WILEY SERIES IN MICROWAVE AND OPTICAL ENGINEERING

KAI CHANG, Editor
Texas A&M University

HISTORY OF WIRELESS

HISTORY OF WIRELESS

Tapan K. Sarkar
Robert J. Mailloux
Arthur A. Oliner
Magdalena Salazar-Palma
Dipak L. Sengupta

With Contributions from:
Duncan C. Baker, John S. Belrose, Ian Boyd, Ovidio M. Bucci,
Paul F. Goldsmith, Hugh Griffiths, Alexei A. Kostenko, Ismo V. Lindell,
Aleksandar Marincic, Alexander I. Nosich, John Mitchell, Gentei Sato,
Motoyuki Sato, and Manfred Thumm

WILEY-
INTERSCIENCE

A JOHN WILEY & SONS, INC., PUBLICATION

Library of Congress Cataloging-in-Publication Data:

History of wireless / Tapan K. Sarkar . . . [et al.] / with contributions from
Duncan C. Baker . . . [et al.].
    p. cm.
  Includes bibliographical references and index.
  ISBN-13 978-0-471-71814-7
  ISBN-10 0-471-71814-9 (cloth : alk. paper)
  1. Radio—History.   2. Wireless communication systems—History.   3.
Electromagnetics—Research—History.   4. Antennas (Electronics)—History.   I.
Sarkar, Tapan (Tapan K.)
  TK6547.H57 2006
  621.384'09—dc22   2005022232

Printed in the United States of America.

10 9 8 7 6 5 4 3 2 1

Contents

Preface

The motivation to write about the History of Wireless comes from Auguste Comte (1798–1857), a French philosopher who is termed the father of positivism and modern sociology [*Les Maximes d'Auguste Comte* (Auguste Comte's Mottos), http://www.membres.lycos.fr/clotilde/]:

On ne connaît pas complètement une science tant qu'on n'en sait pas l'histoire.
(One does not know completely a science as long as one does not know its history.)
Aucune science ne peut être dignement comprise sans son histoire essentielle (et aucune véritable histoire n'est possible que d'après l'histoire générale).
(No science can be really understood without its essential history (and no true history is possible if not from general history.)
L'histoire de la science, c'est la science même.
(The history of science is the science itself.)

and from Marcus T. Cicero (106–43 BC), Roman statesman, orator, and philosopher:

To be ignorant of what occurred before you were born is to remain always a child. For what is the worth of human life, unless it is woven into the life of our ancestors by the records of history?
The causes of events are ever more interesting than the events themselves. History is the witness that testifies to the passing of time; it illuminates reality, vitalizes memory, provides guidance in daily life, and brings us tidings of antiquity.

and enforced by Niccolò Machiavelli (1469–1527), from Florence, Italy:

Whoever wishes to foresee the future must consult the past; for human events ever resemble those of preceding times. This arises from the fact that they are produced by men who ever have been, and ever shall be, animated by the same passions, and thus they necessarily have the same results.

and further elucidated by William Cuthbert Faulkner (1897–1962), the American Nobel Laureate writer:

You must always know the past, for there is no real Was, there is only Is.

and the rationale given by David Hume (1711–1776), the Scottish philosopher and historian:

Mankind is so much the same, in all times and places, that history inform us of nothing new or strange in this particular. Its chief use is only to discover the constant and universal principles of human nature.

and ending in Aristotle (384–322 BC), the Greek philosopher:

If you would understand anything, observe its beginning and its development.

However one has to be careful in writing history, as the British historian Arnold Joseph Toynbee (1989–1975), reminds us that:

"History" is a Greek word which means, literally, just "investigation".

In addition, the French humanist François-Marie Arouet de Voltaire (1694–1778), points out the duties of the historian:

A historian has many duties. Allow me to remind you of two which are important. The first is not to slander; the second is not to bore.

and further reinforced by Pope Leo XIII, born Vicenzo Gioacchino Raffaele Pecci in Italy (1810–1903):

The first law of history is to dread uttering a falsehood; the next is not to fear stating the truth; lastly, the historian's writings should be open to no suspicion of partiality or animosity.

However, in writing about history one has to follow the definition of the American lawyer Noah Webster (1758–1843), in his 1828 dictionary, that states:

History is a narrative of events in the order in which they happened with their causes and effects. A narrative (story) is very different from an annal (a summary listing of dates, events, and definition). Narratives (stories) should be used for teaching history if the student is to gain any understanding. Annals are best used for summary review by one who has already learned the stories as Annals relate simply the facts and events of each year, in direct chronological order, without any observations of the annalist.

For a person to appreciate history, there must be told a story that relates the heart-felt beliefs that led those people to the actions they chose. Without such an understanding of their heart, there is no understanding of the history. To know history is to know what people did and why, that is to know their heart. Cold names without warm understanding of why they did the things they did is no more use to a child than learning the alphabet and not learning to form words. It takes stories from the time to be able to understand the time you are studying. It takes stories leading up to the time, as well as stories of that time.

Therefore to fulfill the requirements of the definition of history according to Webster, we have followed in this book, the two paths as suggested. **The first two chapters provide the annals of wireless, whereas the remaining chapters are narratives of history.**

History is reflected on by the French writer François-René de Chateaubriand (1768–1848), as:

History is not a work of philosophy, it is a painting; it is necessary to combine narration with the representation of the subject, that is, it is necessary simultaneously to design and to paint; it is necessary to give to men the language and the sentiments of their times, not to regard the past in the light of our own opinion.

and history follows the path described by the German philosopher, social scientist, historian, and revolutionary Karl Heinrich Marx (1818–1883):

Men make their own history, but they do not make it just as they please; they do not make it under circumstances chosen by themselves, but under circumstances directly found, given and transmitted from the past.

ending in the words of the American president Abraham Lincoln (1809–1865):

History is not history unless it is the truth.

and those of the Scottish writer Hugh Amory Blair (1718–1800):

As the primary end of History is to record truth, impartiality, fidelity and accuracy are the fundamental qualities of a Historian.

However, it is important to remember that as the American poet and writer Robert Penn Warren (1905–1989), suggests:

History cannot give us a program for the future, but it can give us a fuller understanding of ourselves, and of our common humanity, so that we can better face the future.

and the French historian Numa-Denis Fustel de Coulanges (1830–1889), notes what history is not:

History is not the accumulation of events of every kind which happened in the past. It is the science of human societies.

However, we sincerely hope that in presenting the history of wireless we have paid proper attention to it so that the following quotes do not come true, particularly in the words of the Spanish philosopher, poet, literary and cultural critic, Jorge Augustín Nicolás Ruiz de Santayana y Borrás (known in the United States, where he lived for many years, as George Santayana) (1863–1952):

History is always written wrong, and so always needs to be rewritten.

and enforced by the American jurist Oliver Wendell Holmes, Jr. (1841–1935):

History has to be rewritten because history is the selection of those threads of causes or antecedents that we are interested in.

Finally, we must be failing in our responsibilities if we do not follow the British historian Lord John Emerich Edward Dolberg-Acton (1834–1902):

History, to be above evasion or dispute, must stand on documents, not on opinions.

However, one must remember, as the Jacques Maritain Center points out, what history can and cannot do:

But the truth of history is factual, not rational truth; it can therefore be substantiated only through signs — after the fashion in which any individual and existential datum is to be checked; and though in many respects it can be known not only in a conjectural manner but with certainty, it is neither knowable by way of demonstration properly speaking, nor communicable in a perfectly cogent manner, because, in the last analysis, the very truth of the historical work involves the whole truth which the historian as a man happens to possess; it presupposes true human wisdom in him; it is "a dependent variable of the truth of the philosophy which the historian has brought into play." Such a position implies no subjectivism. There is truth in history. And each one of the components of the historian's intellectual disposition has its own specific truth.

A final remark is that conjecture or hypothesis inevitably plays a great part in the philosophy of history. This knowledge is neither an absolute knowledge in the sense of Hegel nor a scientific knowledge in the sense of mathematics. But the fact that conjecture and hypothesis play a part in a discipline is not incompatible with the scientific character of this discipline. In biology or in psychology we have a considerable amount of conjecture, and nevertheless they are sciences.

Mr. Ferenc M. Szasz (professor of history at the University of New Mexico) collected the above list of quotations about history over the course of his career. The *History Teacher* first published his list in the 1970s. The current list includes scores of new quotations he has come across in the intervening decades. We have also added a few. Readers are welcome to add to the list.

Next comes the definition or meaning of the word "wireless". We follow here the explanation given by J. D. Kraus and R. J. Marhefka in their book on *Antennas for All Applications*, which states:

After Heinrich Hertz first demonstrated radiation from antennas, it was called wireless. And wireless it was until broadcasting began around 1920 and the word radio was introduced. Now wireless is back to describe the many systems that operate without wires as distinguished from radio, which to most people implies AM or FM.

And, finally we provide the roadmap of the book. In Chapter 1 we present a chronology of the developments in magnetism, electricity, and light till the time of Maxwell, who is generally regarded as the greatest physicist of the nineteenth century. The name of Maxwell is synonymous with electromagnetics and electromagnetic waves. Hence we make an attempt to describe who Maxwell was and what he actually did. It is also imperative to point out what was/is his theory as related to wireless. Chapter 2 provides the chronology of the development of wireless up to recent times. The evolution of Continental and British Electromagnetics in the nineteenth century ending in Maxwell is described in Chapter 3. Chapter 4 deals with the genesis of Maxwell's equations. In Chapter 5 it is outlined how the followers of Maxwell redeveloped Maxwell's theory and made it understandable to a broader audience through the experimental verification of Maxwell's results by Hertz. It is interesting to note that the four equations that we use today were not originally developed by Maxwell but by Hertz, who wrote them in the scalar form, followed by Heaviside, who in turn wrote them in vector form. Chapter 6 describes the work of Heaviside and his contributions. The relevant scientific accomplishments in wireless before Marconi is presented in Chapter 7 in detail. Chapter 8 discusses the achievements of Tesla, who holds the first patent for radio in the United States. In Chapter 9 the early experiment of Bose on millimeter waves is described. In fact, many of the artifacts like horn antennas and circular waveguides that he performed experiments with are still in current use. The contributions of Fleming in the development of wireless are presented next in Chapter 10. The many contributions of German scientists to wireless, including the achievements of Hertz, are described in Chapter 11, followed in Chapter 12 by the development of wireless telegraphy and telephony, including the pioneering attempts to achieve transatlantic wireless communications. Chapter 13 presents the evolution of wireless telegraphy in South Africa at the turn of the twentieth century. The development of antennas in Japan is described in Chapter 14, including both the past and the present. The historical background and development of Soviet quasi optics at near-mm and sub-mm wavelengths are illustrated in Chapter 15. Since waveguides are necessary for the circuits that generate, detect and process the waves, it is important to discuss the evolution of electromagnetic waveguides, as done in Chapter 16, from hollow metallic waveguides to microwave integrated circuits. Incidentally, that chapter is the only one that describes the important progress in electromagnetic waves made during and around the World War II period. Finally, in Chapter 17 a history of phased array antennas, and their relations to previous scanning array technology, is provided.

It is important to note that due to the large volume of literature existing on Marconi's work and because his fundamental contributions to the development of wireless communications are widely known and referred to, we explicitly choose to concentrate our attention on most specific and less known aspects and people who also made invaluable contribution to the development of wireless.

Every attempt has been made to guarantee the accuracy of the material in the book. We would, however, appreciate readers bringing to our attention any errors that may have appeared in the final version. Errors and any comments may be e-mailed to tksarkar@syr.edu, regarding all the contributors.

Acknowledgments

We gratefully acknowledge P. Angus, R. H. Corner and M. J. Scmitt for their help and suggestions.

Thanks are due to Prof. Wonwoo Lee, who prepared the front cover for this book, to Prof. Hugh Griffiths, for proofreading the manuscript and to Prof. John Norgard for suggesting ways to improve the readability of the chapters.

We are very grateful to Ms. Christine Sauve, Ms. Brenda Flowers, and to Ms. Maureen Marano from Syracuse University for their expert typing of the manuscript. We would also like to express sincere thanks to Santana Burintramart, Wonsuk Choi, Arijit De, Debalina Ghosh, Seunghyeon Hwang, Youngho Hwang, Zhong Ji, Rucha Lakhe, Mary Taylor, Jie Yang, Nuri Yilmazer, and Mengtao Yuan of Syracuse University, for their help with the book.

1

INTRODUCTION

TAPAN K. SARKAR, *Syracuse University, USA;*
MAGDALENA SALAZAR-PALMA, *Universidad Politécnica de Madrid, Spain;*
DIPAK L. SENGUPTA, *University of Michigan, USA*

1.1   PROLOGUE

This chapter provides an overview of the origin and the developments of magnetism, electricity, and light theories. The chronology is traced up to the time of Maxwell who was the first to link all three together in a formal way even though many conjectured about their interrelations before him. First, an overview of magnetism is provided, followed by that of electricity, and then that of light. The material presented in this chapter is collected from the various references given at the end of the chapter. In addition, the various scientific works done by Maxwell and his legacy are described. Finally, an overview of the theory of electromagnetic waves first developed by Maxwell and how it was subsequently modified by Hertz and Heaviside and later on by Larmor is presented. This is a unique theory in physics where the basic fundamental equations did not change, while their physical interpretations underwent at least two major modifications.

1.2   DEVELOPMENT OF MAGNETISM

The development of magnetism is traced through the last 5000 years.

2637 BC:
  - Emperor Huang-ti of China used the compass in a battle to find the direction along which he should pursue his enemies.
1110 BC:
  - Taheon-Koung, the Chinese minister of state, gave his crew a compass to sail from Cochin, China, to Tonquin.
1068 BC:
  - Chinese vessels routinely navigated the Indian Ocean by compass.
1022 BC:
  - Some Chinese chariots had a floating magnetic needle, the motion of which was communicated to the figure of a spirit whose outstretched hands always indicated the south.

1000 BC:

- Homer of Greece wrote that loadstones were used by the Greeks to direct navigation at the time of the siege of Troy.

950 BC:

- King Solomon (970–928 BC) of Israel knew how to use the compass.

900 BC:

- Magnes, a Greek shepherd, walked across a field of black stones which pulled the iron nails out of his sandals and the iron tip from his shepherd's staff, as suggested by the Italian natural philosopher Giambattista della Porta (1540–1615). The same story had also been told by Gaius Plinius Secundus, better known as Pliny the Elder (23–79AD). This region became known as *Magnesia* in Asia Minor. Probably, the word magnet evolved from this and the iron oxide ore was named as magnetite. Pliny in *Naturalis Historia* also wrote of a hill near the river Indus that was made entirely of a stone that attracted iron.

600 BC:

- First recorded information by Greek philosophers, particularly by Thales of Miletus (624–546 BC), about the magnetic properties of natural ferric oxide (Fe_3O_4) stones. It was also known to the Indians. For example Susruta, a physician in the sixth century BC in India, used them for surgical purposes.

121 AD:

- The Chinese dictionary *Choue Wen* contained an explicit recorded reference of the magnet.

1186:

- Alexander Neckam (1157–1217), a monk and man of science of St. Albans, England, described the working of a compass in the western literature for the first time and he did not refer to it as something new, indicating that it had been in use for some time.

1254:

- Roger Bacon, a philosopher also called Friar Bacon and surnamed Doctor Mirabilis (1214–1294), a Franciscan monk of Ilchester, England, dealt with the magnet and its properties in *Opus Minus*.

1269:

- Petrus Peregrinus or Pierre de Maricourt, a Crusader from Picardy, France, who was a mathematician, aligned needles with lines of longitude pointing between two pole positions of the stone and established the concept of two poles of the magnet. He wrote it in *Epistola de Magnete*.

1400:

- Jean de Jaudun of France wrote about magnets and the problem of action-at-a-distance.

1492:

- Christopher Columbus (1451–1506), from Italy (navigating under the Spanish flag) was the first to determine astronomically the position of a line of no magnetic variation. He observed the compass changes direction as the longitude changes.

1497:
- Portuguese navigator Vasco da Gama (1469–1524) used the compass for his trip to the Indies. He said that he found pilots in the Indian Ocean who made ready use of the compass.

1530:
- Spanish cartographer Alonzo de Santa Cruz produced the first map of magnetic variations from the true north.

1544:
- German technician and physicist Georg Hartmann (1489–1564) also discovered the magnetic dip of the compass.

1558:
- Giambattista della Porta (1540–1615), an Italian natural philosopher, performed experiments with the magnet for the purpose of communicating intelligence at a distance.

1576:
- Robert Norman, a manufacturer of compass needles at Wapping, England, rediscovered the dip or inclination to the Earth of the magnetic needle in London and was the first to measure them.

1590:
- Giulio Moderati Caesare, an Italian surgeon, observed the conversion of iron into a magnet by geographical position alone.

1600:
- Sir William Gilbert (1544–1603), court physician to Queen Elizabeth I, discovered that the Earth was a giant magnet and explained how compasses worked. He gave the first rational explanation to the mysterious ability of the compass needle to point north-south.

1644:
- René Descartes (1595–1650), the French physicist, physiologist, mathematician, and philosopher, in the *Principia Philosophiae*, theorized that the magnetic poles were on the central axis of a spinning vortex of fluids surrounding each magnet. The fluid entered by one pole and leaves through the other.

1687:
- English scientist and mathematician Sir Isaac Newton (1642–1727) estimated an inverse cubed law for the two poles of a magnet. He also published *Principia* that year whose costs and proofreading of the material were carried out by the English astronomer and mathematician Edmund Halley (1656–1742).

1699:
- Halley performed the first magnetic survey showing the variation of the compass.

1716:
- Halley proposed that the magnetic effluvia moving along the magnetic field of the Earth results in the aurora.

1730:

- English scientist Servigton Savery produced the first compound magnet by binding together a number of artificial magnets with a common pole piece at each end.

1740:

- Gowen Knight produced the first artificial magnets for sale to scientific investigators and terrestrial navigators.

1742:

- Thomas Le Seur and Francis Jacquier, of France, in a note to the edition of Newton's *Principia* that they published, showed that the force between two magnets was inversely proportional to the cube of the distance.

1750:

- English scientist John Mitchell (1724–1793) published the first book on making steel magnets. He also discovered that the two poles of a magnet were equal in strength and that the force between individual poles followed an inverse square law.

1759:

- German physicist Franz Maria Ulrich Theodor Hoch Aepinus (1724–1802) published *An Attempt at a Theory of Electricity and Magnetism*, the first book applying mathematical techniques to the subject.

1778:

- Sebald Justin Brugmans (1763–1819), a Dutch professor of natural history, demonstrated the diamagnetic properties of bismuth and antimony. A diamagnetic substance is one that has a permeability of less than one. A bar or a needle of such a substance, when free to move, will tend to be at right angles to the lines of force in a magnetic field.

1785:

- French physicist Charles-Augustin de Coulomb (1736–1806) independently verified Mitchell's law of force for magnets and extended the theory to the law of attraction of opposite electricity. He was the proponent of a two fluid theory proposed in 1759 by the English physicist Robert Symmer based on the ideas of the French physicist Charles François de Cisternay du Fay (1698–1739).

1820:

- French physicists Jean-Baptiste Biot (1774–1862) and Felix Savart (1792–1841) showed that the magnetic force exerted on a magnetic pole by a wire falls off as $1/r$ and is oriented perpendicular to the wire similar to what the Danish physicist Hans Christian Ørsted (1777–1851) had predicted. The English mathematician Edmund Taylor Whittaker (1873–1956) says that "This result was soon further analyzed, to obtain $d\boldsymbol{B} \propto (\boldsymbol{I}d\boldsymbol{s} \times \boldsymbol{r})/r^3$, where $\boldsymbol{B}$ stands for the magnetic flux vector, $\boldsymbol{I}$ for the current, $\boldsymbol{r}$ for the position vector, and $d\boldsymbol{s}$ for the elemental length of current."

1821:

- British scientist Michael Faraday (1791–1867) discovered that a conductor carrying a current would rotate around a magnetic pole and that a magnetized needle would rotate about a wire carrying a current.

- Self-educated British physicist William Sturgeon (1783–1850) made the first electromagnet.
- Physicist Prof. Joseph Henry (1797–1878) of Albany Academy, New York, made an electromagnet with superimposed layers of insulated wires.
- German physicists Johann Solomon Christoph Schweigger (1779–1857) and Johann Christian Poggendorf (1796–1877) constructed independently the first galvanometers.

1824:

- French mathematician Siméon–Denis Poisson (1781–1840) invented the concept of the magnetic scalar potential and of surface and volume pole densities described by the formula

$$F = -\int M \cdot \frac{r-r'}{|r-r'|^3} dV' = \int \frac{\nabla' \cdot M}{|r-r'|} dV' - \int \frac{M \cdot n'}{|r-r'|} dS' .$$

where F is the electric vector potential, M is the magnetic current, r and r' are the field and the source coordinates, respectively, n' is the direction of the outward normal to the surface, dS' and dV' are the elemental surface and volume elements, respectively. He also provided the formula for the magnetic field inside a spherical cavity within magnetized material.

- French physicist Dominique François Jean Arago (1786–1853) demonstrated that a copper disk can be made to rotate by revolving a magnet near it.

1825:

- French mathematician and physicist André-Marie Ampère (1775–1836) published his collected results on magnetism. His expression for the magnetic field produced by a small segment of current was different from that which followed naturally from the Biot-Savart law by an additive term which integrated to zero around a closed circuit. In his memoir one found the result known as *Stoke's theorem* written as $\oint B \cdot ds = \mu_0 I$,

  where μ_0 is the permeability of vacuum. James Clerk Maxwell described this work as *one of the most brilliant achievements in science*.

- Italian physicist Leopoldo Nobili (1784–1835), invented a static needle pair, which produced a galvanometer independent of the magnetic field of the Earth.

1831:

- Henry discovered that a change in magnetism can make currents flow, but he failed to publish this. In 1832 he described self-inductance as the basic property of an inductor. In recognition of his work, inductance is measured in henries. He improved upon Sturgeon's electromagnet, substantially increasing the electromagnetic force. He also developed the principle of self-induction.

1832:

- Karl Friedrich Gauss (1777–1855), the mathematician, astronomer, and physicist from Germany, independently stated Green's theorem (named after the British mathematician George Green, 1793–1841) without proof. He also reformulated Coulomb's law without proof. He formulated separate electrostatic and electrodynamic laws including Gauss's law. All of it remained unpublished till 1867.

1838:

- Wilhelm Eduard Weber (1804–1891), a physicist from Germany, together with Gauss applied potential theory to the magnetism of Earth.

1850:

- Irish-Scottish physicist William Thomson (Lord Kelvin, 1824–1907) invented the idea of magnetic permeability and susceptibility, along with the separate concepts of B, M and H, where H stands for the magnetic field intensity.

1853:

- William Thomson used Poisson's magnetic theory to derive the correct formula for magnetic energy: $U = 0.5 \int \mu H^2 \, dV$. He also gave the formula $U = 0.5 L I^2$, where U is the magnetic energy, μ is the permeability, and L is the self induction parameter.

1864:

- James Clerk Maxwell (1831–1879), the physicist and mathematician from Scotland, published a mechanical model of the electromagnetic field. Magnetic fields corresponded to rotating vortices with idle wheels between them and electric fields corresponded to elastic displacements, hence displacement currents. The equation for H now became $\nabla \times H = 4\pi J_{tot}$, where J_{tot} is the total current, conduction plus displacement, and is conserved, i.e., $\nabla \cdot J_{tot} = 0$. They were all available in scalar form in his paper *On Physical Lines of Force*. This addition completed Maxwell's equations and it now became easy for him to derive the wave equation exactly, and to note that the speed of wave propagation was close to the measured speed of light. Maxwell wrote:

  > *We can scarcely avoid the inference that light is the transverse undulations of the same medium which is the cause of electric and magnetic phenomena. Thomson, on the other hand, says of the displacement current, (it is a) curious and ingenious, but not wholly tenable hypothesis.*

  Maxwell read a memoir before the Royal Society in which the mechanical model was stripped away and just the equations remained. He also discussed the vector and scalar potentials, using the Coulomb gauge. He attributed physical significance to both of these potentials. He wanted to present the predictions of his theory on the subjects of reflection and refraction of electromagnetic waves, but the requirements of his mechanical model kept him from finding the correct boundary conditions, so he never did this calculation. He published his paper *A Dynamical Theory of the Electromagnetic Field* [*Philosophical Trans.*, Vol. 166, pp. 459–512, 1865] – the first to make use of a mathematical theory for Faradays' concept of fields.

1.3   DEVELOPMENT OF ELECTRICITY

Next, the development of electricity is traced back to prehistoric times.

Prehistoric times:
- Early humans were aware of lightning, sting rays, electric eels, and static charges in a dry climate.

600 BC:
- The Greek philosopher Aristophanes was aware of the peculiar property of amber, which is a yellowish translucent resin. When rubbed with a piece of fur, amber developed the ability to attract small pieces of material such as feathers. For centuries this strange, inexplicable property was thought to be unique to amber.
- The Etruscans were known to have been devoted to the study of electricity. They were said to have attracted lightning by shooting metal arrows into clouds threatening thunder and lightning.
- Thales (640–546 BC) of Miletus rubbed amber (*elektron* in Greek) with cat fur and picked up bits of feathers. Unfortunately for posterity he left no writings, and all that we know was transmitted orally until Aristotle (384–322 BC), the great philosopher from Greece, recorded his teachings. So it was not clear whether he discovered the facts himself or learned from the Egyptian priests and others whom he visited on his extensive trips.

341 BC:
- Aristotle wrote about a fish called *torpedo* which gave electrical shocks and paralyzed muscles if touched.

250 BC:
- A Galvanic cell composed of copper and iron immersed in wine or vinegar called the *Baghdad Battery*, was excavated in Baghdad, Iraq, by the German archeologist Wilhelm König in 1938, and was dated back to 250 BC.

1600 AD:
- English physician Sir William Gilbert (1544–1603) proved that many other substances besides amber displayed electrical properties and that they have two electrical effects. When rubbed with fur, amber acquired *resinous* electricity; glass, however, when rubbed with silk, acquired *vitreos* electricity. Electricity of the same kind repels, and electricity of the opposite kind of attracts each other.

1629:
- Italian Jesuit priest Niccolò Cabeo (1585–1650) also observed electrical

repulsion and attraction. Others who also did were the English diplomat and naval commander Sir Kenelm Digby (1603–1665) and the Irish natural philosopher and experimenter Sir Robert Boyle (1627–1691). However, there were some differences in opinion on exactly how it worked!

1646:

- Walter Charlton coined the term electricity. Others say it was coined by the English physician Robert Browne even though he contributed nothing else to science.

1663:

- Otto von Guericke (1602–1686), the German physicist, first published the phenomenon of static electricity and built a machine to produce it.

1665:

- French mathematician, physicist, and astronomer Honoré Fabri (1607–1688) demonstrated the reciprocity of the electric force.

1675:

- Jean Picard (1620–1682), a French astronomer, observed flashes of light from the vacuum space produced in a Torricellian barometer.

1705:

- English physicist Francis Hawksbee (1687–1763) further illustrated this phenomenon and generated light under different environmental conditions.

1729:

- Stephen Gray (1666–1736), a pensioner at the Charter House in London, England, showed that electricity did not have to be made in place by rubbing but can also be transferred from place to place with conducting wires. He also showed that the charge on electrified objects resided on their surfaces.

1733:

- Charles François de Cisternay du Fay (1698–1739), superintendent of gardens of the King of France, also came to the conclusion that electricity came in two kinds, which he called *resinous*(-) and *vitreous*(+).

1742:

- Jean-Théophile Desaguliers (1683–1744), a French-born scientist and Englishman by adoption, since he was brought to England after the revocation of the Edict of Nantes, became a protestant chaplain, continued the work of Gray, and used the names *non-electrics* or *conductors* for materials displaying the corresponding electrical properties.

1745:

- Pieter van Musschenbroek (1692–1761), physicist and professor at Leyden, The Netherlands, invented the Leiden jar, or capacitor, and nearly killed his assistant Andreas Cunaeus in demonstrating his experiment.

- Abbé Jean-Antoine Nollet (1700–1770), member of the court of Louis XV and a physics professor of the French Royal Children, expanded on

Fay's ideas and invented the two-fluid theory of electricity.

1746:

- Sir William Watson (1715–1787), a London apothecary, physicist, physician, and botanist, propounded the doctrine that electrical actions are due to the presence of electric aether and suggested conservation of charge.
- Johann Heinrich Winckler (1703–1770), a professor of philosophy and physics at the University of Leipzig, Germany, was the first to use electricity for telegraphic purposes using sparks.

1747:

- Benjamin Franklin (1706–1790), the American writer, statesman, and scientist, invented the theory of one-fluid electricity in which one of Nollet's fluids existed and the other was just the absence of the first, after observing the performance of some electrical experiments in Boston by a certain Dr. Spence who arrived from Scotland. He also proposed the principle of conservation of charge and called the fluid that existed and flowed "positive". He discovered that electricity can act at a distance in situations where fluid flow made no sense. To demonstrate that, during a thunderstorm in 1752, Franklin flew a kite that had a metal tip and charged a Leyden jar during a thunderstorm demonstrating lightning was an electrical discharge. At the end of the wet, conducting hemp line on which the kite flew, he attached a metal key, to which he tied a non-conducting silk string that he held in his hand. The experiment was extremely hazardous, but the results were unmistakable: when he held his knuckles near the key, he could draw sparks from it. The next two who tried this extremely dangerous experiment were killed.
- Watson passed an electrical charge along a two miles long wire.

1752:

- Johann Georg Sulzer (1720–1779), a Swiss philosopher, published that when two dissimilar metals like lead and silver were placed in touch with the tongue a peculiar taste was observed which did not exist if only one of the metals touched the tongue. This was the forerunner of batteries.

1759:

- German physicist Franz Maria Ulrich Theodor Hoch Aepinus (1724–1802) showed in St. Petersburg, Russia that electrical effects were a combination of fluid flow confined to matter and action at a distance. He also discovered charging by induction. He was assisted in his work by the German physicist Johan Carl Wilcke (1732–1796) when he was working earlier at the Berlin Academy of Science.

1762:

- John Canton (1718–1772), an English physicist, along with Wilcke, demonstrated the principle of electric induction where the near portion of the body acquires an opposite charge to the source near which it was placed whereas the opposite end acquired similar charges. This was also demonstrated by Franklin in 1755.

1764:

- Joseph-Louis Lagrange (1736–1813), the Italian-French mathematician and astronomer, discovered the divergence theorem in connection with the study of gravitation. In 1813 it became known as Gauss's law.

1767:

- Joseph Priestley (1733–1804), a chemist and English Presbyterian minister, acting on a suggestion in a letter from Benjamin Franklin, showed that hollow charged vessels contained no charge on the inside. And based on his knowledge that hollow shells of mass have no gravity inside, he correctly deduced that the law of electric force followed an inverse squared law. Priestley conjectured that the force of attraction followed that of the gravitational forces, and so did the Swiss-Dutch mathematician Daniel Bernoulli (1700–1782) in 1760. Priestley was considered one of the great experimental scientists of the XXVIII century even though he did not take a single formal science course.

1769:

- Physician Dr. John Robison (1739–1805) of Edinburgh, Scotland, determined the force between charges by experiment and found the exponent to be −2.06 that operated on the distance. From this he conjectured that the correct power was the inverse square.

1772:

- English chemist and physicist Sir Henry Cavendish (1731–1810) presented to the Royal Society *An Attempt to Explain Some of the Principal Phenomena of Electricity, by Means of an Elastic Fluid.* Since he was indifferent to publications, his work was published 100 years later at the instigation of William Thomson (Lord Kelvin) and was compiled by James Clerk Maxwell himself when he was the Cavendish Professor later in his life. Cavendish in fact had not only derived the correct form of the inverse square law but also had invented the idea of electrostatic capacity and specific inductive capacity (resistance).

1777:

- Lagrange invented the concept of the scalar potential for gravitational fields.

1780:

- One of Luigi Galvani's (the Italian anatomist and physician, 1737–1798) assistants noticed that a dissected frog leg twitched when he touched its nerve with a scalpel. Another assistant thought that he had seen a spark from a nearby charged electric generator at the same time. Galvani reasoned that the electricity was the cause of the muscle contractions. He mistakenly thought, however, that the effect was due to the transfer of a special fluid, or "animal electricity," rather than to conventional electricity. Experiments such as this, led Luigi Galvani in 1791, to propose his theory that animal tissues generate electricity.

1782:

- Pierre-Simon Laplace (1749–1827), a French mathematician, showed that Lagrange's potential, V, satisfied the equation $\nabla^2 V = 0$.

1785:

- The French physicist Charles-Augustin Coulomb (1736–1806) used a torsion balance to verify that the electric force law had an inverse squared variation. He proposed a combined fluid/action-at-a-distance theory like that of Aepinus but with two conducting fluids instead of one. He also discovered that the electric force near a conductor was proportional to its surface charge density and made contributions to the two-fluid theory of magnetism.

1799:

- Italian physicist Alessandro Guiseppe Antonio Anastasio Volta (1745–1827), professor of Natural philosophy at the University of Pavia, realized that the main factors in Galvani's discovery were the two different metals – the steel knife and the tin plate – upon which the frog was lying. The different metals, separated by the moist tissue of the frog, were generating electricity. The frog's leg was simply a detector. In 1800, Volta showed that when moisture comes between two different metals, electricity is created. This led him to invent the first electric battery, which he made from thin sheets of copper and zinc separated by moist pasteboard (felt soaked in brine) known as an *electrolyte*. He called his invention a *column battery* although it came to be commonly known as the *Volta battery, Voltaic Pile*, or *Voltaic Cell*. Volta showed that electricity could be made to travel from one place to another by wire.

1800:

- English chemist William Nicholson (1753–1815) and the English surgeon Anthony Carlisle (1768–1840) discovered that water may be separated into hydrogen and oxygen by the action of Volta's pile.
- The English chemist Sir Humphrey Davy (1778–1829) developed a theory for the pile based on the contact potentials.

1812:

- Michael Faraday (1791–1867), an English bookbinder's apprentice, wrote to Sir Humphry Davy asking for a job as a scientific assistant. Davy interviewed Faraday and found that he had educated himself by reading the books he was supposed to be binding. He became an assistant of Davy and then gradually became the director of the lab after Davy's death in 1829 and occupied the chair of chemistry from 1833.

1812:

- French mathematician Siméon-Denis Poisson (1781–1840) further developed the two-fluid theory of electricity, showing that the charge on conductors must reside on their surfaces and be so distributed that the electric force within the conductor vanished. This surface charge density calculation was carried out in detail for ellipsoids. He also showed that the potential within a distribution of electricity ρ satisfied the equation $\nabla^2 V = -\rho / \varepsilon_0$, where ε_0 is the permittivity of vacuum.
- Laplace showed that, at the surface of a conductor, the electric field E is perpendicular to the surface and it is given by $E = \rho / \varepsilon_0$, where ρ is the

surface charge density.

- German mathematician, astronomer, and physicist, Karl Friedrich Gauss (1777–1855) rediscovered the divergence theorem of Lagrange which we referred to as the Gauss's divergence theorem. He applied it to derive Gauss's law.

1820:

- Hans Christian Ørsted (also written as Öersted) (1777–1851), professor of philosophy at Copenhagen, Denmark, observed that a current flowing through a wire would move a compass needle placed beside it. This showed that an electric current produced a magnetic field. The French physicist Dominique François Jean Arago (1786–1853) presented these results at the French Academy which excited many scientists to repeat the experiment.

- André-Marie Ampère (1775–1836), a French mathematician and physicist, one week after hearing of Öersted's discovery, showed that parallel currents attract each other and that opposite currents repel. Arago also showed that a wire carrying a current of electricity would attract iron filings.

1821:

- Davy showed that direct current is carried throughout the volume of a conductor and established that $R \propto \ell / A$ for long wires, where R stands for resistance, ℓ for length, and A for the cross-sectional area of a conductor or line. He also discovered that resistance increased with the rise of temperature.

1822:

- Estonian-German physicist Thomas Johann Seebeck (1770–1831) discovered the thermoelectric effect by showing that a current flowed in a circuit made of dissimilar metals if there was a temperature difference between the metals.

1826:

- Georg Simon Ohm (1787–1854), a physicist and a professor of mathematics in Cologne, Germany, established the result $V = I R$ now known as Ohm's law. Using a galvanometer he demonstrated the relation between potential, current, and resistance. The next year he published a book *Die Galvanische Kette, Mathematisch Bearbeitet,* where he proposed the basic electrical law which much later became known as *Ohm's Law.* What Ohm did was to develop the idea of voltage as the driver of electric current. It was not until some years later that Ohm's electroscopic force (V in his law) and Poisson's electrostatic potential were shown to be identical.

1827:

- Felix Savary (1787–1841) was the first experimenter to note oscillatory discharges from capacitors. He did not attach an inductor. He also assisted Ampère in many of his experiments.

1828:

- British baker George Green (1793–1841) generalized and extended the work of Lagrange, Laplace, and Poisson and attached the name *potential* to their scalar function, which was first devised by the Swiss-Russian mathematician Leonhard Euler (1707–1783) in 1744. He showed how to connect the surface and the volume integrals, what is now known as Green's theorem. He became an undergraduate student at Cambridge in October 1833 at the age of 40! He also developed the theory of electrostatic screening.

1831:

- Faraday showed that changing currents in one circuit induced currents in a neighboring circuit and illustrated that they can all be explained by the idea of changing magnetic flux introduced earlier by Niccolò Cabeo and Petrus Peregrinus (see Section 1.2). He, thus, produced electricity from magnetism.
- Ukrainian mathematician and physicist Mikhail Vassilievitch Ostrogradsky (1801–1861) rediscovered the divergence theorem of Lagrange, Gauss, and Green.

1832:

- Antoine-Hippolyte Pixii (1808–1835), an instrument maker of Paris, France, built the first direct current (DC) motor.

1834:

- Jean-Charles Athanase Peltier (1785–1845), a French watchmaker who gave up his profession at the age of thirty to carry out experimental physics, discovered the converse of Seebeck's thermoelectric effect. He found that current driven in a circuit made of dissimilar metals caused the different metals to be at different temperatures.
- Heinrich Friedrich Emil Lenz (1804–1865), a physicist from Estonia, formulated his rule for determining the direction of Faraday's induced currents. In its original form, it was a law for force rather than a law for an induced electromotive force (EMF): *Induced currents flow in such a direction as to produce magnetic forces that try to keep the magnetic flux the same.* Thus, Lenz predicted that if one pulled a conductor into a strong magnetic field, it will be repelled and it will be opposed if one would pull a conductor out of a strong magnetic field.

1836:

- John Frederic Daniell (1790–1845), an English self-taught chemist, proposed an improved electric cell that supplied an even current during continuous operation.

1837:

- Faraday discovered the idea of the dielectric constant.

1838:

- Faraday showed that the effects of induced electricity in insulators are analogous to induced magnetism in magnetic materials. In this way the terms P, D, and ε were realized, where P represents the polarizability, D for the electric displacement, and ε the permittivity of the medium. He formulated his notion of lines of force criticizing action at a distance.

1841:

- English physicist James Prescott Joule (1818–1889) showed that energy was conserved in electrical circuits involving current flow, thermal heating, and chemical transformations.

1843:

- Faraday proved experimentally the conservation of charge.
- The English physicist and inventor Sir Charles Wheatstone (1802–1875) is most famous for the *Wheatstone Bridge*, but he never claimed to have invented it. However, he did more than anyone else to invent uses for it, when he *found* the description of the device in 1843. The first description of the bridge was done by the English mathematician Samuel Hunter Christie (1784–1865) in 1833.

1845:

- German scientist Franz Neumann (1798–1895) connected (i) Lenz's law, (ii) the assumption that the induced emf is proportional to the magnetic force on a current element, and (iii) Ampere's analysis to deduce Faraday's law, and found a potential function from which the induced electric field could be obtained, namely the vector potential A (in the Coulomb gauge), thus discovering the result which Maxwell wrote later on as $E = -\nabla\phi - \partial A/\partial t$, where E stands for the electric field, ϕ for the scalar electric potential, A for the magnetic vector potential, and t for time. He also derived the formula for mutual inductance for equal parallel coaxial polygons of wire.

1846:

- German physicist Wilhelm Eduard Weber (1804–1891) combined Ampère's analysis, Faraday's experiments, and the assumption of the German physicist and philosopher Gustav Theodor Fechner (1801–1887) that currents consist of equal amounts of positive and negative electricity moving opposite to each other at the same speed, to derive an electromagnetic theory based on forces between moving charged particles. This theory has a velocity-dependent potential energy and is wrong, but it stimulated much work on electromagnetic theory, which eventually leads to the work of Maxwell and the Danish physicist Ludwig Lorenz (1829–1891). It also inspired a new look at gravitation by William Thomson (Lord Kelvin, 1824–1907) to see if a velocity-dependent correction to the gravitational energy could account for the precession of Mercury's perihelion.
- William Thomson showed that Neumann's electromagnetic potential A is, in fact, the vector potential from which B may be obtained via $B = \nabla \times A$.

1847:

- German physiologist and physicist Hermann Ludwig Ferdinand von Helmholtz (1821–1894) wrote a memoir *On the Conservation of Force* which emphatically stated the principle of conservation of energy. He described:

> *Conservation of energy is a universal principle of nature. Kinetic and potential energy of dynamical systems may be converted into heat according to definite quantitative laws as taught by Rumford, Mayer, and Joule. Any of these forms of energy may be converted into chemical, electrostatic, voltaic, and magnetic forms.*

He reads it before the Physical Society of Berlin whose older members regarded it as too speculative and rejected it for publication in *Annalen der Physik*. He also suggested electrical oscillation six years before William Thomson theoretically calculated this process, and ten years before the German physicist Berend Wilhelm Feddersen (1832–1918) experimentally verified it.

1848:
- German physicist Gustav Robert Kirchhoff (1824–1887) extended Ohm's work to conduction in three dimensions, gave his laws for circuit networks, and finally shows that Ohm's *electroscopic force* which drives current through resistors, and the old electrostatic potential of Lagrange, Laplace, and Poisson are the same. He also showed that in the steady state electrical currents distribute themselves so as to minimize the amount of Joule heating. He published his circuital laws in 1850.

1854:
- Faraday cleared up the problem of disagreements in the measured speeds of signals along transmission lines by showing that it is crucial to include the effect of capacitance.
- William Thomson, in a letter to the Irish mathematical physicist George Gabriel Stokes (1819–1903), gives the telegrapher's equation ignoring the inductance: $\partial^2 V / \partial x^2 = RC \, \partial V / \partial t$, where x is the spatial variable, R is the cable resistance per unit length, and C is the capacitance per unit length. Since this is the diffusion equation, the signal does not travel at a definite speed.
- German mathematician Georg Friedrich Bernhard Riemann (1826–1866) made an unpublished conjecture about the connection between electricity, galvanism, light, and gravity.

1855:
- Weber and the German physicist Rudolf Hermann Arndt Kohlrausch (1809–1858) determined the value of the speed of light as 3.1×10^8 meters per second based on a comparison of the measures of the charge of a Leyden jar, as obtained by a method depending on electrostatic attraction, and by a method depending on the effects of the current produced by discharging the jar.

1857:
- Kirchhoff derived the equation for an aerial coaxial cable where the inductance is important and derived the complete telegraphy equation, including solving for the circuit parameters and not by components R, L,

C and G as done later by the English physicist and electrical engineer Oliver Heaviside (1850–1925). He observed that the wave propagates with a velocity very close to the speed of light. Kirchhoff noticed the coincidence and is, thus, the first to discover that electromagnetic signals can travel at the speed of light.

1858:

- Riemann generalized Weber's unification of various theories and derived his solution using the wave function of an electrodynamical potential. He also finds the correct velocity of light. He claimed to have found the connection between electricity and optics. The results were published posthumously in 1867.

1859:

- Raymond Gaston Planté (1834–1889), a French physicist, built the first accumulator.

1862:

- Frenchmen Marcel Deprez (1843–1918), an electrical engineer, and Jacques-Arsène d'Arsonval (1851–1940), physicist and a physician, developed the d'Arsonval galvanometer.

1865:

- Scottish physicist and mathematician James Clerk Maxwell (1831–1879) wrote a memoir in which he attempts to marry Faraday's intuitive field line ideas with Thomson's mathematical analogies. In this memoir the physical importance of the divergence and curl operators for electromagnetism first becomes evident. He showed that the entire magnetic field intensity H around the boundary of any surface measures the quantity of electric current passing the surface. The equations $\nabla \cdot (\varepsilon E) = 4\pi\rho$, $\nabla \times A = B$, and $\nabla \times H = 4\pi J$ appear in this memoir only in scalar forms and not in the vector forms, which were first written by Heaviside.

1867:

- Danish physicist Ludwig Lorenz (1829–1891) developed an electromagnetic theory of light in which the scalar and vector potentials, in retarded form, are the starting point. This was also suggested by Riemann in 1858, but his papers were published posthumously in 1867. He showed that these retarded potentials each satisfy the wave equation and that Maxwell's equations for the fields E and H can be derived from his potentials. His vector potential does not obey the Coulomb gauge, however, but another relation now known as the Lorenz gauge. Although he was able to derive Maxwell's equations from his retarded potentials, he did not subscribe to Maxwell's view that light involves electromagnetic waves in the aether. He felt, rather, that the fundamental basis of all luminous vibrations are electric currents, arguing that space has enough matter in it to support the necessary currents.

1870:

- Helmholtz derived the correct laws of reflection and refraction from Maxwell's equations by using the following boundary conditions: D_n, E_t, and B_n, i.e., the normal component of vector D, the tangential component of vector E, and the normal component of vector B, are continuous at material interfaces which are non-conductors. Once these boundary conditions are taken into account Maxwell's theory is just a repeat of the theory of the Irish mathematical physicist James MacCullagh (1809–1847) (see Section 1.4). The details were not given by Helmholtz himself, but appeared rather in the dissertation of the Dutch physicist Hendrick Antoon Lorentz (1853–1928).

1874:

- Irish physicist George Johnstone Stoney (1826–1911) estimated the charge of an electron to be 10^{-20} Coulombs and introduced the term *electron*.

1876:

- American physicist Henry Augustus Rowland (1848–1901) performed an experiment inspired by Helmholtz which showed for the first time that moving electric charge is the same thing as an electric current.

1879:

- Edwin Herbert Hall (1855–1938) performed an experiment that was suggested by Rowland and discovered the *Hall Effect*, including its theoretical description by means of the *Hall term* in Ohm's law.

1880:

- Rowland showed that Faraday rotation can be obtained by combining Maxwell's equations and the Hall term in Ohm's law, assuming that displacement currents are affected in the same way as conduction currents. His earlier work, the first demonstration that a charged body in motion produces a magnetic field, attracted much attention.

1881:

- English physicist Sir Joseph John Thomson (1856–1940) attempted to verify the existence of the displacement current by looking for magnetic effects produced by the changing electric field made by a moving charged sphere.
- Irish mathematical physicist George Francis FitzGerald (1851–1901) pointed out that J. J. Thomson's analysis is incorrect because he left out the effects of the conduction current of the moving sphere. Including both currents made the separate effect of the displacement current disappear.
- Helmholtz, in a lecture in London, pointed out that the idea of charged particles in atoms can be consistent with Maxwell's and Faraday's ideas, helping to pave the way for our modern picture of particles and fields interacting instead of thinking about everything as a disturbance of the aether, as was popular after Maxwell.

1883:

- FitzGerald proposed testing Maxwell's theory by using oscillating currents in what we would now call a magnetic dipole antenna (loop of wire). He performed the analysis and discovers that very high frequencies are required to make the test. Later that year he proposed obtaining the required high frequencies by discharging a capacitor into a circuit.

1884:

- English applied mathematician Sir Horace Lamb (1848–1934) and the English physicist, mathematician and electrical engineer Oliver W. Heaviside (1850–1925) analyzed the interaction of oscillating electromagnetic fields with conductors and discovered the effect of skin depth.

- British physicist John Henry Poynting (1852–1914) showed that Maxwell's equations predict that energy flows through empty space with the energy flux given by $E \times B / (4\pi)$. He also investigated energy flow in Faraday fashion by assigning energy to moving tubes of electric and magnetic flux.

- German physicist Heinrich Rudolf Hertz (1857–1894) asserted that E made by charges and E made by changing magnetic fields are identical. Working from dynamical ideas based on this assumption and some of Maxwell's equations, Hertz was able to derive the rest of them. He showed in the limit Helmholtz's theories become Maxwell's equations. He wrote Maxwell's equations in *scalar form* (12 in number, instead of Maxwell's 20 equations in scalar form) by discarding the concept of *aether* introduced by Maxwell and starting from the *sources* rather than the *potentials* as Maxwell did.

1885:

- The Ganz Company of Budapest patented the *electric transformer*, following joint research by the Austrian-Hungarian electrical engineer Károly Zipernowski (1835–1942), the Hungarian electrical engineer Miksa Déri (1854–1938), and the Hungarian mechanical engineer Ottó Titusz Bláthy (1860–1939).

1886:

- Heaviside expressed Maxwell's equations in vector form using the notation of gradient, divergence, and curl of a vector. The comment of FitzGerald about them is:

  > *Maxwell's treatise is cumbered with the debris of his brilliant lines of assault, of his entrenched camps, of his battles. Oliver Heaviside has cleared these away, has opened up a direct route, has made a broad road, and has explored a considerable trace of country.*

  Heaviside introduced the term *impedance* as the ratio of voltage over current.

1887:

- Hertz found that ultraviolet light falling on the negative electrode in a spark gap facilitates conduction by the gas in the gap. This was the first demonstration of photo-electricity. Hertz established experimentally the existence of radio waves.

1888:

- Hertz discovered that oscillating sparks can be produced in an open secondary circuit if the frequency of the primary is resonant with the secondary. He used this radiator to show that electrical signals are propagated along wires and through the air at about the same speed, both about the speed of light. He also showed that his electric radiations, when passed through a slit in a screen, exhibit diffraction effects. Polarization effects using a grating of parallel metal wires were also observed. This is one of the most beautiful experiments performed in physics and it established Maxwell's theory, and prediction and generation of electromagnetic waves. Hertz produced, transmitted, and detected electromagnetic waves of wavelength 5 m and 50 cm. He used reflectors at the transmitting and receiving positions to concentrate the wave into a beam.
- German physicist Wilhelm Conrad Röntgen (1845–1923) showed that when an uncharged dielectric is moved at right angles to E, a magnetic field is produced.

1889:

- Hertz presented the theory of radiation from his oscillating spark gap. He was the first to obtain a solution of Maxwell's equation with an electric source: this was the introduction of what we call today the *Hertz vector*.
- Heaviside found the correct form for the electric and magnetic fields of a moving charged particle, valid for all speeds $v < c$, where c stands for the velocity of light.
- FitzGerald suggested that the speed of light is an upper bound for any possible speed.
- English physicist John William Strutt (Lord Rayleigh, 1842–1919) presented a model of radiation in terms of wave propagation.

1890:

- FitzGerald used the retarded potentials of Lorenz to calculate the electric dipole radiation from Hertz's radiator.

1892:

- Dutch physicist Hendrik Antoon Lorentz (1853–1928) presented his electron theory of electrified matter and the aether. This theory combines Maxwell's equations, with the source terms ρ and J, with the Lorentz force law, and the acceleration f of charged particles as: $m f = q E + q v \times B$, where m is the mass of a particle having a charge q and a velocity v. Lorentz concluded that the "null" result obtained by the Prussian-American physicist Albert Abraham Michelson (1852–1931) and the American chemist Edward Williams Morley (1838–1923) (see Section 1.4) was caused by an effect of contraction made by the aether on their apparatus and introduced the length contraction equation.

1895:

- The reciprocity theorem for electric waves was clearly stated and proved by Lorentz.

1.4   DEVELOPMENT OF THE THEORY OF LIGHT

The development of the theory of light is traced back to prehistoric times.

12,000 BC:
- Earliest known use of oil burning lamps.

2000–1000 BC:
- Mirrors found in Egyptian tombs.

900–600 BC:
- The Babylonians made convex lenses from crystals.

~500 BC:
- Pythagoras (582–500 BC) of Samos, Ionia, a Greek philosopher and mathematician, put forth the particle theory of light that assumed that every visible object emits a steady stream of particles that bombarded the eye. He said light consists of rays that acting like fillers travel in straight lines from the eyes to the object.
- Empedocles (492–432 BC) of Acragas, Sicily, a Greek philosopher and poet, postulated an interaction between rays from the eyes and rays from a source such as the sun.

~400 BC:
- Plato (427–347 BC) of Athens, Greece, a philosopher, belonged with respect to this matter to the Pythagoras school of thought.
- Aristotle (384–322 BC) of Stageria, Macedonia, a philosopher, concluded that light was like a wave and therefore differed from the prevalent school of the Pythagoras school of thought.

~300 BC:
- Euclid (325–265 BC) of Alexandria, Egypt, a Greek mathematician, wrote among many other works, *Optica,* dealing with vision theory and perspective. He defined the laws of reflection and refraction and stated that light travels in straight lines.

280 BC:
- The Egyptians built the first lighthouse.
- The Chinese used optical lenses and were the first to use corrective lens.

150 BC:
- Convex lenses were in existence at Carthage during this time.

1st century AD:
- Hero of Alexandria, a Greek inventor, published *Catoptrica* where the law of reflection relating the angle of incidence equal to the angle of reflection was demonstrated.

2nd century:
- Claudius Ptolemy (100–170) from Alexandria, a Greek astronomer and geographer, wrote on optics, derived the law of reflection from the assumption that light rays traveled in straight lines (from the eyes), and tried to establish a quantitative law of refraction.

1000:

- Abu Ali al-Hasan Ibn al-Haitham (Alhazen) (965–1038) of Basra, Persia, an Arab mathematician, wrote *Kitab al-Manazir* [translated into Latin as *Opticae Thesaurus Alhazeni* in 1270 by the Polish friar, theologian and scientist Erazm Ciolek Witelo (1230–1280)] on optics, dealing with reflection, refraction, lenses, parabolic and spherical mirrors, aberration and atmospheric refraction. Al-Haitham argued that sight is due only to light entering the eye from an outside source and there were no beam emanating from the eye itself. He gave an example of the pin-hole camera. He used spherical and parabolic mirrors and was aware of spherical aberration.

1093:

- Chinese philosopher, astronomer and mathematician Shen Kua (1031–1095) wrote *Meng ch'i pi t'an* (Brush Talks from Dream Brook) where he discussed concave mirrors and focal points. He noted that the image in a concave mirror is inverted.

1220:

- English theologian and scientist Robert Grosseteste (1168–1253) wrote on optics and light, experimenting with lens and mirrors. In his work *De Iride* (On the Rainbow) and *De Luce* he wrote:

  > *This part of optics, when well understood, shows us how we may make things a very long distance off appear as if placed very close, and large near things appear very small, and how we may make small things placed at a distance appear any size we want, so that it may be possible for us to read the smallest letters at incredible distances, or to count sand, or seed, or any sort or minute objects.*

1270:

- English philosopher Roger Bacon (1214–1294), a student of Grosseteste, was the first to apply geometry to study optics. He considered that the speed of light is finite and propagated in the medium similar to that of sound. In *Opus Malus*, he discusses the magnifying glass, eyeglasses and the telescope. He followed the work of Alhazen and spoke of convex and concave lenses. He postulated that the color of the rainbow was due to the reflection and refraction of sunlight through the rain drops.
- Polish theologian and scientist Erazm Ciolek Witelo (1230–1280) wrote *Perspectiva* treating geometric optics, including reflection and refraction, which became the standard text for several centuries. He also reproduced the data given by Ptolemy on optics, though he was unable to generalize or extend the study.
- Eyeglasses, convex lenses for the far-sighted, were first invented in or near Florence (as early as the 1270s or as late as the late 1280s – concave lenses for the near-sighted appeared in the late 15th century).

1275:

- German Dominican scholar, scientist, philosopher, and theologian Albert Magnus (later St. Albertus Magnus) studied the rainbow effect of light and speculated that the velocity of light is extremely fast but finite.

1304:

- German Dominican monk Theodoric (Dietrich) of Freiberg proposed the hypothesis that each raindrop in a cloud makes its own rainbow. He verified his hypothesis by observing the diffraction of sunlight through a circular bottle. The French physicist, physiologist, mathematician, and philosopher René Descartes (1595–1650) presents a nearly identical theory roughly 350 years later.

1500:

- Italian renaissance painter, architect, engineer, mathematician, musician, inventor, anatomist, sculptor, and philosopher Leonardo da Vinci (1452–1519) mentioned diffraction in his notebooks and studied reflection, refraction, and mirrors.

1590:

- Dutch lens makers, Hans Janssen and his son Zacharias Janssen (1580–1638) invented the first compound (twin lens) microscope.

1593:

- Giambattista della Porta (1540–1615), a natural philosopher, described the use of convex lens in order to improve the formation of images.

1604:

- Austrian mathematician and astronomer Johannes Kepler (1571–1630) developed the inverse square law for light and claimed that the speed of propagation is infinite.

1609:

- The Italian astronomer and physicist Galileo Galilei (1564–1642) from Pisa, developed a telescope modeled after the German-Dutch lens maker Hans Lippershey (1520–1619).

1611:

- Kepler discovered total internal reflection.

1618:

- Italian Jesuit physicist, mathematician, and geometer Francesco Maria Grimaldi (1618–1663) discovered diffraction patterns of light and became convinced that light was a wave-like phenomenon. In 1665, in a posthumous report, it was found that he gave the name of diffraction to the bending of light around opaque bodies.

1621:

- Lawyer and mathematician Willebrord van Roijen Snell (1580–1626) from Leiden, The Netherlands, experimentally determined the law of angles of incidence and reflection for light and for refraction between two media. He did not publish his discovery, which remained unknown till 1703 when it was published by the Dutch mathematician and physicist Christiaan Huygens (1629–1695).

1630:

- Vincenzo Cascariolo, a shoemaker in the city of Bologna, Italy, discovered fluorescence.

1637:

- French physicist, physiologist, mathematician, and philosopher René Descartes (1596–1650) published *La Dioptrique* explaining the formation of the rainbow and the laws of reflection and refraction. He theorized that light is a pressure wave flowing through the second of his three types of matter of which the universe was made. He invented properties of this fluid that made it possible to calculate the reflection and refraction of light. The *modern* notion of the aether was born. He did not believe in action at a distance and also thought that the velocity of light was infinite.

1638:

- Galileo Galilei attempted to measure the speed of light by a lantern relay between distant hilltops. He got a very large answer.

1657:

- French lawyer and mathematician Pierre de Fermat (1601–1665) showed that the principle of least time was capable of explaining refraction and reflection of light. Fighting with the Cartesians began. This principle for reflected light was anticipated long ago by Hero of Alexandria.

1667:

- English inventor, natural philosopher, experimental scientist, physicist, and architect Robert Hooke (1635–1703) reported in his *Micrographia* the discovery of the rings of light formed by a layer of air between two glass plates. These were actually first observed by the Irish philosopher, physicist, and chemist, Robert Boyle (1627–1691). These rings are now known as Newton's rings.

1669:

- Danish scientist and physician Rasmus Bartholin (1625–1698) discovered polarization by observing light through Iceland spar, which is a naturally occurring transparent crystal (optical quality calcite, $CaCO_3$). It separates an image into two displaced images when looked through along certain directions. Bartholin, not only saw double, but also performed some experiments and wrote a 60-page memoir about the results. This was the first scientific description of a polarization effect (the images are polarized perpendicular to each other).

1671:

- English scientist and mathematician Sir Isaac Newton (1642–1727) destroyd Hooke's theory of color by experimenting with prisms to show that white light is a mixture of all the colors and that once a pure color is obtained it can never be changed into another color. Newton argued against light being a vibration of the aether, preferring that it be something else that was capable of traveling through the aether. He did not insist that this something else consisted of particles, but allowed that it may be some other kind of emanation or impulse.

1672:

- Newton was the first to publish that white light was composed of different colors that were refracted at different angles of the prism.

1675:

- Ole Christensen Römer (1644–1710), the Danish astronomer, demonstrated the finite speed of light via observations of the eclipses of the satellites of Jupiter. He calculated the speed as 225,000 km per second.

1678:

- Dutch mathematician and physicist Christiaan Huygens (1629–1695) introduced his famous construction and principle defending the wave theory of light. He discovered the polarization of light by double refraction in calcite. To him light was still a longitudinal wave. He also accepted the presence of an all pervading medium called aether. He did not publish his work till 1690.

1690:

- Huygens, in *Traité de Lumiere*, provided the first numerical value for the speed of light of 2.3×10^8 m/s as opposed to Römer, who estimated it from observations.

1704:

- Newton published *Opticks* arguing that the corpuscles of light created waves in the aether.

1720:

- Wilhelm Jacob s'Gravesande (1688–1742), a Dutch physicist, was the proponent of the corpuscular theory of light and spread it after Newton's death.

1728:

- English astronomer James Bradley (1693–1762) showed that the orbital motion of the Earth changed the apparent motions of the stars in a way that was consistent with light having a finite speed of travel.

1800:

- Sir Frederick William Herschel (also known as Wilhelm Friedrich Herschel) (1738–1822), a German born British astronomer, discovered the infrared region of sunlight.

1801:

- English physicist Thomas Young (1773–1829) correctly concluded that light must be a transverse wave. His wave theory also explained the interference of light.
- German astronomer Johann Georg von Soldner (1776–1833) made a calculation for the deflection of light by the sun assuming a finite speed of light corpuscles and a non-zero mass. The result, 0.85 arc-sec, was rederived independently by English physicist Sir Henry Cavendish (1731-1810) and in 1911, by the German-Swiss-American scientist and physicist Albert Einstein (1879–1955), but went unnoticed until 1921.
- German physicist Johann Wilhelm Ritter (1776–1810) found that sun emits invisible ultraviolet radiation.

1803:

- Thomas Young explained the fringes at the edges of shadows by means of the wave theory of light. He explained the formation of colored bands in soap films:

> *Thus, when a film of soapy water is stretched over a wine glass, and placed in a vertical position, its upper edge becomes extremely thin, and appears nearly black, while the parts below are divided by horizontal lines into a series of colored bands ...*

and Newton's rings:

> *... and when two glasses, one of which is slightly convex, are pressed together with some force, the plate of air between them exhibits the appearance of colored rings, beginning from a black spot at the center, and becoming narrower and narrower, as the curved figure of the glass causes the thickness of the plate of air to increase more and more rapidly. ...*

establishing that there was an 180° change of phase when light was reflected from the surface of a denser medium, e.g., light traveling in air reflecting from the surface of glass or metal.

1808:

- French mathematician Pierre-Simon Laplace (1749–1827) gave an explanation of double refraction using the particle theory, which Young attacked as improbable.
- French physicist Étienne Louis Malus (1775–1812) discovered that light reflected at certain angles from transparent substances as well as the separate rays from a double-refracting crystal had the same property of *polarization*. In 1810, he received the prize of the French Académie des Sciences and embolded the proponents of the particle theory of light because no one saw how a wave theory can make waves of different polarizations.

1811:

- French physicist Dominique François Jean Arago (1786–1853) showed that some crystals alter the polarization of light passing through them.
- The French physicist Augustin-Jean Fresnel (1788–1827) and Arago discovered that two beams of light polarized in perpendicular directions do not interfere.

1812:

- French physicist Jean-Baptiste Biot (1774–1862) showed that Arago's crystals rotate the plane of polarization about the direction of propagation.

1814:

- Fresnel independently discovered the interference of light and explained it in terms of the wave theory.

1815:

- Scottish scientist and writer David Brewster (1781–1868) established his law of complete polarization upon reflection at a special angle now known as Brewster's angle.

1816:

- Arago, an associate of Fresnel, visited Young and described to him a series of experiments performed by Fresnel and himself which shows that light of differing polarizations cannot interfere. Reflecting later on this curious effect, Young saw that it can be explained if light is transverse instead of longitudinal. This idea is communicated to Fresnel in 1818 and he immediately saw how it clears up many of the remaining difficulties of the wave theory.

1819:

- French Académie des Sciences proposed as their prize topic for the 1819 Grand Prix a mathematical theory to explain diffraction. Fresnel wrote a paper giving the mathematical basis for his wave theory of light and in 1819 the committee, with Arago as chairman, and including Poisson, Biot, and Laplace, met to consider his work. It was a committee which was not well disposed to the wave theory of light, most believing in the corpuscular model. However, Poisson was fascinated by the mathematical model which Fresnel proposed and succeeded in computing some of the integrals to find other consequences. He wrote

> *Let parallel light impinge on an opaque disk, the surrounding being perfectly transparent. The disk casts a shadow – of course – but the very centre of the shadow will be bright. Succinctly, there is no darkness anywhere along the central perpendicular behind an opaque disk (except immediately behind the disk).*

This was a remarkable prediction, but Arago asked that Poisson's predictions based on Fresnel's mathematical model be tested. Indeed the bright spot was seen to be there exactly as the theory predicted. Arago stated in his report on Fresnel's entry for the prize to the Académie des Sciences :

> *One of your commissioners, M. Poisson, had deduced from the integrals reported by [Fresnel] the singular result that the centre of the shadow of an opaque circular screen must, when the rays penetrate there at incidences which are only a little more oblique, be just as illuminated as if the screen did not exist. The consequence has been submitted to the test of direct experiment, and observation has perfectly confirmed the calculation.*

Fresnel was awarded the Grand Prix and his work was a strong argument for a transverse wave theory of light.

1823:

- Physicist Joseph von Fraunhofer (1787–1826), from Germany, published his theory of diffraction.

1825:

- Fresnel showed that combinations of waves of opposite circular polarization traveling at different speeds can account for the rotation of the plane of polarization.

1827:

- Fresnel published a decade of research on the wave theory of light. Included in these collected papers are explanations of diffraction effects, polarization effects, double refraction, and Fresnel's sine and tangent laws for reflection at the interface between two transparent media.
- French engineer and physicist Claude-Louis Marie Henri Navier (1785–1836) published the correct equations for vibratory motions in one type of elastic solid leading to the well known Navier-Stokes equations (the second name refers to the Irish mathematical physicist George-Gabriel Stokes, 1819–1903). This begins the quest for a detailed mathematical theory of the aether based on the equations of continuum mechanics.

1828:

- French mathematician Augustine-Louis Cauchy (1789–1857) presented a theory similar to Navier's, but based on a direct study of elastic properties rather than using a molecular hypothesis. These equations are more general than Navier's. In Cauchy's theory, and in much of what follows, the aether is supposed to have the same inertia in each medium, but different elastic properties. He wrote 789 mathematical papers!

1837:

- British mathematician George Green (1793–1841) attacked the elastic aether problem from a new angle. Instead of deriving boundary conditions between different media by finding which ones give agreement with the experimental laws of optics, he derived the correct boundary conditions from general dynamical principles. This advance makes the elastic theories not quite fit with light.

1839:

- Irish mathematical physicist James MacCullagh (1809–1847) invented an elastic aether in which there are no longitudinal waves. In this aether the potential energy of deformation depends only on the rotation of the volume elements and not on their compression or general distortion. This theory gives the same wave equation as that satisfied by E and B in Maxwell's theory.
- Irish-Scottish physicist William Thomson (Lord Kelvin, 1824–1907) removed some of the objections to MacCullagh's rotation theory by inventing a mechanical model which satisfies MacCullagh's energy of

rotation hypothesis. It has spheres, rigid bars, sliding contacts, and flywheels.

- Cauchy and Green presented more refined elastic aether theories. Cauchy does it by removing the longitudinal waves by postulating a negative compressibility, and Green using an involved description of crystalline solids.

1845:

- British scientist Michael Faraday (1791–1867) discovered that the plane of polarization of light is rotated when it travels in glass along the direction of the magnetic lines of force produced by an electromagnet (Faraday rotation).

1846:

- English astronomer Sir George Biddell Airy (1801–1892) modified MacCullagh's elastic aether theory to account for Faraday rotation.
- Faraday, inspired by his discovery of the magnetic rotation of light, wrote a short paper speculating that light might be electromagnetic in nature. He thought it might be transverse vibrations of his beloved field lines.

1849:

- French physicist Armand–Hippolyte-Louis Fizeau (1819–1896) repeated Galileo's hilltop experiment (9 km separation distance) with a rapidly rotating toothed wheel and measured the velocity of light, as $c = 3.15 \times 10^8$ m/s.
- Irish mathematical physicist George Gabriel Stokes (1819–1903) published a long paper on the dynamical theory of diffraction in which he showed that the plane of polarization must be perpendicular to the direction of propagation.

1850:

- French applied physicist Jean-Bernard-Léon Foucault (1819–1868), improved on Fizeau's measurement and used his apparatus to show that the speed of light is less in water than in air. He also measured the velocity of light, finding a value that is within 1 percent of the true figure.
- Stokes's law relating to the curl of a vector is stated without proof by Lord Kelvin. Later, Stokes assigns the proof of this theorem as part of the examination for the Smith's Prize. Presumably, he knew how to do the problem. Maxwell, who was a candidate for this prize, later remembers this problem, traces it back to Stokes, and calls it the Stokes's theorem.

1867:

- French physicist and mathematician Valentin Joseph Boussinesq (1842–1929) suggested that instead of aether being different in different media, perhaps the aether is the same everywhere, but it interacts differently with different materials, similar to the modern electromagnetic wave theory.
- German mathematician George Friedrich Bernhard Riemann (1826–1866) proposes a simple electric theory of light in which Poisson's

equation is replaced by $\nabla^2 V - (1/c^2)\partial^2 V / \partial t^2 = -4\pi\rho$.

1872:

- French physicist Éleuthère Élie Nicolas Mascart (1837–1908) looked for the motion of the Earth through the aether by measuring the rotation of the plane of polarization of light propagated along the axis of a quartz crystal. No motion was found with a sensitivity of $v/c \approx 10^{-5}$.

1873:

- Scottish physicist and mathematician James Clerk Maxwell (1831–1879) published his *Treatise on Electricity and Magnetism*, which discusses everything known at the time about electromagnetism from the viewpoint of Faraday. He states:

  > *In several parts of this treatise an attempt has been made to explain electromagnetic phenomena by means of mechanical action transmitted from one body to another by means of a medium occupying the space between them. The undulatory theory of light also assumes the existence of a medium. We have now to show that the properties of the electromagnetic medium are identical with those of the luminiferous medium.*
  >
  > *But the properties of bodies are capable of quantitative measurement. ... If it should be found that the velocity of propagation of electromagnetic disturbances is the same as the velocity of light, and this not only in air, but in other transparent medium then we have strong reasons for believing that light is an electromagnetic phenomenon, and the combination of the optical with the electrical evidence will produce a conviction of the reality of the medium similar to that which we obtain, in the case of other kinds of matter, from the combined evidence of the senses.*

- Maxwell also produced the first color picture.

1881:

- Prussian-American physicist Albert Abraham Michelson (1852–1931) and the American chemist Edward Williams Morley (1838–1923) attempted to measure the motion of the Earth through the aether by using interferometry. They found no relative velocity. Michelson interpreted this result as supporting Stokes's hypothesis in which the aether in the neighborhood of the Earth moves at the velocity of Earth.

1.5   WHO WAS MAXWELL?

The name of James Clerk Maxwell is synonymous with the development of modern physics. He laid the basic foundation for electricity, magnetism, and optics [12-14]. The theory on electromagnetism is one of the few theories where the equations have not changed since its original conception, whereas their interpretations have gone through revolutionary changes at least twice [15-17]. The first revolution was by Hertz and Heaviside and the second by Larmor, as discussed in the next section. Maxwell's work on electromagnetic theory was only a small part of his research. As the English physicist and mathematician Sir James Hopwood Jeans (1877–1946) [18] pointed out: *In his hands electricity first became a mathematically exact science and the same might be said of other larger parts of Physics.* In whatever area he worked, he brought new innovation. He published five books and approximately 100 papers. Maxwell can be considered as one of the world's greatest scientists even if he had never worked on electricity and magnetism. His influence is everywhere, which surprisingly is quite unknown to most scientists and engineers! As the German physicist Max Karl Ernst Ludwig Planck (1858–1947) [18] said:

> *His name stands magnificently over the portal of classical physics and we can say this of him: by his birth James Clerk Maxwell belongs to Edinburgh, by his personality he belongs to Cambridge, by his work he belongs to the whole world.*

Here we provide **a cursory overview of some of his achievements**, which are still in vogue today. Hopefully, this will make one more familiar with Maxwell, as electromagnetics [23], the physical basis of wireless, took shape under him. As pointed out in [12], the name of Maxwell invokes according to well known scientists: *One scientific epoch ended and another began with James Clerk Maxwell* – Albert Einstein, and *From a long view of the history of mankind – seen from, say, ten thousand years from now – there can be little doubt that the most significant event of the nineteenth century will be judged as Maxwell's discovery of the laws of electrodynamics* – Richard Feynman.

His first publication *On the Description of Oval Curves, and those having a plurality of foci* was done at the age of 14 in 1846. It dealt with drawing curves on a piece of paper using pins, string, and pencil. He varied the number of times he looped the string around each pin to generate various egg shaped graphs. Earlier, René Descartes had discovered the same set of bi-focal ovals, but Maxwell's results were more general and his construction method much simpler [11]. It turned out that such constructions have significant practical applications in optics. He also wrote in 1855 on *Description of a new form of Platometer*, an instrument for measuring areas of plane figures drawn on a paper.

In 1855, he published the first-part of his paper *On Faraday's Lines of Force*, and the second part in the following year. His objective was: *to find a physical analogy which will help the mind to grasp the results of previous investigations without being committed to any theory founded on the physical science from which that conception is borrowed so that it is neither drawn aside from the subject in the pursuit of analytical subtleties nor carried beyond the truth by a favorite hypothesis.* The laws of electricity were compared with the properties of an incompressible fluid, the motion of which was retarded by a force proportional to the velocity, and the fluid is supposed to posses no inertia.

In the latter part of the paper he proceeded to consider the phenomenon of Electromagnetism and showed how the laws discovered by Ampère lead to conclusions identical with that of Faraday. *On Physical Lines of Force* he further speculated that the magnetic field were occupied by molecular vortices the axes of which coincide with the lines of force.

In 1855, he published a paper on *Experiments on Colour as Perceived by the Eye*. He wrote two papers on the sensitivity of the retina to color. To achieve this he developed equipments to look into the eye which was a modification of the ophthalmoscope originally developed by Helmholtz [13]. At the point of the retina where it is intersected by the axis of the eye there is a *yellow spot*. Maxwell observed that the nature of the spot changes as a function of the quality of the vision. The macular degeneration of the eye affects the quality of vision and is the leading cause of blindness in people over 55 years old. Today, the extent of macular degeneration of the retina is characterized by this *Maxwell spot test*. Maxwell also worked on the generation of white light by mixing different colors and in 1860, published the paper *On the Theory of Compound Colors and its Relations to the Colors of the Spectrum*. In this, he extended the work of Young who first postulated only three colors, red, green and violet are necessary to produce any color including white and not all the colors of the spectrum are necessary as first illustrated by Newton. Maxwell created a color triangle and illustrated that any color can be generated with a mixture of any three primary colors and that a normal eye has three sorts of receptors as illustrated in his 1861 paper *On the Theory of Three Primary Colors*. He chose the three primary colors as red, green, and blue colored light. Today, color television works on this principle, but Maxwell's name is rarely mentioned [11]. However, other choices for the primary colors are equally viable. He demonstrated these principles by pasting different color papers on a spinning top. He provided a methodology of generating any color represented by a point inside a triangle whose vertex represented the three primary colors. The new color is generated by mixing the three primary colors in a ratio determined by the respective distance of the point representing the new color from the vertices of the triangle. In the present time, this triangle is called a chromatist diagram and differs in details from his original. He took the first color photograph in 1861. He also developed a *color box* to replace his spinning top to standardize his experiments which could generate any color by impinging sunlight through various slits of different colors. He used polarized light to reveal strain patterns in mechanical structures and developed a graphical method for calculating the forces in the framework. He was the first to show that in color blind people, their eyes are sensitive only to two colors and not to three as in normal eyes. Typically, they are not sensitive to red. He suggested using spectacles with one red lens and one green for color blind people. He was the first to develop the fish eye lens [11].

On the General laws of Optical Instruments published in 1858, Maxwell laid down the conditions which a perfect optical instrument must fulfill, and showed that if an instrument produce perfect image of an object, i.e., image free from astigmatism, curvature and distortion, for two different positions of the object, it will give perfect image at all distances.

His paper in 1858 was on the nature of Saturn's rings, which won him the Adams prize [named after the English astronomer John Couch Adams (1819– 1892), the discoverer of Neptune]. It took him two years to do this work. The problem that he addressed was under what conditions the rings of Saturn would be stable if they were (1) solid, (2) fluid, or (3) composed of many separate pieces of matter. Maxwell showed that a solid ring could be stable except in one arrangement where about 80% of its mass was concentrated at one point on the circumference and the rest distributed throughout. Since telescopic observations reveal otherwise, this cannot be a possible solution. He also used a Fourier analysis to show that the ring would break up if it were of a single fluid, as waves would be generated. The only solution is that the ring consisted of separate bodies. He showed that it could vibrate in four different ways as long as its average density was low enough compared with that of Saturn. When he considered only two rings, one inside the other, he found that some arrangements were stable, but others not. Also, because of friction, he predicted that the inner ring will move inward and the outer ring outward on a very long time scale. Here, we have almost the only example of a piece of work that Maxwell himself completed and left in perfect form. Airy said that this was one of the most remarkable applications of mathematics he has ever seen [18]. His conclusions received observational confirmation 38 years later when the American astronomer James Edward Keeler (1857–1900) obtained spectroscopic proof that the outermost portions of the rings were rotating less rapidly than the inner portions [18]. He also built a hand cranked model to demonstrate the motion of the rings. Some say this research led him directly to the next topic.

Next, he worked on the kinetic theory of gases and published a paper in 1860. He tried to determine the speed with which the smell fills the air when a perfume-bottle is opened. In order to address this problem he derived the Maxwell distribution for molecular velocities. The distribution turned out to have the well-known bell-shaped curve now known as the normal distribution. This was done to represent various motions of the molecules in a single equation, representing a statistical law [11]. His discovery opened up an entirely new approach to physics, which led to statistical mechanics, to a proper understanding of thermodynamics, and to the use of probability distributions in quantum mechanics. No one before Maxwell ever applied a statistical law to a physical process. He also predicted a new law of viscosity for gases in 1860, which was a magnificent piece of work, but was not devoid of flaws. His work inspired the Austrian mathematician, physicist, and inventor of statistical mechanics Ludwig Boltzmann (1844–1906), and their works led to the *Maxwell-Boltzmann* distribution of molecular energies. In 1866, he wrote *Dynamical Theory of Gases* and produced the first statistical law of physics. He wrote a paper on Boltzmann's theorem in 1879 on the *Average Distribution of Energy* in a system of material points. This enabled people to explain the properties of matter in terms of the behavior, en masse, of its molecules. One of the ideas in the paper was the method of ensemble averaging, where the whole system is much easier to analyze, rather than dealing with individual components. Interestingly, this mode of analysis is quite prevalent nowadays in most signal processing and communications theory applications. However, his result for the specific heat of air was off from the measurement. Instead of trying to explain this discrepancy in his theory by ingenious attempts, he said: *Something essential to the complete statement of the physical theory of molecular encounters must have hitherto escaped us, and that the only thing to do was to adopt the attitude of thoroughly conscious ignorance that is the prelude to every real advance in science* [11]. He was right. The explanation came 50 years later from quantum theory. He was also the first to suggest using a centrifuge to separate gases, which is still being used in modern times.

He developed a coherent set of units of measurement of electricity and magnetism in 1863. They were later adopted almost unchanged as the first internationally accepted system of units, which became known misleadingly as the Gaussian system, which is a combination of the Electrostatic units and the Electromagnetic units. He also introduced the dimensional analysis in physics which is used currently and yet nobody wonders who first thought about it. He also produced the first standard of electrical resistance in 1868.

Sir Ambrose Fleming wrote [18]: *In electricity course, he gave us a new and powerful method of dealing with problems in networks and linear conductors. Kirchoff's corollaries of Ohm's Law had provided a means only applicable in the case of simple problems in which one could foresee the direction of flow of current in each conductor. But that was not possible in complicated networks. Maxwell initiated a new method by considering the actual current in each wire to be the difference of two imaginary currents circulating in the same direction round each mesh of the network. In this way, the difficulty of foreseeing the direction of the real currents was eliminated. The solution of the problem was then reduced to the solution of a set of linear equations and the current in any wire could be expressed as the quotient of two determinants. After Maxwell's death, in 1885 I communicated a paper to the Physical Society of London in which the method was extended so as to give an expression for the electrical resistance of any network between any two points.* One would immediately recognize this as a method for writing the loop equations that are currently available in all undergraduate textbooks dealing with electrical circuits, and yet no mention is made of Maxwell, the inventor of this technique!

When creating his standard for electrical resistance, he wanted to design a governor to keep a coil spinning at a constant rate. He made the system stable by using the idea of negative feedback. He worked out the conditions of stability under various feedback arrangements. This was the first mathematical analysis of control systems [11]. This work did not get any attention till 1940, when gun control radars were in demand during the Second World War. After the war the American mathematician Norbert Wiener (1894–1964) took things further and developed the science of cybernetics, based on his paper *On Governors* in 1868.

He wrote the *Dynamical Theory of Electromagnetism* using the Quaternion convention. However, he wrote the final equations in the scalar form even though he used the terms "curl", "convergence" and "gradient". Nowadays, convergence is replaced by its negative, which is called divergence, and the other two are still in the standard mathematical literature. These are available *On the Mathematical Classification of Physical Quantities.* Maxwell identified light with electromagnetic waves and introduced the concept of aether as the basic medium of the electromagnetic field to retain the possibility of a mechanical interpretation. He presented his first paper on the new theory before the Royal Society in 1864 and published the comprehensive *Treatise on Electricity and Magnetism* in 1873 [15].He also showed that radiation pressure from the sun exists, which has a force of 4 pounds per square mile.

In 1868 [23], *On a Method of making a Direct Comparison of Electrostatic with Electromagnetic Force; with a note on the Electromagnetic Theory of Light,* he measured the speed of light by using 2,600 batteries to produce 3,000 volts. The goal was to balance the electrostatic attraction between two charged metal plates against the magnetic repulsion between two current carrying coils and built a balance arm to do this. He got a result of 288,000 km/sec as compared to the current accepted value of 299,792.5 km/sec.

He, along with the English biologist Thomas Henry Huxley (1825–1895), was the joint scientific editor of the 9th edition of Encyclopedia Britannica. Maxwell always delivered scientific lectures for the common people using models. He also was very prolific in writing limericks, as we will see later.

He wrote a book on The *Theory of Heat* in 1871 and provided a completely new formulation of the relationships between pressure, volume, temperature, and entropy and expressed them through differential equations known as Maxwell's relations [11]. He introduced the "Maxwell demon", as termed by Lord Kelvin, i.e., the molecule-sized creature which was going to defy the second law of thermodynamics [19]. The goal was that this demon can separate low velocity molecules from high velocity ones by opening an aperture and, thus, making heat flow from a colder region to a hotter one defying the second law of thermodynamics. Hence, the demon is generating a perpetual motion machine: the machine will keep on working till the temperature difference between the two regions fell back to zero and then we would be back to where we started. This cannot really happen. And why it cannot happen, the answer according to Maxwell, is:
  (1)   The second law of thermodynamics is a statistical law. So that if one throws a bucket of water into the sea, one cannot get the same bucket out again as the law applies to molecules in masses and not to individuals.
  (2)   If we are sufficiently nimble fingered like the demon, we could break the second law. The reason we cannot is because we are not clever enough.
The Austro-Hungarian physicist Leo Szilard (1898–1964) in 1929 showed that the very act of acquiring information about a system increases its entropy proportional to the amount of information gathered. Through such work of Szilard and others, Maxwell's demon helped the creation of information theory, now an essential part of communication and computing [11].

He also wrote a paper *On Hills and Dales* in 1870, extending the work of Cayley published in 1859. The surface of the Earth has high areas or hills and low areas with a bottom point. There are also ridges, valleys, or dales, and passes. He thought that the numbers of each of these features must be somehow related by mathematical rules. His original ideas about the Earth's surface have now evolved into a branch of topology called global analysis [11].

Even though Maxwell has influenced development in many areas of physical sciences and had started a revolution in the way physicists look at the world, he is not very well known, unfortunately, outside some selected scientific communities. In fact, when the Royal Society of London held its tercentenary celebration, Queen Elizabeth II presided and praised a number of former Fellows – presumably listed by the Society. Inexplicably, Maxwell was not among them [11,12]. He has been more widely commemorated elsewhere, even in countries without a strong scientific tradition. For example, the governments of Mexico, Nicaragua, and San Marino are among those who have issued postage stamps in his honor. Some claim that the reason Maxwell was not so well known (he was not even knighted) is because he was too humble and never proselyte his theory. There may be some truth to that. In addition, there may be other reasons as outlined below.

J. J. Thomson, speaking at the centenary celebration of Maxwell, said that, according to his teacher, the mathematician William Hopkins (1793–1866) at Cambridge,

> Maxwell was unquestionably the most extraordinary man he had met with in the whole course of his experience; that it appeared impossible for Maxwell to think wrongly on any physical subject, but that in analysis he was far more deficient. ... The public lectures were however read and he was compelled by the manuscript to keep to the track.

In the same celebration, Sir James Jeans succinctly summarized his contributions

> as of purely abstract in nature. As a consequence, Maxwell did not alter the face of civilization as Faraday did, or at least did not alter it so immediately or in a manner obvious to the eye. It could hardly have been written of him during his lifetime, as it was of Faraday, that our life is full of resources which are the results of his labours; we may see at every turn some proof of the great grasp of his imaginative intellect. Faraday had used his clear vision and consummate skill as an experimenter to explore those strata of nature which lie immediately under our hands; Maxwell used his clear vision and consummate skill as a theorist to explore the deeper strata in which the phenomenon of the upper strata have their origin.

However, as Sir James Jeans pointed out [18]:

> It was not keeping with Maxwell's methods that he should finish off a piece of work so completely that nothing could be added to it, his plan was rather to open up wide vistas which

could provide work in their detailed exploration for the whole generations yet to come. ... It was his power of profound physical intuition coupled with adequate, although not outstanding mathematical technique, that lay at the basis of Maxwell's greatness.

Sir Horace Lamb [18] said:

He had his full share of misfortunes with the blackboard, and one gathered the impression, which is confirmed I think by the study of his writings, that though he had a firm grasp of essentials and could formulate great mathematical conceptions, he was not very expert in the details of minute calculations. His physical instincts saved him from really vital errors.

However, there may be an additional reason that he was not recognized during his life time for his work, which is that he had an eccentric side to his personality. He had a tendency to make insulting remarks, largely in the form of sardonic limericks [12]! During the 1874 British Association meeting at Belfast he delivered himself of several poems, in one of which he refers to members of this highly respected body in this way [20, p. 637]:

So we who sat, oppressed with science,

As British asses wise and grave,

Are now transformed to Wild Red Lions...

[The Red Lions are a club formed by members of the British association to meet for relaxation after the graver labors of the day].

And after hearing a lecture by the Scottish physicist Peter Guthrie Tait (1831–1901) in 1876, he wrote [20, p.646]:

Ye British asses, who expect to hear

Ever some new thing,

I've nothing to tell but what, I fear,

May be a true thing.

For Tait comes with his plummet and his line,

Quick to detect your

Old bosh new dressed in what you call a fine

Popular lecture.

And again in 1878 where he devoted the Rede lecture on the invention of the telephone as:

> *One great beauty of Professor Bell's invention is that the instruments at the two ends of the line are precisely alike...The perfect symmetry of the whole apparatus – the wire in the middle, the two telephones at the ends of the wire; and the two gossips at the ends of the telephones may be very fascinating to a mere mathematician, but would not satisfy the evolutionist of the Spencerian type, who would consider anything with both ends alike, to be an organism of the very low type, which must have its functions differentiated before any satisfactory integration can take place.*

It is rather amusing to note that in recent times this statement of Maxwell in many places has been used to claim that Maxwell was against the evolutionary theory of Darwin, which the British philosopher and sociologist Herbert Spencer (1820–1903) popularized [21]. However, according to Sir John Ambrose Fleming [18], who attended the lecture: *It was a brilliant discourse, illustrated by flashes of wit, apt analogies and much learning but it was not of the type most useful to convey to unscientific hearers an idea of the mode in which a telephone operates.*

Maxwell's personality often showed two additional unusual characteristics [11]. The first is that he could return to a subject, often after a gap of several years, and take it to new heights using an entirely new approach. He did this twice with electromagnetism. The second is that often his intuition led him to correct results even when he had made mistakes along the way. He was tolerant of the mistakes of others, but was very critical of the failure to be honest and open to the readers. He rebuked Poisson for telling lies about the way people make barometers and Ampère for describing only perfected experiments to demonstrate his law of force and hiding the poor experiments by which he had originally discovered the law.

1.6   WHAT WAS/IS MAXWELL'S ELECTROMAGNETIC THEORY?

Interestingly there is no clear cut answer to this question as we will see. This section is derived from the writings of several scientists who themselves modified his theory. First, we describe what Maxwell presented, followed by the modifications made by others. As elucidated by Sir James Jeans [18]

> *Maxwell pictured electromagnetic theory of light in terms of a medium whose properties could be specified completely in terms of a single mathematical constant. He saw that if the value of this constant could once be discovered, it ought to become possible to predict all phenomena of optical theory with complete mathematical precision. Maxwell showed that the constant in question ought to be merely the ratio of electromagnetic and electrostatic units of electricity and his first calculations suggested that this was in actual fact equal*

to the constant of the medium which measured the velocity of light.

> *The first mention of the great discovery comes in a letter, which he wrote to Michael Faraday on 19th October 1861: I suppose the elasticity of a sphere to react on the electrical matter surrounding it, and press it downwards. From the determination by Kohlrausch and Weber of the numerical relation between static and the magnetic effects of electricity, I have determined the elasticity of the medium in air, and assuming that it is the same with the luminiferous aether, I have determined the velocity of propagation of transverse vibrations. The result is 193,088 miles per second. Fizeau has determined the velocity of light as 193,118 miles per second by direct experiment.*

> *Even though the two numbers quoted above agreed to within 30 miles per second, oddly enough both are in error by more than 6,000 miles a second. When Maxwell came to publish his paper,* A Dynamical Theory of the Electromagnetic Field, *probably the most far reaching paper he ever wrote, he gave the two velocities in terms of kilometers per second Kohlrausch and Weber (310,740,000 meters a second) and Fizeau (314,858,000 meters a second) and is nowhere near the figure. Happily he seems to have realized that the velocity of light was not at all accurately known, and so he did not allow himself to be deterred, as Newton had been, by a substantial numerical disagreement in his law of universal gravitation.*

According to English physicist Sir Oliver Lodge (1851–1940):

> *Maxwell perceived that a magnetic field was wrapped around a current, and that a current could equally well be wrapped around a magnetic field, that in fact the relation between that was reciprocal, and could be expressed mathematically by what he subsequently called* curl. *When the two fields coexisted in space, the reaction between them could be expressed as the curl of a curl, and this he simplified down to the well known wave equation, the velocity of propagation of the wave being the reciprocal of the geometric mean of an electric and a magnetic constant. This velocity for a time he called the number of electrostatic units in a magnetic unit, and proceeded to devise experiments whereby it could be measured. Experiments made by him at King's College resulted in some near approach to the velocity of light, so that thenceforth, in his mind light became an electromagnetic phenomenon. He was assisted in his ideas by an imaginative constructive model of the aether, a model with rolling wheels and sliding particles, which subsequently he dropped, presumably as being too*

complex for reality and remained satisfied with his more abstract equations, which were reproduced in his great Treatise *in 1873.*

This provides one reason why Maxwell's theory was so difficult to follow. By never identifying his physical pictures with reality, Maxwell left himself free to discard one picture and adopt another as often as expediency or convenience demanded. He described his method of procedure in the following words:

If we adopt a physical hypothesis, we see the phenomenon only through a medium, and are liable to that blindness to facts and rashness in assumption which a partial explanation encourages. We must therefore discover some method of investigation which allows the mind at every step to lay hold of a clear physical conception, without being committed to any theory founded on physical science for which that conception is borrowed.

Maxwell considered Faraday's line of force similar to the lines of flow of a liquid. Maxwell himself wrote to Lord Kelvin [13]:

I suppose that the "magnetic medium" is divided into small portions or cells, the divisions or cell walls being composed of a single stratum of spherical particles these particles being "electricity". The substance of the cells I suppose to be highly elastic both with respect to compression and distortion and I suppose the connection between the cells and the particles in the cell walls to be such that there is perfect rolling without slipping between them and that they act on each other tangentially. I then find that if the cells are set in rotation, the medium exerts a stress equivalent to a hydrostatic pressure combined with a longitudinal tension along the lines of axes of rotation. ... Thus there will be a displacement of particles proportional to the electro-motive force, and when this force is removed, the particle will recover from displacement. I have calculated the relation between the force and the displacement on the supposition that the cells are spherical and that cubic and linear elasticities are connected as in a perfect solid. I have found from this the attraction between two bodies having given free electricity on their surfaces. And then by comparison with Weber's value of the statical measure of a unit of electrical current I have deduced the relation between elasticity and density of the cells. The velocity of the transverse undulations follows from this directly and is equal to 193,088 miles per second, very nearly that for light.

Even though the final results of the original Maxwell's theory are valid even today, however, the intermediary steps used to arrive at the conclusion was in question [16]. The first problem was associated with the definition of the charge. For example, consider a positively charged ball that is placed inside an infinite dielectric medium. Modern theory defines the positive surface charge on the surface of the conductor as seen in Figure 1.1. This charge creates an electric field which produces polarization throughout the dielectric. Next we divide the dielectric using an imaginary surface C. One part (A) of the dielectric lies between the conducting sphere and C; the other part (B) lies between C and goes to infinity. The innermost boundary of A which actually touches the sphere, according to modern theory will carry a negative polarization charge which is smaller in magnitude than the conduction charge on the sphere. The outermost boundary of A, characterized by the surface C bears a positive polarization charge numerically equal to the negative polarization charge on the inner surface of A. The charge on the outermost boundary of A is exactly compensated by a negative polarization charge on the innermost boundary of B, i.e., by a charge on surface C considered as the inner boundary of B. Accordingly no space charge at all exists and we have only the positive conduction charge and the numerically smaller negative polarization charge on the surface of the dielectric which is immediately adjacent to it as shown in Figure 1.1 [16]. Now if we use Maxwell's interpretation of the same situation using his quote: *The charge therefore at the bounding surface of a conductor and the surrounding dielectric; which in the old theory was called the charge of the conductor, must be called in the theory of induction (Maxwell's theory) the surface charge of the surrounding dielectric* [15, Vol. I, Art 111].

In Maxwell's theory, we begin with a displacement **D** which exists throughout the dielectric and which points away from the center of the sphere. Since the displacement points away from the center of the sphere, it enters B's inner boundary in a direction parallel to that boundary's inward-directed normal. According to Maxwell's definition of charge, the inner boundary of B has a positive charge on it numerically equal to **D** as shown in Figure 1.2. In addition, the innermost boundary of A (surface C) coincides with the innermost boundary of B. Therefore, the outermost boundary of A has on it a negative charge equal

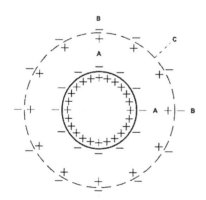

Figure 1.1. Modern picture for a conducting sphere embedded in a dielectric.

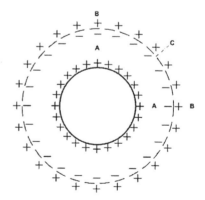

Figure 1.2. Maxwellian picture for a conducting sphere embedded in a dielectric.

and opposite to the positive charge on the inner boundary of B as shown in Figure 1.2. Since the boundaries coincide, no net charge can exist anywhere in the dielectric. At the surface of the sphere and the innermost boundary of the dielectric the displacement enters the dielectric boundary parallel to its inwardly directed normal and so we have on this surface a positive charge. But since no displacement at all exists within the sphere, its surface is unchanged. Consequently, the positive charge on the inner surface of the dielectric is uncompensated. The result is that what modern theory calls the positive surface charge of the conductor, Maxwell's theory called the positive surface charge of the inner surface of the dielectric [16].

Maxwell considered magnetism as a phenomenon of rotation and electric currents as a phenomenon of translation. So in any magnetic field the medium is in rotation about the lines of magnetic force. Maxwell's theory sought unity through a highly plastic set of field equations coupled to Hamilton's principle, named after the Irish mathematician William Rowan Hamilton (1805–1865). In current bearing linear circuits, Maxwell thought that the currents were linked by rigid constraints to an intervening medium called the aether. Thus, the most difficult concepts for the modern reader to grasp in the original Maxwell's theory are the concepts of "charge" and "current". In modern theory, charge is the source of the electric field and current is the source of the magnetic field. In his theory, charge is produced by the electric field; current, in the usual sense, is the rate of change of charge over time, and is only indirectly related to the magnetic field. Therefore, in Maxwellian theory, charge is a discontinuity of the displacement D and not in E. Maxwell's goal was to create a theory of electromagnetism which made no use whatsoever of the microstructure of the matter. To Maxwell, the conduction current was effectively a continuous series of charging and discharging. The conduction current then is the process and growth of displacement. Maxwell proposed all currents are closed. In this fashion, he introduced the displacement current. Maxwell quite explicitly limited electric polarization to the boundary conditions on the flux characteristics of electric displacement and magnetic induction. Maxwell did not apply boundary conditions to explain the phenomenon of reflection and refraction. This was achieved by Hertz and Heaviside.

In the European continent, physicists of the period were intimately familiar with field equations. However, each of them attempted to obtain Maxwell's field equations as a limiting case of Helmholtz's (entirely non Maxwellian) polarization theory of aether and matter. Maxwell's theory was based on a property of vortex notion, which Helmholtz had deduced (1858). Even those who no longer reached the Maxwell equations via Helmholtz (like H. A. Lorentz after 1892 or Hertz after 1890) continued to bear unmistakable marks of the Helmholtz polarization theory. However, there is a marked distinction between the two theories due to the difference between those who viewed electricity as a by-product of the field processes from those who did not [16]. Helmholtz's theory makes a distinction between conduction and polarization charge, with free charge being their sum. Helmholtz's theory consisted of three components [16]:

(1) expression for electromagnetic potentials and the forces derived from them
(2) a continuity equation linking charge and current, and
(3) a model for an electrically and magnetically polarizable medium.

The conflict between Helmholtz's and Maxwell's theories occurs where one would expect to find it, i.e., in the continuity equation. In Helmholtz's theory, all fields involve interaction between charge densities, and these interactions are not in fact propagated [16]. Only the polarizations propagate, charge interactions are always instantaneous. The basic difference is that in Maxwell's theory charge is the discontinuity in the electric displacement whereas in Helmholtz's theory it is with the discontinuity in the electric field. Thus, the problem with Helmholtz's theory is then at material interfaces where there will be charges as the electric field is discontinuous and where the displacement is not. Jules Henri Poincaré (1854–1912) [19], the French mathematician, theoretical scientist, and philosopher of science, said on Maxwell's *Treatise* that was published in 1873: *I understand everything in this book except what is meant by a body charged with electricity.*

In summary, Maxwell's *Treatise* provided an immense fertile theory, but in a form which was awkward, confusing, and on some points simply wrong [22, p. 108]. It was especially ill-tuned to handling the propagation problems that were coming into increasing prominence in the 1880s in connection with advances in telegraphy, telephony, and the study of electromagnetic waves. Before Maxwell's theory could gain wide acceptance and come into general use, it required substantial revision and clarification; both its physical principles and their mathematical expressions had to be put into a simpler and more easily grasped form. The most important steps in this process were taken in the mid 1880s by Poynting, FitzGerald, and Heaviside on the flow of energy for an electromagnetic field [22].

Maxwell was able to show [15, Art. 783 & 784; Equations 8 & 9] that in free space $(d^2 J)/(d t^2) + \{d(\nabla^2 \Psi)\}/(d t) = 0$, where ψ is the scalar electric potential. He then made an important assertion, for which he provided no real justification: " $\nabla^2 \Psi$ ", he said, *which is proportional to the volume density of the free electricity, is independent of t*, i.e., he claimed that the electric potential is determined solely by the spatial distribution of charge which in a non-conductor does not change. This is the assumption usually made in electrostatics, and Maxwell simply extended it to general electromagnetic theory without alteration or explanation. Time independence implied that the electric potential adjusted instantaneously across all space to any changes in the positions or magnitudes of the charges. It also implied $(d^2 J)/(d t^2) = 0$, so that as Maxwell wrote J *must be linear function of t, or a constant, or zero and we may therefore leave J and ψ of account in considering wave disturbances.* In practice, Maxwell generally $J = 0$ and so worked out what we now call the *Coulomb Gauge*, a gauge 'ited to electrostatic problems but with the serious drawback in treating ' fields that it requires the electric potential to be propagated instantaneously [22]. FitzGerald differed from Maxwell on this point and instead of assuming the two potentials to be independent as Maxwell did, FitzGerald put $J = -(d\Psi)/(dt)$ or equivalently: $\nabla \bullet A + (d\Psi)/(dt) = 0$. This *Lorenz Gauge*, as it was later called, is much better suited to treating propagation phenomenon than was Maxwell's *Coulomb Gauge* with $J = 0$. This new gauge thus eliminated the question of the instantaneous propagation of the electric potential. However, FitzGerald found that Heaviside had independently done it already [22]! As FitzGerald and Rowland put it in 1888, *That ψ should be murdered from treating propagation problems* [22]. Soon after, the same fate happened to *A*, as Heaviside puts it, *not merely the murder of Maxwell's ψ, but of that wonderful three legged monster with a scalar parasite on its back, the so called electrokinetic momentum at a point – that is the vector potential itself* [22]. That is the first complete modification of Maxwell's theory done by Heaviside and Hertz to get rid of the potentials and to start the problem with the sources, i.e., currents and charges.

Poynting in 1883 showed how the energy from an electric current passes from point to point, i.e., by what paths and according to what law does it travel from the part of the circuit where it is first recognizable as electric and magnetic to the parts where it changed into heat and other forms. In addition, Heaviside in 1884–1885 cast the long list of equations that Maxwell had given in his *Treatise* into the compact and symmetrical set of four vector equations now universally known as *Maxwell's equations*. Heaviside independently developed Poynting's theorem six months later. Reformulated in this way, Maxwell's theory became a powerful and efficient tool for the treatment of propagation problems, and it was in this new form (*Maxwell Redressed*, as Heaviside called it) that the theory eventually passed into general circulation in the 1890s. Heaviside's distance from the mainstream of British mathematical physics made it easier for him to dispense with the potentials and the Lagrangian methods favored by members of the Cambridge school and to approach the problem from a new and, as he believed through his vector operational calculus, into more fruitful directions [22].

After Maxwell's books on the treatise were published, in the European Continent, the Berlin Academy of Science announced a prize for research on the following problem: *To establish experimentally any relation between electromagnetic forces and the dielectric polarization of insulators, that is to say, either an electromagnetic force exerted by polarizations in non-conductors, or the polarization of a non-conductor as an effect of electromagnetic induction.* At that time Hertz was working with Helmholtz at the Physical Institute in Berlin. Helmholtz suggested to Hertz that, should he address this problem, the resources of the institute will be available to him. However, at that time Hertz gave up the idea because he thought a solution was not possible, as he found no adequate sources for generation of high frequencies. However, he continued thinking about this problem. Hertz addressed this problem when he went to the Technical High School at Karlsruhe. However, before he moved there he was at the University of Kiel. There he did not have much experimental equipment. So he worked on the theoretical aspects of the Maxwell's theory. In his book, he himself addresses the question as to what exactly is Maxwell's theory. In his words:

> *Maxwell left us as the result of his mature thought a great treatise on Electricity and Magnetism; it might therefore be said that Maxwell's theory is the one propounded in that work. But such an answer will scarcely be regarded as satisfactory by all scientific men who have considered the question closely. Many a man has thrown himself with zeal into the study of Maxwell's work, and even when he has not stumbled upon unwanted mathematical difficulties, has never the less been compelled to abandon the hope of forming for himself an altogether consistent view of Maxwell's ideas. I have fared no better myself. Notwithstanding the greatest admiration for Maxwell's mathematical conceptions, I have not always felt quite certain of having grasped the physical significance of the statements. Hence, it was not possible for me to be guided directly by Maxwell's book. I have rather been guided by Helmholtz's work, as indeed may plainly be seen from the manner in which the experiments are set forth. But unfortunately, in the special limiting case of Helmholtz's theory which lead to Maxwell's equations, and to which the experiments pointed, the physical basis of Helmholtz's theory disappears, and indeed it does, as soon as action-at-a-distance is disregarded. I therefore endeavored to form for myself in a consistent manner the necessary physical conceptions, starting from Maxwell's equations, but otherwise simplifying Maxwell's theory as far as possible by eliminating or simply leaving out of consideration those portions which could be dispensed within as much as they could not affect any possible phenomena... To the question, "What is Maxwell's Theory?" I know of no shorter or more definite answer than the following: - Maxwell's theory is Maxwell's system of equations. Every theory which leads to the same system of equations, and therefore comprises the same possible phenomena, I would consider as being a form or special case of Maxwell's theory; every theory which leads to different equations, and therefore to different possible phenomena, is a different theory. Hence, in this sense, and in this sense only, may the two theoretical dissertations in the present volume be regarded as representations of Maxwell's theory. In no sense can they claim to be precise rendering of Maxwell's ideas. On the contrary, it is doubtful whether Maxwell, were he alive, would acknowledge them as representing his own views in all respects.*

What Hertz missed was the core idea of the discontinuity in the displacement. The quandary of Hertz forced him into an uneasy compromise with the traditional Helmholtzian view on charge. Hertz distinguished the free electricity from which one calculates forces, and which is alterable by non-conducting means, from the true electricity, which is alterable only by conduction. So, though Hertz referred the measure of true charge to the divergence of the displacement, he preserved Helmholtzian wording because he had not seen how to avoid it. Whereas, a Maxwellian would write apparent charge as $\nabla \cdot E$, Hertz wrote of free charge and felt it necessary to retain the idea of bound charge to grant free charge physical significance, though he refused to consider why such a thing as bound charge exists. We see now the significance of Hertz's famed rejection of the Maxwellian distinction between electric intensity and displacement in the free aether. Without this distinction, it is impossible, in Maxwellian theory, to understand the existence of a charged surface in vacuum because charge is due to the discontinuity in the displacement. The fact that in free aether D reduces to E is merely a mathematical artifact. This is due to the definition of capacity of the aether being unity. The conceptual and physical distinction between displacement and intensity is still essential. Not knowing or understanding this distinction goes to the heart of Maxwell's theory. Hertz felt free to ignore it where it seemed mathematically to make no difference. In addition, Hertz started from the sources of the fields which were charges in electrostatics and currents in magnetostatics and not treat the potentials as fundamental quantities in analogy to a mechanical model as Maxwell did. Hertz showed that at a dielectric boundary the tangential components of the electric fields are continuous and the normal component is discontinuous. He obtained similar boundary conditions for a magnetic media.

In short [22], Hertz defined the electric and magnetic constants as unity in free space, thus eliminating displacement as a primary quantity. While this somewhat simplified the mathematical structure of Maxwell's theory, it was fatal to its dynamical basis. Heaviside raised this point with Hertz in 1890 after reading his first paper *On the Fundamental Equations of Electrodynamics.* He asked, *Can you conceive of a medium for electromagnetic disturbances which has not at least two physical constants, analogous to density and elasticity? If not, is it not well to explicitly symbolize them, leaving to the future their true interpretation?* Heaviside was calling on Hertz not to sacrifice for the sake of a small and perhaps illusory mathematical simplification, the dynamical aether, filled with stresses, strains, and stored energy on which Maxwell had built his theory. The British Maxwellians took this dynamical aether much more seriously than did their Continental counterparts, insisting that even on those points where Maxwell's theory required clarification and correction, this could and should be done without reducing the theory to a mere set of equations. The important modifications Heaviside had made to Maxwell's theory, including his abandonment of the potentials in favor of his four equations in the vector form that we call Maxwell's equations today, he said, meant as sustentative changes or new departures, but were directed solely at bringing out the leading points of the theory more clearly than had Maxwell himself. Both Heaviside and FitzGerald drew a careful distinction between *Maxwell's Treatise* and *Maxwell's theory* and said that Maxwell's book gave only an imperfect account of the real nature of the theory [22].

The Maxwellian and Continental ideas were so profound that only Lorenz was able to retain the substantial character of charge while incorporating certain Maxwellian elements that did not violate basic preconceptions. Lorenz's theory computes interparticle actions directly by using retarded forces and it employs careful microphysical averaging. However, Lorenz's theory very much obscures the difference between ionic motions and the basic field equations.

In 1886, Hertz observed while examining some of the apparatus used in lecture demonstrations that the oscillatory discharge of a Leyden jar or induction coil through a wire loop caused sparks to jump a gap in a similar loop a short distance away. He recognized this as a resonance phenomenon and saw that such sparking loops could serve as very sensitive detectors of oscillating currents and, thus, of electromagnetic waves. This provided him with the proper experimental tool which had eluded Lodge, FitzGerald, and others.

Heaviside, on the other hand, differentiated between the absolute and the relative permeability and permittivity, defining the relative quantities as the ratio of the absolute value for a medium and that for free space. Heaviside rewrote Maxwell's equations in the vector form that we use today using the above modifications. There are conductors and non-conductors or insulators and since the finite speed of propagation in the non-conducting space outside conductors was unknown, attention was almost entirely concentrated upon the conductors and an assumed field which was supposed to reside upon or in them, and to move about, upon, or through them. And the influence on distant conductors was attributed to instantaneous action-at-a-distance, ignoring an intermediate agent. Maxwell explained these actions of the intermediate agent of an intervening medium being transmitted at finite speed. In Heaviside's words:

> *What is Maxwell's Theory or what should we agree to understand by Maxwell's theory? The first approximation to the answer is to say there is Maxwell's book as he wrote it; there is his text and there are his equations, together they make his theory. But when we come to examine it closely, we find that this answer is unsatisfactory. To begin with, it is sufficient to refer to papers by physicists, written say during the twelve years following the first publication of Maxwell's Treatise, to see that there may be much difference of opinion as what his theory is. It may be, and has been, differently interpreted by different men, which is a sign that it is not set forth in a perfectly clear and unmistakable form. There are many obscurities and inconsistencies. Speaking for myself, it was only by changing its form of presentation that I was able to see it clearly, and so as to avoid inconsistencies. It is therefore impossible to adhere strictly to Maxwell's theory as he gave it to the world, if only on account of its inconvenient form... But it is clearly not admissible to make arbitrary changes in it and still call it his. He might have repudiated them utterly. But if*

we have good reason to believe that the theory as stated in his treatise does require modification to make it self-consistent and to believe that he would have admitted the necessity of the change when pointed out to him, then I think the resulting modified theory may well be called Maxwell's.

Maxwell defined the Ampère's law defining the electric current in terms of magnetic force. This makes the electric current always circuital implying that electric current should exist in perfect non-conductors or insulators. It is the cardinal feature of Maxwell's system. But Maxwell's innovation was really the most practical improvement in electrical theory conceivable. The electric current in a non-conductor was the very thing wanted to coordinate electrostatics with electrokinetics and consistently harmonize the equations of electromagnetic. It is the cardinal feature of Maxwell's system when dealing with Faraday's law where the voltage induced in a conducting circuit was conditioned by the variation of the number of lines of force through it. Instead of putting this straight into symbols, they went in a more round about way and expressed an equation of electro-motive force containing a function called the vector potential of the current and another potential, the electrostatic, working together but not altogether in the most harmoniously intelligible manner – in plain English muddling one another. It is I believe a fact, which has been recognized that not even Maxwell himself quite understood how they operated in his general equations of propagation. We need not wonder, then, that Maxwell's followers have not found it a very easy task to understand what his theory really meant, and how to work it out. It was Maxwell's own fault that his views obtained such slow acceptance; and in now repeating the remark, do not abate one of my appreciations of his work, which increases daily.

To understand the problem, one must introduce the Maxwellian concept of a current. It was apparently not possible to incorporate conductivity into the dynamical structure of the Maxwellian theory. Maxwell did not himself incorporate conductivity directly into the dynamical structure of the theory. Rather, he treated Ohm's law as an empirical, independent fact, and subtracted the electromagnetic intensity it requires from the induced intensity.

It is important to point out that Heaviside's duplex equations, as he called them, can be found in Maxwell's own writing in 1868, *Note on the Electromagnetic Theory of Light*. Maxwell gave a clear statement of the second circuital law relating the magnetic current to the curl of the electric force, crediting it to Faraday and asserting that it afforded "the simplest and most comprehensive" expression of the facts. But despite the advantages of this simple equation, particularly in treating electromagnetic waves, Maxwell did not use it at all in his *Treatise* in 1873 and it remained unknown even to his close students till 1890 when Maxwell's scientific papers [23] were published. It is rather amusing to note that even Lodge (one of the Maxwellians) actually credited this equation to Heaviside in his Presidential Address to the Physical Society in February 1899, when the Irish physicist and mathematician Joseph Larmor (1857–1942) pointed out to him that it was in the writing of Maxwell himself [22]!

In summary, why Maxwell's theory was not accepted for a long time by contemporary physicists was that there were some fundamental problems with his theory. Also, it is difficult to explain using modern terminology as to what those problems were. As stated in [16] *Maxwellian theory cannot be translated into familiar to the modern understanding because the very act of translation necessarily deprives it of its deepest significance, and it was this significance which guided British research. It assumed that field and matter can always be treated as a single dynamical system, subject to modification according to the circumstances.* In modern days, we reject this very basis of the Maxwellian theory! Maxwell's paper in 1862 *On Physical Lines of Force* described an elaborate mechanism for aether. He was deeply attached to the mechanism despite certain problems with it and remained throughout his life till 1879 strongly committed to the principle to model building. Yet, in 1864, his paper on *A Dynamical Theory of the Electromagnetics*, avoided specifying the structure of the aether, but nevertheless, presumed the field to be governed by what he called dynamical laws. This is done by treating H as a velocity and D as the curl of the corresponding mechanical displacement. Through such assumption, it does not mean that the true structure of the aether is fully understood. Maxwell's theory was based on the assumption that all electromagnetic phenomenon, including boundary conditions, can be obtained by applying Hamilton's principle to suitably chosen field energy densities which contain the appropriate electrical parameters like permittivity and permeability for the material medium. Modern theory implies that this can at best work on certain occasions as the macroscopic fields (D, H) are not a simple dynamical system but a construct obtained by averaging over the true state and the combined field vectors (E, B) with material vectors (P, M). Hence, where modern theory introduces the electron, the Maxwellian theory invented new forms of energy. This was possible because the Maxwellians were quite willing to invent modifications to the basic equations governing the electromagnetic field – as long as the results held up experimentally. Modern theory seeks unified explanations in an unmodifiable set of field equations coupled through the electron motions to intricate microphysical models. Maxwellian theory sought unity through a highly plastic set of field equations coupled to Hamilton's principle. In Maxwell's theory, the goal was to create a theory of electromagnetism which made no use whatsoever of the microstructure of matter. Hence, the basic problem in understanding Maxwell's theory by a modern reader is to decipher what exactly did he mean by the words *charge* and *current*! Some of this confusion depends on Maxwell's having altered at least once his choice for the sign for the charge density in the equation which links it to the divergence of the electric flux D. The conduction current in Maxwell's theory cannot be simply explained! The details of these subtleties may be found in [16]. Finally, as pointed out by Hunt, [22]: *When the 1880s began, Maxwell's theory was virtually a trackless jungle. By the second half of the decade, guided by the principle of energy flow, Poynting, FitzGerald and above all Heaviside have succeeded in taming and pruning that jungle and in rendering it almost civilized.*

In modern electrodynamics, we do not regard the field itself as a material structure, so we do not consider the stresses that may act upon it. Electromagnetic radiation, for example, transports energy and momentum but stresses arise only when the radiation impinges on material structures. The Maxwellians did not think in this way. For them energy inhomogeneity, whether matter is present or not, implies stress. Indeed after the discovery of Poynting's theorem, they realized that the free aether must be stressed when transmitting radiation, and so must move. In Maxwellian theory, the electromagnetic field transmits stress and is itself acted upon by stress. In modern theory, the field only acts; it is not acted upon [16].

In summary, the great difficulty with Maxwellian theory is that one cannot setup correspondences between mechanical and field variables which lead to consistent results unless one ignores conductivity, as illustrated by Heaviside. Heaviside's demonstration of this fact was based on the Green's potential function. He suggested using E as the velocity instead of H in the mechanical model and this requires a complete reconstruction of Maxwellian theory, as Maxwell did it the other way around.

From 1894 to 1897, the British electromagnetic theory abandoned the basic principles of Maxwellian theory and the entire subject was reconstructed on a new foundation – the electron, developed by Joseph Larmor in consultation with George FitzGerald [16]. Larmor had effected a revolution in the Maxwellian theory, one in which the electron had become the fundamental for generation of the field. Heaviside felt that Larmor's ideas of the electron as a nuclear of intrinsic twist was not sufficiently fundamental. The major impact of Larmor's theory was the destruction of the idea that continuum theory can serve as a sufficient basis for electromagnetism. Maxwell's theory with its fundamental assumption that the electromagnetic field can be subjected to precisely the same type of analysis as the material continuum was an artifact after 1898. This was the second conceptual modification of Maxwell's theory after Hertz, and Heaviside even though the basic equations still remained the same. Larmor's student, John Gaston Leathem, connected mathematically displacement with polarization by writing $D = E + P$ and thus effectively destroying the Maxwell-FitzGerald theory of using Hamilton's principle as a starting point based on a continuous energy function [16].

From this time onwards, future work on electromagnetism, in Britain and in the Continent, depended directly on microphysics based on the electron rather on the macrophysics. Hence there was a complete divorce between the matter and the field [16]. Thus, the nature of conduction, the stumbling block in the Maxwell's theory, led to the final modification by Larmor by incorporating the microscopic view of matter through the introduction of the electron as the source of charge, the flow of which results in a current. These sources now produce the fields.

In short, the current perception of Maxwell's theory consists of Maxwell's system of equations supplemented by his radical concept of displacement current.

1.7   CONCLUSIONS

This chapter provided an introduction of the developments made up to the time of Maxwell in the creation of the theory of electromagnetism based on merging the different theories related to magnetism, electricity, and light. This is a unique theory where the fundamental equations did not change even though the philosophy went through two complete modifications. We have provided a brief outline of who Maxwell was as his name is synonymous with electromagnetic waves and that is the basis of wireless. The four of his equations that we use today under the heading of Maxwell's equations were written by Oliver Heaviside who first put them in the vector form. We have also provided what was/is his theory that resulted in the explosion of wireless.

REFERENCES

[1]   Herbert Meyer, *A History of Electricity and Magnetism*, MIT Press, Cambridge, MA, 1971.

[2]   Heinrich Hertz, *Electric Waves*, (Translated by D. E. Jones), Macmillan and Company, New York, 1900.

[3]   Edmund Taylor Whittaker, *A History of the Theories of Aether and Electricity,* Harper Books, New York, 1960, first published by Thomas Nelson and Sons, England in 1910. Vol. I and II.

[4]   Jeff Biggus, *Sketches of a History of Classical Electromagnetism (Optics, Magnetism, Electricity, Electromagnetism).* http://history.hyperjeff.net/Electromagnetism.html.

[5]   Ross Spencer, http://maxwell.byu.edu/~spencerr/phys442/node4.html.

[6]   The Electricity Forum, http://www.electricityforum.com/electricity-history.html

[7]   http://www.geocities.com/CapeCanaveral/8341/

[8]   http://www-groups.dcs.st-and.ac.uk/~history/

[9]   http://www.thespectroscopynet.com/Educational/newton.htm

[10]   http://hjem.get2net.dk/Hemmingsen/Rainbow/history.htm

[11]   Basil Mahon, *The Man Who Changed Everything – The Life of James Clerk Maxwell*, John Wiley and Sons, New York, 2003.

[12]   Ivan Tolstoy, *James Clerk Maxwell*, University of Chicago Press, Chicago, 1981.

[13]   Joseph Larmor, *Origins of James Clerk Maxwell's Electric Ideas*, Cambridge University Press, 1937.

[14]   Oliver Heaviside, *Electromagnetic Theory*, Vol. I and II, The Electrician Printing and Publishing Company, London, 1893.

[15] James Clerk Maxwell, *A Treatise on Electricity and Magnetism*, Vol. I and II, Oxford University Press, London, 1873.

[16] Jed C. Buchwald, *From Maxwell to Microphysics: Aspects of Electromagnetic Theory in the Last Quarter of the Nineteenth Century*, University of Chicago Press, Chicago, 1985.

[17] R. T. Glazebrook, *James Clerk Maxwell and Modern Physics*, Macmillan & Co., London, 1896.

[18] *James Clerk Maxwell – A Commemoration Volume*, Cambridge University Press, 1931.

[19] Martin Goldman, *The Demon in the Aether – The Story of James Clerk Maxwell*, Paul Harris Publishing, Edinburgh, Scotland, 1983.

[20] Lewis Campbell and William Garnett, *The Life of James Clerk Maxwell – With a Selection from his Correspondence and Occasional Writing and a Sketch of his Contributions to Science*, Macmillan and Company, London, 1982.

[21] http://www.charlespetzold.com/etc/MaxwellMoleculesAndEvolution.html

[22] Bruce J. Hunt, *The Maxwellians*, Cornell University Press, Ithaca, NY, 1991.

[23] *The Scientific Papers of James Clerk Maxwell*, Vol. 1 & 2.

[24] http://www.acmi.net.au/AIC/

[25] http://www.hpo.hu/English/feltalalok/

[26] http://www.answers.com/

[27] http://www.cartage.org.lb/en/themes/Sciences/Physics/Optics/brief histoy/briefhistory.htm

[28] http://www.3rd1000.com/chronoatoms.htm

[29] http://www.ee.umd.edu/~taylor

[30] http://kr.cs.ait.ac.th/~radok/

[31] http://www.encyclopedia.com/

[32] http://www.britannica.com/eb/

[33] http://libserv.aip.org:81/

[34] http://octopus.phy.bg.ac.yu/web_projects/giants/

[35] http://nobelprize.org

[36] http://www.aip.org/history/

[37] http://en.wikipedia.org/

[38] http://52.1911encyclopedia.org/

[39] http://chem.ch.huji.ac.il/~eugeniik/history/

[40] http:// scienceworld.wolfram.com

[41] http://reference.allrefer.com/encyclopedia

[42] http://ac6v.com/history.htm

[43] http://www.carc.org.uk/

[44] http://www.utm.edu/research/iep/

[45] http://encyclopedia.laborlawtalk.com

[46] http://www.mpoweruk.com/history.htm

[47] http://www.bbc.co.uk/

[48] http://www.infoscience.fr/

[49] http://www.slcc.edu/schools/hum_sci/physics/whatis/famous.html

[50] http://www.geocities.com/bioelectrochemistry/

[51]   http://www.newadvent.org/cathen/
[52]   http://wise.fau.edu/~jordanrg/bios/
[53]   http://www.seds.org/messier/xtra/Bios/
[54]   http://www.maths.tcd.ie/pub/HistMath/People/
[55]   http://www.connected-earth.com/Galleries/
[56]   http://www.eresie.it/
[57]   http://www.nationmaster.com/encyclopedia/
[58]   http://www.bookrags.com/
[59]   http://www.absoluteastronomy.com
[60]   http://understandingscience.ucc.ie/
[61]   http://www.gesource.ac.uk/
[62]   http://www.wsu.edu:8080/~dee/
[63]   http://www.aip.org/history/
[64]   http://chemistry.mtu.edu/~pcharles/SCIHISTORY/
[65]   http://www.iihr.uiowa.edu/products/history/
[66]   http://www.sparkmuseum.com/
[67]   http://maxwell.byu.edu/~spencerr/phys442/
[68]   http://www.code-electrical.com/historyofelectricity.html
[69]   http://www.rare-earth-
       magnets.com/magnet_university/history_of_magnetism.htm
[70]   http://www.tcd.ie/physics/schools/what/materials/magnetism/top.html
[71]   http://earlyradiohistory.us/1963hwa.htm
[72]   http://www.ieee.org/organizations/history_center/
[73]   http://www.cosmovisisons.com/
[74]   http://turnbull.mcs.st-and.ac.uk/~history/
[75]   http://www.infopage.org/
[76]   http://www.zoomschool.com/inventors/
[77]   http://www.factmonster.com/encyclopedia/
[78]   http://www.infoplease.com
[79]   http://www.peak.org/~danneng/chronbio.html
[80]   http://americanhistory.si.edu/
[81]   http://www.iee.org/theiee/research/archives/histories&biographies/
[82]   http://www.philosophypages.com
[83]   http://www.bartleby.com
[84]   http://www.nrao.edu/whatisra/
[85]   http://plato.stanford.edu/
[86]   http://encyclopedia.thefreedictionary.com/
[87]   http://en.wikibooks.org/wiki/Electronics:History_chapter_1(to_chapter_7)
[88]   http://www.speed-light.info/measure/
[89]   http://www.psigate.ac.uk/newsite/
[90]   http://homepage.newschool.edu/het/
[91]   http://www.luminet.net/~wenonah/history/
[92]   http://history.hyperjeff.net/
[93]   http://www.hps.cam.ac.uk/starry/
[94]   P. Dunsheath, *A History of Electrical Engineering*, Faber and Faber,
       London, 1962.

2

A CHRONOLOGY OF DEVELOPMENTS OF WIRELESS COMMUNICATION AND SUPPORTING ELECTRONICS

MAGDALENA SALAZAR-PALMA, *Universidad Politécnica de Madrid, Spain;*
TAPAN K. SARKAR, *Syracuse University, USA;*
DIPAK L. SENGUPTA, *University of Michigan, USA*

2.1   INTRODUCTION

The goal of this chapter is to present a chronology of the developments of wireless communication and supporting electronics that evolved during the last three centuries. Quite often, it is found that, for example, the invention of radio is delegated to one or two persons, the names of whom vary from country to country, depending on the country of origin of the author. This chapter aims to illustrate that simultaneous development was going on all over the world and that each invention provided a solution to the portion of the puzzle. This presentation is in no way complete because the subject area is extraordinarily vast. Thus, only some of the crucial events will be highlighted and facts will be reflected without giving further details.

2.2   ACKNOWLEDGMENTS

The voluminous work of compiling this chronology was started by Professor Moti Chand Mallik (Figure 2.1) from the Indian Institute of Technology at Kharagpur, Bengal, India [1]. It is our privilege to take his initial results on which he spent five years, almost twenty years ago, and bring it to the interested readers, as he could not finish it himself. We are very grateful to him for introducing us to his Herculean task. Some of the figures in this chapter are taken from [105].

Figure 2.1. M. C. Mallik.

2.3   BACKGROUND

The history of Wireless Communication, explicitly involving electric signals[1], starts with the understanding of electric and magnetic phenomena (already observed during the very early days of the antique Chinese, Greek and Roman cultures) and the experiments and inventions carried out during the last half of the seventeen century and the eighteen century (see Chapter 1) by scientists like the German physicist Otto von Guericke, the French astronomer Jean Picard, the English scientist and mathematician Sir Isaac Newton, the English electrical experimenter Stephen Gray (Section 7.2.1), the French physicist Charles François de Cisternay du Fay, the German cleric and physicist Ewald Jürgen von Kleist, co-inventor of the *Leyden Jar* together with the Dutchmen Pieter van Musschenbroek (Figure 2.2), physician and physicist, and his assistant Andreas Cunaeus, the American writer, statesman, inventor and scientist, Benjamin Franklin and his friends, the English-American electrician Ebenezer Kinnersley and the English chemist Joseph Priestley, the British physicist, physician, and botanist Sir William Watson, the Swiss philosopher Johann Georg Sulzer, the Swiss-born French mathematician and physicist Georges-Louis Lesage (to whom belongs the honour of being the first to transmit intelligible signals over a system of wires: he produced an *Electrostatic Telegraph* using 24 wires for each letter of the French alphabet and 24 pith ball electroscopes at the far end, Figure 2.3), the English chemist and physicist Henry Cavendish, the French physicist Charles Augustin Coulomb (Figure 2.4), the Italian anatomist and medical scientist Luigi Galvani (Figure 2.4, Section 7.4), the Italian physicist Alessandro Giuseppe Antonio Anastasio Count Volta, the Spaniards Francisco Salvá i Campillo (Figure 2.4), physician, physicist and engineer (who proposed the *Spark Conduction Telegraph*, Section 7.2.1), and Agustín de Betancourt y Molina (Figure 2.4), engineer, and many other from different countries [2-5].

Figure 2.2. Left, Pieter van Musschenbroek (1692–1761); right, Leyden jars kept at the Borhaave Museum, Leiden, The Netherlands.

---

[1] French inventor and engineer Claude Chappe, with the help of the French watchmaker Abraham-Louis Bréguet, developed in 1792 the first *Visual (semaphore) Signaling System* or *Mechanical-optical Telegraph* that found systematic use in various countries (with variations) over about fifty years until it was gradually replaced by telegraphy using electric signals.

It may be also pointed out that the origins of digital communications, on use nowadays in most services utilizing wireless transmission, may be traced back to the early work on binary codes done by the Englishmen Sir Francis Bacon, a philosopher, and John Wilkins, a scholar, during the seventeen century, and the German philosopher and mathematician Gottfried Wilhelm Leibniz at the beginning of the eighteen century. Binary codes are also at the basis of the electronic computer [6-7].

Figure 2.3. George-Louis Lesage's Electrostatic Telegraph.

Figure 2.4. From left to right, Charles Augustin Coulomb (1736–1806), Luigi Galvani (1737–1798), Francisco Salvá i Campillo (1751–1828), and Agustín de Betancourt y Molina (1758–1825).

2.4   SOME CRUCIAL EVENTS OF THE NINETEENTH CENTURY

1800:

- Volta presented the first electric battery that he constructed in 1799 extending the research of Galvani. It consisted of a series of alternating copper and zinc rings in an acid solution kwon as *electrolyte*. He called his invention a *column battery* although it came to be commonly known as the *Volta battery, Voltaic Pile* or *Voltaic Cell* (Figure 2.5) [4-6].

- English chemist William Nicholson, and his friend the English surgeon Anthony Carlisle, constructed a cell similar to the voltaic pile and discovered the electrolytic decomposition of water [4].

- American Jonathan Grout obtained the first optical telegraph patent in the United States [1].

1807:

- French mathematician Jean Baptiste Joseph Fourier discovered what is now called Fourier's theorem. This earned him his Baron title (Figure 2.5) [2].

Figure 2.5. Left and middle, Alessandro Giuseppe Antonio Anastasio Count Volta (1745–1827) and a voltaic pile kept at Tempio Voltiano (Volta Museum), Como, Italy; right, Baron Jean Baptiste Joseph Fourier (1768–1830).

1809:

- German anatomist Samuel Thomas von Sömmering demonstrated to the Munich Academy of Science the use of the development of hydrogen bubbles on the negative electrode as an indicator of a transmitted signal, an idea for an *Electrochemical Telegraph* already suggested independently by Salvá i Campillo [4].

1811:

- English chemist Sir Humphrey Davy discovered that an electrical arc passing between two poles produced light. In fact he invented the bright arc light sending a powerful current of electricity between to carbon terminals [5,6].

1817:

- Swedish natural scientist, physician, and chemist Baron Jöns Jakob Berzelius discovered *selenium* in the residue resulting from the distillation of pyrites [5,8].

1820:

- Danish physicist Hans Christian Ørsted (Figure 2.6) announced his discovery of the electromagnetic field caused by electric current. The French physicist Dominique François Jean Arago (Figure 2.6) showed that a wire became a magnet when current flowed through it [2,4].

- French mathematician and physicist André-Marie Ampère (Figure 2.6) discovered *electrodynamics* and proposed an *Electromagnetic Telegraph* based on electrodynamics that was never constructed [4].
- French physicist Antoine César Becquerel, father of Alexandre Edmond Becquerel, and grandfather of the Nobel Prize winner Antoine Henri Becquerel, was the first person to observe piezoelectric effects and, seven years later, diamagnetism [9].
- German professor of physics, Johann Salomon Christoph Schweigger, presented the first *galvanometer*. Johann Christian Poggendorf, a student of physics also from Germany, presented independently a similar device called *multiplicator* [4].

Figure 2.6. From left to right, Hans Christian Ørsted (1777–1851), Dominique François Arago (1786–1853), and André-Marie Ampère (1775–1836).

1822:

- The *Difference Engine,* designed already in 1820 by the English mathematician Charles Babbage, was presented as the first programmable computer and won an award from the Astronomical Society. Babbage never perfected it because he conceived a much grander device, a fully functional programmable analytic mechanical computer, including *store* (memory), *mill* (central processor unit) and *control* (computer programs). He never finished such a computer, but his concepts were almost 100 years ahead of their time [5,10].
- Fourier completed his work on the flow of heat using his theorem and published it in a book entitled *Théorie Analytique de la Chaleur* [11].

1824:

- Berzelius discovered *silicon* [2,5].

1825:

- Italian physicist Leopoldo Nobili invented a static needle pair, which produced a galvanometer independent of the magnetic field of the earth [4].

1828:

- American physicist Joseph Henry improved Sturgeon's electromagnet substantially increasing the electromagnetic force [4].

- British mathematician George Green published an *Essay on the Application of Mathematical Analysis to the Theories of Electricity and Magnetism*, where he coined the term *potential*, developed the theory of *Electrostatic Screening*, and introduced the concepts that are now known as Green's theorem and Green's functions [8].

1830:

- William Ritchie, a British physicist, demonstrated *Electrical Telegraphy*, i.e., the transmission of electric signals over a distance (in his case, of 20 to 30 m), using a *torsion galvanometer* [4].
- Henry discovered the principle of self-induction [2].

1831:

- Another self-educated British scientist, Michael Faraday (Figure 2.7), discovered *electromagnetic induction*. He predicted the existence of electromagnetic waves [2,4,5].
- Faraday proved that vibrations of metal could be converted to electrical impulses. This was the technological basis of the telephone but no one used yet such system to transmit sound [6].
- Based in his previous work with electromagnets, Henry invented the *Electromagnetic Telegraph* [13]. He signaled over one mile using a horseshoe magnet, a swinging pivot, and an electric bell (Figure 2.7).

Figure 2.7. Left, Michael Faraday (1791–1887); middle and right, Joseph Henry (1797–1878) and a simplified sketch of his Electromagnetic Telegraph.

1833:

- German mathematician, astronomer, and physicist, Karl Friedrich Gauß (normally written as Gauss) and his colleague, Wilhelm Eduard Weber (Figure 4.3), physicist, built an electromagnet relay for sending signals between Gauss laboratory and their two observatories, a distance of 5,000 feet. In fact it was a telegraph, but they did not consider the invention of sufficient importance to publicize it [4,5,12].
- British scientist and mathematician Samuel Hunter Christie invented the *electric bridge* and used it for experimental determination of the

laws of magnetoelectric induction [14].

- Faraday observed that silver sulphide has a negative temperature coefficient of resistance [16].

1834:

- American artist and inventor Samuel Finley Breese Morse (Figure 2.8), invented the code for telegraphy named after his name [4,15].

- English lady Augusta Ada Byron, Countess of Lovelace (Figure 2.8), met Babbage and started cooperating with him towards the completion of the analytic mechanical computer that was never finished. She could be called the first computer programmer. In fact the United States Department of Defense gave the name *Ada*, in her honor, to a fail-safe computer language for use in unattended systems developed in 1979 [5].

Figure 2.8. Left, Samuel Finley Breese Morse (1791–1892); right, Augusta Ada Byron, Countess of Lovelace (1815–1852).

1835:

- Henry constructed the first relay [4].
- Morse demonstrated the *Recording Telegraph* [4].
- Munk Roschenschold, a German scientist, discovered the rectifying action of semiconductors over alternating current under certain conditions [3].

1836:

- Karl August von Steinheil (Figure 7.2), a German physicist, mathematician and astronomer, developed a recording telegraph and a telegraphic code [4].

1837:

- Englishmen Sir William Fothergill Cooke, inventor, and Sir Charles Wheatstone, physicist and inventor, applied for a patent for a telegraph system with 5 magnetic needles each of which could be deflected by an electric current as in Ørsted's original experiment. By deflecting the needles successively in pairs any one of the 20 letters of the alphabet could be selected in turn. The deflection of one needle could be used to point to a numeral. British patent (GBP) 7,390 was granted in 1841. Later on, they developed two-needles and one-needle improved systems (Figure 2.9) [4,17].

Figure 2.9. Upper part: left, Sir Charles Wheatstone (1802–1875); right, Sir William Fothergill Cooke (1806–1879). Lower part: left and middle, sketch and photograph of a five-needles system; right, a two-needles system under operation.

- Morse perfected the telegraph and his code (Figure 2.10). He demonstrated it at the University of the City of New York. The telegraph is the forerunner of digital communications in that the Morse code may be considered a variable-length ternary code using an alphabet of four symbols: a dot, a dash, a letter space and a word space; short sequences represented frequent letters, whereas long sequences represented infrequent letters. This type of signaling is ideal for manual keying [4,18].
- Submarine cable industry started on the Thames [3].

1838:

- German physicist and engineer Moritz Herman Jacobi, working in St. Petersburg, Russia, constructed a 28-feet boat propelled by an electric motor with a large number of battery cells [3].

- Morse demonstrated his one-needle telegraph to American President Martin Van Buren [6].

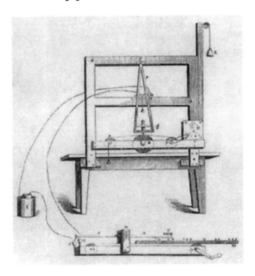

Figure 2.10. Upper part: Contemporary drawing of Morse's first telegraph equipment, 1837. Lower row: left, Morse's sounder widely use on telegraph lines after 1855; right, Morse's receiving magnet.

1839:

- French physicist Alexandre Edmond Becquerel, son of Antoine César Becquerel and father of Nobel Laureate Antoine Henri Becquerel, discovered the *Photovoltaic Effect* [6].
- British physicist Sir William Robert Grove devised and electric cell making use of hydrogen and oxygen. He demonstrated that water in contact with a strongly heated electric wire would absorb energy and break up into hydrogen and oxygen [2].
- British physician Sir William Brooke O'Shaughnessy, director of East India Company, erected the first telegraph line in India (21 miles long including 7,000 feet under the Hooghly River). The Kolkata-Diamond Harbor line started operation in 1851 [1,3,4].

- Electrical telegraph line using Cooke-Wheastone's five-needle telegraph operated between London-Paddington and West Drayton in England [4].
- Wheastone and Cooke applied for a patent for their "step-by-step letter showing" pointer telegraph, also called the *ABC Pointer Telegraph*. GBP 8,345 was granted next year [4].

1840:

- English physicist James Prescott Joule worked out the formula known as *Joule's Law* [5]. Joule provided also the breakthrough theory for the solution to electric light that lay not in an electrical arc in open space but in electricity passed through a filament. In fact Joule theorized that electrical current, if passed through a resistant conductor, would glow white-hot with heat energy turned to luminous energy. The problem was devising the right conductor, or filament, and inserting it in a container, or bulb, without oxygen, because the presence of oxygen would cause the filament to burn [6].
- Faraday experimentally verified electrostatic screening [8].

1841:

- Experimental *arc lights* were installed as public lighting along the Place de la Concorde in Paris. Other experiments took place in Europe and America, but the arc light proved impractical because it burned out too quickly [6].
- The Scottish inventor Alexander Bain made the first electric clock.

1842:

- In his first successful attempt on space communication at Princeton Campus, Henry discharged a battery of Leyden jars through a ground wire at a distance of several hundred feet and deflected a magnetic needle [1].

1843:

- Bain authored the first recorded proposal and patent for the electrical transmission of pictures (Figure 2.11). He called this device an *Electro Chemical Recording Telegraph* and was the first idea of facsimile transmission [7,17]. However, the first apparatus for transmission of pictures was made four years later.
- Cooke-Wheastone's one-needle telegraph was used successfully in the extended line Paddington-West Drayton-Slough, in England [4].

1844:

- With the transmission of the words "What hath God wrought?" through Morse's electric telegraph (Figure 2.12) between Washington, D.C., and Baltimore, Maryland, a completely revolutionary means of real-time, long-distance communication was triggered (Figure 2.13) [7,12].

1845:

- German physicist Franz Ernst Neumann derived the formula for mutual inductance of equal parallel coaxial polygons of wire [19].
- Cooke and Wheastone obtain GBP 10,655 for their one-needle telegraph. The next year they sold their patents to the *Electric*

Telegraph Company, which started the construction of a public telegraph network [4].

- Scottish mathematician James Bowman Lindsay suggested building a world-wide-wave of sea cables (Section 7.2.3).

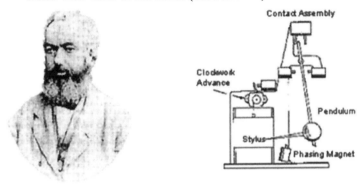

Figure 2.11. Left, Alexander Bain (1811–1877); right, Bain's telegraph scheme.

Figure 2.12. Photograph of the original tape received in Baltimore: under the code for each letter the corresponding letter was written by Morse forming the sentence "What hath God wrought?". In the upper part it is written: "This sentence was written by me at the Baltimore terminal at 8^h $45^{min.}$ A. M. on Friday May 1844, being the first ever transmitted from Washington to Baltimore, by Telegraph and was indited by my much loved friend Annie G. Ellsworth. Sam. F. B. Morse, Superintendent of Elec. Mag. Telegraphs."

Figure 2.13. Left, photograph of Morse's *embosser* utilized in 1844 as recording device. In 1845 Bain's chemical recorder, shown in the right photograph, substituted it.

1847:

- German physiologist and physicist Hermann Ludwig Ferdinand von Helmholtz (Figure 2.14) suggested electrical oscillation six years before the Irish-Scottish physicist William Thomson (future Lord Kelvin) (Figure 2.14) theoretically calculated this process [20], and ten years before the German physicist Berend Wilhelm Feddersen experimentally verified it [4].

- English mathematician George Boole (Figure 2.14) established *Boolean Algebra* in his work *The Mathematical Analysis of Logic* [21].
- French physicist and watchmaker, Louis-François-Clement Bréguet, grandson of Abraham-Louis Bréguet (see footnote 1 on p. 54) constructed a one-wire step-by-step telegraph similar to the Wheastone and Cooke pointer telegraph [4].
- German physicist Werner von Siemens constructed a gutta-percha press for seamless insulation of copper wires [4].
- British physicist Frederick Collier Bakewell made the world first successful image transmission in London using an *Image Telegraph* [4,20].

1848:
- Louis Napoléon Bonaparte (future Emperor Napoléon III) ordered the construction of a national electrical telegraph network [4].

1849:
- The first demonstration of telephone was made by Antonio Santi Giuseppe Meucci (1808-1896) (Figure 2.18 illustrating the unofficial and the official inventor of the telephone) at Havana, Cuba. He also demonstrated that inductive loading of the circuit improves the performance. His is an extraordinary episode in American history in which justice was perverted. In 1871, unable to obtain $250 for obtaining a definite patent he filed the patent *caveat* 3335 which was mysteriously lost from the patent office! Ultimately, the United States House of Representative passed the resolution HR 269 on 11 June 2002, recognizing him as the inventor of the telephone and not Alexander Graham Bell. The parliament of Canada retaliated by passing a bill recognizing the Canadian immigrant Bell as the only inventor of telephone.
- Englishman J. W. Wilkins suggested a sea conduction wireless telegraph between England and France (Section 7.2.3).

1850:
- The first under-sea telegraph cable was laid between Dover and Calais [17].
- German physicist Gustav Robert Kirchhoff (Figure 2.14) first published his circuit laws [12].

1851:
- German technician Heinrich Daniel Rühmkorff applied the induction principle to generate very high voltages. He constructed a coil with two windings galvanically insulated form each other. For this coil, later called *Rühmkorff Coil*, which became the basis for future experiments with spark bridges, he received the Volta Prize from Emperor Napoléon III [4].
- The first private telegrams were sent in France [4].

1852:

- Grove reported the first observation of metal deposits (thin film) sputtered from the cathode of a glow discharge [22].
- American lawyer Erasmus Peshine Smith coined the word *telegram* [4].
- British brothers Edward and Henry Highton started experimenting with a sea telegraph between England and France. Two bare metal wires were laid in the water with transmitter and receiver connected at their ends (Section 7.2.3).

Figure 2.14. Upper part: left, Hermann Ludwig Ferdinand von Helmholtz (1821–1894); right, William Thomson, Lord Kelvin (1824–1907). Lower part: left, George Boole (1815–1864); right, Gustav Robert Kirchhoff (1824–1887).

1853:

- Viennese physicist Julius Wilhelm Gintl invented *Duplex Telegraphy* [15,17].
- William Thomson (future Lord Kelvin) calculated the period, damping and intensity as a function of the capacity, self-inductance and resistance of an oscillatory circuit [23].
- French physicist Armand-Hippolyte-Louis Fizeau developed the electrolytic condenser as a means for increasing the efficiency of induction coils [1].

1854:

- David Edward Hughes (Figure 7.12), a London-born music professor in Kentucky, invented a letter printing telegraph with 52-symbol keyboard in which each key caused the corresponding letter to be printed at the distant receiver (Figure 2.15). His achievement was significant because the typewriter had not been invented at that time. He developed his plain-language writing telegraph during the next year. The *Teleprinter*, the *Telex System*, and present-day computer keyboards/visual displays are direct descendents of Hughes's invention [4,17].
- Charles Bourseul, a French telegraph official born in Brussels, proposed electrical speech transmission [4].
- Lindsay filed a patent on wireless sea telegraph, emphasizing the need of huge batteries (Section 7.2.3).

Figure 2.15. Hughes's Type-printing Telegraph machine.

1855:

- French engineer Jean-Mothée Gaugain studied the rectifying action between two metal balls in an evacuated chamber, producing the first *Electric Valve* [1].
- German instrument maker Heinrich Geissler developed a gas discharge tube known later on as *Geissler tube* [24].

- The first international sea cable was laid connecting the telegraph networks of Sweden, Denmark and Germany [4].
- Wheatstone developed the *ABC Automatic* (no operators required) *Telegraph*, also known as *Communicator*, using incoming and outgoing dials [4].

1857:

- Feddersen verified experimentally the resonant frequency of a tuned circuit as suggested by Helmholtz in 1847 [23].
- French mathematician Jules Antoine Lissajous developed the principle of his figures [8].
- Kirchhoff deduced capacitance and self-inductance of wires in telegraphy [25].

1858:

- The first submarine cable was laid in India between the mainland and the island of Ceylon (now Sri Lanka) [4].
- First transatlantic submarine cable was laid between the United States and England [15], which was opened with a 90-word message from Queen Victoria to American President James Buchanan. The analysis of William Thomson showed that a faint current would be carried through the 3,000 miles cable. He constructed and patented a very sensitive telegraph receiver that responded to the reduced signal. However his view did not prevail against others who insisted that the signal would be stronger. The line failed after few months of service [4,5,6,12].
- William Thomson developed a mirror galvanometer [4].

1859:

- French physicist Gaston Planté invented the *lead acid storage battery* [6].
- German mathematician and physicist Julius Plucker discovered *Cathode Ray* and observed that magnetic effects affected phosphorescence [26].

1861:

- German self-taught inventor and school teacher Johann Phillip Reis (1834–1874) (Figure 11.1) built a simple apparatus that changed sound to electricity and back again to sound, i.e., a telephone. However, it was a crude device, incapable of transmitting most frequencies, and it was never fully developed [4,6].
- Weber suggested a fundamental system for electrical and magnetic measurements [3].
- The American transcontinental telegraphy line was inaugurated [4].

1862:

- Italian physicist Giovanni Caselli built a machine to send and receive images over long distances using telegraph that he called *pantelegraph* and was based on Bain's invention. It was used by the French Post/Telegraph agency between Paris and Marseille from 1856 to 1870 [6].

1864:
- In France Hughes's telegraph replaced Bréguet pointer telegraph [4].

- Scottish mathematician and physicist James Clerk Maxwell (Figure 2.16) formulated the electromagnetic theory of light and developed the general equations of the electromagnetic field. He formulated twenty equations in terms of twenty variables, not the four *Maxwell's equations* as we know them today. Maxwell showed the existence of displacement currents and of electromagnetic vibrations identical to those of light. However, the statement that Maxwell predicted the existence of radio waves is only half-truth and its implication inaccurate [2,7,12,15,18,27,28] (see Chapters 1, 3-5, 7).
- American dentist Dr. Mahlon Loomis (Figure 2.17) wrote the earliest description of a wireless transmission system. He demonstrated and patented it in 1866. This was the world's first patent on wireless telegraphy [4,29].

Figure 2.16. Two well-known portraits of James Clerk Maxwell (1831–1879).

1865:
- The *International Telegraph Union* (ITU) was founded and met for the first time in Paris [4,30].
- Pantelegraph image transmission began on Paris-Lyon line [4].

1866:
- William Thomson directed as chief engineer the laying of a transatlantic cable between the United States and England. The line showed to be reliable following Thomson's techniques. Queen Victoria knighted him for his efforts (Lord Kelvin). A frenzy of cable laying began between continents. Lord Kelvin held the patents on designs of the cables and receivers [5,12].

- Loomis demonstrated the transmission of signals between two mountains in the Blue Ridges range, a distance of 22 km. He used a kite held by a wire loop. A rectangular copper wire aerial of 40 cm by 40 cm was attached to the kite and connected with two upper ends of the wire loop with a length of 180 m. At the lower side of the wire loop, one end was connected with earth and the other to a galvanometer. One such arrangement, which he called an *Aerial Telegraph*, was placed on top of Coshocton Mountain and a second on top of Beorse Deer Mountain. Disconnecting and connecting to the wire to the galvanometer at one station caused a clear deflection on the galvanometer at the other station [4] (Figure 2.17) (Section 7.2.4).

Figure 2.17. Mahlon Loomis (1826–1886) and a drawing of his Aerial Telegraph set up.

1867:

- American publisher, politician, and philosopher Christopher Latham Sholes invented the typewriter. In 1873 he sold the manufacturing rights of his typewriter to firearm, sewing machine, and farm toolmakers *E. Remington & Sons* [4].

- Wheatstone presented his *cryptograph,* the first example of mechanical coding of a telegraph text, at the world exhibition in Paris [4].

1868:

- Maxwell developed the theory of feedback systems [31].

1870:

- Maxwell deduced (a) the relations between the mean diameter and axial length of a square section coil for maximal inductance, and (b) the increase of resistance and decrease of inductance of a coil due to skin effect [32]. He also deduced the shape of the electrostatic field about a wire grid placed between two electrodes [33].

- British physicist John Tyndall explained and demonstrated the principle of guiding light through a transparent conductor (first reference to optical fibers) [4].

- French instrument maker Jean G. Bourbouze suggested the use of river Seine as medium for conduction wireless telegraphy (Section 7.2.3).

1871:

- British electrical engineer Willoughby Smith discovered the photoconductive effect. It was reported in 1873 [34].

- The Society of Telegraph Engineers is organized in London, UK [12].

- Maxwell published the electromagnetic screening by perfect conductors [8].

1872:

- French telegraph operator Bernhad Meyer developed the first telegraph multiplexer [4].

- Loomis received a patent for his aerial telegraph [4].

1873:

- Englishman George Little used electrolytic condenser in his automatic telegraph [1].

- English physicist and chemist Frederick Guthrie observed that a negatively charged electroscope was discharged when a metal ball heated to dull red was brought near it, but positively charged electroscope was not discharged unless the ball was heated to higher temperatures. This lead to the foundation of vacuum tubes. An electroscope charged either positively or negatively became discharged when the ball was at a temperature of white heat [33].

1874:

- Sir Arthur Schuster, an English physicist born in Germany, observed that contacts between tarnished and untarnished copper wire did not obey Ohm's Law [16].

- Scottish-American Alexander Graham Bell, a teacher of the deaf, conceived the *Magneto Telephone* in Brantford, Ontario. It was born next year, in Boston, Massachusetts, USA. Bell patented it in 1876 (Figure 2.18). The telephone made real-time transmission of speech by electrical encoding and replication of a sound a practical reality. The first version of the telephone was crude and weak, enabling people to talk only over short distances. The quality and range was greatly enhanced by the invention and development of the carbon microphone and the induction coil during the years 1877–1890 [7,12,15,18].

- American physicist Elisha Gray built the first steel diaphragm/electromagnet receiver. Gray accidentally discovered that he could control sound from a self-vibrating electromagnetic circuit and in doing so invented a basic single note oscillator. The *Musical Telegraph* used steel reeds whose oscillations were created and transmitted, over a telephone line, by electromagnets. Gray also built a simple loudspeaker device in later models consisting of a vibrating diaphragm in a magnetic field to make the oscillator audible. In fact he could have been known as the telephone inventor. Incredibly Gray filed for a caveat patent the same day in 1876 that Bell did, but two hours later! On the basis of its earlier filing time and on the subtle distinctions between a caveat and an actual patent application, the US Patent Office awarded Alexander Bell, not Elisha Gray, the patent for the telephone (Figure 2.19) [6].

- Jean-Maurice-Émile Baudot, officer in the French Telegraph Service, invented a system based on the use of a *five unit binary code* where each letter of the alphabet was represented by a unique combination of the five elements. He combined the use of such code with time-division multiplexing. This system was officially adopted by the French Telegraph system in 1977. His invention was particularly significant because it involved two principles: First the representation, as with Morse, of a symbol by a code, and second, the concept of time division multiplexing. The first resulted in the pulse-code modulation, while the second was the forerunner of modern digital techniques. Baudot's name, in the shortened form *Baud*, symbols per second, is used nowadays as a unit of the speed of data transmission [4,7,17,18].

- German professor of Physics at the University of Strasbourg Karl Ferdinand Braun (Figure 2.20), observed that contact between metals and various sulphides (e.g., galena) and between metals and various pyrites showed rectifying action [16].

1875:

- English physicist, mathematician, and electrical engineer Oliver Heaviside (Figure 2.20) and the American inventor Thomas Alva Edison (Figure 2.20) improved telegraphy leading to *Quadruplex Telegraphy* [17].

- A worldwide network of telegraph cables had been laid linking England, the United States, India, the Far East and Australia by the Victorian telegraph pioneers of England and America [19].

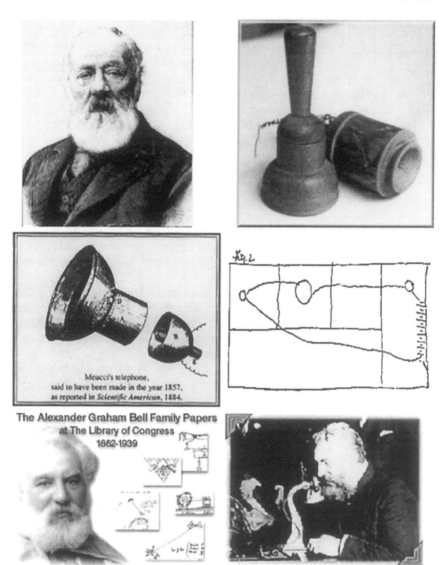

Figure 2.18. The unofficial and the official originator of the telephone. Top right: Antonio Santi Giuseppe Meucci (1808–1889), the real inventor of the telephone. Top left: Meucci's original phone used in Havana in 1849. Midddle right: Meucci's improved phone as published in the *Scientific American*. Middle left: Layout of Meucci's first experiment in Havana, performed in 1849. The first three rooms shown in top of the figure, were part of Meucci's apartment. The batteries (about sixty Bunsen batteries, yielding a total emf of about 114 volts) were located in the theater's workshop (shown at the right side of the figure). The big circle in the second room indicates a reel of wire. Bottom part: right, a poster from the Library of Congress showing a photograph of Alexander Graham Bell (1847–1922); left: a photograph of him using his telephone.

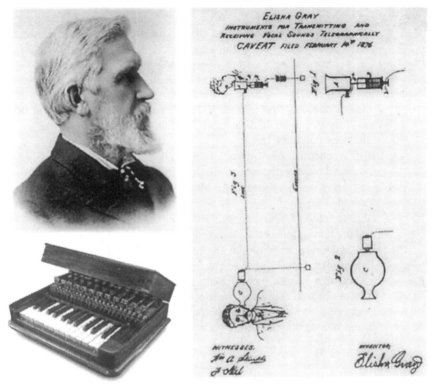

Figure 2.19. Left: upper part, Elisha Gray (1835–1901); lower part, a photograph of one of his Musical Telegraph. Right: drawing for his patent caveat.

Figure 2.20. From left to right, Karl Ferdinand Braun (1850–1918), Oliver Heaviside (1850–1925), and Thomas Alva Edison (1847–1931).

- American civil servant and inventor George R. Carey proposed to imitate the human eye by mosaic consisting of a large number of selenium cells [1].

- Scottish physicist John Kerr discovered a link between electricity and optics. He showed that when dielectrics are subjected to powerful electrostatic force, they acquire the property of double refraction [25].

1876:

- Bell fabricated the electromagnetic telephone transmitter (instrument for converting acoustic power into electric power). The term *microphone* was first used in 1878. He submitted a patent application for "Improvements in Telegraphy" which resulted in the world's first telephone patent [4,15].
- British chemist and physicist Sir William Crookes (Figure 7.16) measured radiation pressure of light using radiometer [8].
- German physicist Eugene Goldstein introduced the name *Cathode Rays* [1].
- German physicist Kart Rudolf Koenig found ultrasonic vibrations by making tuning form capable of producing 90 kHz [1].
- English-American electrical engineer, inventor and entrepreneur Elihu Thomson and the American electrical engineer Edwin James Houston proved that sparks were actually oscillatory high-frequency electric currents, which could not be detected by DC equipment. Then Edison gave up in his idea of spark telegraphy (Section 7.4.2)
- Braun demonstrated the rectification effect of a metal-semiconductor contact (Section 11.2.4).

1877:

- Edison invented the carbon transmitter (mouth piece) for telephone [3, 4].
- Edison invented the *Phonograph*. The French Charles Cros, independently, presented a similar device this same year. There is no evidence that the Cros phonograph actually worked. However, Edison, who called his phonograph the *Talking Machine*, made a recording of himself (i.e., the first recording) reciting, "Mary had a little lamb" that still survives today. Edison phonograph was powered by an electric motor in 1878 [4,6,35].
- The first telephone system, known as *Exchange*, was installed in Hartford, Connecticut [6].
- First permanent telephone line was installed between Boston and Somerville, Massachusetts [4].
- The company Siemens & Halske (initiated by Siemens and the German mechanic Johann Georg Halske) started telephone production in Germany [4].
- The French electrical engineer Georges Leclanché invented the dry cell battery [6].
- The earliest known example of dielectric non-linearity was the discovery of the *Electro-Optical Effect* by Kerr [15,16].

1878:

- English biologist Prof. Thomas Henry Huxley communicated to the London Royal Society Hughes's investigation about the effect of contact pressure between elements of an electric circuit on the value of current flowing through it. Huxley gave the name *Microphone* to the instrument he was using because it magnified the weak sounds [3].

- Bell successfully demonstrated voice modulation of a light beam by a microphone and detected it at a distance of 500 yards using a selenium (photo) cell and a parabolic reflector [15].
- French engineer and inventor Hippolyte Fontaine and the Belgian electrician and inventor Zénobe Théophile Gramme invented the alternator, a device for turning direct current (DC) into alternating current (AC) [6].
- Manual telephone switchboard operated in New Haven, Connecticut [4].
- *Bell Telephone Company* opened telephone service in London [4].
- Louis-François-Clement Bréguet and his son, Antoine Bréguet, started telephone production in Paris, France [4].

1879:

- American physicist Edwin Herbert Hall discovered the effect named after his name [16].
- The *electric light* was invented simultaneously by Edison in the United States and Sir Joseph Wilson Swan, English physicist and chemist. Edison sent electricity through the loop of carbonized cotton thread and devised a host of subsidiary equipment to keep banks of incandescent lights burning at constant levels [5,6]. Swan lamps gained popularity in Great Britain [2].
- The first *telephone exchange* outside USA was constructed in London, UK [6].
- Dutch physicist Hendrik Antoon Lorentz provided the formula for a single layer of solenoid [36].
- Hughes published methods for preventing cross talk in telephone [1].
- English inventors William Edward Ayrton, physicist, and John Perry, engineer, produced the first practical portable ammeter [3].
- English Professor Walter Bailey demonstrated a rotating electromagnetic field without the aid of any mechanical motion [3].

1880:

- Ayrton and Perry proposed a system using selenium for transmission of pictures [20].
- American engineer and inventor Charles Fritts made possible to have variable density and variable width sound recording on photographic film. He filed a patent that was finally granted some years after his death, in 1916 (USP 1,203,190) [15].
- By the end of this year Edison had produced a 16-watt bulb that could last for 1500 hours [7].
- French physicist Pierre Curie, while working with his brother, Jacques, observed the phenomenon of piezoelectricity [2].

- German physicists Julius Elster and Hans Friedrich Geitel enclosed a lamp filament and metal plate in vacuum and found that a tiny current flowed from the filament to the plate [10].
- American scientist Dr. H. E. Licks invented a system of transmitting color by telegraph that he called *Diaphote* (from the Greek *dia*, through, and *photos*, light) [1].
- French physicist and engineer Ernest Mercadier rechristened the Bell's photo phone as *radiophone* and this was the first time that the word *radio* was used in the sense we use it today [1].
- American Professor John Trowbridge, of Harvard University, was the first to apply Bell's telephone receiver to wireless communications between ships and stations on the shore (Section 7.2.5)

1881:
- Two separate channels of transmitters (microphones) and earphones were used in Paris opera for stereophonic recording [1].
- International telephone operation started between Detroit, USA, and Windsor, Ontario, Canada [4].
- Bell patented the metallic circuit for telephone line [4].

1882:
- American Amos Emerson Dolbear (Figure 2.21), professor of physics and astronomy, was granted a patent for a wireless transmission and reception system using an induction coil, a microphone, a telephone receiver and a battery [1].
- English physicist John William Strutt, Lord Rayleigh (Figure 2.21), found that induction between two coils decreased when a copper sheet was placed between them [1].

Figure 2.21. Left, Amos Emerson Dolbear (1837–1910); right, John William Strutt, Lord Rayleigh (1842–1919).

- Nathan B. Stubblefield, a Kentucky melon farmer, transmitted audio signals without wires. In 1892, he demonstrated transmission of human voice using his wireless telephone, which was ultimately patented in 1907 as USP 887,357. Even though the details of his experiments are very scant, it appeared he used an Earth battery (USP 600,957, obtained in 1898) and used the principle of induction to transmit human voice to within a radius of half mile. This appears to be the first report of transmission of human voice available on the web!

- Scottish meteorologist and geophysicist Balfour Stewart explained certain phenomenon of terrestrial magnetism by postulating a conducting layer in the upper atmosphere [1].

1883:

- Edison observed that the bulb of a carbon filament lamp became coated with black deposit. He also discovered that when a metal plate is suspended inside the bulb and connected externally through a galvanometer to the positive end of the filament a steady current flowed into the plate as long as the filament was heated [3]. This flow of electrons in vacuum was called the *Edison Effect*, the foundation of the electron tube [12].

- French scientist Lucien Goulard and his colleague the British inventor John Dixon Gibbs developed the AC transformer [4].

- French electrical engineer Léon Charles Thévenin developed the theorem that bears his name [37].

- The first telephone exchange system connecting two major cities was established between New York and Boston [6].

- American electrical engineer Frank Julian Sprague, an assistant of Edison, used thermoionic emission from an incandescent filament to monitor the supply voltage of his illuminator circuit (vacuum tube voltmeter) [1].

- Irish physicist and chemist George Francis FitzGerald (Figure 7.15) published a formula for the power radiated by a small loop antenna (Section 7.4.7).

1884:

- Edison used and also exhibited the emission in a bulb to monitor the supply voltage of his illumination circuit (vacuum tube voltmeter). He had also observed that the current through the galvanometer was a highly sensitive indication of the filament voltage and in a sense he anticipated the equation that the English physicist Sir Owen Williams Richardson will develop later on [2,38].

- Italian engineer Arturo Malignani patented the use of arsenic and iodine as *vaporizable reagents* to take up residual gasses in an incandescent lamp (*getters*). The English chemist Frederick Soddy patented in 1906 the use of calcium and magnesium to clean up the gasses in vacuum chambers and used them as getters to improve vacuum of hard valves in 1916 [34].

- Russian-German inventor Paul Gottlieb Nipkow, when he was a student, filed the first patent on a complete system of television, *Mechanical Television*, using selenium cells (Figure 2.22) [7,13].
- Fritts discovered the barrier-layer photo voltaic cell while working with the selenium cell [39].
- German physicist Heinrich Rudolf Hertz (Figure 2.23) wrote Maxwell's equations in *scalar form* (12 in number, instead of Maxwell's 20 equations in scalar form) by discarding the concept of *aether* introduced by Maxwell and starting from the *sources* rather than the *potentials* as Maxwell did [28].
- The *American Institute of Electrical Engineers* (AIEE) was formed [12].
- German physicist Johann Wilhelm Hittorf observed that incandescent wire in vacuum acts as cathode [1].

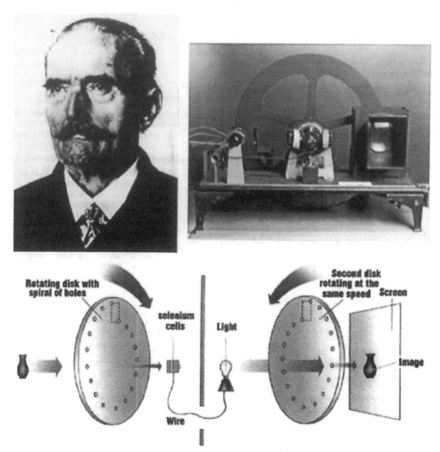

Figure 2.22. Upper part: left, Paul Gottlieb Nipkow (1860–1940); right, a photograph of his disk. Lower part, a sketch of how his system worked.

1885:

- Edison patented a system of wireless communication by electrostatic induction. He placed two high masts at a distance apart and fixed a metal plate on top of each. One metal plate was charged to a high voltage producing an electric field, which could be detected on the plate of the receiving mast. The two plates on the high masts were the first electrical aerials. The patent for the aerials was sold to the Italian electrical engineer Guglielmo Marconi (future Marchese) in 1903 [35].
- American inventor William Seward Burroughs developed the commercial adding machine [4].
- Irish-American electrical engineer Marmaduke Marcelus Michael Slattery described a high resistance, comprising carbon and a refractory insulator molded into a cylinder with carbon termination at each end [1].
- Welsh electrical engineer Sir William Henry Preece (Figure 7.9) pointed out that reversal of anode potential resulted in decrease of Edison current [1]

1886:

- Germanium was isolated from argyrodites by Clemens Alexander Winkler, a German chemist, who gave it the name *Germanium* [2].
- Edison, in partnership with the Americans Charles Sumner Tainter and Chichester Bell, cousin of A. G. Bell, improved the phonograph [6].
- Strowger Company, owned by the American undertaker Almon Strowger, develops the telephone-dialling disk [4].
- French electrical engineer Maurice Leblanc published the principle of multiplex telephony [1].
- Elihu Thomson introduced electric welding [3].
- Heaviside introduced the term impedance as the ratio of voltage over current [3].
- Hertz initiated his experiments to demonstrate the existence of radio waves. He completed his work during the next year and published his paper in 1888 [7,10,18].

1887:

- English physicist Sir Oliver Joseph Lodge (Figure 2.23) discovered *Sympathetic Resonance* (i.e., standing waves) in wires [15].
- Hertz discovered the *Photo Emissive Effect*. He observed that the maximum length of the spark between two electrodes increases when ultraviolet light falls on the negative electrode of the spark gap [34].
- While the Edison phonograph recorded sound on cylinders, a German-born American inventor Émile Berliner invented a machine, that he called *Gramophone*, which used flat disks. This system became the standard one [6].
- Heaviside made iron powder cores with wax as binder and insulator [40]. He also developed the principle of loading in transmission lines to make them distortion-less.

- Lord Rayleigh published a theoretical explanation of the oscillation of a system in which the stiffness parameter was periodically varied [1].

Figure 2.23. Heinrich Rudolf Hertz (1857–1894) and Sir Oliver Joseph Lodge (1851–1940).

1888:

- German physicist Wilhelm Hallwachs proved that photo emissive effect involved the loss of negative electricity from the illuminated surface [34].
- Hertz produced, transmitted, and detected electromagnetic waves of wavelength 5 m and 50 cm. He used reflectors at the transmitting and receiving positions to concentrate the wave into a beam [4,12,15].
- Italian mathematician and physicist Galileo Ferrari and the Croatian-American electrical engineer Nikola Tesla, independently, produced rotating fields using 2-phase currents [3].
- Hertz discovered also the principle underlying *RADAR (Radio Detection and Ranging)* [6].
- Heaviside (a man who did not complete college) wrote Maxwell's equations in vector form: the four equations that we use today [27].
- Austrian botanist Friedrich Reintzer discovered the liquid crystal [41].
- American scientist, inventor, and industrialist Oberlin Smith published an article suggesting the magnetic recording of sound [6,42].
- Tesla (Figure 2.24) obtained 13 USA patents for the design of a Polyphase Alternate Current (AC) Motor and Power Transmission [4].

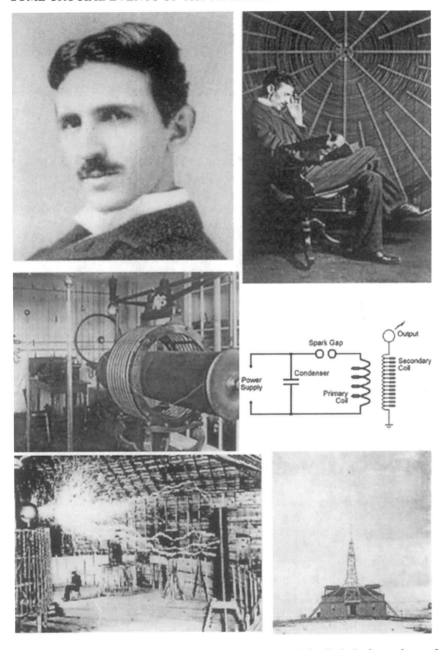

Figure 2.24. Upper part: left, Nikola Tesla (1856–1943); right, Tesla in front of one of his coils. Middle part: left, a photograph of a Tesla transformer; right, Tesla transformer schematic. Lower part: left, a double exposure photograph showing Tesla under his coils; right, Tesla's Colorodo Springs Laboratory.

- Austrian engineer Ernst Lecher established the relation between frequency, wire length, velocity of wave propagation, and the electrical constants of the wire [43].
- Elisha Gray developed the *Telautograph* facsimile machine [4].
- Fritts made the first photovoltaic cells from selenium covered with a thin gold film [6].
- Fifty pairs of cables were first used in telephone [1].

1889:

- Lead sulphide diode was invented [23].
- American engineer and mathematician Herman Hollerith developed the punch card system [4]
- American undertaker (and funeral director) Almon Brown Strowger invented an automatic telephone switching system that dispensed with operators [4,17].
- German physicist Louis Carl Heinrich Friedrich Paschen formulated the law that bears his name relating ignition voltage, distance between two electrodes and gas pressure of a gas-filled enclosure containing two electrodes [34].
- American inventor William Gray patented the coin-operated *Pay Telephone*. The first coin box telephone was installed at Hartford, Connecticut [4,6].
- The *Institute of Electrical Engineers* (IEE) was formed from the Society of Telegraphs Engineers in London [12].
- On being asked by Mr. H. Huber, a friend of Hertz, Hertz said there was no prospect of his discovery of electromagnetic waves being used for telephony as the wavelength would be 300 km: "One has to construct a mirror as large as the continent to succeed with the experiment", he said [3].
- German scientists Julius Elster and Hans Friedrich Geisel stated that the current in Edison's tube was due to negative ions (now called electrons) [1].
- Joubert demonstrated that when the frequency of the current applied to a muscle of a frog was raised to a certain level it could not cause the muscle to contract [44].
- Ernest Mercadier proposed voice-frequency telegraphy [4].

1890:

- Elster and Geitel predicted and proved that alkali metals should be the most sensitive photoelectric emitters of all metals and found that sodium and potassium respond to visible light also [34].
- English electrical engineer Sir John Ambrose Fleming (Figure 2.25) published a paper "On electric discharges between electrodes at different temperatures in air and in high vacuum" (emission in a diode) [45].
- Lecher used standing waves produced in parallel wires to measure frequency [15].

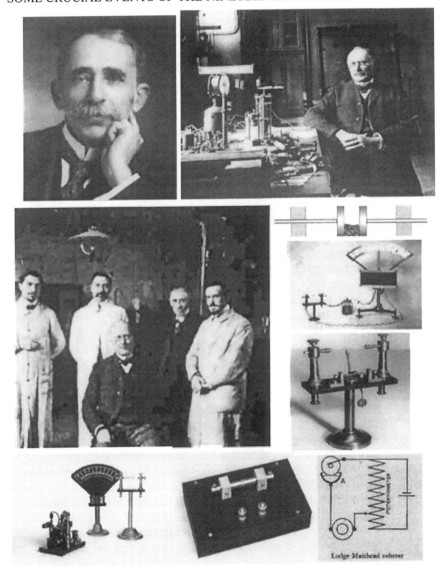

Figure 2.25. Upper photographs: left, Sir John Ambrose Fleming (1849–1945); right: Edouard-Eugène-Desiré Branly (1844–1875) in his laboratory. Middle part: left, Branly and his team; right, from up to bottom, a simplified drawing of Branly coherer, the first receiver made by Branly himself, a photograph of Branly's coherer. Lower part: left, coherer with decoherer system and coherer with galvanometer; middle, Lodge's coherer; right, Lodge-Muirhead coherer sketch.

- Tesla introduced high frequency (HF) currents in therapeutics. He observed that currents of high frequency were capable of raising the temperature of living tissues. The French physicist and physician

Jacques-Arsène d'Arsonval used high ampere low voltage high frequency current for treatment of certain diseases and demonstrated that HF current could be used for coagulation of proteins [1].

- French engineer Édouard Bouty constructed the mica condenser [1].
- Tesla patented his *Tesla Coil* (USP 433702), which was used later in every spark gap generator for generation of high frequencies (Figure 2.24) [4].
- Auburn prison in New York State electrocuted a prisoner to demonstrate the ugly side of AC as opposed to Edison's DC.
- Heaviside, the British electrical engineer and inventor Sebastian Ziani de Ferranti, and others, reported the practicality of induction heating [46].
- Heinrich Rubens and R. Ritter made a very sensitive bolometer which measured the intensity of electromagnetic waves by means of the heat they generated in an extremely thin wire [25].
- Australian physicist and chemical engineer Richard Threlfall suggested use of electromagnetic waves in communications (Section 7.4.8).

1891:

- Captain Henry Jackson (later Sir Henry Jackson and First Sea Lord) of the British Admiralty sent Morse-codes by wireless over a few hundred yards [17,35].
- French physicist and physician Professor Édouard Eugène Desiré Branly (Figure 2.25) reported the results of the work he initiated the previous year, namely, the rediscover of the cohesion effect on small particles under the influence of electricity, i.e., an electric spark occurring near an ebonite tube of metal fillings increased their conductivity appreciably. Such a tube of metallic powder does not normally pass current. Branly received the Nobel Prize in Physics in 1921. The cohesion effect had been noticed by the Frenchman Guitard in 1850, by the English engineer Samuel Alfred Varley in 1866, by Lord Rayleigh in 1879, and by the Italian physicist Temìstocle Calzecchi-Onesti in 1884–1885, but had not yet been applied [4] (Section 7.4.5). In 1894 Lodge in a lecture given to the Royal Institution in London about his experiments for detection of Hertzian waves publicly demonstrated that the conductivity of loosely packed iron fillings, contained in a glass tube closed by a metal plug at each end was increased when electric radiation fell upon it. He gave the name *Coherer* to the device, which was a practical version of Branly's coherer (Figure 2.25) [3,4].
- Strowger patented the first *Automatic Telephone Exchange*. It was installed next year but manual switchboards remained in common use until the middle of the twentieth century [6]. Of all the electromechanical switches devised over the years, the *Strowger step-by-step switch* was the most popular and widely used [7,18].

- Frenchmen Maurice Hutin and Leblanc reviewed the theory of resonance in an inductively coupled circuit [1]. The first submarine telephone cable from England to France was in use [3].
- Englishman Alexander Pelham Trotter suggested the use of electromagnetic waves for ship to shore communications (Section 7.4.8).

1892:

- Heaviside showed in his famous electrical papers that a circuit had four fundamental constants, the resistance, inductance, capacity and leakage [3].
- Crookes (Figure 7.16) published in the Fortnightly Review an article on the possibilities of Hertzian waves, i.e., wireless communications such as: "wave lengths can be used from few thousand miles to a few feet", "telegraphy without wires", "directional transmission", "receiver could be adjusted to receive a particular wavelength by turning a screw, this would give *sufficient secrecy* and this could be increased by using a code" (Section 7. 4.8) [35].
- Preece signaled between two points on the Bristol Channel, Loch Ness, Scotland, employing both induction and conduction to affect one circuit by the current flowing in another. Using loops of wire several hundred feet long detected current interruptions in one with the other [35].
- Zender used triggering of gas filled tube for demonstrating to a large audience the presence of Hertzian waves [1].
- Sennett showed means of communication with light vessels by using submerged cables from which signals were received on board light ship by means of multiturn loops and telephone receiver [1].
- Englishman C. A. Stevenson patented means for navigating vessels over electrically energized cables. He also suggested that telegraphic communications could be established between ships by coils of wire, "the larger the diameter the better to get induction at a great distance." [1].
- Strowger telephone exchange operated in La Porte, Indiana [4].
- Stubblefield demonstrated the first broadcast of a human voice using his wireless telephone attached to a land aerial.

1893:

- Serbian-American inventor Michael (Mihailo) Idvorsky Pupin suggested the use of tuned circuits for harmonic analysis [34].
- Joseph John Thomson, an English physicist, successor of Lord Rayleigh as both Professor of Physics in Cambridge, and director of the Cavendish Laboratory, published the first theoretical analysis of electric oscillations within a conducting cylindrical cavity of finite length suggesting the possibility of wave propagation in hollow pipes (waveguides) [1,2,47].
- Elster and Geitel made selenium cells, which converted light to electricity [6].
- Tesla developed a wireless system for transmitting intelligence [1].

- Born in Colaba, India, American electrical engineer Arthur Edward Kennelly published the use of complex notation in Ohm's law for AC circuit [1].
- Hertz conducted experiments of electromagnetic (EM) shielding and for coaxial configuration [4].
- Italian physicist Augusto Righi improved Hertz's oscillator devising his spark oscillator, which he would further improve during the next few years (Section 12.1).
- Tesla constructed an AC power plant at the World's Fair (Columbian Exposition), Chicago (Figure 2.26).

Figure 2.26. Tesla's AC power plant at World's Fair (Columbian Exposition), Chicago, 1893 showing four of the twelve 1000 horse power two-phase generators.

1894:

- Lodge demonstrated at the British Association meeting at Oxford in September 1894 the possibility of Morse signaling by wireless, i.e., wireless communication. As suggested by the English telegrapher Alexander Muirhead, he sent dot and dash signals over a relatively short distance of 180 feet through two stonewall. He used a Kelvin dead-beat mirror galvanometer to make long and short sweeps of a beam of light on scales (Section 7.4.7) [12,18,35].
- German Professor Erich Rathenau developed the conduction wireless telegraph to its peak by extending its range to the distance of about 5 km over water. In 1896 Strecker achieved wireless reception at a distance of 17 km (Section 7.2.6).

1895:

- Alexander Stepanovich Popoff (or Popov) (Figure 2.27), the Russian inventor, demonstrated at the convention of the Russian Physical and Chemical Society in Petersburg his *Thunderstorm Recorder* using an aerial, a coherer, and an electromagnetic relay. He reported sending and receiving a wireless signal across a 600 yards distance [4,8,48].

- French brothers Louis and Auguste Lumière developed the *Cinematograph* [4].
- Marconi (Figure 2.28) transmitted and received a coded message at a distance of 1.75 miles near his home at Bologna, Italy [6,15].
- Indian physicist, Sir Jagadis Chunder Bose (Figure 2.29) generated and detected wireless signals of 6 mm wavelength at Presidency College, Calcutta, India. He also produced a fantastic variety of devices and techniques: wave-guides, horn antenna, cut-off grating, dielectric lens, microwave reflectors, double-prism directional coupler, polarimeter, interferometer, dielectrometer, and so on. He studied the effects of treating the contacts of coherer with chemicals (impurity) and its response to a flash of radiation (pulse) using a home made recorder (Chapter 9) [38,49].
- American Engineer John W. Howell, an associate of Edison, pointed out at a meeting of AIEE that rectification takes place in an Edison tube [1].

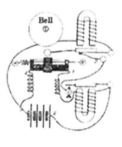

Figure 2.27. From left to right, Alexander Stepanovich Popoff (1859–1906), Popoff's radio receiver, and its electrical scheme.

Figure 2.28. A well-known photograph of Guglielmo Marconi (1874–1937).

Figure 2.29. Sir Jagadis Chunder Bose (1858–1937) at his Friday Evening Discourse before the Royal Institution.

1896:

- Tesla obtained 8 patents for producing currents of high frequency (Figure 2.30).
- Marconi applied in England for the first patent in wireless, covering the use of a transmitter with a coherer connected to a high aerial and to earth [4,35]. He also used a copper mirror to project a beam of electric radiation along certain directions.

1897:

- Marconi obtained GBP 12,039 for *Improvements in transmitting electrical impulses and signals and in apparatus there-for* [12].
- Marconi demonstrated a radio transmission to a tugboat over an 18-mile path at the Bristol Channel, England. This was the official trial of Marconi's radio system [4,50].
- The first wireless company, *Wireless Telegraph and Signal Company Ltd.,* was founded; the company bought most of Marconi's patents. Its name was changed to *Marconi's Wireless Telegraph Co. Ltd.,* in 1900 [35].
- J. J. Thomson discovered the electron [3]. In 1906, he was awarded the Nobel Prize in physics for his work on the electron and in 1908 he was knighted. Subsequently, no fewer of seven of his research assistants were to win Nobel Prizes, among them, the New Zealand-born English mathematician and physicist Ernest Rutherford and the English chemist and physicist Francis William Aston.
- Lodge patented the fundamental method of tuning transmitters and receivers of electric wave: *Improvement in Synchronized Telegraphy without Wires, Syntonic Wireless* [35].

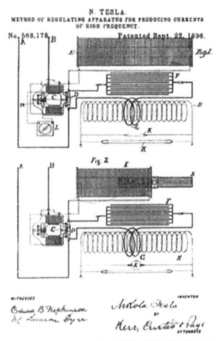

Figure 2.30. Tesla's drawings for USP 568,178 "Method of regulating apparatus for producing currents of high frequency".

- Englishman William du Bois Duddell, a student at the City Guilds College, and later an electrical engineer invented a practical oscillograph using a galvanometer with an extremely small periodic time and a viewing device which would spread out the deflections on a time scale in such a way so as to display the shape of a current wave repeating itself 50 to 100 times a second. The French physicist André-Eugène Blondel suggested the idea in 1892 [3].
- German Professor Adolf Karl Heinrich Slaby and Count George Wilhelm Alexander Hans von Arco, used a spark oscillator, a coherer, and a Morse telegraph which operated at about 250 MHz. Contrary to Marconi, who installed the antenna isolated from Earth, Slaby and Arco connected the upper end of the antenna to Earth via a coil. The receiver was also connected to the same antenna arrangement in which the coil reduced the influence of atmospheric disturbances. Experiments were made with transportable stations on land and on warships. With a 300-m-long wire as antenna connected to a balloon, a distance of 21 km was achieved [4].
- Braun invented the cathode-ray tube. He used a fluorescent screen, such that the screen produced visible light when struck by a beam of electrons (for long time the tube was known as *Braun's Tube*) [3,4,13].
- A. G. Davis devised the electrolytic condenser [1].

- Lord Rayleigh suggested that for small wave lengths, when the outer conductor of a coaxial transmission line exceeds a critical fraction of the wave length, the centre conductor can be removed and although it, electromagnetic waves may still propagate down the pipe by back and forth diagonal reflections between the walls (first conception of a waveguide) [1].
- Englishmen E. Wilson and C. J. Evans operated a radio wave controlled boat in the Thames River. In 1900 they were granted the USP 663,400 [51].
- Lord Rayleigh published an analysis of propagation through dielectrically filled waveguide [1].
- T. E. Gambell, in England, disclosed the dispersion of carbon in a liquid medium containing a binder and volatile solvent, which applied to an insulating base, is heated to produce a resistive film [1].
- Lodge wanted to measure long wave radiation from the sun by using coherer, but could not do it. J. C. Bose predicted that sun might emit electric radiation, which is absorbed by solar or terrestrial atmosphere [1].

1898:

- Danish electrical engineer and inventor Valdemar Poulsen built the first magnetic recorder of sound. He called his invention the *Telegraphone* [6,13].
- Lodge patented the spherical dipoles, square plate dipoles, biconical dipoles and *bow-tie* antennas [106].
- X. Boucherot devised the paper condenser [1].
- Braun improved the Slaby-Arco radio system by developing a spark oscillator circuit, which was connected to the antenna inductively instead of galvanically. This significantly delayed the damping of the transmitter. The coherer circuit was also connected to the antenna through a transformer [4]. He thus introduced tuning circuits in the transmitter and the receiver.
- Tesla demonstrated a radio controlled boat in Madison Square Garden and obtained a patent for it (USP 613,809) (Figure 2.31).
- French inventor Eugène Ducretet and his colleague Ernest Roger transmitted radio signals from the Eiffel tower to a receiver located near the Panthéon [4,6].
- German physicist Max Abraham calculated the radiation resistance of a simple vertical antenna situated over a plane conducting earth [1].
- AEG Company in Germany started production of the radio equipment developed by Slaby [4].

1899:

- Pupin read his paper before the American Institution of Electrical Engineers on the propagation of long electric waves and took out a patent for distributing inductance along the length of a conductor, i.e., the *Pupin Coil*. The American Telephone and Telegraph Company

(AT&T) bought the patent [3]. The use of loading coils made telephony possible over distances extending up to 2000 miles [7].

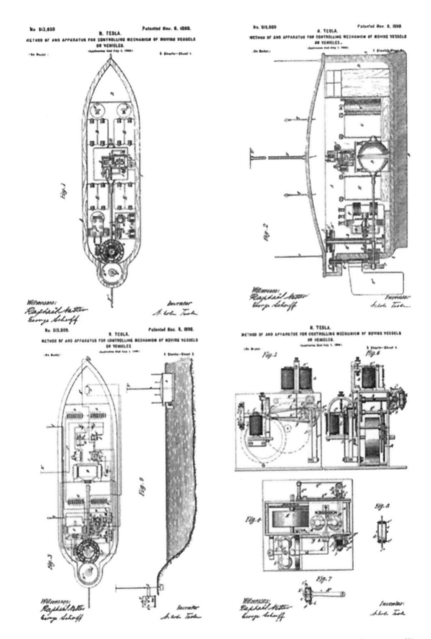

Figure 2.31. Tesla's drawings for USP 613,809 "Method of and apparatus for controlling mechanism of moving vessels or vehicles".

- Braun used a loop aerial for transmission and reception of wireless signal [15].
- Duddell discovered that arc lamps could also be used for the generation of frequencies up to about 1 MHz [4].
- Marconi sent the first international wireless message from Dover, England, to Wimereux, France, a distance of 50 Km [6].
- German physicist Arnold Sommerfeld published a paper describing a surface wave guided by a cylindrical conductor of finite conductivity [1].
- American-born British electrical engineer Sidney George Brown proposed the use of vertical aerials spaced half a wavelength apart for directional transmission of wireless signals. He patented parabolic reflectors with the transmitting aerial at the focus [52].
- Scottish electrical engineer James R. Erskine-Murray erected two vertical aerials with the transmitter connected between them in order to concentrate the radiation in one direction [36].
- Tesla built a gigantic coil resonant at 150 kHz and fed 300 kW to it to demonstrate the transmission of power without wires [44].
- Marconi installed wireless telegraph on the ship "St. Paul" [1].
- Elihu Thomson suggested methods for directional reception [36].
- Kennelly worked out the Delta-Y or Pi-T transformations [53].
- American electrical engineer John S. Stone filed a patent application for directional transmission and reception, using aerials spaced half a wavelength apart aligned in the direction of transmission [2,30]. He obtained the patent in 1902.
- First radio emergency call was made from a ship to England [4].

2.5   SOME CRUCIAL EVENTS OF THE TWENTIETH CENTURY

1900:

- Tesla obtained patents USP 645,576 and 649,621 on *System of Transmission of Electrical Energy*, submitted in 1897, which United States of America Supreme Court recognized to be the first patents on *Radio* (Figure 2.32 and upper part of Figure 8.5).
- Canadian-American Reginald Aubry Fessenden did the first speech transmission (over 25 miles) using a spark transmitter. The carrier frequency was 10 kHz (Figure 2.33) [4,15].
- Duddell discovered that when an inductance and a condenser are connected in series across an ordinary DC arc a steady oscillation sets up on the circuit for a certain ratio of inductance and capacitance. Maximum frequency obtained by Duddell was 10,000 Hz. He patented it as a *Singing Arc*, a method of producing alternating currents from direct current [36].
- Russian scientist Constantine Perskyi coined the word *Television* [10].

- Tesla patented a security system (ECCM) for remote control, using coincidental transmission (wireless) on two channels, a forerunner of the AND circuit [51].
- Marconi submitted his first US patent (November 10, 1900). He continued to submit patent applications on Radio. They were all turned down. In 1903 the US Patent Office made the following remarks: "Many of the claims are not patentable over Tesla patents # 645,576 and 649,621, of record, the amendment to overcome said references as well as Marconi's pretended ignorance of the nature of a *Tesla Oscillator* being little short of absurd ... the term Tesla Oscillator has become a household word on both continents (Europe and North America)".
- Tesla was the first person to describe a system of determining the location of a moving object using radiowaves, i.e., a radar system [6].
- German physicist Max Karl Ernst Ludwig Planck demonstrated that energy at the atomic level exists in indivisible packets or quanta [17].
- German mathematician, natural scientist, physicist and electrical engineer, Jonathan A. Zenneck demonstrated the screening of a vertical aerial in certain direction by the erection of additional vertical wires [52].

1901:

- German physicist Max Karl Werner Wien developed a bridge for measurement of imperfect condenser and cable testing [32].

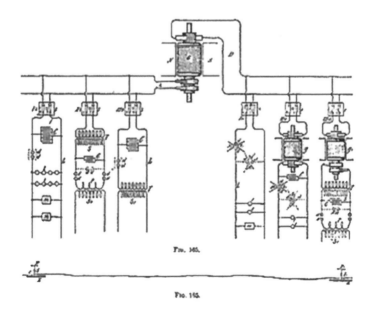

Figure 2.32. Tesla's drawings for USP 645,576 which was the first patent on a radio system.

 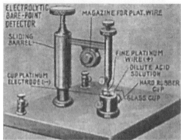

Figure 2.33. Upper photographs: left, Reginald Aubrey Fessenden (1866–1932); right, a photograph with two of his collaborators, transmitting radio messages (he is at the right). Lower part: left, one of his first alternators; right, his electrolytic detector (1903 USP 727,331).

- O. W. Richardson developed the theory of emission of electricity from hot bodies, *Negative Radiation from Hot Platinum Wire* [17,34]. He received the Nobel Prize in 1928, and was knighted for his work.
- On December 12, the first long distance wireless message (the letter S in Morse code) was transmitted from Poldhu, Cornwall, England, and received by Marconi in Signal Hill, Newfoundland, Canada. The distance covered was 1,700 miles across the Atlantic Ocean. The way was thereby opened toward a tremendous broadening of the scope of communications. For the experiments Fleming, a Marconi collaborator at that time, designed a 20 HP engine to drive an alternator of 2,000 V stepped-up by a transformer to 20 kV, and a spark gap of 2 inches (immersed in oil). The transmitting aerial height was 219 feet, and the receiving aerial was suspended from a kite flying at 400 feet at a site called Signal Bay. The wavelength of transmission was 800 m and the cost of the experiment was £ 40,000 [7,18,35].
- Bose filed a patent for his *Point Contact Diode Using Galena* (Galena detector). It was given to him in 1904 [49] (Figure 2.34).

- Italian physicist Giovanni Giorgi proposed the MKS system [3].
- Fessenden patented the idea of using a saturable reactor in a *Wireless Signaling* system as a magnetic modulator [54].
- Fessenden received the first patent of the world for radiotelephony [4].
- Maher first used measurement of current flowing through a circuit instead of voltage across a capacitor to determine resonances [1].
- Marconi installed in a car the earliest form of mobile radio [55].

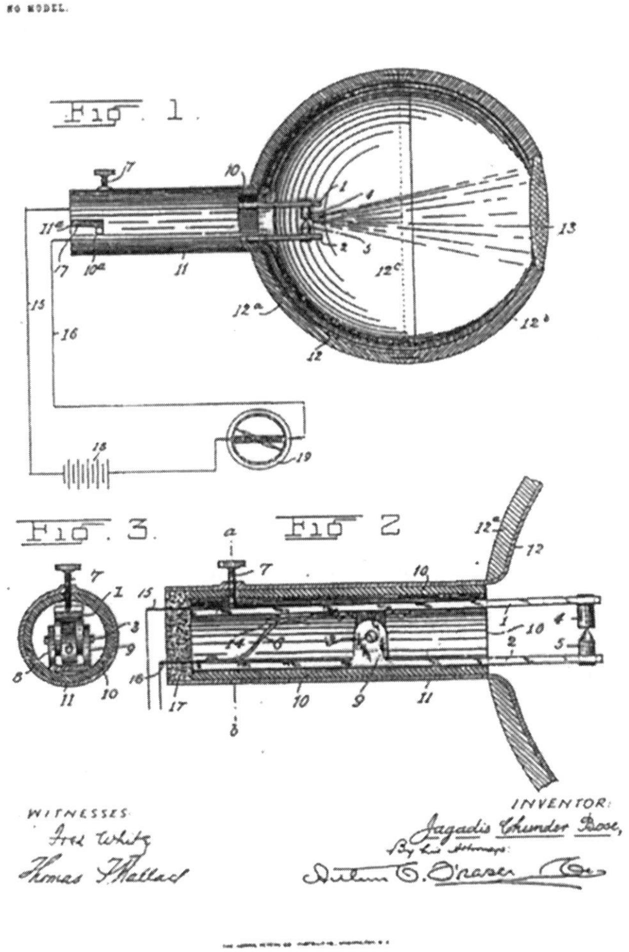

Figure 2.34. Bose's drawings for patent "Point contact diode using galena".

1902:

- American scientists Louis Winslow Austin and H. Starke first identified secondary emission [46].

- German physicist Arthur Korn developed a facsimile machine suitable for transmission of photographs. He was the first to introduce a facsimile machine with optical scanning in the transmitter and photographic reproduction in the receiver [4,6,17].
- O. W. Richardson published experimental results confirming his theory (published in 1901) on the emission of electrons from hot bodies [56].
- Righi worked out the theoretical expression for frequency of glow tube relaxation oscillator [34].
- Heaviside and Kennelly suggested that the upper layers of the earth's atmosphere could have conducting layers. In fact they predicted the existence of an ionized layer in the upper atmosphere [4,15].
- J. J. Thomson pointed out the effect of the conductivity of the gaseous ions in the ionosphere on the propagation of electromagnetic waves [1].
- Fessenden patented the *Heterodyne* method of reception [15].
- Austrian physicist Philipp Eduard Anton von Lenard used a third electrode in a vacuum tube for the purpose of studying photoelectric effect [33]. He received the Nobel Prize in 1905.
- American physicist Albert Hoyt Taylor reported ignition interference from a two-cylinder automobile [1].
- American Cornelius D. Ehret filed two patents covering the transmission and reception of coded signals or speech (*Frequency Modulation*: FM). So far as it is known this was the first disclosure to describe any system of modulation by name. The method of modulation consisted of varying the resistance or reactance of an oscillator [1].
- Poulsen was first to develop a continuous-wave transmitter [4].
- American electrical engineer Greenleaf Whittier Pickard developed a loop aerial [1].
- Weber proposed an interesting physical interpretation for the fact that the wave velocity in the tube (waveguide) is less than light [44].
- Danish telegraph engineer Carl Emil Krarup introduced transversal inductance cable [4].

1903:

- Marconi purchased the patent of Edison's antenna [35].
- Marconi established a transmission station in South Wellfleet, Massachusetts. The dedication ceremonies included an exchange of greetings between American President Theodore Roosevelt and British King Edward VII [6].
- J. E. Taylor suggested the effect of the ionization of air by sunlight on radio wave propagation [1].
- An international conference was held in Berlin to discuss radio communications; the *International Radiotelegraph Union* emerged from the suggestion of a second conference in 1906 [30].

- Pickard filed a patent application for a crystal detector, in which a thin wire was in contact with silicon [41].
- To produce high frequency for wireless telegraphy and keep the arc stable Poulsen found that the following modifications were necessary: (a) use of copper positive electrode, water cooled and slowly rotating carbon negative electrode, (b) to maintain the arc in a hydrogenous atmosphere, and (c) a transverse magnetic field across the arc [36].
- Blondel showed that, provided the aerials were fed in correct phase (suggested by Brown in 1899) they need not be half-a-wave length apart. Italian engineer E. Bellini had also suggested that twenty Blondel pairs would radiate a beam in one direction, but the suggestion proved premature because such phasing of radio frequency (RF) currents was not possible at that time and the distance between the aerial rendered it impractical for long waves (concept of end-fire array) [15].
- Spanish civil engineer, mathematician and inventor Leonardo Torres Quevedo patented the *Telekine*, a wireless remote control that was demonstrated on tricycles and engine-driven boats. He created a family of different and easily readable code words by using the signal generated with a common wireless telegraph transmitter. He built a completely new type of receiver, which was able to react in a different way to each codeword sent by the transmitter (Figure 2.35) [104].
- W. M. Miner patented the application of time division multiplex methods to telephone channels [1].
- American engineer Malcolm Rorty of AT&T developed the queuing theory out of telephone traffic problems [1].
- American Charles L. Krum, vice-president of the *Western Cold Storage Company*, obtains the first teleprinter patents [4].

1904:

- French engineer and writer Louis Marie Édouard Estaunié, director of the Ecôle Supérieure des Postes et Télégraphes de France and member of the Académie Française, coined the word *telecommunication* [4].
- German engineer Christian Hülsmeyer patented in England an obstacle detector and ship navigational device by Hertzian-wave projecting and receiving devices, based on Tesla's radar idea [57]. It was called a *Telemobiloscope*. In the 1920s Gregory Breit and Merle A. Tuve used a similar system to measure the depth of the ionosphere [6].
- Korn's Telautograph was tested and the first telegraphic transmission of pictures on a Munich-Nuremberg-Munich loop took place [4,58].
- Maxwell calculated the shape of electrostatic field about a wire grid placed between two electrodes as a problem in pure electrostatics [33].
- American engineer Frank J. Sprague developed the idea of the printed circuit [59].

Figure 2.35. Left, Leonardo Torres Quevedo (1852 – 1936); right, his Telekine.

- Fleming was the first to suggest the rectifying action of *vacuum-tube diode* (electric valve) for detecting high frequency oscillation. He used a carbon filament as the incandescent electrode and a metal plate as the cold electrode (anode). He called this device the *Oscillation Valve*. It also became known as *diode* (two electrodes) and *rectifier* (because it converted alternating current into direct current). The diode was also a better detector of high frequency oscillating current than the crystals used in radio receivers. Fleming's valve detected radio waves and converted them to weak direct current that could be played through a radio headset. It was the first practical radio tube. The vacuum-tube diode paved the way for the invention of the vacuum-tube triode (Figure 2.36) [5,6,18,33,36].
- The Wireless Telegraphy Act was passed to grant license to amateurs in UK [15].
- The patent on *Point contact diode using galena* was granted to Sir J. C. Bose. He applied for it in 1901 [49].
- German physicist Arthur Wehnelt invented the oxide-coated emitters, *Wehnelt cathode*, for vacuum diode. They were used in valves in 1914 [3].
- The United States Patent Office reversed itself and gave the Radio patent to Marconi (the Marconi Company stock increased, the millionaire Andrew Carnegie invested in American Marconi, and Edison was consulting engineer). In 1943, two months after Tesla's death, the US Supreme Court upheld Tesla's patent 645,576 for invention of *Radio*. The court had selfish reasons for doing so. Marconi Company was suing the American government for using its patents in World War I. The court simply avoided the action by resorting priority to Tesla's patent over Marconi.
- Marconi entered into an agreement with the Cunard steamship line to create the first ship-to-shore communication system [6].
- F. K. Vreeland published that fading was due to anomalous attenuation of electromagnetic wave while passing through the ionized air through the influence of sunlight [1].

- Fessenden achieved radiotelephony over 40 km [4].

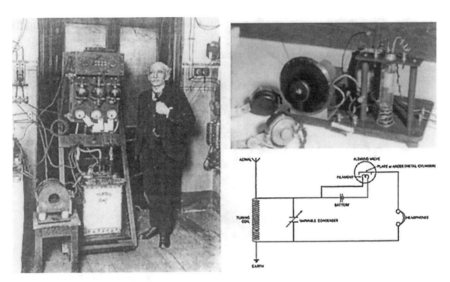

Figure 2.36. Left, Fleming with some of his valves. Right: upper part, an early model of his diode valve; lower part, schematic of his diode valve in operation.

1905:

- German physicist Albert Einstein (Figure 2.37) explained the photoelectric emission under the action of radiation. He proposed to apply Planck's quantum theory of radiation to photoelectric studies (second law of Photoelectric emission) [34,45].
- Wehnelt suggested coating of platinum filament (of cathode-ray tube) with lime (BaO and SrO) so that the necessary emission for practical purposes could be achieved at a reasonable voltage and many experimenters became interested. At about 1911 the American researcher Harris J. Ryan brought together a number of ideas to control the deflection of a stream of electrons. A cathode-ray tube without filament was called cold-cathode cathode-ray tube and required 10 kV or more for its operation and use; the one with filament was called hot-cathode cathode-ray tube. Even in 1922 papers were published in *Proceedings of the Institute of Radio Engineers* explaining operation and use of cold-cathode type cathode-ray tube [3].
- Wien increased the efficiency of the damped spark gap oscillators by developing a special spark bridge consisting of a series of copper disks separated from each other by a gap of 0.5mm which produced very short silent sparks of high energy that was extinguished automatically on the first zero level. This caused a pulse excitation in the secondary circuit, which then produced a wave of constant frequency at slowly diminishing amplitude [4].

- Radio controlled torpedoes designed by the American engineer Harry Shoemaker and built in the United States, were sold to Japan [51].
- Stone patented circuits for transmission and reception of multiplex radiotelephony and telegraphy [1].
- Duddell and J. E. Taylor measured first the electric field intensity by using a thermo galvanometer [1].
- Fessenden demonstrated wireless telephony by transmitting speech and music over a radio channel [12,18].
- Fessenden invented the *superheterodyne* circuit [4].
- Tokyo-Sasebo (1,583 km) record phone call was achieved [4].
- Swedish immigrant and electrical engineer Ernst Frederik Werner Alexanderson working for General Electric in the United States obtained six patents in the field of motor and generators [36].

Figure 2.37. Left, Albert Einstein (1879–1955); right, Einstein's office and blackboard.

1906:

- Marconi determined the radiation curves of an *inverted L type aerial* fed at one end and found that the aerial was strongly directional for both transmitting and receiving [36].
- American physicist Lee de Forest patented the general principle of omni-range using a rotating radio beam keyed to identify the sector forming 360° sweep of the beam [1].
- De Forest invented the three-electrode valve or *vacuum tube triode*. The original device was known as *Audion* valve (Figure 2.38). It had a tantalum filament, nickel wire grid on one or both sides of the filament and an anode consisting of one or two nickel plates connected together after the grid. He found that when a microphone transformer was connected between the grid and the filament of the audion valve, the sound in the earpiece was much louder than when the earpiece was connected across the transformer (i.e., his discovery represents the first use of the three electrode valve as an amplifier). He published "Audion, a New Receiver for Wireless Telegraphy" in Proc. AIEE,

Oct. 1906 [12,15,36]. The discovery of the triode was instrumental in the development of transcontinental telephony in 1913 and signaled the dawn of wireless voice communications. Indeed, until the invention and perfection of the transistor, the triode was the supreme device for the design of electronic amplifiers [18].

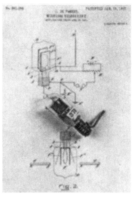

Figure 2.38. From left to right, Lee de Forest (1873–1961), the Audion patent drawing with a real Audion above it, and a photograph of the first triode "Audion".

- Robert von Lieben, an Austrian physicist working in Vienna on the amplification of telephone signals, added a third electrode to Fleming's diode, in the form of a wire mesh grid between the electron emitting cathode and the anode. In fact, he discovered the *vacuum tube triode*, independently from De Forest. However, von Lieben's work was not fully exploited, perhaps because of his early death and the outbreak of war in 1914 [4,17].
- Fessenden did the first advertised Radio Broadcast in the United States. [29].
- German engineers Max Dieckmann and Gustav Glage proposed and performed an experimental use of cathode ray tube for facsimile reproduction [34].
- The *International Radio Telegraph Union* was formed [4,30].
- Pickard filed a patent for shielding of antenna against static using a Faraday cage [52].
- Pederson patented electroplating of ductile ferromagnetic material over a brass wire [1].
- French inventor Eugène Lauste recorded sound on photographic film and reproduced from the film [60].
- French scientist Édouard Belin developed a telephoto machine, which he called the *Telegraphoscope* for telephotography. It was smaller than a typewriter and capable of being connected to an ordinary telephone. In 1913 he developed a portable version called the *Belinograph* [4].

1907:

- Radio telephony apparatus was installed in 24 battle ships for reception at a distance of 25 miles [15].
- Italian engineers Bellini and A. Tosi developed an aerial for directional transmission and reception [15].
- Pickard developed direction finding using a vertical aerial [15].
- German engineer Otto Scheller patented his *course-setter* (for navigation) using interlocked A/N (Morse code) signal [52].
- Poulsen transmitted music by wireless using an arc transmitter with 1 kW input power and 200-feet-high aerial. The music was heard at a distance of 300 miles [36].
- Fessenden used in an article for the first time the term *Modulation* and also talked about atmospheric absorption of wireless signals [1].
- English scientist Captain Henry Joseph Round first reported observing light emission from beneath a *Cat Whisker* [61].
- Korn sent the first intercity fax sending a photograph from Munich to Berlin [6].
- The *Society of Wireless Telegraph Engineers* was formed in the United States [12].
- Patent for cold-cautery (radio knife) using Poulsen arc generator was issued to De Forest and was demonstrated to surgeons in Paris and Berlin in 1908 [1].
- Morkrum Company produced the first practical teleprinter. The name was changed to *Teletype* in 1925 [1].
- Zenneck showed that a plane interface between two semi-infinite media such as ground and air could support an EM wave, which is exponentially attenuated along the direction of propagation [1].

1908:

- American physicist Edward B. Rosa modified the formula suggested by H. A. Lorentz for calculation of the inductance of an air core coil [36].
- Alexanderson constructed in the United States 100 kHz, 2 kW and 22 kHz, 200 kW alternators (Figure 2.39) [36].
- De Forest installed a transmitter at the top of the Eiffel Tower for broadcasting music from a gramophone. The transmission was heard up to 500 miles [15].
- Fleming found that, for valves, tungsten filament was better than carbon filament and patented it [15]. At about 1903 the Austrian scientists Alexander Just and Franz Hanaman first made tungsten filament for lamps [33].
- Scottish electrical engineer Alan Archibald Campbell-Swinton publishes basic ideas of television broadcasting [14]. He proposed, and the Russian scientist Boris L'vovich Rosing patented in England, the use of cathode ray tube as television picture-receiver [20].
- Transatlantic radiotelegraph service was opened to the public [1].

- Fessenden reported successful radiotelephone communication between Brant Rock and Washington, DC, a distance of 600 miles using a 70 kHz, 2.5 kW alternator [1].
- Fessenden found that night transmission over the north Atlantic was severely depressed during magnetic disturbances [1].

Figure 2.39. Left, Ernst Frederik Werner Alexanderson (1878–1975); right, his alternators.

1909:
- American physicist Louis Winslow Austin and Louis Cohen, engineer, developed an empirical formula to measure range of wireless stations and checked it up to a range of 1100 miles over the Atlantic [36].
- A. C. and L. S. Anderson of Denmark proposed color television using mechanical scanning, selenium cell and dispersive prism [62].
- Japanese scientist J. Nagaoka provided the inductance coefficient of solenoids [63].
- First electrocardiograph was developed [3].
- Physicist and inventor William David Coolidge of General Electric (GE) Company of Schenectady, New York, developed a sintering process for tungsten which improved the ductility of the metal [15].
- The *Wireless Institute* was established in the United States [12].
- A. Russel worked out the transmission of alternating current over a concentric system of conductors [1].
- Sommerfeld published his paper on radiation of a Hertzian dipole over a plane earth [1].
- Marconi and Braun shared the Nobel Prize for Physics for their contributions to the physics of electric oscillations and radiotelegraphy.

1910:
- W. R. Ferris transmitted wireless telegraphy from an aeroplane [1].

- Before 1910, the terms used for modulation were "control" or "vary" or "mould" or "modify". Fleming and the American communication engineer and inventor George Washington Pierce used the term *modulation* in books published in 1910 [1].
- The origin of optical waveguides may be traced back to the work on dielectric rod started this year by D. Hondros and P. Debye [64].
- Coolidge invented the *tungsten filament*, which further improved the longevity of the light bulb [6].
- De Forest sought to demonstrate the potential of radio by arranging the first ever radio broadcast featuring the great tenor Enrico Caruso singing from the Metropolitan Opera House in New York City [6].
- Von Lieben applied for a patent on the so-called *"Lieben Tube"*(Section 11.2.6)
- T. Baker demonstrated a method of transmitting photography by wireless using a coherer [11].

1911:

- The early German transmitting valve, called the von Lieben's transmitting triode, was perfected with an oxide coated filament, perforated aluminum disc as grid and aluminum coil as anode. It was a mercury vapor valve with a small pocket for mercury. Softness of the valve could be increased by heating the mercury [15,36].
- A. Blondel demonstrated the guidance of aircraft by radio telegraphy [1].
- K. W. Wagner developed an earthing device to eliminate the effect of capacitance between the headset, null detector and observer's head [32].
- Von Lieben and Eugen Riesz developed a cascade amplifier [1].
- Campbell-Swinton patented an electronic scanning picture transmitting and receiving system based on cathode ray tubes [17].
- Coolidge introduced a small quantity of Thorium into the tungsten and made thoriated-tungsten filament, which increased its ductility and gave better light efficiency. The filament became standard for lamps and valves [15].
- American physicist Clement Dexter Child demonstrated the law named after him relating plate current and plate voltage of a high vacuum diode with ample supply of electrons from the cathode for parallel-electrode surfaces [33,56].
- American physicist John Milton Miller demonstrated the equivalent circuit of a vacuum triode [34].
- British physicist William Henry Eccles reported semiconductor oscillating crystals, which predate vacuum tube oscillators [10].
- H. Gerdien patented the practical application of the motion of electrons in perpendicular electric and magnetic field (as in the magnetron) [45].

- Hugo Germsback, an American science fiction novelist, envisaged the concept of pulse RADAR in one of his works where he proposed the use of a pulsating polarized wave, the reflection of which was detected by an *actinoscope* [65].
- Round, working for the British Marconi Company, commenced work on the design of a diode valve (Section 11.2.6)

1912:

- De Forest patented the reaction in a valve circuit [15]. He sold the patent to AT&T [4].
- Engineers were coming to realize that the triode had other uses besides detection of radio waves. De Forest, Fritz Loewenstein, and Irving Langmuir in the United States as well as von Lieben and Otto von Bronk in Germany realized that it could be used in a transmitter and could work as an oscillator. These functions were soon put to use. The three-electrode vacuum tube was included in designs for telephone repeaters in several countries (Section 11.2.6).
- American Edwin Howard Armstrong, a student of electrical engineering at Columbia University in New York City, found that he could obtain much higher amplification from a triode by transferring a portion of the current from the anode back to the signal going to the grid (regenerative receiver). He also found that increasing this feedback beyond a certain level made the tube into an oscillator, a generator of continuous waves (CW) (Section 11.2.6).
- Austrian physicist Alexander Meißner (or Meissner), working at Telefunken in Berlin, discovered and patented an electronic high frequency (HF) generator by feeding-back part of the amplified signal to the grid of the Audion [4].
- D. C. Prince developed a loop aerial modified from the Bellini-Tosi aerial [35].
- Sinding and Larsen transmitted television (TV) by wireless using three channels, one for sound, one for picture, and the third for synchronization [62].
- Two wireless operators went down with the *Titanic,* sending wireless signals. This saved 712 passengers [35].
- G. W. O. Howe measured self-capacity and natural frequency (self resonant frequency) of an inductance coil using a variable frequency generator (alternator) and a wave meter [36].
- F. Keitz published the History of Earth Aerials, i.e., the first history related with wireless [52].
- G. A. Campbell developed guided wave filters [32,66].
- C. L. Krum and his son, Howard Krum, along with the American Joy Morton (owner of *Morton Salt Company*) produced the first start-stop teletypewriter. A start signal was transmitted immediately preceding the code of each character. Similarly a stop signal was transmitted at the end of the code of each character. The code employed for the characters was a five-unit code similar to that of Baudot in 1874 [4].

- Italian physicist Quirino Majorana found another interesting solution for the modulation of the Poulsen transmitter. He patented a four-electrode valve with two grids, used as a full-wave detector [1,4]. The four other tetrodes (excluding screen grid tube) developed were, the *Space-Charge Tetrode* by the American physicist and chemist Irving Langmuir, in 1913 [4,34]; the *Negatron*, a negative resistance device developed by John Scott-Taggart in 1921 [36]; the *Quadrode* developed by Marconi Company in 1924 [1]; and the *Wanderlich Tube* developed by H. A. Pidgeon and J. O. McNally in 1930 [53].
- The *Institute of Radio Engineers* (IRE) was formed in the United States from the Society of Wireless Telegraph Engineers and the Wireless Institute [12].
- Eccles pointed out the possible influence of upper layer and suggested diurnal variations of signal intensity were due to change in degree of ionization produced by the solar radiation [1].
- German scientist Siegmund Strauss invented and patented the feedback principle [1].
- American scientist John H. Hammond suggested radio relaying by change of the carrier frequency [51].
- The Federal Wireless Company started wireless communication between San Francisco and Honolulu (2,100 miles) using Poulsen arc [1].
- Campbell made a wave filter using inductance and capacitance. In 1934, the American electrical engineer Warren P. Mason made it using quartz crystal [32,66].
- Austin described a method of measuring the total equivalent resistance of antennas using a buzzer for exciting the antenna [67].
- First rotary exchange is installed in Landskrona, Sweden [4].
- *Western Union* starts the teletypewriter service [4].

1913:

- After the development of the air pump done by the American scientist Saul Dushman the first hard (high vacuum) valve to come into prominence was the *Kenotron* (the Greek word "Kenos" means empty). The *Pliotron,* a three-electrode valve in which no blue glow appeared, followed the two-electrode valve [36].
- Fessenden patented the use of the vacuum triode as oscillator [15].
- Alexanderson patented the radio frequency amplifier in cascade [15].
- Australian-American electrical engineer Cyril F. Elwell made possible *Frequency Shift Keying* of a Poulsen arc by short-circuiting a portion of the tuning inductance [36].
- Langmuir showed the effect of space-charge and residual gasses on thermoionic currents in high vacuum diode [45].
- Transmission of time signal by wireless for ships was started from the Eiffel Tower [15].

- Eccles suggested that ultraviolet light from the sun could ionize the gases in the upper atmosphere to produce layers of gas containing free electrons [15].
- Coolidge developed the X-ray tube [1].
- Round patented a neutralizing circuit for RF amplifier [15].
- Round patented the idea of heterodyning in radiotelephony [1].
- Marconi receiver with Round's C valve was available (Section 11.2.6).
- Transcontinental telephony came into being with the use of electronic amplification, thanks to vacuum tubes [7].
- The term *radio* taken from *radii* was originally meant to indicate that waves were broadcast. Standard Committee of IRE standardized the term *Radio* [1].

1914:

- Carl R. Englund first set up the equation of a modulated wave (*Amplitude Modulation*: AM). He also discovered the frequencies related to the sidebands [1].
- A repeater for landline telephony was constructed [33].
- German physicist Walter Schottky (Figure 11.20) discovered the phenomenon named after him, i.e., the *Schottky Effect* (effect of electric field on the rate of electron emission from thermoionic-emitters) [45].
- Langmuir was the first to observe the reduction of electron affinity of tungsten as the result of introduction of a small amount of thorium [34].
- Frequency modulation of carrier was proposed to accommodate more number of channels within the available wavebands (only medium and long waves) using very small frequency deviations [68].
- Fleming discovered the atmospheric refraction and its bearing in the transmission of EM wave round the surface of the Earth. [52].
- Edison invented the *alkaline storage battery* [6].
- English physicist Edmund Eward Fournier d'Albe made the *Optophone,* the first known machine for reading print using selenium cells [1].
- Howe published methods for the calculation of capacitance of elongated flat top antennas composed of parallel wires [1].
- The first application of automatic pilot developed jointly by Sperry Gyroscope Company and Hammond Laboratory was installed in the boat *NATALIA* and was put to test run for 60 miles [51].
- Hammond used crossed loops in the ship Dolphin for direction finding [51].

1915:

- American electrical engineer John Harold Morecroft invented the unipotential thimble shaped tungsten filament for electron tubes [56].
- Arlington to Paris and Honolulu radiophone services was opened [1,4].

- American electrical engineer Edwin Howard Armstrong received in the United States a signal from England using a single valve tuned-anode as tuned-grid receiver [15] (Figure 2.40).
- Bell System completed an American transcontinental telephone line [12].
- Direction finding equipment was used to locate sources of atmospherics [15].
- Western Electric Company made push-pull audio frequency power amplifiers [15].
- American electrical engineer Raymond A. Heising of Western Electric Company of patented the use of vacuum tube (diode) to measure voltage [69].
- J. R. Carson developed the mathematical theory of modulation and various forms of amplitude modulations including the single sideband and suppressed carrier mode of transmission [17]
- Zenneck developed a rotating omnirange with an identification signal [1].
- AT&T engineers developed permalloy cable [4].
- Western Electric engineers John G. Roberts and John N. Reynolds, obtained the first patent for crossbar switching [4].
- Schottky started work on developing a space-charge-grid tube and a screen grid tube or *Tetrode,* in which he achieved good amplification by placing a screen grid between the grid and the anode so that a tetrode was created [4].

Figure 2.40. Left, Edwin Howard Armstrong (1890–1954); right, a photograph of him showing the first portable radio.

1916:

- American scientist Charles V. Logwood modulated an oscillator by impressing audio signal on the grid [1].
- O. E. Buckley used a triode for measurement of ionization [33].
- J. H. Roger used EM wave for sub-surface communication [1].
- Schottky finished his work on the tetrode [3]. He patented it in 1919 [34]. The American physicist Albert Wallace Hull and the Englishman Captain Henry Joseph Round made it in 1926, independently [3,45].
- French-American physicist Léon M. Brillouin and the French physicist Georges A. Beauvais patented the R-C coupled amplifier [15].
- Alexanderson and S. P. Nixdrof constructed a magnetic amplifier for radiotelephony, used as modulator with Alexanderson high frequency alternator [36].
- American engineer G. S. Meikle, working for GE, developed the hot-cathode argon-gas filled rectifier [45].
- C. H. Sharp and E. D. Doyle patented in the United States the *Crest Voltmeter* using vacuum diode [45].
- F. Cutting designed radio telegraphic transformer [56].
- E. Leon Chaffee deduced and experimentally verified the relative amplitudes and phases of currents in a coupled circuit [56].
- Hammond filed for a patent for transmitting and receiving electromagnetic waves for secret radio communication using a rotary spark gap transmitter [1].
- GE developed automatic signaling for railways, which was installed by Union Switch and Signal Company and was tested in 1923 [1].
- S. M. Aisenstein in Russia used a 300 kW – 25 kHz spark transmitter to communicate with a 20 feet submerged submarine fitted with a loop aerial [1].
- Scheller, working for Lorenz Company of Germany, patented the *Goniometer* [1].
- F. Adcock used open vertically spaced aerials for direction finding in aircrafts [1] and was granted the British Patent 130,490 in 1919.

1917:

- Marconi-Osram marketed 150 W anode dissipation glass valves [70].
- R. E. Russell developed the *Tunger* (mercury vapor) rectifier [45].
- Englund was first to suggest the measurement of field strength by voltage substitution method [69].
- Morecroft constructed an artificial transmission line, resembling a long antenna using ten capacitors of 18.3 microfarad each; nine inductances of 0.0415 H and 0.702 ohm each. Using wattmeter, ammeter, voltmeter and generator the resonance characteristics of the line were determined. The line showed three series and three parallel resonances between 12 and 152 cycles [56].

- Elwell designed and constructed 714 feet high wooden lattice mast for supporting aerial; the mast was erected at Rome and was the highest wooden structure in the world [36].
- R. S. Kruse generated 750 MHz, the highest frequency of continuous wave triode oscillator up to 1932 [1].
- C. S. Franklin designed and constructed mirrors made of wires for reflecting waves of 3 to 15 m. Radio communication was carried out with short wave transmitters of 200 W between London and Birmingham [15].
- Frenchmen Henri Abraham and Eugène Bloch developed the *Astable Multivibrator Oscillator*, which generated a harmonic-rich 1 KHz signal [15].
- Valures analyzed RC coupled amplifier and obtained the expression for output voltage [15].
- An automatic transmitter for distress signal was developed [1].
- NPL of UK first used direction-finding equipment in an aircraft [13].
- Englund first suggested the measurement of field strength using the voltage substitution method [69].
- Einstein theorized that if there were more photons than electrons in a form of radiation (such as light), the energy difference could *stimulate* an electron to jump to a lower energy level, causing the *emission* of another photon, hence *amplification*. This laid the principle for the invention of the *Maser* and *Laser* [6].
- Radio controlled torpedoes from an airplane flying at a height of 5,000 ft were demonstrated [1].
- Hammond invented an early form of automatic volume control by shunting the input of detector triode with another triode biased from the detector whose resistance decreases with increase in detector voltage [1].
- Frederick Schwers studied the effect of water vapor on the propagation of EM waves [1].
- Morecroft developed an artificial transmission line by connecting inductors and capacitors [56].

1918:

- Armstrong invented the *Superheterodyne Radio Receiver* using eight valves. Even to this day, almost all radio receivers are of this type [1].
- American scientist A. M. Nicholson, working for Bell Telephone Laboratories, employed Rochelle salt crystal as the control element in a vacuum tube oscillator [54].
- Davis of Dunwoody Industrial Institute of Minneapolis, Minnesota, patented a signaling system for the teaching of radio signals to students in a classroom [1].
- Ryan developed the equivalent circuit of a quartz crystal during war research on submarine detection [53].
- Langmuir patented the feedback amplifier [45].

- E. H. O'Shaughnessy development of direction finding was one of the key weapons in England during the First World War. Thirty-feet-high Bellini-Tosi aerials were installed around the coast to locate transmission from ships and aircrafts [36].
- English natural scientist and physicist Sir Edward Victor Appleton first suggested a transconductance bridge to measure the transconductance of a valve [34].
- The *Dynatron*, first advocated by A. W. Hull, was a three-electrode tube consisting of spiral tungsten filament, perforated metal cylinder anode surrounding the filament, and a third electrode, called plate or target, surrounding anode. The tube was used as a negative resistance oscillator [36].
- Schottky predicted theoretically that due to current fluctuations in emitted electron stream of thermoionic vacuum tube, disturbances would set up in any tuned circuit connected to the plate. He called the phenomenon *Schroteffekt* or "small shot effect" [33].
- Schottky also first called the attention to the fluctuation of current in a conductor due to thermal agitation of electrons (thermal noise). J. B. Johnson and Harry Nyquist (independently) verified such fluctuations experimentally in 1928 [33].
- Schottky also invented, independently from Armstrong, the superheterodyne detection principle (Section 11.2.11).
- Miller developed a dynamic method for determining the characteristics of three electrode vacuum tubes [63].
- F. H. Millenner developed radio communication with moving trains [1].
- G. N. Wattson deduced the diffraction of electric waves by earth [1].
- American electrical engineer and physicist Louis Alan Hazeltine invented the neutrodyne circuit with tuned RF amplifier with neutralization [1].
- Tigerstedt suggested a method of secrecy in speech transmission using magnetic recording on endless iron tape, a number of electromagnets and commutator for scrambling the speech before passing it on to the telephone line [15].
- H. H. Arnold and C. F. Kettering worked out the basic principles of a pilot-less aircraft, which the US Navy demonstrated in 1946 [1].
- Bjornsthal observed dynamic scattering in Nematic liquid crystal (used in display since 1967) under the influence of an electric field [71].
- Four-channel multiplex was used on Baltimore-Pittsburgh line [4].
- Britain-Australia radiotelegraphy was made over 17,700 km [4].
- Radio transmission was made from the Eiffel tower [3].

1919:

- F. K. Richtmeyer made metal film resistor by sputtering [1].

- F. Kruger patented the idea of increasing the value of deposited carbon resistance by cutting a helical groove in the film [1].
- A. H. Taylor described the possibilities of concealed receiving system [1].
- Schottky and H. Abraham developed the relation between amplification factor and electrode dimensions of a triode [34].
- Using a triode as detector in a receiver H. J. Van der Bijl, of the Western Electric Engineering Department, USA, detected a three picowatt input signal [56].
- Miller first described bridge circuits for the dynamic measurements of tube factors [34].
- H. Abraham developed the formula for radiation resistance of vertical loop aerials in terms of area, number of turns, and wavelength [36].
- The *Silica Valve* was shown to get up to 1 kW of anode dissipation [70].
- Marconi-Osram Company developed the U-5 twin-anode full-wave rectifier [15].
- R. Bartelemy used AC to heat the filament of valves used in receiver. Vellette used an old triode receiving-valve to rectify AC together with a condenser as filter to smooth the high voltage [15].
- A 500 kW Poulsen Arc converter was constructed in UK [56].
- American mathematician Joseph Slepian, research engineer at Westinghouse Electric and Manufacturing Company, filed a patent application for a vacuum tube electron multiplier [62].
- J. J. Stevens and others detected infra-red radiation using thermopile detector located at the focus of a parabolic mirror [56].
- J. H. Dellinger published variation of induction and radiation fields of aerials and coil (loop) antennas with distance. R. R. Ramsey measured such variations in 1928 at distances from 0.05 wavelength [56].
- Based on Abraham-Bloch multivibrator, Britishmen Eccles and F. W. Jordan invented a binary circuit capable of either one of two stable states at any given time. The invention became known as the *flip-flop* circuit [6,72].
- Schottky patented the screen grid valve [34].
- The cross-modulation in the Ionosphere was discovered [1].
- Jwett suggested talking with a flying airline pilot [1].
- Robinson published direction finding for aircrafts using two rotating coils fixed at right angles and free to rotate about a vertical axis [1,36].
- Hollingworth developed a new form of catenary-trailing wire antenna for aircrafts [1].
- Hammond published aero-radio surveying and mapping [1]
- J. E. Taylor first observed the existence of errors at night in the apparent observed direction of a radio station [1].

- Scottish physicist Sir Robert Alexander Watson-Watt patented a device concerned with radiolocation by means of short-wave radio waves. This was the forerunner of the Radar system [5,6].

1920:

- C. E. Prince installed valve transmitters in aeroplanes for wireless telephony [36].
- L. B. Turner devised the *Kallitron*, a type of regenerative resistance-coupled amplifier suitable for use as DC amplifier [33].
- Miller published the *Miller Effect*, i.e., dependence of the input impedance of a triode upon the load in the plate circuit [45].
- S. G. Brown developed the *Ossiphone*, hearing aid for deaf persons without eardrum [15].
- A blow-by-blow account of a boxing match, broadcasting from the ringside, was first made in England [15].
- Marconi Company constructed a 15 kW valve transmitter at Poldhu. Nillie Melba sang into a microphone at Chelmsford. The song was heard in Newfoundland 3,000 miles away [15].
- O. J. Zobel of AT&T Company perfected the wave filter designs which made it possible to put more single-sideband voice channels to be packed into the frequency band available on a wire or cable system [17].
- G. Vallauri measured the field strength of a signal produced at Leghorn from Annapolis USA, and found that power of 10 pico watts could be detected by using eight-valve amplifier [36].
- G. W. O. Howe measured AC/DC resistance ratio of RF coils using a thermocouple soldered to the center of the coil, for measuring the temperature rise of the coil [36].
- Heising developed plate modulation (AM), known as constant current modulation, for aircraft transmitters [73].
- Morecroft observed and photographed impulse excitation of a RLC circuit with change in pulse length [56].
- Breit developed square-law detection of AM signal [34].
- L. B. Turner developed push-pull negative-resistance R-C oscillator using triode [34].
- R. T. Beatty and A. Gilmour found that high values of grid-leak resistance and coupling capacitors in RF oscillator are the reasons for *motor boating*, i.e., objectionable feedback or even low frequency oscillation [34].
- G. Vallauri formulated the mathematical operation of the triode [3].
- Dutchman Hugo Alexander Koch obtained a patent for an automatic ciphering machine. However, it was the German engineer Albert Scherbius who, in 1923, built the first machine, which he called *Enigma* [4].

- Westinghouse Company started the first regularly scheduled radio programming, which begun with the broadcasting of the presidential election results from Pittsburg, Pennsylvania, US [6,12].
- Radio Broadcasting was instituted in UK [3].
- Radio Research Board was established in UK [3].
- American J. R. Carson of Bell Laboratories applied sampling to communication [12].
- Harvey L. Curtis calculated the proximity effect in two parallel cylindrical conductors [1].
- German physicists Heinrich Georg Barkhausen and K. Kurz proposed *velocity-modulated vacuum tubes* or *transit-time tubes* in order to generate signals of frequency higher that 10 MHz (that was the maximum achievable frequency with Meissner's HF generator). They developed an oscillator of frequency 300 MHz [4,39].

1921:

- E. S. Purington made the all-electronic frequency modulator [51].
- J. Valasek first discovered ferrolectricity in Rochelle salt [1].
- S. P. Shackleton filed a patent for a testing apparatus for vacuum tubes to check if they are operating or not. This instrument was marketed by Jefferson Electric Co. in 1930 [1].
- A. W. Hull invented the *Magnetron* oscillator, frequency 30 kHz, power output 8 kW, and efficiency 69 percent [57].
- A. Crossly developed piloting vessels by electrically energized cables [1].
- The *All-British Wireless Exhibition* took place [15].
- John Scott-Taggart developed the *Negatron,* negative resistance tetrode [36].
- Mullard Company produced 25 kW Silica Valve [36].
- L. M. Hull used a cathode ray tube to measure frequency by Lissajous pattern method [43].
- Austin did and developed the measurement of wave-front angle in radiotelegraphy [56].
- E. H. Colpitt and O. B. Blackwell developed modulation of an audio frequency carrier by signals of lower audio frequency for carrying telephony over wires [34].
- A police car radio dispatch with 2 MHz frequency band was in use at the Detroit Police Department [50].
- Stone used vertically stacked dipoles for vertical directivity and patented it [1].
- Jones patented an underground and underwater antenna system [1].
- A coaxial structure was first used for telephone and telegraph service in a submarine installation between Key West and Havana [1].
- S. Butterworth published a classic paper on high frequency (HF) resistance of single coil considering skin and proximity effect [15].

1922:

- Appleton developed the automatic synchronization of triode oscillators [53].
- Englishmen P. P. Eckersley and T. L. Eckersley first started entertainment broadcast [17].
- Korn successfully transmitted pictures from Italy to the United States by radio [74].
- American physicist Walter Guiton Cady invented the *piezoelectric (Quartz) crystal oscillator* [53].
- A. H. Taylor and L. C. Young detected wooden ships using continuous wave (CW) interference radar with separated receiver and transmitters, wavelength 5 m [57].
- J. R. Carson in his paper 'Notes on Theory of Modulation' showed that frequency modulation requires more bandwidth than amplitude modulation [45].
- E. B. Moulin used the anode bend detector (triode) as vacuum tube voltmeter for measuring small RF voltage [36].
- V. Bush and G. C. Smith developed the cold cathode [1].
- Armstrong invented the super-regenerative receiver circuit [15].
- C. S. Franklin sets up *Wireless Lighthouse*, a short-wave transmitting station with rotating parabolic reflector, sending distinctive signal at every half and full rotation [36].
- British Broadcasting Corporation (BBC), a private limited company, was registered in England [15]. It broadcasted its first news program on November 14 [6].
- Broadcasting as an advertising media was started in the United States [15].
- J. B. Johnson and H. J. van der Bijl developed *sealed-off* cathode-ray tube with small amount of inert gas for *gas focusing* of electron beam (first attempt for focusing) [1].
- Marconi suggested using different transmitter and receiver for obstacle detection using electromagnetic wave [57].
- AT&T made available *toll broadcasting* at $100 for 10 minutes [76].
- Watt developed how it is possible to locate direction of rainfall area [56].
- Bush and Smith developed the use of diode as rectifier [1].
- T. W. Case developed infrared telephone using modulated acetylene flame [1].
- Carrier telephone was applied to power systems [1].
- W. G. Housekeeper in the United States filed a patent application for passing cooling fluid around the anode of a high power tube [1].

1923:

- A form of automatic-gain control was obtained by using a triode biased from the detector valve to shunt the aerial circuit [77].
- T. Spooner introduced the term *Incremental Permeability* [32].

- The *decibel* (1/10th of a *bel*, after A. G. Bell, inventor of the telephone sounder) was used to express loss in a telephone cable [70].
- H. W. Nichols developed point-to-point communication using single sideband communication [36].
- In England the cost of war surplus of 0.005-microfarad air dielectric variable condenser was 2.10 s. Amateurs used them to make small variable condensers [15].
- N. V. Kipping used *Neon Lamp* (cold-cathode gas diode) to produce time base (for low voltage cathode-ray oscillograph) by charging and discharging a variable capacitor. Before the time base circuit was developed, time base for cathode ray tubes was obtained by spinning a permanent magnet or current carrying coil at a suitable rate near to the base of the tube or by applying to the deflecting plates a unidirectional voltage through a potential divider of circular form whose moving contact was rotated at a uniform speed [32].
- A. W. Hull suggested for heating filament using AC, complete electrical insulation of thimble-cathode from filament [56].
- H. D. Arnold and L. Espenchied used eliminated carrier frequency form of AM signal in transatlantic radio *Telephony Balanced Modulator* [56].
- D. C. Prince demonstrated that power output of (class A) oscillator is greater than power output from the same tube when used as class A amplifier. He also presented the analysis of class C amplifiers [34].
- Langmuir described the action of positive ions on the grids of cold-cathode gas triode [34].
- French engineer Antoine Barnay developed the first *Rotary Dial Telephone* [6].
- Scottish engineer John Logie Baird built and patented the first practical TV, an eight-scan-line system. He demonstrated his *Televisor* in 1926, which was used for the first public television broadcast in UK [6] (Figure 2.41).
- Watson-Watt perfected his radiolocation device by displaying radio information on a cathode ray oscilloscope, so that the radar operator could tell the direction, distance and velocity of the target [5,6].
- American electronics researcher Ralph Vinton Lyon Hartley showed that the amount of information that can be transmitted at a given time is proportional to the bandwidth of the communication channel [1].
- H. D. Arnold and G. W. Elmen developed *Permalloy* material of very high permeability [53].
- W. Stephenson and G. Walton proposed interlaced scanning for TV [62].
- H. Flurschein filed a patent application on a radio warning system for use on vehicles. The arrangement consisted of a receiving system installed on a moving vehicle for warning the occupant of the first vehicle of his approaching to a position of danger [1].

- Portable receivers were marketed in England [15].
- P. Mertz patented the use of FM for facsimile [51].
- Automatic long-distance telephony began in Germany [4].
- Pierce introduced three different designs for quartz oscillators [61].

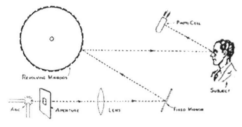 

Figure 2.41. Upper part: Two photographs of John Logie Baird (1888–1946) at work. Lower part: left, Baird's television scheme; right, transmitting portion of the original experimental Baird's television apparatus.

1924:

- The crystal-controlled oscillator was first used in a broadcast station in the United States [54].
- The principle of electroencephalograph evolved from the study of the electrical activity of the brain by Prof. Hans Berger, an Austrian psychiatrist. [1].
- A. C. Carlton used coagulatory property of RF for removal of tonsils [1].
- German electrical engineer Siegmund Loewe (Figure 11.15), owner of Loewe Radio Company in Berlin-Steglitz, produced metal film resistance for radio using vacuum evaporation [1].
- J. R. Carson in his classical paper "Selective circuit and static interference" showed that the energy absorbed by a receiver is directly proportional to its bandwidth [53].
- Carson extended Lorentz's reciprocity theory of electromagnetic fields to antenna terminals [1].

- A. Glagowela-Arkadeiwa generated by spark gap damped submillimeter waves of wavelength 82 micrometer [1].
- C. S. Franklin patented a system of wires erected as two vertical arrays across the direction of transmission. One row of wires, used as aerial, was energized by a special feeder system in same phase. The other row placed half-a-wave length behind acted as reflector (*broadside array*) [13,15].
- R. H. Barfield first suggested the use of electrostatic shielding for coil (loop) antenna to reduce errors in bearing [73].
- R. M. Foster formulated his reactance theorem [73].
- American scientist Lloyd Espenschied invented the first radio altimeter [10].
- First patent of a moving coil loudspeaker was issued to C. W. Rice and E. W. Kellog [15].
- Hopkinson suggested the use of baffle to mount a loudspeaker [15].
- Western Electric Co. constructed 14 kW, water-cooled metal valves [36].
- First use of Public Address system during election in England [15].
- It was suggested to use *Bel* and *Neper* to compare loudness [15].
- Amateurs made wireless communication halfway round the world [15].
- During a study of optical dispersions in materials the investigators observed that instead of absorbing the incident radiations (light) some of the materials were emitting radiation (light). The investigators dubbed them as *negative oscillators* (stimulated emission of radiation). The effect was again observed in 1930 [78].
- Synthetic lead sulphide crystal (Hertizite) was fabricated [15].
- L. Theremin produced *Elect-one*, musical instrument using one fixed frequency and one variable frequency oscillator, a detector, and a loudspeaker [15].
- R. L. Smithrose demonstrated the reliability of radio direction finding for navigational purpose using electromagnetic shielding [52].
- BTH Co. developed *twin-triode valve* [15].
- J. B. Johnson developed the *cathode follower* connecting in a cascaded triode RF amplifier a tuned circuit between cathode and HT negative, and HT positive to anode [15].
- A. W. Hull measured the voltage gain of a cascaded RC coupled amplifier using screen grid tube and found 120 dB gain at 50 kHz and 80 dB at 10 MHz [56].
- *International Amateur Radio Union* is formed [15].
- A photo-radiogram was sent from New York to London and back to New York [1].
- Dr. C. Stille, a German scientist, started work on magnetic recording of sound and introduced steel tape for recording [60].
- Commercial selenium rectifiers were available [3].

- W. J. Brown was the first to give rules for the choice of load resistance in a triode power amplifier with transformer coupled-load [34].
- Kipping developed the circular sweep for Cathode Ray Oscilloscope (CRO) and frequency measurement producing toothed wheel pattern [34].
- J. Taylor and W. Clarkson measured time intervals using electron tubes [34].
- A. Kalin devised the oscillographic determination and recording of tube characteristics [34].
- R. L. Wegel and C. R. Moore designed a portable harmonic analyzer based upon the tuned filter (only one) principles [34].
- The *Mobile Telephone* was invented by Bell Telephone Company, USA, and introduced into New York City police cars [6].
- Russian-American electronic engineer Vladimir Kosma Zworykin invented the *Iconoscope*, which is the direct ancestor of the *picture tube* used in modern TV sets [6].
- Heising patented a method of transmission of speech using pulse length modulation [1].
- *US Federal Communications Commission* was formed to minimize the mutual interference, which was due to a chaotic evolution of more than 1000 stations operating in the United States to deliver speech, and music, as suggested by David Sarnoff of RCA Company [17].
- Appleton and M. A. S. Barnett in England used an FM radar technique to measure the height of the ionosphere.
- O. V. Lossev from Russia applied cat whisker to radio [3].
- German electrical engineers Manfred von Ardenne (Figure 11.15) and H. Heinert developed and patented at Loewe's Company the so-called Loewe-3 fold tube 3NF (3 triodes) in which the audion (receiver), the resistive amplifier (RC amplifier), the output amplifier and the coupling capacitors and anode resistors where integrated into a single vacuum tube. Only a few months later they developed the Loewe 2-fold tube 2 HF (2 triodes with common space-charge grid) for broadband amplifiers (in 1926) (Section 11.2.8).
- Scientists Erich Habann in Jena and Napsal August Zázek in Prague independently investigated the magnetron for high frequency oscillations (Section 11.2.9).

1925:

- Morecroft and A. Turner described shielding of electric and magnetic fields [1].
- Dr. Petroff observed that small tadpoles were killed when brought near the coil of a 2-metre oscillator of about 15 W output [1].
- R. Mesny generated polyphase oscillations by means of electron tubes [73].
- E. W. Kellog designed non-distorting (class A) audio frequency power amplifier (for triode) [73].

- Vannever Bush and his colleagues constructed a machine capable of solving differential equation, the first analog computer [2].
- Heising developed single sideband modulator circuit for transatlantic Radio Telephony [73].
- Gregory Breit and Merle A. Tuve developed the first application of pulse technique for the measurement of distance and height of the Ionosphere [57].
- L. O. Grondahl and P. H. Geiger constructed copper oxide rectifier [79].
- Rectifier power supply units for receivers were made [15].
- Sullivan developed triggering of astable multivibrator by external signal [15].
- Bell Telephone Co. transmitted pictures by wireless [15].
- First conference on frequency allocation was held in Geneva [15].
- Master-Oscillator power amplifier (MOPA) to avoid oscillator frequency change due to movement of aerial by wind was suggested [15].
- Appleton and Barnett showed that signals reflected from the upper atmosphere were elliptically polarized. They gave the first experimental proof of the existence of an ionosphere [73].
- Using pulse technique for height measurement Appleton found another reflecting layer above the Heaviside layer [70].
- Gramophone Co. did the first electrically produced gramophone record [15].
- W. W. Brown and J. E. Love suggested reasonable power factors for RF coils at 200 kHz, 0.3 percent, and at 2 MHz, 1 percent [56].
- G. A. Ferrie and others used a triode with free grid connected with a photocell to measure the energy coming from the stars [56].
- Zworykin filed a patent application for Iconoscope (camera for TV), which was issued on Nov. 1928, and for all-electronic color TV [1].
- English mathematician Sir Ronald Aylmer Fisher made significant contributions to *Estimation Theory*, in particular *Maximum Likelihood Estimation* [7].
- W. J. Brown developed the theory of lossless/lossy rejector circuit (parallel resonant) in receivers [1].
- Russian-American engineer Joseph Tykocinski-Tykociner demonstrated that the characteristics of a full sized antenna can be replaced with sufficient accuracy from measurements made on a small scale model about 1/100 of the full size excited by a proportionally short wave in the range of 3 to 6 m. G. H. Brown and R. W. P. King also used in 1934 a small-scale model of antenna excited by HF source to investigate the characteristics of larger antenna [1].
- M. Czerny published his first observation of pure rotation spectra of hydrogen chloride. This is regarded as the beginning of the far-infrared molecular spectroscopy [44].

- E. Merritt found rectification between a metal point contact and germanium leading to the establishment of the Germanium diode [16].
- Respondek filed a patent application for connection of vacuum tubes using a Wheatstone bridge circuit, where a spare tube will be put into the circuit automatically replacing a tube, which had failed in the normal course of operation [1].
- CCIT (*Consultative Committee International on Telegraphy*) and CCIF (*Consultative Committee International on Long Distance Telephony*) were founded [4].
- AT&T introduced telephoto service [4].

1926:

- J. P. Maxfield and H. C. Harrison made electrical recording of sound on disk (Gramophone Record) [1].
- The Post Office high power transmitter at Rugby used 729 strands of 36 SWG wire for its coils [15].
- V. Bush demonstrated that the action of non-uniform electromagnetic and electrostatic fields upon accelerated electron beams was similar to the action of lenses upon light [45].
- A. V. Longhern and J. C. Warner developed graphical methods for solving vacuum-tube problems based upon the plate or transfer families of characteristics [34].
- A. W. Hull made the screen grid valve [45].
- Round invented the *tetrode* [3].
- Dutch physicist Bernardus Dominicus Hubertus Tellegen, working at Philips Research Laboratories (founded in 1914 by Holst and Oosterhuis) developed the *pentode* to minimize the production of undesired electrons from the anode when using a tetrode [4].
- B. Van der Pol developed the relaxation oscillator using pentode [34].
- E. Peterson developed an AC watt-meter using a balanced modulator with a triode [34].
- G. Belfils developed fundamental suppression type harmonic analyzer using a resonance bridge [34].
- H. A. Wheeler developed the *Automatic Volume Control Circuit* [1].
- F. M. Colebrook deduced that the dynamic resistance of a tuned circuit is proportional to the Q of the circuit [15].
- *British Broadcasting Corporation* (BBC) was formed [15].
- Baird and C. F. Jenkins, from the United States, demonstrated TV by mechanical scanning without using cathode ray tube [12,15].
- Loewe and von Ardenne constructed multiple valves – Two triodes with load, grid leak, and coupling capacitor inside one glass envelope [15].
- Woodruffe developed the gramophone pick-up suggested previously by Round [15].
- First use of scrambled speech in transatlantic wireless telephone service. Cost of first three minutes was £5 [15].

- J. E. Lilienfield patented the theory of *Field-Effect Transistor* (FET). W. Shockley experimentally demonstrated it in 1952 [80].
- A. S. Eve and others demonstrated experimentally propagation of EM waves in tunnels in the Mount Royal tunnel of the Canadian National Railways, Montreal, Canada [1].
- Japanese engineers Hidetsugu Yagi and Shintaro Uda developed the so-called *Yagi Antenna* or *Yagi-Uda Antenna,* a row of aerials consisting of one active antenna and twenty undriven members as *wave canal.* The beam width at half height was about 8 degree (Figures 2.42 and 14.7) (Chapter 14) [4,81,82].
- Gabor constructed the first high speed CRO and developed the first trigger circuit for automatic viewing of transients [1].
- Hulsenback & Company first patented identification of buried objects using continuous-wave (CW) radar, based on the German patent 489,434.

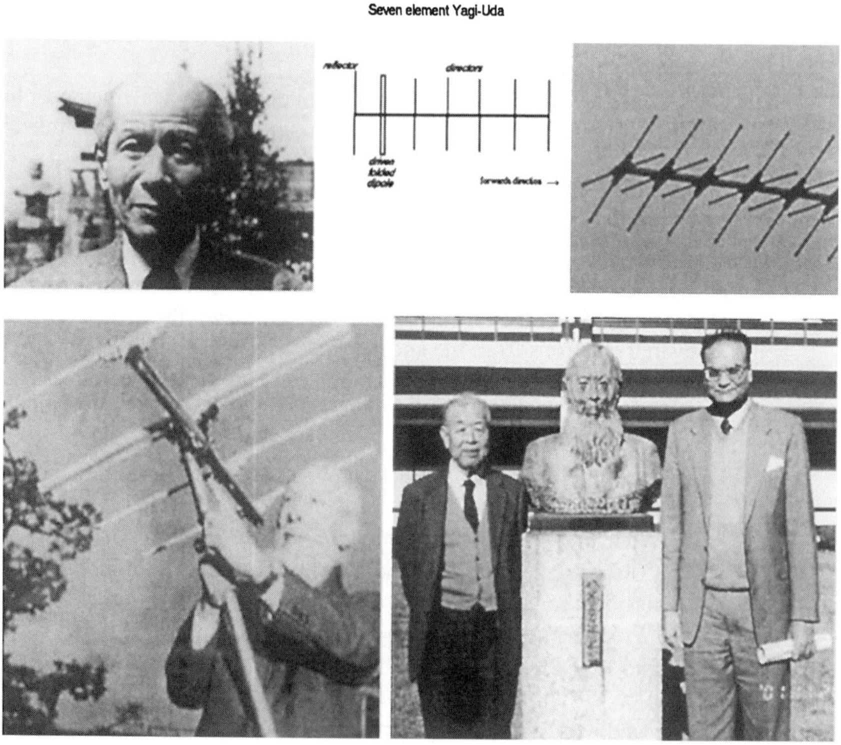

Figure 2.42. Upper part: left, Hidetsugu Yagi (1886–1976); middle, example of a seven element Yagi-Uda antenna; right, photograph of a tridimensional Yagi-Uda antenna. Lower photographs: left, Yagi presents the Yagi-Uda antenna; right, Professors Gentei Sato and Tapan K. Sarkar besides the monument dedicated to Yagi.

1927:

- R. V. L. Hartley developed the mathematical theory of communication [83].
- The International Radio Telegraph Convention, Washington DC, established the International Technical Consulting Committee on Radio (CCIR) for the purpose of studying technical and related questions pertaining to radio communication [1].
- E. D. McArthur of GE first observed body temperature rise and headache of technicians while testing 3 kW transmitting tubes [1].
- Using pulse technique of height measurement Appleton found another reflecting layer (F) above the Heaviside layer [1].
- H. Bush invented the use of short magnetic coil surrounding the electron beam (neck of Cathode ray tube) for focusing [84].
- C. Davison and L. H. Germer demonstrated that the diffraction of electrons by a crystal of nickel is similar to the diffraction of light waves by grating [45].
- Warner Brothers showed the first full-length picture with talking, singing, and music, 'The Singing Fool' in October 27 at New York. The sound recording was sound on-disc [15].
- 100 kW single sideband transmitter for transatlantic radiotelephony used thirty 10 kW valves in parallel [70].
- In Bell Laboratories it was experimentally observed that for a given amount of distortion the power output from two tubes in push-pull connection is much greater than twice the power output from a single tube [56].
- J. E. Anderson found that the presence of impedance in the B-supply of a multistage amplifier or in common lead between the voltage supply and the plates or other electrodes may result in objectionable feedback or even low frequency oscillation, called *motor boating*. In 1931 W. Cocking suggested RC *decoupling* filters to avoid this oscillation [34].
- C. Davison and L. H. Germer explained the wave behavior of electrons [45].
- American electrical engineer Philo T. Farnsworth filed a patent for *Image Dissector* for TV (Figure 2.43) [17,76].
- Harold Stephen Black of Bell Laboratories conceived the negative feedback amplifier [12,17].
- H. J. Reich and F. Bedell proposed the use of a commutator to observe two or more waves in cathode ray oscilloscopes (switching) [34].
- C. R. Moore and A. S. Curtis developed the heterodyne type harmonic analyzer using mechanical filters [34].
- Gary, Horton, and Mathes of Bell Labs gave the first theoretical discussion about the influence of the bandwidth on the quality of TV pictures and were able to fix the minimum bandwidth requirement [1].
- The *Varistor*, a voltage dependent resistor, was produced [85].

Figure 2.43. Philo Taylor Farnsworth (1906–1971).

- J. A. O'Neill introduced a magnetic recording system in the United States that used ribbons coated with iron oxide [6].
- Bell Telephone Company broadcasted TV images from Washington to New York via telephone lines [6].
- The *Federal Radio Commission* (FRC) was created in the United States [12].
- A. de Hass studied fading and independently developed diversity reception system near Bandung, Java [1].
- W. B. Roberts developed the linear detector [34].
- Eckersley used the reciprocity theorem to calculate the radiation from transmitting antennas over an imperfect earth [1].
- Bartlett developed the relationship between lattice and bridged T networks [1].
- AT&T developed the picture telephone [13].
- H. Diamond and J. S. Webb developed the procedure to display the response curve of an amplifier or other network in a cathode-ray oscilloscope [34].

1928:

- Baird conducted the first transatlantic TV broadcast from London to New York using shortwaves [6].
- Baird built the first *color television* system using Nipkow disc. Bell Laboratories devised a parallel system in the United States in 1929 [6,13].
- C. R. Englund published the characteristics of horizontal, isolated and ungrounded half wave antenna also called "Hertz" antenna and a matching device with transmission line [73].
- Farnsworth demonstrated the first all-electronic TV system [12,18].

- Two crystal oscillators in temperature-controlled ovens from US Bureau of Standards were taken to different countries of Europe to compare with their standards [56].
- W. E. Benham suggested that for operation at wavelength of few meters or less transit time of electrons from cathode to plate should be considered [34].
- G. Koehler developed the analysis of transformer coupled amplifier considering leakage inductance [34].
- J. H. Horton and W. A. Marrison made suggestion for driving a synchronous clock from a-f output of a chain of frequency dividers (multivibrators) the first one controlled by a crystal-controlled RF oscillator [34].
- C. C. Shangraw developed *Radio Beacons* for transpacific flight [79].
- Nyquist published a classic paper on the theory of signal transmission in telegraphy. In particular, Nyquist developed criteria for the correct reception of telegraph signals transmitted over dispersive channels in the absence of noise. Much of Nyquist's early work was applied later to the transmission of digital data over dispersive channels [7,18].
- J. B. Johnson and Nyquist did the experimental verification of thermal noise [33].
- R. R. Ramsey measured the variation of radiation and induction fields of loop antennas with distance [56].
- H. Forestier developed ferrite [86].
- C. S. Franklin patented the coaxial cable in England by to be used as an antenna feeder [4].
- A. W. Hull constructed the *Thyratron* (hot cathode gas filled triode) [45].
- H. H. Beverage and H. O. Peterson developed spaced aerial for diversity reception [1].
- Fritz Pfleumer, from Germany, introduced a paper strip system for the recording of sound. He sold his idea to the company AEG. They in turn sold it to BASF Company, which replaced the paper with cellulose-acetate strips. In 1932 BASF produced a plastic-backed magnetic recording tape similar to the tapes in use today [6].
- Gresky carried out an experimental investigation on the operation of cylindrical, parabolic and plane reflectors at 2.98m wavelength and found that the operation characteristics were dependent on the physical dimension of the reflectors [1].
- A. D. Blumlein developed the principle and practical advantages of toroidal transformer as bridge arms [14].

1929:
- L. Cohen proposed circuit tuning by wave resonance (resonant transmission line) and its application to radio reception [66].
- A. S. Eve and others discovered the penetration of rock by EM waves [1].

- H. O. Siegmund developed the aluminum electrolytic condenser [66].
- Sound-on-Film recording – RCA variable width systems were available [15].
- H. J. Walls developed four-course radio range as radio-aids to air navigation [73].
- F. Aughtie developed single stage RC coupled phase inverter circuit to feed push-pull stage [34].
- *Thyratron* sweep oscillator was used for cathode-ray oscilloscope [34].
- M. C. McCurdy and P. W. Blye used two tuned filters with amplifiers and thermocouple meter for measurement of the amplitude of a small-amplitude component whose frequency did not differ greatly form that of a large-amplitude component [34].
- I. Koros used glow-discharge tube (cold cathode) as a voltage stabilizer [34].
- A. C. Rockwood and W. R. Ferris developed microphonic improvement in vacuum tubes [73].
- R. V. L. Hartley theoretically suggested an early ancestor of today's *Parametric Amplifier*. In 1936 E. Peterson experimentally verified it by using prongs of a tuning fork as variable capacitor [38].
- F. Bellow developed the *Beam Power Pentode* [34].
- H. A. Affel and L. Espenscheid of AT&T Company/Bell Laboratories created the concept of a coaxial cable for a frequency division multiplexed (FDMA) multi-channel telephony system [17].
- Spark transmitters were forbidden except for on very small ships [15].
- R. C. Hitchcock developed a direct reading heterodyne type radio-frequency meter (with crystal oscillator for calibration) [1].
- M. Czerny developed the first infrared image device [1].
- F. A. Cowan suggested the possibility of program transmission over a telephone line [1].
- K. Okabe made possible the breakthrough in generation of cm-waves by magnetrons when operating his slotted-anode magnetron (5.35 GHz) at Tohoku University in Sendai, Japan (Section 11.2.9).
- German electrical engineer Hans Erich Hollmann anticipated the operation principle of a reflex klystron. He patented his idea as a "*double-grid retarding-field tube*" (Section 11.2.9).
- Schottky did the experimental verification of barrier layer in metal-semiconductor contact (Schottky Barrier) (Section 11.2.11).
- W. H. Martin proposed the Decibel as transmission unit [79].

1930:

- Wolf and Hart filed a patent application for radio-pulse altimeter [1].
- Sisir K. Mitra from Calcutta provided electrical measurements for the existence of the E-layer [1].
- J. C. Schelleng and J. Groszkowski filed a patent application for frequency division of sinusoidal oscillator [53].

- L. A. Hyland of Naval Research Laboratory of England performed (accidentally) detection of aircraft flying at a distance of 2 miles using the wave interference effect at frequency 33 MHz [57].
- K. B. McEachron first developed the use of *Thyrite*, a voltage dependent resistor, as a lightning arrester [56].
- For broadcasting purpose only 550 KHz to 1000 KHz band was assigned [56].
- Photocell control of traffic was made in 20 lanes of Detroit bridge [1].
- Remote control of radio receivers was done by cable [1].
- Experimental radio remote control of tanks using short wave was made in Tokyo Military College [1].
- Jefferson Electric Company developed the vacuum tube checker [1].
- A squadron of British bombers was controlled from ground within a radius of 400 miles; the pilots aboard kept their hands off from the controls [1].
- L. Rolla and L. Mazza proposed infrared radiation telephony [66].
- Easu and Hahnemann experienced no fading or atmospherics during telephone communication with aeroplanes (within line of sight) at 100 MHz [56].
- F. B. Llewellyn deduced the magnitude of the shot effect when the space-current is limited by space-charge [33].
- Uda worked on short wave transmission from wavelengths of 4.4 meters to 0.5 meters [56].
- S. Ballantine suggested that the performance of power amplifiers should be specified by power sensitivity [34].
- Devry Corporation made synchronised sound-on-disk for 16mm camera cinema film [1].
- F. E. Terman and N. R. Morgan showed the advantage of linear grid detector over linear diode detector [34].
- D. D. Knowels developed control of power by grid-glow tubes [34].
- J. A. Stratton showed effects of rain and fog on propagation of very short wave [1].
- French electrical engineer André G. Clavier developed an experimental microwave telephone link in New Jersey using a 10 feet (3 m) parabolic dish antenna [1].
- Lal C. Verman and L. A. Richards demonstrated regulation of generator (DC) voltage using current amplifiers [34].
- BBC began broadcasting regular television programming when Baird's improved Televisor went on the market. Although Baird's invention was not an immediate commercial success, by 1939 there were 20,000 television viewers in UK [6].
- J. H. Hammond placed TV on plane for safe landing in fog [1].

1931:
- L. E. Barton obtained satisfactory cent percent AM using class B modulator [1].

- A. G. Clavier and his team at LCT (Laboratoire Central des Telecommunicaccations, an affiliate of IT&T) experimentally demonstrated a microwave link over a 21-mile path between Dover and Calais, radiated power 0.5 W, and frequency 1.72 GHz [4,10].
- 500 kW demountable valve was used in GPO single sideband transmitter at UK [70].
- W. J. Polydrof proposed ferro-inductors and permeability tuning [1].
- F. Vacchiacchi developed oscillations in the circuit of a strongly damped triode (Blocking oscillator) [73].
- R. Jouaust was first to point out variation of dielectric constant of atmosphere with increase in height [73].
- H. H. Beverage and H. O. Peterson devised a diversity receiving system using three spaced aerial (to avoid fading) for radio telegraphy [56].
- E. Karplus proposed communication using quasi-optical wave (0.001 mm to 10 m) [66].
- F. B. Llewellyn developed a rapid method of estimating the signal-to-noise ratio of a high gain receiver [34].
- Aeolight lamp hot cathode gas diode was developed for sound recording on motion picture film [34].
- Approximate cost of 50kW broadcasting station was $296,000 [56].
- H. J. Reich and C. S. Marvin developed current amplifier to use with electromagnetic oscillograph [34].
- Herd obtained voltage amplification approximating amplification factor of pentode by using the plate resistance of a second pentode as load resistance (cascode) [34].
- H. Diamond and F. W. Dunmore conceived a Radio Beacon and Receiving System for Blind Landing of Aircraft (IIS) [73].
- It was shown by Barton that the high power output and efficiency of class B operation could also be attained without excessive distortion in audio frequency amplifiers [34].
- Although Carson showed in 1922 that frequency FM cannot be used in radiotelegraphy, H. Roder showed that FM transmitters might be operated at much higher efficiency than AM [34].
- S. Ballantine demonstrated that a RC-coupled amplifier with automatic gain control might be used as a logarithmic voltmeter [34].
- W. R. Blair and H. M. Lewis developed radio tracking of meteorological balloons [1]
- R. F. Field used Wien's bridge for frequency measurement [34].
- Russian-American mathematician Sergei Alexander Schelkunoff, working at Bell Labs, provided theoretical studies for losses and dispersion of cables based on electromagnetic theory. O. J. Zobel also of Bell Labs solved the problem of equalizing the various cable losses with frequency by suitably designed networks [17].
- Teletypewriter service was initiated [12].

- F. Guarnaschelli and F. Vacchiacci made a direct reading frequency meter using vacuum tube amplifier, capacitor and DC meter, where the current was directly proportional to the frequency value [34].
- C. W. Hansell and others developed directive transmitting (V) antennas [73].
- Jansky and Bailey calculated broadcast station coverage by field intensity measurement [1].
- Wired Radio Inc. planned to produce broadcast program over electric line and De Forest Company installed at Lincoln Hotel in New York radio program distribution over lighting circuit with receivers in each room [1].
- E. Bruche and H. Johansson of Germany demonstrated an electron microscope of magnification factor of 15 [1].
- M. Michelson of Telefunken made an infrared sensitive resistance using tellurium [1].
- H. E. Hollmann built and operated the first decimeter transmitter and receiver at the Heinrich Hertz Institute in Berlin. The unit worked by using a symmetric opposed, resonant Lecher circuit excited by a so-called *"hammer"* retarding field tube. He called his device magnetron. Later only reflex klystrons were used as local oscillators. The so-called *Vircator* (Virtual Cathode Oscillator) is a modern version of the retarding-field tube (Section 11.2.7).

1932:
- V. J. Andrue first suggested that the simplest type of discriminator for reception of FM radio signals is a series resonance circuit slightly off resonance with the carrier [73].
- W. A. Fitch found that incidental phase or frequency modulation was often present in the output of a telephone or telegraph transmitter as a by-product of the modulation or keying [73].
- Professor Schroter commented "None of the proposed cathode-ray scanning systems show promise of development in the foreseeable future". However, RCA demonstrated with an improved cathode-ray tube for receiver "all electronic TV". On 20th August BBC did a first experimental TV broadcast in London [20,67].
- Required frequency stability of transmitters was increased by FRC ±500 Hz to ±50 Hz [1].
- J. D. McGee and W. Tedham of THORN EMI developed a photosensitive mosaic for TV camera [3].
- The word *Telecommunication* was coined and the *International Telecommunications Union* was formed [30].
- American physicists George C. Southworth and J. F. Hargreaves developed the circular waveguide, whose theoretical properties were described by S. P. Mead and S. Schelkunoff [17].
- Karl Jansky accidentally discovered *Weak Static* (radio noise) from outer space (namely, from constellation Sagittarius). The array used by Jansky was a horizontal group of vertical half-wave elements with a

similar reflecting group a quarter wavelength away designed by E. Bruce and H. T. Friis of AT&T Company [17]. In 1921 Marconi expressed serious interest in radio waves "coming apparently from outer space" [35,58]. This gave birth to radio astronomy.

- Harry Nyquist developed the feedback without oscillation *Regeneration Theory* [73].
- R. Darbord developed the UHF (*Ultra High Frequency*) Antenna with parabolic reflector [73].
- H. M. Lane demonstrated that the drooping of the frequency-response curve of a RC-coupled amplifier at low frequency might be prevented by shunting a portion of the coupling resistor with a condenser [34].
- P. H. Osborn studied class B and class C amplifier tank circuits [34].
- H. D. Brown developed the grid-controlled (thyratron) rectifier [34].
- H. A. Robinson modulated RF using tetrode [34].
- C. E. Winn-Williams developed high-speed scale-of-eight counter using hot cathode gas triode [34].
- W. W. Garstang developed the voltage quadrupler circuit [34].
- W. B. Lewis developed the analysis of condenser input filter circuit for high vacuum and gas filled rectifier [34].
- M. Knoll and E. Ruska developed the electron microscope [34].
- F. S. Dellenbaugh and R. S. Quinby deduced and experimentally proved that a "critical" value of inductance of an inductance input filter for rectifier was required [34].
- The use of 4/3 effective earth's radius to account for the refraction of radio waves predated radar, and because of its convenience was widely used in radio communication and radar [57].
- Watson-Watt suggested the term ionosphere connoting the ionized region of upper atmosphere [1].

1933:

- Harvey Fletcher of Bell Laboratories, USA, obtained a large audience for stereophonic reproduction [1].
- C. E. Cleeton and N. H. Williams generated the highest frequency of CW oscillator, 30 GHz, using a split-anode magnetron [53].
- Slepian and the engineer Leon R. Ludwig invented the *ignitron* [79].
- Allen B. Dumont developed the *Cathautograph*, an electronic pencil for writing on the screen of a CRT [1].
- The *hexode* (vacuum tube) converter for receiver was developed [63].
- A. W. Kishpaugh demonstrated linear grid bias modulation [34].
- P. H. Macneil developed the infrared *Fog-Eye* to detect iceberg [1].
- Morecroft reported variation of resistance and reactance of a coil with frequency [56].
- L. B. Snoddy of GE Research Labs reported breakdown (ionization) time of thyratron of a few microseconds [34].
- C. H. Willis developed counter-EMF inverter (using thyratron) in DC power transmission [34].

- Zworykin developed the theory of electrostatic focusing of cathode-ray tube [84].
- Cabot S. Bull and Sidney Rodda of EMI patented the beam-tetrode [1].
- J. C. Schelleng and others described duct propagation of ultra short waves. Schelleng also did the correction of the radius of the Earth due to refraction of radio waves [57].
- H. Z. Sewing developed an electronic switch for observation of two or more waveforms in a single beam cathode ray oscilloscope [34].
- Reflex amplifier used to amplify two frequencies usually IF (intermediate frequency) an audio were fairly common in the United States [63].
- Armstrong demonstrated another revolutionary concept, namely, a modulation scheme that he called *Frequency Modulation* (FM). Armstrong's paper making the case for FM radio was published in 1936 [12,18].
- Englishmen J. Neyman and E. S. Pearson did significant contributions to *Statistical Decision Theory* [7].
- EMI introduced in England *stereophonic* disks (two-track sound) for the gramophone [6].
- T. Minohra and Y. Ito experimentally proved that meteors affect propagation of radio waves [1].

1934:

- The *Federal Communication Commission* (FCC) was created from the Federal Radio Commission in the United States [12].
- The FCC authorized four radio channels in the 30 to 40 MHz range [50].
- G. Fayered calculated the speed of propagation of EM waves by measuring the phase difference between modulated signal and detected signal, received and retransmitted from another transmitter [1].
- Bergamann and Kruegel measured field inside a short hollow metal cylinder and the radiation from its open end when a half-wave coaxial antenna was properly excited [1].
- W. B. Roberts developed the vibrator (AC) power supply from dry cells [53].
- American scientist Clarence Melvin Zener invented the diode that is named after him [87].
- W. L. Everitt obtained the optimum operating conditions for class C amplifier [73].
- Warren P. Mason developed wave filters using quartz crystal [66].
- F. E. Terman demonstrated transmission line as a resonant circuit [73].
- L. F. Gaudernack demonstrated how to measure the degree of modulation (in AM) by measuring the rectified envelope [73].
- I. G. Mallof and D. W. Epstein developed the theory of electron gun (cathode ray tube) [84].

- O. W. Livingston and H. T. Maser added a fourth electrode in hot-cathode gas triode, which acts as shield [34].
- Saul Dushman developed a hot cathode arc tube (gas diode) that could be use as light source using mercury vapor, sodium vapor, and so on [34].
- E. D. Scott demonstrated how transmission lines might be used as frequency modulators [73].
- The triode hexode converter and the heptode frequency changer were developed [63].
- B. Salzberz designed and used the "acorn" triode (a special triode for UHF), with maximum frequency 700 MHz and a power output of 0.5W [1].
- Microwave commercial telephone links using parabolic reflectors were installed [10].
- Associated Press introduced the first system for routinely transmitting *wire photos* [6].
- German physicist Oskar Ernst Heil (Figure 11.17) applied for a patent on "Improvements in, or relating to electrical amplifiers and other control arrangements and devices" which can be seen as the theoretical invention of the capacitive current control in field effect transistors (FETs) (Section 11.2.10).

1935:

- W. J. de Haas and J. M. Casimir-Jonker first tested superconducting switching element [1].
- G. L. Beers developed automatic selectivity control of receiver [79].
- L. C. Wallers and P. A. Richards developed the magic-eye tuning indicator for receiver [34].
- C. Travis used vacuum tubes as variable impedance elements, for automatic frequency control of local oscillator of receiver [34,77].
- Cross-modulation between two strong radio waves while passing through the ionosphere (*Luxemburg effect*) was observed during transmission from Radio Luxemburg; the effect was also observed in the United States in 1919 [1].
- C. J. Frank of Boonton Radio Corporation demonstrated Q-meter at the fall meeting of IRE [116]. The ratio of reactance to resistance of a coil as its 'Quality factor' was first suggested at about 1926 [15].
- Space-radio was applied to detect concealed weapons in jail [1].
- Coaxial cable for operating up to one MHz was developed [1].
- Armstrong demonstrated (on the 7 meter band) that wide-band FM is remarkably effective in suppressing noise, interference and static [1].
- Telefunken of Germany revealed details of 10 cm "mystery ray" system capable of locating position of aircraft through fog, smoke, and cloud [1].
- A. B. Dumont developed the procedure to measure the degree of modulation (AM) using oscilloscope [1].

- P. A. McDonald and E. M. Campbell measured 0.0002 picoamp using ordinary amplifier tubes [34].
- H. E. Iams and B. Salzberg developed the electrostatic electron multiplier phototube [45].
- D. E. Harnett and N. P. Case studied the waveguide behavior at wavelengths greater than cut-off [88].
- A. C. Bartlett published intermodulation in audio frequency amplifiers [34].
- T. G. Castner developed a heterodyne harmonic analyzer using quartz crystal filter [34].
- G. W. O. Howe reported the behavior of high resistance at high frequency [63].
- A signal synchronized sweep circuit for oscilloscope was developed [66].
- In France an experimental pulse-modulated 600 MHz radio-relay link was installed near Lyon, between Saint-Genis-Laval and Le Col de la Faucille. This appeared to be the first attempt to use digital modulation techniques for microwave radio transmission [4,7].
- A French television transmitter was installed atop Eiffel Tower. RCA installed one atop New York's Empire State Building in 1936. Because of World War II, television did not become a common fixture in private homes until the late 1940s and early 1950s [6].
- Watson-Watt of the British National Physical Laboratory developed and patented the first practical radar for use in the detection of airplanes. It was installed at a secret military site at Bawetsey Manor on the east coast of England. A workable system was deployed before World War II. In Germany Rudolf Kuhnold at Telefunken had actually made progress toward a similar system by 1938. However further development was halted and resumed only in 1943. Today radar is an essential part of air traffic control at airports, navigation, speed measurements, and planetary observation [5,6,12,57].
- American electrical engineer Hendrik Wade Bode used the true null balance property of bridged-T [1].
- E. G. Bowen suggested the development of the pulse position indicator, which was first used in 1940 [1].
- Russian-German physicist Agnesa Arsenjeva-Heil and his husband Oskar Ernst Heil (Figure 11.17) published their historical paper on velocity modulation of electrons in a short gap and the consequent bunching of them as they traversed in a drift tube. This publication provided the first description of the fundamental principles behind modern high power linear beam microwave electron tubes [44].
- H. E. Hollmann filed a patent application on the multi-cavity magnetron. US Patent 2,123,728 was granted on July 12, 1938, well ahead of J. Randall's and H. Boot's work in 1939 (Section 11.2.9).

1936:

- H. E. Iams patented the addition of E. S. memory to CRT [1].

- O. S. Puckle reported Saw-tooth wave generation using Eccles-Jordan circuit [34].
- W. H. Doherty developed a new high efficiency power amplifier for modulated waves [73].
- W. L. Barrow and G. C. Southworth discovered independently the practical possibility of waveguides as transmission systems. Initially they were thought for VHF (*Very High Frequency*) applications [73].
- English engineer Paul Eisler devised the printed circuit [56].
- First radar echo at 200 MHz from a range of 12 miles was observed using duplexing system with common antenna for both transmitters and receiver [57].
- N. H. Jack patented the semi-rigid coaxial cable using thin soft copper tube as outer conductor [44].
- Harold Wheeler used two flat coplanar strips side-by-side to make a low loss transmission line that could be rolled to save space [44].
- G. H. Brown developed a turnstile antenna, used for TV transmission from the Empire State Building in 1939 [73].
- H. E. Edgerton and K. J. Germeshausen developed the *Strobotron*, a cold-cathode arc tube (tetrode), using the tube self-excited and multivibrator for a controlled stroboscope [34].
- W. R. Ferris described the transit-time effect on input conductance of triode [34].
- J. D. McGee and H. D. Lubszynski developed a sensitive *Super Emitron* [3].
- The British Broadcasting Corporation (BBC) did the first TV broadcast in UK [3,12].
- H. T. Friis and A. C. Beck invented the horn-reflector antenna with dual polarization operating from 2-10 GHz based on the earlier work of Southworth of Bell Labs [17].
- The first commercial tape recording was made by BASF Company [6].
- J. H. Denlinger proved the synchronous occurrence of daytime radio fadeout and solar eruptions at intervals of fifty four days [1].
- R. M. Foster and S. Seely developed the FM detector (discriminator) [1,34].

1937:

- 12 channel carrier and coaxial cable in telephone was used in UK [3].
- J. R. Woodyard developed the detection of FM signal using (a) artificial delay line and (b) auto synchronized oscillator [73].
- Grote Rober was the first to construct a *Radiotelescope* [2].
- H. T. Friis and C. B. Feldman developed multiple unit steerable antennas for short-wave reception [73].
- George Stibitz of AT&T developed the first digital computer using telephone switches and electromechanical relays.
- G. Grammer developed electronic voltage stabilizers for rectifiers [34].

- J. R. Day and J. B. Russel developed the negative-feedback class B audio frequency amplifier [73].
- R. L. Freeman and J. D. Schantz developed video-frequency (wideband) amplifiers [73].
- W. R. Blair patented the first anti-aircraft fire control radar (SCR-568) as *Object Locating System* [57].
- H. H. Beverage reported field strength calculation of UHF transmitters [1].
- American physicist Russell H. Varian and his brother Sigurd Varian, a mechanic, together with the American physicist William Hansen, developed the reflex Klystron, name derived from the Greek word "Klyzein" expressing the braking of waves in a beach (Figure 2.44) [1,61].
- W. N. Tuttle devised a new instrument and new circuit (Bridged T) for coil and condenser checking [73].
- British radio engineer Alex H. Reeves, while working in Paris, invented *Pulse-Code Modulation* (PCM) for the digital encoding of speech signals (Figure 2.45). Next year he applied for his first patent in France, a second in England in 1939, and a third in the United States. The technique was developed during World War II to enable the encryption of speech signals. At the end of the war American military used in the field a full-scale, 24-channel system. However, PCM had to await the discovery of the transistor and the subsequent development of large-scale integration of circuits for its commercial exploitation [7,12,18].
- American engineer Howard Aiken, assisted by the American mathematician and physicist Grace Brewster Murray Hopper, designed the first large-scale computer, IBM Mark I, the grandfather of today's *mainframe computers* (Figure 2.46) [6].

1938:

- H. H. Scott described a parallel LRC resonant circuit [45].
- Harry F. Olson developed a multiple-coil multi-cone loudspeaker [1].
- Spectrum analyzer using swept local oscillator was made for modulation measurement [1].
- E. L. Chaffee determined the optimum load for class B amplifiers [34].
- IRE published standards on transmitters, receivers and antennas.
- W. W. Hansen developed a type of electrical (cavity) resonator [73].
- L. J. Chu and W. L. Barrow studied the propagation of EM waves in hollow metal pipes of (a) elliptical and (b) rectangular cross sections [73].
- F. A. Lindemann proposed a double-sided mosaic for TV [20]. The octode frequency changer was developed [63].
- The FCC established rules for regular radio service [50].
- A. Preisman developed the cathode-coupled (cathode-follower) amplifier [73].

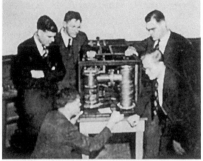

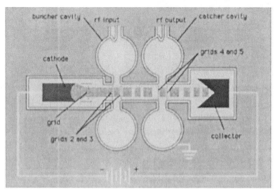

Figure 2.44. Upper photographs: Left, Russell H. Varian (left, 1898–1959) and Sigurd F. Varian (right, 1901–1961) with a high-powered klystron. In the palm of his hand, Russell Varian holds a smaller type of klystron used for radar, aircraft instrument landing, and microwave communications. Right, Stanford researchers in 1939 examine their invention, the klystron. Standing from left to right are Sigurd Varian, physicists David Webster and William Hansen, and in the front are Russell Varian, left, also a physicist, and John Woodyard, an engineering graduate student. Lower part: Two-cavity klystron scheme.

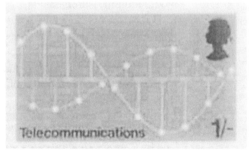

Figure 2.45. Left, Alec Harley Reeves (1902–1972); right, a stamp commemorating the invention of Pulse Code Modulation.

Figure 2.46. Computer memory and logic of the first computer IBM Mark I.

- M. J. O. Strutt and A. Van der Zeil suggested the causes for the increase of the admittance of modern high-frequency amplifier tubes on short waves [73].
- American Claude Elwood Shannon (Figure 2.47) recognized the parallels between Boolean algebra and functioning of electrical switching system [89].
- O. H. Schmitt's trigger circuit was developed [1].
- Direct current was used to vary the inductance of a high frequency tuning device [155].
- R. Hilsch and R. W. Pohl working with alkali-halide crystals made active solid-state triode. The cut-off frequency of this experimental device was about one cycle [16].
- During this year and the next one Schottky developed the space-charge and edge-sheet theory of crystal rectifiers (Section 11.2.11).

Figure 2.47. Claude Elwood Shannon (1916–2001).

1939:

- By this year the British Broadcasting Corporation (BBC) was broadcasting TV on a commercial basis [18].
- P. S. Carter improved Lodge's biconical antennas with a tapered feed. N. E. Lindenblad of RCA developed a coaxial horn antenna [106].
- W. R. Hewlett developed the Wien-bridge (RC) oscillator [45].
- P. H. Smith at RCA developed the well known *Smith Chart* [88].
- R. E. Mathes and J. N. Whitaker developed radio facsimile using sub-carrier FM [68].

- W. L. Barrow and F. D. Lewis developed sectorial electromagnetic horns [73].
- Englishmen John Turton Randall and Henry Albert Boot developed at Birmingham University, the first practical magnetron, the *Cavity Magnetron* magnetron suitable for radar (10 cm and 100 kW peak power). Their work is based on various other forms of magnetron developed in the 1920s and 1930s. Secretly brought to the United States, the Randall-Boot magnetron becomes the central component of radar systems [57].
- Aircraft- interception radars (200 MHz) were installed on aircrafts for detecting hostile aircrafts [57].
- W. S. Duttera developed the high pressure coaxial transmission line [66].
- Reeves patented pulse code modulation [61].
- H. A. Affel and others proposed the "Volume unit" – VU meter [69].
- L. Espenschied and R. Newhouse developed an absolute altimeter [73].
- Albert Rose and Harley Jams of RCA developed the *Orthicon* TV camera tube [1].
- Andreï Nikolaevich Kolmogorov, from the Soviet Union, developed a treatment of the linear prediction problem for discrete-time stochastic processes [7].
- Terman developed variable selectivity wave analyzer using negative feedback [70].
- Byrne developed polyphase broadcasting of AM [1].
- John D. Kraus described the operation of a folded dipole and other multi-wire doublet antennas [1].
- L. Espenschied patented in the United States FM altimeters [10].

1940:

- J. de Gier developed the post acceleration cathode ray tube [73].
- Paul Eisler was granted the patent for printed circuit board [56].
- The *Thermistor* was developed [85].
- Albert Alford and A. G. Kanodian developed the *UHF Loop Antenna* [79].
- The US Navy officially adopted RADAR as code word by for *Radio Echo Equipment* [57].
- L. J. Black and H. J. Scott developed the low frequency amplifier using HF carrier [34].
- M. G. Crossby developed the reactance tube (FM) modulator [72].
- M. G. Crossby measured frequency deviation in FM [66].
- M. Wald devised noise suppression in receivers using amplitude limiters [73].
- R. F. Guy performed frequency modulation field tests [1].
- C. W. Carnahan applied FM to TV system [1].
- C. E. Strong developed the inverted (grounded grid) amplifier [73].

- Harvey Fletcher of Bell Lab demonstrated stereophonic reproduction of music and speech for motion picture films [1].
- Using field sequential system developed by Peter Goldman full color TV broadcast was made by CBS [1].
- C. A. Lovell and D. B. Parkinson of Bell Laboratories suggested control of the movements of guns by computers [1].

1941:

- W. Fischsig in France patented the principle of shadow-mask grid type color picture tube [61].
- W. C. Godwin developed the direct-coupled push-pull amplifier with inverse feedback [34].
- D. R. Griffin showed that bats and porpoises use ultrasound echolocation principles. The bats emit a series of ultrasonic frequency modulated rectangular pulses about 2 msec in width, repetition frequency varying from 5 to 60 Hz and carrier frequency varying from 39 kHz to 78 kHz. The ears of the bats are directional antennas and can detect obstacles as close as 5 cm [37,57,90].
- E. L. Ginzton and L. M. Hollingsworth developed the RC phase-shift oscillator [34].
- Commercial 50 kW FM broadcast station [68].
- The company Siemens & Halske made the Germanium diode [16].
- Sidney Warner realized a two-way police FM radio [68].
- R. S. Ohl made the silicon junction diode [16].
- J. H. Scaff filed a patent application for n-type and p-type silicon [16].
- J. K. Hellard elaborated a distortion test by intermodulation method [66].
- Alan Turing and M. H. A. Neuman built at the University of Manchester, England, the first truly *electronic computer*, Colossus I, for the British Government [6].
- The FCC authorized TV broadcasting in the United States [12].
- John V. Atanasoff invented the computer at Iowa State College [12].
- Germans used powdered iron core frame aerial of 12 inches long and 3 inch in diameter for direction finding in their aircrafts whose performance was almost equal to that of a larger structure [1].
- A. W. Horton of Bell Labs patented the AND circuit [1].
- A. D. Blumlein in the United States patented the principle of slotted cylindrical antennas [1].
- The operation of automatic *lock-and-follow* ground radar with an aircraft as target was successfully demonstrated [1].
- C. S. Franklin tested and filed a patent application for helical aerials in England [1].

1942:

- Traffic control using FM was used in Pennsylvania [68].

- J. A. Van-Allen invented the *Proximity Fuse*, a CW radar for shells of anti-aircraft. The electronic circuit was fabricated on a printed circuit board. This was the first time they were used [2,57].
- J. G. Brainerd developed in the United States the first automatic computer [58].
- L. Espenschied, from Bell Laboratories, applied for a patent covering switching of pulsed voice signals [4].
- American mathematician Norbert Wiener formulated the continuous-time linear prediction problem and derived an explicit formula for the optimum prediction. He also considered the filtering problem of estimating a process corrupted by an additive noise process. The explicit formula for the optimum estimate required the solution of an integral equation known as the Wiener-Hopf equation. The second name was after his colleague, the American the Austrian mathematician Eberhard Frederich Ferdinand Hopf [7].
- W. H. T. Holden of Bell Labs invented the OR circuit [1].
- Eddy current tuning of VHF receivers was reported [1].
- A. Philips developed a moving target indicator (MTI) system and displayed on 'A' scope (PPI was not known to him). In 1946 A. G. Emslie developed MTI using Pulse Position Indicator (PPI) display [1,57].
- The first electronic digital computer was built at Iowa State University by John V. Atanasoff and Clifford Berry. In 1973 US Federal court gave the patent to them over the *ENIAC*.
- A. P. King built a probe fed conical horn antenna [106].

1943:

- L. W. Alverez and R. M. Robertson developed waveguide linear array scanners [57].
- H. J. Finden developed the frequency synthesizer [1].
- Austrian engineer Rudolf Kompfner developed the *Travelling wave tube* while working at University of Birmingham under the advice of Prof. L. M. Oliphant.
- Randall and Boot developed the *Multi-cavity Pulsed Magnetron* [1].
- C. K. Chang developed frequency modulation of RC oscillator [31].
- First submerged telephone repeater between Holyhead and Isle of Man was installed [3].
- W. J. Polydrof developed the iron core loop antenna [29].
- H. Salinger designed dummy dipole network (dummy antenna) for receiver measurement [3,69].
- J. P. Blewett and others developed a multilobe tracking system [57].
- H. Wallman developed the stagger-tuned amplifier [1,72].
- R. M. Sprague developed a frequency shift radiotelegraph and teletype system [68].
- C. F. Edwards developed the microwave mixers [88].

- Electronically controlled milling machines for automatic operation were developed [1].
- H. T. Friis developed noise figures of radio receivers [69].
- D. O. North devised the concept of the matched filter for the optimum detection of a known signal in additive white noise. J. H. Van Vleck and D. Middleton obtained, independently, a similar result in 1946, The later ones coined the term *matched filter* and also described the effect of noise on pulsed signal [7,18,57].
- An automatic test set up using AC and DC bridges which makes static comparison between circuits of electronic equipment coming off an assembly line was published [1].
- Slepian and Thomas E. Browne improved circuit breakers.

1944:

- F. Shunaman of Bell Labs developed pulse position modulation [1].
- Harold Goldberg suggested pulse frequency modulation [1].
- J. R. Whinnery, H. W. Jamieson and T. E. Robbins analyzed discontinuities in coaxial lines [88].
- J. T. Potter described a scale-of-ten counter using feedback in flip-flop circuit [1].
- E. C. Quackenbush of Amphenol developed the VHF coaxial connectors [44].
- Paul Neil of Bell Labs developed type *N* connectors, which were named after him (also called Navy type connector) [44].
- V. H. Rumsey and W. H. Jamieson developed a flat strip coaxial line for power division network in antenna design [44].
- T. E. Robins published the distortion of electric and magnetic fields due to discontinuities in transmission lines [1].
- George Westinghouse developed an electronic balancing machine for rotors to detect vibrations as small as 25 millionths of an inch [1].
- E. Karplus developed the *Butterfly* circuit (UHF resonator) [69].
- Phantastron time-delay circuit was developed for radar [1].
- Bell Labs designed, and Western Electric Co. made, electronic gun-director for antiaircraft guns, which computes the path of the aircraft and aims the gun for sure hit considering wind velocity direction, and gravity [1].
- R. K. Luneberg developed the lens at Brown University, which still carries his name [82].

1945:

- Frenchman Maurice Deloraine, together with P. R. Adams and D. H. Ransom, applied for patents covering *switching by pulse displacement* a principle later defined as *time-slot interchange*, in fact, the basic idea on digital switching [4]. Thus *Time-Division Multiplexing* (TDMA), a natural for digital carrier systems was invented. The first application was in integrated military tactical telephone networks [7]. Deloraine and E. Labin also invented Pulse-time modulation [91].

- IRE published standards on Radiowave propagation and guided waves [88].
- C. F. Kettering and G. C. Scott experimentally proved that electrons are the carrier of current by measuring m/e of electrons in a current carrying conductor [1].
- Robert Adler developed the phasitron modulator tube for FM [1].
- R. G. Peters developed FM ratio detector [1].
- John Robinson Pierce developed the theory of reflex klystron [88].
- E. D. McArthur invented *Lighthouse* or *Disk-seal* tube to reduce the lead inductances of ordinary tube at UHF [45].
- Radio Research Laboratory developed apparatus for Radar counter measures (jamming) in the range of 25 MHz to 6 GHz. [1].
- C. E. Nobles suggested mixers for use in broadcasting [39].
- Nobles suggested that a network of airplanes flying at 40,000 feet might be used for microwave broadcasting and relaying [39].
- Microwave band was released for amateurs on Nov. 15th. Merchant and Harrison communicated the same day using 5,300 MHz. [1].
- The British science-fiction writer Arthur C. Clarke proposed the idea of using *earth-orbiting* satellites as relay point for communication between earth stations [4,7]. The International Astronomical Union has named the geostationary orbit the *Clarke Orbit* [4].
- Ceramic capacitors for VHF circuits were introduced [1].
- S. Pickles published description of portable Instrument Landing System (ILS) consisting of localizer, glide path and marker beacons during Second World War [79].
- C. Chapin Cutler developed dummy antenna and power meter (up to 100W) for VHF using carbon resistance and thermocouple type meter [1].

1946:

- Aram developed the folded dipole antenna for FM broadcasting [1].
- Silicon photocell was developed [88].
- V. H. Regener developed the decade counting circuit [1].
- Winston E. Kock developed metal lens antenna for microwave band [88].
- Radar countermeasures were developed [1].
- H. S. Black, J. W. Beyer, T. J. Grieser and F. A. Polkinghorn developed a multichannel microwave radio relay system [88].
- Arthur Samuel and others developed a Gas-discharge transmit-receive tube (TR and ATR) at Bell Labs [88].
- Sharbaugh and L. C. Wallers reported communication over a distance of 800 ft at 21 GHz [1].
- E. V. Appleton discovered that sunspots emit radio wave [58,79].
- M. Katzin patented a probe fed rectangular horn antenna [106] which was first introduced by J. C. Bose in 1895.

- Detection of radar echoes from the Moon using 111.5 MHz radar of 3 kW peak power [57].
- S. L. Ackerman and G. Rappaport developed a radio control system for guided missiles [79].
- G. L. Fredendal and others suggested pulse width modulation for the sound channel in TV [39].
- The first *commercial mobile telephone service* became available in St. Louis, Missouri. However, the mobile telephone would not become common for another four decades [6].
- H. C. A. van Duuren, during World War II, devised an idea that was to be widely used in computer communications, the *automatic repeat-request* (ARQ). It was published this year. It was used to improve radio-telephony for telex transmission over long distances [18].
- The electronic digital computer *ENIAC (Electronic Numerical Integrator & Computer)*, using 18,000 vacuum tubes was built at the Moore School of Electrical Engineering of the University of Pennsylvania, under the technical direction of John Presper Eckert and John W. Mauchly. However, John von Neumann's contributions were among the earliest and most fundamental to the theory, design and application of digital computers, which go back to the first draft of a report written in 1945. Computers and terminals started communication with each other over long distances in the early 1950s. The links used were initially voice-grade telephone channels operating at low speeds (300 to1200 b/s) [12,18,92].
- Bell Telephone Laboratories produced a simplex system [1].
- E. M. Williams developed the radio frequency spectrum analyzer [93].
- Andrew observed demodulation of commercial broadcasting stations by superconducting material [1].
- Bell Lab developed foil type polystyrene capacitor with Q of 10000 [1].
- V2 rockets fitted with pulse-time modulated telemetering system were used for upper atmosphere research [1].

1947:

- G. E. Mueller and W. A. Tyrrel developed the dielectric rod antenna.
- John D. Kraus invented the helical antenna [88].
- W. V. Smith and others developed frequency stabilization of microwave oscillator by spectrum lines [31].
- W. M. Goodall developed telephony by pulse-code modulation [79].
- W. Tyrell proposed hybrid circuits for microwaves [88].
- R. S. Mantner and O. H. Schade developed high-voltage RF supplies for TV circuits [88].
- J. H. Van Vleck reported absorption of microwaves by uncondensed water vapor and oxygen [57].
- R. W. Masters proposed inverted triangular (diamond) dipoles [106].
- H. E. Kallaman of MIT constructed the voltage standing wave ratio

(VSWR) indicator meter [1].

- V. A. Kotel'nikov, from the Soviet Union, developed the geometric representation of signals [7,18].
- Walter H. Brattain, John Bardeen, and William Bradford Shockley at Bell Laboratories invented the *point-contact transistor*, although it was not announced until next year [12].
- Steve O. Rice developed statistical representation for noise at Bell Laboratories [12].
- Thick film printed conductors and resistances were used in miniaturized transmitters [1].

1948:

- W. H. Brattain, J. Bardeen, and W. Shockley of Bell Labs built the *junction transistor*. In 1956 they were jointly awarded the Nobel Prize in physics for their research on semiconductors and their invention of the transistor [5,6,18]. A patent application for the junction transistor was filed. The first n-p-n junction transistor was demonstrated in 1950 [43]. This invention was to spur the application of electronics to switching and digital communications. The motivation was to improve reliability, increase capacity, and reduce cost [18].
- L. N. Brillouin patented the omni-directional and directive coaxial horn antennas [106].
- Long-playing record systems (LP), at 33 revolutions per minute (rpm), were introduced by Columbia Broadcasting Corporation (CBS), as an improvement of the previous 78 rpm record system [1,6].
- New York - Boston microwave repeater, delayed due to war was installed by Bell Lab [1].
- F. T. Haddock reported scattering and attenuation of microwave radiation through rain [57].
- B. D. H. Tellegan suggested use of Faraday rotation of microwaves in ferrites and called it *Microwave Gyrator* [1,25].
- William Webster Hansen (1900–1949) designed, constructed and tested optical radar for surveying, using pulses of light and retro-directive reflector [1].
- E. L. Ginzton and others developed distributed wideband amplifier using pentodes in parallel [88].
- Dennis Gabor (1900–1979) invented holography and obtained the Nobel Prize in Physics in 1971 for it [38].
- L. A. Meacham and E. Peterson developed an experimental pulse-code modulation system of toll quality [88].
- W. W. Hansen reported surveying with pulsed-light radar [1].
- Bernard Oliver (1916–1995) and others reported the philosophy of pulse code modulation [1].
- Englishman Thomas H. Flowers applied for patents on switching of pulse amplitude modulation signals [4].
- George Sinclair reported patterns of slotted-cylinder antennas [88].

- The first port radar system in the world was installed in Liverpool [3].
- Wang developed static magnetic recording, storage and delay line [1].
- Air-Force portable instrument landing system (surplus of World War II) was installed at Dum Dum (Calcutta) airport, first use in civilian airport in the World [1].
- Frank Hamilton developed the first electronic computer with stored memory, *IBM Selective Sequence Electronic Calculator (SSEC)* [6].
- Shannon laid the theoretical foundations of digital communications in a paper entitled "A Mathematical Theory of Communication", his famous work on information theory [7,12,18].
- Arthur Charles Clarke discussed cm and optical frequency bands for communication, navigation and other related purposes [1].
- The BALUN – a transmission line transformer for transforming a coaxial line into a balanced twin wire line using a quarter wavelength sleeve was described by Paine [1].
- Printed conductor (zinc) and resistance (graphite) were used for mass production of radio in England [1].

1949:

- Shannon's paper was reproduced in the form of a book, authored by C. E. Shannon and W. Weaver with the title "The Mathematical Theory of Communication" [7,18].
- Shannon programmed a computer to play chess [1].
- Vannever Bush developed a 'rapid selector' for microfilms in a library using photocells [1].
- Anon developed stable time and frequency standard using ammonia resonator [31].
- J. H. Scaff and co-workers coined the names 'Acceptor' and 'Donor' [16].
- G. Tiemann developed *Ediphor* for TV projection [1].
- E. J. Barlow published the principle of operation of Doppler radar [1].

1950:

- J. M. L. Janssen developed stroboscopic (sampling) oscilloscope [1].
- The phototransistor evolved from the observation made by J. N. Shieve that the conductivity of the diode changed when light was incident on the p-n junction [45].
- Henry Booker and William E. Gordon developed the theory of radio scattering in troposphere [88].
- L. A. Manning and co-workers studied the meteoric echo of upper atmosphere winds [88].
- Single crystal germanium was produced in laboratory [29].
- D. R. Griffin reported ultrasonic cries of bats [57].
- P. K. Weimer and co-workers developed *Vidicon* (RCA) [20].
- First generation digital computer was introduced [61].
- TDMA was applied to telephony [12].

- During the following decade microwave telephone and communications links were developed [12].
- Penteman constructed electronic decade counter with neon indicator [1].
- Magnetic Tape as memory capable of storing 64 bits was made for Mark III computer [1].
- George Goubau invented and displayed surface wave transmission line by coating a single wire with special insulations which can replace coaxial lines and waveguides for many applications [1].
- N. A. Bergovitch and A. R. Margodien of Hughes Aircraft Co. reported for the first time the original analysis of strip transmission line [44].
- The phonetic-steno-sonograph, which converted speech into a written record was advertised in Germany [1].

1951:

- K. L. Henry designed the *Tinkertoy* system [94].
- The American physicist Charles Hard Townes published the principle of *MASER* (*Microwave Amplification by Stimulated Emission of Radiation*), based in 1917 Einstein theory, but using microwaves instead of light. In 1954 Townes demonstrated a working model at 23.87 GHz. Townes received the 1964 Nobel Prize in physics for his invention [6,61].
- At the Laboratoire Central des Télécommunications (LCT), in Paris, the first mock-up model of a time-division multiplex highway connecting subscriber lines by electronic gates handling amplitude-modulated pulses (PAMs) was done [4].
- John Presper Eckert (1919–1995) and John Mauchly designed the first commercially successful electronic computer, the *Sperry Universal Automatic Computer* (*UNIVAC*) [6].

1952:

- C. L. Hogan demonstrated microwave circulator using Faraday rotation in a circular waveguide [1].
- Geoffrey W. Dummer developed the concept of semiconductor integrated circuit at Royal Signals and Radar establishment in Malvern, UK [93].
- Andy Kay developed the digital voltmeter [94].
- Unipolar field-effect transistor was developed by Shockley [80].

1953:

- G. Deschamps first proposed the concept of microstrip radiators. However, the first practical antennas were not developed until 1972 by J. Q. Howell [95].
- NTSC color TV was introduced in the United States [12].
- Hagelberger of Bell Lab built a readout device using a single node neon tube and ten stacked wire-cathodes in the form of numerals [1].

- The first transatlantic telephone cable (36 voice channels) was laid [12].

1954:

- Charge controlled storage CRT was developed by M. Knoll and others [20].
- Transistor radio set was first developed [94].
- Gordon Pearson and Calvin Fuller at Bell Laboratories, USA invented the silicon p-n junction photovoltaic cell. This invention, together with the advent of high-budget orbiting spacecraft, made the *Silicon Cell* a practical device. Since the 1980s solar cells are also used on Earth to power telephones, water heaters, automobiles and even airplanes [6,61].
- J. P. Gordon, H. J. Zeiger, and C. H. Townes produced the first successful maser [14].

1955:

- Ian M. Ross filed a patent for MOS [92].
- R. H. DuHamel and D. E. IsBell developed the log periodic antenna.
- R. Braunstein reported infrared emission from Gallium Arsenide (GaAs) [94].
- William Ross Aiken devised a flat CRT for TV [97].
- John R. Pierce proposed the use of satellites for communications [18].
- Sony Corporation marketed the first *transistor radio*, the TR-55. Within a decade, transistor radios and electronics had replaced their vacuum tube counterparts by transistors [6].
- Indian scientist Narinder S. Kapany designed the first practical use of an optical fiber incorporating it in an endoscope [6].

1956:

- First Transatlantic Cable-Telephone was inaugurated [3].
- Murray Hopper developed the FLOW-MATIC compiler that had near English-like statements. She was one of the first people to realize that programming languages should be written for people who were neither mathematicians nor computer experts. She helped formulate the specifications for COBOL, the common oriented business language [5].

1957:

- Japanese scientist Leo Esaki developed Germanium Tunnel-diode [61].
- B. Kazan and F. H. Nicoll developed Solid-State light amplifier [1].
- Jordel Bank's Radio Telescope (250 ft. diameter dish) was completed. In 1960 it received signal from a 5 W transmitter of Pioneer V deep space probe covering a distance of 22.5 million miles. The Radio Telescope was the only antenna capable for sending signal to the space probe through the same distance to detach its carrier rocket [1].
- V. H. Rumsey introduced the concept of frequency independent antennas.

- Soviet Union launched the satellite Sputnik I, which transmitted telemetry signals for about five months [7,12,18].
- Gordon Gould of Columbia University described a *LASER (Light Amplification by Stimulated Emission of Radiation)* system. He did not file for a patent until 1959 after other similar designs had been filed, and his patent was not issued until 1977 [6].
- German physicist Herbert Kroemer (Figure 11.21) originated the concept of the *heterostructure bipolar transistor* (HBT) (Section 11.2.12).

1958:

- C. H. Townes (Figure 2.48) and his brother-in-law, Arthur Leonard Schawlow of Bell Labs, advanced the possibility of a visible light maser, a *laser*. Townes was awarded the Nobel Prize in physics for his fundamental work that led to the construction of masers and lasers. He shared the prize with two Russian scientists who had done similar work but not as well publicized [1,5,12].
- Ampex developed the video tape recorder [94].
- J. D. Andrea patented the thick film resistors [96].
- Robert Noyce of INTEL and Jack Kilby from Texas Instruments produced the first silicon *integrated circuit* (IC). The transistor and this landmark innovation in solid state devices and integrated circuits led to the development of *Very-Large-Scale Integrated* (VLSI) circuits single-chip *microprocessors*, and with them the nature of the telecommunication industry changed forever [18,29].
- A first working model of a digital switch was made at the Bell Laboratories, under the direction of H. Earle Vaughan who started a project in 1956 known as ESSEX (Experimental Solid-State EXchange). The goals to improve reliability, increase capacity, and reduce cost in switching of digitized telephone were achieved. However, economical industrial production of the required complex technology was not yet possible [4,7].

Figure 2.48. Charles Hard Townes (born 1915).

- The satellite Explorer I was launched by the United States [4].
- Seymour Cray designed the *transistorized computer* for Control Data Corporation [6]
- Australian physicist R. Q. Twiss publishes a paper which was followed by another paper the next year from the German physicist Jürgen Schneider (Figure 11.22), which were the first publications on the quantum electronic (QE) model of the electron cyclotron resonance maser interaction (Section 11.2.13).

1959:

- Tantalum Thin film capacitor was developed at Bell Labs [94].
- RCA developed the *Nuvistor* high frequency tube as an alternate to the transistor [61]. It was probably developed by Albert Rose.
- The Moon was used as a reflector for radiotelephony test between Jordell Bank and Massachusetts [1].
- American physicist Theodore Harold Maiman invented *laser* (light amplification by stimulated emission of radiation) and developed it during the next year [4,18].
- The IBM 1401 computer replaced mechanical business machines in large numbers in business and industry [6].

1960:

- R. L. Carbery developed video transmission over telephone cable by seven digit PCM [97].
- The first commercial telephone service with digital switching began in Morris, Illinois [6,18].
- Maiman built and operated the first *Ruby Rod Laser* at Hughes Research Laboratory in California. Other lasers were demonstrated thereafter by four other groups independently also in 1960 [5,6,12,61].
- American electrical engineer Nick Holonyak, Jr., first began work on the visible-spectrum light emitting diode (LED) while working at General Electric. He had discovered that the wave length of the GaAs diode could be shifted from the infrared to the visible spectrum by merely changing the chemical composition of the crystal itself to GaAsP. Holonyak developed the first practical visible-spectrum LED in 1962 [61].
- J. W. Allen and B. E. Gibbone developed *Light Emitting Diode* [94].
- Passive communication satellite ECHO was launched into a 1000 mile-high orbit of a 100ft diameter light-weight balloon satellite by NASA [1].
- Second generation (transistorized) computer was introduced [97].
- D. Khang and M. M. Atalla developed silicon-silicon dioxide field induced surface devices (MOS) [61].

1961:

- E. Pfeiffer and O. Gentner developed Earthed-Collector amplifier (*emitter follower*) [1].
- Robert Biard and Gary Pittman of Texas Instruments (TI) developed Gallium Arsenide (GaAs) diodes, using an infrared microscope, and found that they emitted significant light in the infrared region. Based on their findings, TI immediately began a project to manufacture these diodes and announced the first commercial product, the SNX-100, in October, 1962. Infrared LEDs continue to be used today as transmitters in fiber optic data communication systems [61].
- Digital Equipment Corporation (DEC), USA developed *Minicomputer* [94].
- Fernando J. Corbató demonstrated *Compatible Time-sharing System* (CTSS) of computer [92].
- The *helium neon gas laser* was developed [6].
- Stereo FM broadcasts begin in the United States [12].

1962:

- P. Vogel and co-workers developed *Electronic Timepiece* [94].
- Robert N. Hall and co-workers developed semiconductor LASER [61].
- G. Robert-Pierre Marié patented a wide band slot antenna [106].
- S. R. Hofstein and F. P. Heiman developed MOS IC [94].
- P. K. Wiemer developed *Thin Film Transistor* (TFT) [1].
- In continuation of successful experiments with the passive satellite ECHO I in 1960, a major experimental step in communication satellite technology was taken with the launching of TELSTAR I. The satellite was capable of relaying TV programs across the Atlantic. This was made possible only through the use of maser receivers, large antennas, reliable transistors, solar cells, microwave resonators, and low-noise receiver circuitry [1,4,7,12,18].
- Bell Laboratories installed the first *T-1 carrier digital system transmission* in the United States [18].

1963:

- Bell Punch and Co., UK, developed the *Electronic Calculator* [94].
- W. S. Mortley and J. H. Rowen developed *surface acoustic wave* (SAW) devices [94].
- Hughes Aircraft Company launched a satellite of 78 lbs sponsored by NASA under the project SYNCOM in the geo-synchronous orbit at a height of 22,300 miles. It consisted of a single repeater receiving signals from ground at 7.4 GHz and transmitting on 1.8 GHz with a 4 watt TWT amplifier [17].
- N. F. Foster and coworkers developed the *thin film piezoelectric transducer* [22].

- John B. Gunn of IBM demonstrated microwave oscillations in Gallium Arsenide and Indium Phosphide diodes [17].
- American scientists Richard Williams and George Heilmeier at the David Sarnoff Research Center, RCA laboratory in Princeton, New Jersey, first published a suggestion for using liquid crystal materials for display. Heilmeier went on to head a group at the lab, including Nunzio Luce, Louis Zanoni, Joel Goldmacher, Joseph Castellano and Lucian Barton, to investigate the use of liquid crystal displays (LCD) for a "TV-on-a-wall" concept, a dream of David Sarnoff himself [61]. The LCD display was also independently developed by Frank Leslie at the Royal Signals and Radar Establishment, Malvern, UK.
- The Dutch electronic company Philips NV introduced into the mass market *Compact audio cassettes* [6].
- Bell System introduced the touch-tone phone. It is also known as *Dual-Tone Multiple Frequency* (DTMF) signaling [12].
- The *Institute of Electrical and Electronic Engineers* (IEEE) was formed from the merging of the IRE and AIEE [12].
- During 1963 and up to 1966 error-correction codes and adaptive equalization for high-speed error-free digital communication were developed [12].

1964:

- Highest frequency, up to 1974, 60 GHz, was generated at MIT using laser beam and sapphire crystal [29].
- R. L. Johnston, B. C. De Loach and B. G. Cohen developed the IMPATT diode oscillator [61].
- R. L. Stermer made the first screened (thick film) capacitor [64].
- The multinational organization INTELSAT was formed. Its purpose was to design, develop, construct, establish, and maintain the operation of the space segment of a global commercial communication satellite system [7].
- Xerox Corporation introduced *Long Distance Xerography* (LDX) [6].
- The electronic telephone switching system (No. 1 ESS) was placed into service [12].
- Scientists at the Institute of Applied Physics of the Russian Academy of Sciences in Gorky operated the first *gyrotron* (mode TE_{101}, rectangular cavity, power 6 W, CW).
- Kroemer first explained the Gunn Effect (Section 11.2.12).

1965:

- The third generation of computers using IC was introduced [97].
- The American Communications Satellite Corporation (COMSAT) in cooperation with INTELSAT started launching a series of communications satellites that were the initial building blocks in the global network of international communications satellites. The first, designated INTELSAT F-I, and known as *Early Bird*, a geostationary communication satellite, was launched on April 6, 1965. In a period of

seven years four generations of satellites (INTELSAT I to IV) were placed in commercial operation. Capacity increased from 240 telephone circuits and 1 TV channel to 6000 telephone circuits and 12 TV channels. This growth was possible thanks to the feasibility of increased power, antenna gain, and transponder bandwidth, together with smaller size components [6,7,12].

- Robert Lucky pioneered the idea of *adaptive equalization*, which was to be instrumental for the support of high data transmission rates by telephone channels [7].
- The geometric representation of signals was brought to full fruition in a book by J. M. Wozencraft and I. M. Jacobs [7,18].

1966:

- In the United Kingdom, Charles K. Kao and George A. Hockham, working under Reeves, predicted that fibers drawn from extremely pure glass would be an ideal support for the transmission of modulated light waves. They predicted that these *optical fibers* would have a very low intrinsic loss [4,7,12,18,94].
- German scientist M. Boerner arrived to findings similar to those of Kao and Hockham. He obtained a German patent as well as British and American patents, but he died soon thereafter [4].
- Medium Scale Integrated Circuits were developed [98].
- Xerox introduced *Magnafax Telecopier*, an improved fax machine. In the late 1970s Japanese companies produced new generations of faster, smaller, and more efficient fax machines. By the late 1980s compact fax machines had revolutionized everyday communication around the world [6].
- Kroemer invented the double heterostructure laser (Section 11.2.12).

1967:

- Heilmeier developed the *Liquid Crystal Display* (LCD) using Dynamic Scattering Mode effect (DSM) [61].
- The BBC commenced its color transmission on the 625-line PAL (*Phase Alternation by Line*) system and so did Western Europe. The United States and Japan selected the 525-line NTSC (*National Television System Committee*), whilst France and the USSR chose SECAM (*Système Électronique Couleur avec Mémoire*) [17].
- A. C. Traub propose the use of varifocal mirror for stereoscopic display [20].

1968:

- T. de Boer developed a TV display using gas discharge panel [61].
- Pulsating radio source (*Pulser*) was discovered [58].
- RCA developed C-MOS Integrated Circuits [94].
- Sony Company produced the *Trinitron Color Television System*. With this improvement color TV became commercially available [6].
- Cable TV systems were developed [12].
- W. Stohr patented an ellipsoidal monopole and dipole antennas [106].

1969:

- Large Scale Integrated Circuits were introduced [57].
- American physicist James Ferguson, while Associate Director of the Liquid Crystal Institute at Kent State University in Ohio, discovered the twisted pneumatic field effect of liquid crystals, which led to an improved version of LCD [61].
- The INTERNET began as ARPANET, the American government-supported *Advanced Research Project Agency Network*, which searched for means to avoid disruption in telecommunication networks caused by enemy action. It was based on a computer protocol (software instructions) that enabled each computer to co-operate with others in the network, which involved data packet switching. [17].
- Agustar and co-workers developed semiconductor memory [94].
- George Smith and Willard Boyle developed charge coupled devices [61].
- The first digital radio-relay system went into operation in Japan. It was operated in the 2 GHz band with a transmission capacity of 17 Mbps, corresponding to 240 telephone channels. Digital radio-relay equipment appeared on the market rapidly in the 1970s and gradually stopped the production of analog equipment [4]. Renewed interest in microwave radio transmission or digital communications was largely due to the introduction of digital switching during these years [7].

1970:

- S. K. Deb and R. F. Shaw patented an electrochemical display [99].
- Robert Maurer, Peter Schultz, Donald Keck, and Felix Kapron from Corning Glass Works achieved the goal of less than 20-dB/Km loss (in fact, 16 dB/Km) predicted by Kao and Hockham for optical fibers, with an industrially produced fiber with a core of titanium-doped silica glass and a cladding of pure silica glass. Nowadays, transmission losses as low as 0.2 dB/Km are achievable. Optical fibers have been used since this year to transmit telex, telephone, and TV (cable TV) signals much more efficiently than was ever possible with metal wire [4,6,7,18].

1971:

- Ted Hoff at Intel Corporation invented the silicon microprocessor that was instrumental for the development of personal microcomputer [6,12].
- C. A. Burrus of Bell Labs produced the first light emitting diodes [19]
- Patent application for microcomputer was filed by Michael Cochran and Gary Boone of Texas Instrument [98].
- Statek, an American company started in 1970 by the German-American physicist Juergen Staudte, began manufacturing and marketing quartz oscillators. Saudte had invented and patented before a process for mass producing miniature quartz crystal oscillators, which used a photolithographic process that is similar to the way integrated circuits are made [61].

- K. Izuka and coworkers developed *Hologram Matrix Radar* [94].
- During 1950-1970, various studies were made on *computer networks*. However, the most significant of them in terms of impact on computer communications was ARPANET (Advanced Research Project Agency Network), now popular worldwide as INTERNET, first put into service in this year of 1971. This pioneering work in *packet switching* was sponsored by the Advanced Research Projects Agency of the USA Department of Defense [4,7,18].

1972:

- Kees Hart and Arie Slob developed *integrated injection logic* [94].
- T. J. Rodgers developed V-MOS [94].
- INTELSAT V was launched. It provided the first international use of *time-division-multiple-access* (TDMA) digital technique [7].
- Philips Company of The Netherlands marketed the first videodiscs, followed by Thomson-CSF I France, JVC in Japan, and RCA in the United States. The JVC system was the first to use an optical tracking signal linked to the video signal. Since the 1990s videodiscs enjoy large-scale consumer acceptance [1].
- Motorola demonstrated the *cellular telephone* to the FCC [12].

1973:

- American space probe Pioneer 10 transmitted TV pictures from within 81,000 miles of Jupiter (1200 million km) [58].
- William E. Good and Thomas T. True developed projection color TV display [100].

1974:

- 16-Bit single chip microprocessor was introduced [94].
- E. Spiller and A. Segmuller developed *Light-pipe* for guiding X-rays to desired location [29].
- The first step to extend the digital network to the telephone service was made. The concept of integrated digital transmission and switching was proposed in the late 1950s, about the same time as the deployment of the first digital switching system. The move toward and *Integrated Digital Network* (IDN) was motivated by expectations of continued reductions in component and interface costs [7].

1975:

- The *Cray I*, first member of the fourth generation of computers, the supercomputers, was developed by the company formed by Cray. Cray I was superseded by Cray II introduced in 1985. Both of them where superseded by CM-200, from Thinking Machines, Inc., introduced in 1991 [6].
- William Gates and Paul Allen (Figure 2.49) produced the first microcomputer software that helped to launch the *Personal Computer* (PC) revolution [17].
- *Very-large-scale-integrated-circuits* (VLSI) were developed [98].

Figure 2.49. Paul Allen (born 1953), left, and William H. Gates (born 1955), right.

- Western Electric and Bell Labs in Atlanta, Georgia, carried out the first field trial of an optical system. They used 144 optical fibers with a loss of less than 6dB/km, permitting a spacing of 10km between regenerative receivers. Each fiber operated at a bit rate of 45 Mbit/sec using a GaAs laser light source of 0.82 microns wavelength and had a capacity exceeding 100,000 telephone circuits [17].
- F. K. Reinhart and R. A. Logan developed integrated optical circuits [94].
- M. Fukushima and coworkers developed color TV display using gas discharge panels [61].

1976:

- Steve Jobs and Steve Wozniak, in the Silicon Valley, created a homemade microprocessor computer board called *Apple I* [6,12].

1977:

- Pocket TV receivers were made [94].
- Jobs and Wozniak founded Apple Computer, Inc., and introduced Apple II, the world first desktop computer or PC [6].

1978:

- University College of London developed the *Acoustic Microscope* [101].
- Westinghouse made TV *Liquid Crystal Displays* (LCD) with Thin Film Transistors (TFT) [20].
- American Telephone and Telegraph (AT&T) Bell Laboratories began testing a mobile telephone system based on hexagonal cells (*Cellular Telephone*) [6].

1979:

- Sony Co., Japan, and Philips Co., The Netherlands, collaborated in introducing the *Compact disc* (CD) technology. Since the 1990s CDs enjoy large-scale consumer acceptance [6].

- 64-kbit random access memory ushers in the era of VLSI circuits [12].

1980:

- T. Migoshi and K. Iwasa made electrochromic display for watches [100].
- CW performance of GaAs MESFET (Metal-Semiconductor FET) reached 10 w at 10 GHz [102].
- Bell System FT3 fiber optic communication system was developed [12].
- ATLAS I electromagnetic pulse simulator was built for testing large aircrafts on a wooden trestle 400m long, 75 m high and 105 m wide. It was the largest wooden structure in the world as shown in Figure 2.50 and does not have any metal nails!

Figure 2.50. The ATLAS I EMP Simulator with a full scale B1B (upper photograph) and a C130 (lower photograph) undergoing test (photos provided by Carl E. Baum, AFRL).

1981:

- W. A. Crossland and P. J. Ayliffe developed a dyed phase-change liquid crystal display over a MOSFET switching array [100].
- Laser was used for recording and reproduction of Video discs [67].
- *Voice control* and *voice recognition* systems were installed in cars of Toyota and Nissan in Japan [1].
- Coverage of the United States by the cellular telephone system [6].
- IBM Corporation introduced its personal computer (PC), which became familiar to computer users all over the world under the acronym MS-DOS (Microsoft Disc Operating System) [17]. By the 1990s, Apple and IBM divided the market, and microcomputers had become a familiar fixture in business, schools and homes throughout the industrialized world spurred by the miniaturization of transistors, microprocessors and electronics in general [6,12].

1982:

- Sony Corp. of Japan, developed filmless 35 mm camera using light sensitive material also called CCD [103].
- Wrist watch TV set was available in Japan [1].
- Stored picture of Saturn taken by *Voyager I* from a distance of about 100 million km were received in earth covering more than 1200 million km from a 20 W transmitter after it has passed Saturn [103].
- G. Ungerboeck pioneered efficient modulation techniques that together with adaptive equalization have contributed to a dramatic increase in data transmission rates in digital communication systems [18].
- The US Air Force began work on the Super Cockpit or Visually-Coupled Airborne Systems Simulator (VCASS), continuing the line of research on flight simulators started in World War II. This is one example of *Virtual Reality* (VR) applications. In 1985 Jaron Lanier coined the term VR. By the early 1990s VR systems were being use for everything from simulated racquetball games to architectural design, including a virtual map of the human body. Future applications of VR in science, industry and entertainment seem virtually unlimited [6].
- AT&T agreed to divest its 22 Bell System telephone companies [12].

1983:

- A laser armed US Air Force plane knocked off air-to-air missiles by destroying their guidance systems [1].

1984:

- Apple introduced its Macintosh computer [12].

1985:

- Fax machines became popular [12].

1986:

- The European Community in a bid to bypass Japanese dominance in the consumer electronics field sponsored the development of an HDTV-MAC system. The HDTV (*High Definition Television*) pioneered by the Japanese Broadcasting Authority NHK and SONY proposed a new TV standard using 1125-lines at 30 frames a second, thus requiring new designs of the transmitting and receiver equipments. The picture format proposed a 9 to 16 ratio, compared with the present day 3 to 4, and 1250 lines compared with 625, the wider aperture being more appropriate for viewing stage and field events. The present day wide-screen TV does not in general incorporate high definition capability. The MAC (Multiplexed Analogue Components) system, primarily suited for satellite transmission, was devised by Independent Broadcasting Authority in the UK, sought an evolutionary development from the 625-line, 25 frames per second standard. MAC minimized the effect of the higher levels of noise at the upper end of the video spectrum (where the color information is transmitted), characteristic of transmission by satellite, by sending each line of color information in a time-compressed form immediately before its monochrome component. [17].

- The first international optical fiber submarine cable link between UK and Belgium (led by British Telecom in 80 km deep water) provided 11,500 telephone circuits. [17].

1988:

- The first trans-Atlantic optical fiber cable system between Europe and America, TAT 8, capable of carrying 40,000 simultaneous telephone conversations, was designed by AT&T Co, British Telecom and France Telecom. It doubled the previous capacity [17].

1989:

- F. Lalezari invented the broad band notch antenna [106].
- Motorola introduced the *pocket* cellular telephone [12].

1990:

- Scientists working in AT&T Bell Laboratories built the first *digital optical processor* for computers [6].
- In December, the *World Wide Web* (www), a standardized communication system between users, servers and data banks, was created in CERN (European Center for Nuclear Research), Geneva, Switzerland, as the result of a project initiated in 1989 by Sir Timothy John Berners-Lee. He implemented the first server, using *html* (Hypertext Markup Language), *http* (Hypertext Transfer Protocol) and URL (Uniform Resource Locator) developed by himself [17].

2.5   EPILOGUE

The events of the 1980 decade paved the way for an extraordinary evolution and revolution of wireless systems from 1990 until nowadays and for the years to come. A rapid development and extension (globalization) of in-door and out-door communications and telecommunications has taken place. The digital era has bloomed with digital signal processing using microprocessors, digital oscilloscopes, megaflop workstations, DVDs (*Digital Video Disc*), digital audio recording, digitally tuned receivers, spread spectrum systems, ISDN (*Integrated Service Digital Network*), HDTV and so on. PCs, portable PCs and PDAs

(*Personal Digital Assistant*) of unprecedented high memory and storage capability are available. *Flash* cards and drives have been made possible by advances in transistor logic [12]. Analog cellular mobile systems, commonly referred to as first generation systems, have been substituted by digital second generation systems, like those based on GSM (*Global System for Mobile Communications*) in Europe, PDC (*Personal Digital Cellular*) in Japan, IS-95 or *cdmaone* in America and Korea, IS-136 or US-TDMA in the United States, which enabled voice communication, text messaging and access to data networks to go wireless in many of the leading markets. The work to develop the third generation mobile systems started when the World Administrative Radio Conference (WARC) of the ITU (*International Telecommunications Union*), at its 1992 meeting, identified the frequencies around 2 GHz that were available for use by future third generation mobile systems, both terrestrial and satellite. Air interfaces like WCDMA (*Wideband Code Division Multiple Access*), EDGE (*Enhanced Data rates for GSM Evolution*), and multicarrier CDMA (*cdma2000*), and other wireless technologies like GPRS (*General Packet Radio Service*), *Bluetooth*, WLAN (*Wireless Local Area Network*), WI-FI (*Wireless Fidelity*) have been developed and UMTS (*Universal Mobile Telecommunication System*) is now offering new services. The fourth generation is nowadays under development. It will provide unprecedented connectivity via wireless and high-speed or optical fiber lines between users of a wide variety of handsets, data banks, computers and so on. Other wireless services initially restricted to some users, like GPS (*Global Positioning System*), are now becoming popular.

REFERENCES

[1] M. C. Mallik, "Chronology of Developments of Wireless Communication and Electronics", *IETE Technical Review*, vol. 3, n. 9, Sept. 1986, pp. 479-522.

[2] I. Asimov, *Asimov's Biographical Encyclopedia of Science and Technology*, 2nd Revised Edition, Doubleday & Company, Inc., Garden City, New York, 1982.

[3] P. Dunsheath, *A History of Electrical Engineering*, Faber and Faber, London, 1962.

[4] A. A. Huurdeman, *The Worldwide History of Telecommunications*, John Wiley & Sons, Hoboken, NJ, 2003.

[5] J. H. Tiner, *100 Scientists who Shaped World History*, Bluewood Books, 2000.

[6] B. Yenne, *100 Inventions that Shaped World History*, M. Grosser, Consulting Ed., Bluewood Books, 1993.

[7] S. Haykin, *Digital Communications*, John Wiley & Sons, Inc., New York, NY, 1988.

[8] W. L. Bragg and G. Porter, Ed., *Physical Sciences*, The Royal Institution Library of Sciences, Vol. 1-5, Elsevier Publishing Co. Ltd., Essex, 1970.

[9] R. A. Heising, *Quartz Crystals for Electrical Circuits, Their Design and Manufacture*, D. Van Nostrand Co. Inc., New York, 1946.

[10]   L. W. Turner, *Electronics Engineers' Reference Book*, Butterworth and Co. Ltd., London, 1976.

[11]   E. Eastwood, *Wireless Telegraphy,* Royal Institution Library of Science, Applied Science Publishers Limited, London, 1974.

[12]   L. W. Couch II, *Modern Communications Systems. Principles and Applications*, Prentice-Hall, Upper Saddle River, NJ, 1995.

[13]   W. Benton, *The New Encyclopedia Britannica*, London, UK, 1973.

[14]   R. Hague, *Alternating Current Bridge Methods*, Sixth Edition, Pitman Publishing, 1971.

[15]   W. M. Dalton, *The Story of Radio*, Vol. I, II and III, Adam, Hilger, London, 1975.

[16]   G. L. Pearson and W. H. Brattain, "History of Semiconductor Research", *Proceedings of the Institute of Radio Engineers*, p.1794, 1955.

[17]   J. Bray, *Innovation and the Communications Revolution*, IEE, London.

[18]   S. Haykin, *Communications Systems*, John Wiley & Sons, Inc., New York, NY, 1994.

[19]   F. W. Grover, *Inductance Calculations, Working Formulas and Tables*, D. Van Nostrand Co., New York, 1946.

[20]   T. P. McLean and P. Schagen, *Electronic Imaging*, Academic Press Ltd., London, 1979.

[21]   M. P. Marcus, *Switching Circuits for Engineers*, 2nd ed., Prentice-Hall of India Pvt. LTD., New Delhi, 1969.

[22]   L. I. Maissel and R. Glang, *Handbook of Thin Film Technology*, McGraw-Hill, New York, 1970.

[23]   P. Lenard, *Great Men of Science*, G. Bell and Sons Ltd., London, 1958.

[24]   G. Parr and O. H. Davie, *The Cathode Ray Tube and Its Applications*, Reinhold Publishing, New York, 1959.

[25]   R. Taton, *Science in the Nineteenth Century*, Thames and Hudson, London, 1965.

[26]   J. J. Thompson, *Conduction of Electricity through Gases*, Cambridge University Press, 1933.

[27]   G. R. M. Garratt, *The Early History of Radio from Faraday to Marconi*, IEE, London, 1995.

[28]   H. Hertz, *Memoirs: Letters, Diaries*, arranged by J. Hertz, 2nd edition enlarged and edited by M. Hertz and C. Susskind, San Francisco Press, 1977.

[29]   N. McWhirter, *The Guinness Book of Answers*, General Editor, Guinness Superlatives Ltd., 2, Cecil Court, Middlesex, 1976.

[30]   W. Fraser, *Telecommunications*, Macdonald, London, 1960.

[31]   W. A. Edson, *Vacuum-Tube Oscillators*, John Wiley and Sons, Inc., New York, 1953.

[32]   E.W. Golding, *Electrical Measurements and Measuring Instruments*, 3rd ed., Sir Isaac Pitman and Sons Ltd., London, 1944.

[33]   E. L. Chafee, *Theory of Thermoionic Vacuum Tubes*, McGraw-Hill Book Co., New York, 1933.

[34]   H. J. Reich, *Theory and Applications of Electron Tubes*, 2nd ed., McGraw-Hill Book Co., 1944.

[35] J. G. Growther, *Six Great Inventors*, Hamish Hamilton, London, 1961.

[36] W. Greenwood, *A Text-Book of Wireless Telegraphy and Telephony*, University Tutorial Press Ltd., New Oxford, 1925.

[37] H. A. Thompson, *Alternating Current and Transient Circuit Analysis*, McGraw-Hill Book Co., New York, 1955.

[38] J. Ramsey, *Electronic Spectrum Before 1900,* The Raman Laser, Z.H. (Herb) Haller IEEE Spectrum, April and Aug. 1964.

[39] L. B. Arguimbau and R. B. Adler, *Vacuum Tube Circuits and Transistors*, Wiley, New York, 1957.

[40] W. D. Jones, *Fundamental Principles of Powder Metallurgy*, Edward Arnold Ltd., London, 1960.

[41] *Silicon Rectifier Handbook*, Sarkes Tarzian Inc., Bloomington, IN, 1960.

[42] C. E. Lowman, *Magnetic Recording*, McGraw-Hill Book Co., New York, 1972.

[43] H. A. Brown, *Radio Frequency Electrical Measurements*, McGraw-Hill Book Co. Inc., London, 1938.

[44] *IEEE Transactions on Microwave Theory and Techniques*, 1984, Vol. 9, Historical Perspectives of Microwave Technology.

[45] T. S. Gray, *Applied Electronics*, Asia Publishing House, Calcutta, 1958.

[46] W. I. Bendz, *Electronics for Industry*, John Wiley & Sons, New York, 1947.

[47] K. S. Packard, "The Origin of Waveguides: A Case of Multiple Rediscovery," *IEEE Trans. Microwave Theory Tech.*, vol. MTT-32, no. 9, Sept. 1984, pp. 961-969.

[48] V. Barkan and V. Zhadanov, *Radio Receivers*, Foreign Languages Publishing House, Moscow.

[49] V. Mukherjee, "Some Historical Aspects of J.C. Bose's Microwave Research During 1895-1900", *Indian Journal of Science*, Vol. 14, No. 2, 1979.

[50] D. P. Agrawal, *Introduction to Wireless and Mobile Systems*, Brooks/Cole-Thomson Learning, Pacific Grove, CA, 2003.

[51] J. H. Hammond Jr. and E. S. Purington, "History of Some Foundations of Modern Radio-Electronic Technology", *Proc. IRE*, Sept. 1957, p. 1191.

[52] R. Keen, *Wireless Directions Findings*, 4th ed., Iliffe and Sons Ltd., London, 1947.

[53] F. E. Terman, *Radio Engineering*, 2nd ed., 3rd impression, McGraw-Hill Book Co. Inc., New York, 1937.

[54] W. A. Geyger, *Magnetic-Amplifier Circuit*, McGraw-Hill Book Co. Inc., New York, 1957.

[55] J. Martin, *Future Development in Telecommunication*, 2nd ed., Prentice-Hall Inc., Englewood Cliffs, New Jersey 1977.

[56] J. H. Morecroft, *Principles of Radio Communication*, 3rd ed., Chapman and Hall Ltd., London, 1933.

[57] M. I. Skolnik, *Introduction to Radar Systems*, International Student Edition, Kogakusha Co., Ltd., Tokyo, 1952.

[58] B. Grun, *The Time Tables of History*, Thames and Hudson Ltd., London, 1975.

[59]   T. D. Schalabach and D. K. Rider, *Printed and Integrated Circuitry, Materials and Processes*, McGraw-Hill Book Co., Inc., London, 1963.

[60]   A. Wood, *Acoustics*, Blackie and Son LTD., London and Glasgow, 1940.

[61]   *IEEE Transaction on Education*, Vol. 7, 1976, Historical Notes on Important Tubes and Semiconductor Devices.

[62]   *The Focal Encyclopedia of Film and Television Techniques*, Focal Press, London, 1968.

[63]   F. Langford-Smith, *Radio Designer's Handbook*, 4th ed., Iliffe Books Ltd., London, 1963.

[64]   T. Tamir, *Integrated Optics*, Spring-Verlag, New York, 1975.

[65]   T. I. Williams, *A Bibliographical Dictionary of Scientists*, Adams and Charles Black, London, 1976.

[66]   E. K. Sandeman, *Radio Engineering*, 2nd ed., Chapman and Hall LTD., London, 1953.

[67]   R. C. Newnes, *Newnes Dictionary of Dates*, London, 1962.

[68]   C. E. Tibbs, *Frequency Modulation Engineering*, Chapman and Hall LTD., London, 1947.

[69]   F. E. Terman and J.M. Pettit, *Electronic Measurements*, International Student Edition, Kogakusha Co., Ltd., Tokyo, 1952.

[70]   *Admiralty Handbook of Wireless Telegraphy*, Vol. II, His Majesty's Stationary Office, London, 1938.

[71]   G. Meir, E. Sackmann and J. G. Grabnair, *Applications of Liquid Crystals*, Springer Verlag, Berlin, 1975.

[72]   S. Seely, *Electron-Tube Circuits*, Asian Students Edition, Kogakusha Co. Ltd., Tokyo, 1958.

[73]   F. E. Terman, *Radio Engineers' Handbook*, McGraw-Hill Book Co. Inc., 1943.

[74]   D. M. Cosligan, *Electronic Delivery of Documents and Graphics*, Van Nostrand Reinhold Co., New York, 1978.

[75]   J. H. Ruiter, *Modern Oscilloscopes and Their Uses*, Murray Hill Books, New York, 1949.

[76]   J. Greenfield, *Television, the First Fifty Years*, Harry N. Abrams, Inc., New York, 1977.

[77]   K. R. Sturley, *Radio Receiver Design*, Part II, Chapman and Hall, Ltd., London, 1949.

[78]   M. Brotherton, *Masers and Lasers*, McGraw-Hill Book Co., New York, 1964.

[79]   J. D. Ryder, *Engineering Electronics*, Kogakusha Co., LTD., Tokyo, 1957.

[80]   F. C. Fitchen, *Transistor Circuit Analysis and Design*, East-West Student Edition, Affiliated East-West Press Pvt. Ltd., New Delhi, 1966.

[81]   E. B. Moulin, *Radio Aerials*, Oxford University Press, London, 1949.

[82]   H. Jasik, *Antenna Engineering Handbook*, McGraw-Hill Book Co., New York, 1961.

[83]   L. Marton, *Advances in Electronics,* Ed. III. Academic Press Inc., New York, 1951.

[84] I. G. Maloff and D. W. Epstein, *Electron Optics in Televisions*, McGraw-Hill Book Co., New York, 1938.

[85] *Physical Design of Electronic Systems*, Vol. III, Bell Telephone Laboratories, Prentice-Hall Inc., Englewood Cliffs, NJ, 1972.

[86] E. C. Snelling, *Soft Ferrites, Properties and Appl*ications, Iliffe Book, London, 1969.

[87] W. P. Mason, *Physical Acoustics, Principles and Methods*, Vol. I – Part B, Academic Press, New York, London, 1964.

[88] F. E. Terman, *Electronic and Radio Engineering*, Asian Students' Edition, Kogakusha Co. Ltd., Tokyo, 1955.

[89] J. D. Ryder, *Electronic Fundamentals and Applications*, 5th ed., Prentice-Hall of India, New Delhi, 1976.

[90] D. G. Fink, *Radar Engineering*, McGraw-Hill Book Co., New York, 1947.

[91] K. Henny, *Radio Engineer's Handbook*, McGraw-Hill Book Co., New York, 1959.

[92] J. M. Rosenberg, *The Computer Prophet*, Collier-Macmillan Ltd., London, 1969.

[93] A. B. Glaser and G. E. Subak-Sharpe, *Integrated Circuit Engineering*, Addison-Wesley Publishing Co., London, 1977.

[94] G. W. A. Dummer, *Electronics Inventions and Discoveries*, 2nd ed., Pergamon Press, Oxford, 1978.

[95] I. J. Bahl and P. Bhartia, *Microstrip Antennas*, Artech House Inc., 1980.

[96] C. A. Harper, *Handbook of Thick Film Hybrid Microelectronics*, McGraw-Hill Book Co., New York, 1974.

[97] R. M. Kline, *Digital Computer Design*, Prentice-Hall, Englewood Cliffs, NJ, 1977.

[98] J. Millman, *Microelectronics, Digital and Analog Circuits and Systems*, McGraw-Hill Book Co., 1979.

[99] A. R. Kametz and F. K. Von Willison, *Nonemissive Electrooptic Displays*, Plenum Press, New York, 1976.

[100] G. F. Weston and R. Bittlestone, *Alphanumeric Displays, Devices, Drive Circuits and Applications*, Granada Publishing Ltd., London, 1982.

[101] H. V. Hodosn, (ed.), *The Annual Register, World Events in 1978*, Longman, 1979.

[102] G. D. Venedelin, *Design of Amplifiers and Oscillators by the S-Parameter Method*, John Wiley and Sons, New York, 1972.

[103] *1983 Year Book of Science and the Future*, Encyclopaedia Britannica Inc., 1982

[104] A. Pérez-Yuste, M. Salazar-Palma, "Scanning our past from Madrid: Leonardo Torres Quevedo", *Proc. IEEE*, vol. 93, no. 7, July 2005, pp. 1379-1382.

[105] Eugenii Katz, Home page given by http://chem.ch.huji.ac.il/~eugeniik/.

[106] H. G. Schantz, "A Brief History of UWB Antennas", www.uwbgroup.ru/ /pdf/37schantz.pdf.

3

EVOLUTION OF ELECTROMAGNETICS IN THE NINETEENTH CENTURY

I.V. LINDELL, *Helsinki University of Technology, Finland*

3.1   INTRODUCTION

The path leading to the present-day electromagnetic theory made its greatest leaps in the 19th century. From modest beginnings in the 1600s with experiments which might have created interest only in the most curious philosophers, the knowledge started to grow and the experiments became more and more conspicuous. In the 1700s electricity became the main topic in physics or "natural philosophy" as it was then called. Finally, at the end of the 18th century it gained the status of exact science after the force law was formulated by Coulomb. In the 19th century the development took two paths which will be labeled as Continental and British Electromagnetics. The former was based on Newton's action-at-a-distance concept, proceeded using mathematical tools and ended in Weber's force law. The British branch started by Faraday's field concept, proceeded using physical tools, and finally ended in Maxwell's mathematical formulation. In this chapter a short overview is given on the two processes.

3.1.1   Early Experiments

William Gilbert (1544–1603), physician to Queen Elizabeth in London, appears to be the first to have made systematic studies of electricity and magnetism. In his book *De Magnete* (1600) he listed a number of media that could be electrified by friction and studied properties of permanent magnets [9]. One of Gilbert's aims was to emphasize the difference between electric and magnetic forces. He erroneously claimed that there is no repulsive electric force and that the attractive force is not reciprocal: An electrified body attracts a non-electrified body but not the other way around. In contrast, he showed that a magnet attracts a piece of iron in the same way as the iron attracts the magnet. This was because the electric forces were so small that they could only attract very light objects. The repulsive electric force was first described in 1629 by Niccolò Cabeo (1585–

1650), in Ferrara, and more reliably in 1663 by Otto von Guericke (1602–1686) in Magdeburg. The reciprocity ("mutuality") of the electric force was demonstrated in 1665 by Honoré Fabri (1607–1688), in Florence, through more sensitive experiments by hanging the electrified body on a silk thread [12].

The electric experiments did not interest too many people at this stage. When the glass tube and the glass globe were introduced to produce frictional electricity in greater amounts in the early 1700s, the experiments became more interesting. In 1729 Stephen Gray (1666–1736) found that electric force can be transmitted over a distance of 270 meters along a wire. He also defined a number of conducting and insulating materials and started the electrification of human bodies. In 1733 Charles François de Cisternay du Fay (1698–1739), in France, found that there are two different kinds of electricity. A charged object repels an object charged with the same kind and attracts an object charged with the other kind of electricity. This started the two-fluid concept of electricity (vitreous and resinous electricity). Independently, Benjamin Franklin (1706–1790) noticed that a capacitor (Leyden jar) separates two kinds of electricity which completely annihilate each other. To explain this, he introduced the single-fluid theory of electricity (positive and negative electricity). To explain the force, he introduced the qualitative concept of "electric atmosphere" as a kind of field surrounding electrified objects.

3.1.2   Coulomb's Force Law

The first mathematical theory for electric and magnetic forces was constructed by the German Franz Maria Ulrich Theodor Hoch Aepinus (1724–1802) in his treatise [1] of 1759. He considered both attractive and repulsive electric and magnetic forces being similar to Newton's action-at-a-distance gravity force with no medium effect. He assumed the force law in the form $1/r^n$ with an unspecified value for n. However, qualitative properties could be explained without knowing the exact value of n. Following Franklin's suggestion, Joseph Priestley (1733–1804) found in 1767 that there is no electric force inside a hollow conducting sphere and reasoned that the electric force most probably obeys Newton's $1/r^2$ law, which was known to imply vanishing of the gravity force inside a hollow sphere. More careful zero-force experiments were done in 1772 by Henry Cavendish (1731–1810). Direct although crude measurements of the electric or magnetic force were also performed by John Robison (1739–1805) in 1769 and by John Michell (1724–1793) in 1750. The results gave reason to assume the inverse square law for each force.

Convincing measurements were finally made starting in 1785 by Charles-Augustin Coulomb (1736–1806), in France, with a sensitive torsional balance developed by him. After this, the inverse square law for point charges in electrostatics and for single poles in magnetostatics was generally accepted. Coulomb had the view that all electric phenomena can be explained by assuming "two electric fluids and an inverse square force law for fluid particles, repulsion between similar and attraction between different fluid particles" [32]. He also showed that electricity is confined at the surface of a metal obstacle and the force at the surface is proportional to its surface density. He was able to determine surface densities for some geometries by measuring the force at the surface. Coulomb had a firm view that electricity and magnetism were two separate phenomena with no coupling between them, which probably damped the interest to find a connection between them. It was easy to share this view after making multiple futile experiments with charged bodies and the most powerful magnets.

Coulomb's law for the force between two static point charges (Figure 3.1), defined by the charge density functions [modern system international (SI) notation and vector symbols are used throughout]

$$\rho_1(\mathbf{r}) = Q_1 \delta(\mathbf{r} - \mathbf{r}_1), \quad \rho_2(\mathbf{r}) = Q_2 \delta(\mathbf{r} - \mathbf{r}_2), \tag{3.1}$$

can be expressed as:

$$\mathbf{F}_{12} = \mathbf{u}_{12} \frac{Q_1 Q_2}{4\pi\varepsilon_o r_{12}^2} = -\mathbf{F}_{21} \tag{3.2}$$

with

$$r_{12} = \sqrt{\mathbf{r}_{12} \cdot \mathbf{r}_{12}}, \quad \mathbf{r}_{12} = \mathbf{r}_1 - \mathbf{r}_2, \quad \mathbf{u}_{12} = \frac{\mathbf{r}_{12}}{r_{12}}. \tag{3.3}$$

The force is repellant for $Q_1 Q_2 > 0$ and attractive for $Q_1 Q_2 < 0$.

3.1.3   Galvanism and Electromagnetism

In 1780 Luigi Galvani (1737–1798) started a new branch in electricity, in Bologna, by making electric experiments with dissected frogs, which gave rise to an interest in electricity created by material contacts. After experimenting with different combinations of metals and other conducting materials, Alessandro Volta (1745–1827) introduced in 1800 a chemical source of electricity, the Volta pile.

William Hyde Wollaston (1766–1828) described, in 1801, the electricity obtained from Volta piles when compared to static electricity as "less intense but being produced in much larger quantity." This launched the somewhat vague concepts of "intensity" and "quantity" of electricity, which stayed in use for two

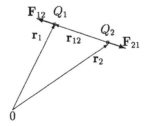

Figure 3.1. Coulomb's force between two static charges.

decades until the concepts "voltage" and "current" with more exact definitions were introduced after Ohm's studies. In particular, the electric quantity (current) was not a clear concept because of the competing two-fluid and single-fluid theories. Electric current in a conductor could be considered either as a one-way movement of a single fluid or a symmetric movement of two fluids in opposite directions. Two-way movement was often favored because it explained the behavior of the electrolytic solution between the electrodes of a Volta pile. Before the electron was discovered in 1897, it was not, however, possible to decide between the two models [32].

The connection between electricity and magnetism was first reliably shown in 1820 by Hans-Christian Öersted (1777–1851) in Copenhagen. In his interpretation, a closed circuit containing a Volta pile caused a "conflict of electricity" kind of magnetic-field concept, which could deflect the magnetic needle. This discovery offered a possibility of detecting the electric current in a conducting wire by a magnetic needle, which eventually developed to the galvanometer. Öersted was first to describe the strange nature of the magnetic force around the current conductor (right-hand rule). He also showed that a magnet exerts a force on a current loop.

3.1.4   Electromagnetic Induction

After Öersted's discovery became known in 1820, Ampère showed that a solenoid of electric current behaves like a bar magnet. This gave him the idea to interpret all magnetic phenomena in terms of electric currents. To find out whether these currents pre-existed in unmagnetized iron or were induced when the iron was magnetized, he made an experiment to see if an electric current can be induced in a conductor by another electric current. In the experiment he tried to induce a current in a copper ring by a coplanar and concentric current loop. He expected to see a stable deflection of the ring caused by a nearby magnet. The test was made twice, in 1821 and 1822; in the latter experiment Ampère noticed a slight transient motion. Since it was against his expectations, he did not study or report the experiment and lost a major discovery [6].

The principle of energy conservation was given a concrete mathematical form in the 1840s. Before that there was a general idea that physical forces could be transformed from one form to another. For example, it was known that electricity could generate heat and light through incandescence. Also, heat could generate electricity through Seebeck's thermoelectric element. This general idea had inspired Oersted to find out whether electricity could generate magnetism. The problem remained to show that magnetism could also generate electricity. It was known that an iron bar inside a solenoid became a magnet when the solenoid was connected to a battery. The converse was tried many times during the 1820s: a magnet was put inside a solenoid and a galvanometer was watched for flowing current. Because the experiment did not show the expected effect, it was slowly believed that there was no such effect. In 1831, after a handful of unsuccessful attempts over many years, Michael Faraday (1791–1867) finally succeeded in observing a transient electromagnetic induction effect when current was switched on or off. Almost simultaneously, Joseph Henry (1797–1878) made a similar observation but was late to publish. After these basic experiments there was a need for a unifying theory which could mathematically explain all the known electromagnetic phenomena. Construction of such a theory took somewhat different paths on the European continent and on the British Isles.

3.2   CONTINENTAL ELECTROMAGNETICS

3.2.1   Electrostatics and Magnetostatics

Coulomb's law had made electricity and magnetism part of the exact sciences. It inspired research based on the Newtonian action-at-a-distance concept and use of mathematics. Since this was mainly done in the Continent in contrast to the field concept and physical reasoning mainly done in Great Britain, it will be referred to as Continental Electromagnetics.

In 1812 Siméon-Denis Poisson (1781–1840) read a memoir to the French Academy on solving boundary problems in electrostatics. Considering electric matter as a fluid he determined its thickness on a conducting ellipsoid and on two conducting spheres by requiring vanishing of the electric force inside conducting objects. The results confirmed Coulomb's experimental data. Lagrange had in 1777 introduced the concept of gravitational potential V and Laplace had in 1782 proved that outside the gravitating matter it satisfies $\nabla^2 V=0$. Poisson showed that inside electric matter the electrostatic potential satisfies the inhomogeneous equation:

$$\nabla^2 V = -\rho / \varepsilon_o , \tag{3.4}$$

and V is constant on a conducting object. The potential function corresponding to the Coulomb force (3.2) is:

$$V = \frac{Q}{4\pi\varepsilon_o r}, \quad \mathbf{F}_{12} = -Q_1 \nabla V_2 = -\nabla \frac{Q_1 Q_2}{4\pi\varepsilon_o r_{12}} . \tag{3.5}$$

In 1824 Poisson published a corresponding mathematical theory for magnetostatics. He introduced the concept of magnetic moment for magnetic sources, and defined polarization of magnetic material and equivalent magnetic surface sources for magnetized objects. Poisson's approach was generalized by George Green (1793–1841), in an article of 1828, by introducing what is now known as the Green function and Green's theorem. However, his privately published article went practically unnoticed during Green's lifetime.

3.2.2   Ampère's Force Law

In 1820 Öersted had shown that a force was exerted on a magnet by a current and the converse was also valid. In addition, of course, a magnet exerts a force on another magnet. These do not, however, imply that a current would exert a force on another current (to see this, replace "current" by "unmagnetized iron" in the preceding sentences). This effect was shown after Öersted's experiment in 1820 by André-Marie Ampère (1775–1836) who demonstrated that there exists a force between two parallel current conductors, attractive for currents in the same direction and repellent for currents in opposite directions. Jean-Baptiste Biot (1774–1862) and Felix Savart (1791–1841) showed by measurements that the magnetic force exerted by a line current obeys $1/r$ law when r is the distance from the current line.

In 1825, Ampère formulated the basic force law between two differential current elements, $I_1 d\mathbf{c}_1$ and $I_2 d\mathbf{c}_2$, denoted by the current density functions as:

$$d\mathbf{J}_1(\mathbf{r}) = I_1 d\mathbf{c}_1 \delta(\mathbf{r}-\mathbf{r}_1), \quad d\mathbf{J}_2(\mathbf{r}) = I_2 d\mathbf{c}_2 \delta(\mathbf{r}-\mathbf{r}_2), \qquad (3.6)$$

in the form,

$$d^2\mathbf{F}_{12} = -\mathbf{u}_{12}\frac{\mu_o I_1 I_2}{4\pi r_{12}^2}\left[2d\mathbf{c}_1 \cdot d\mathbf{c}_2 - \frac{3}{r_{12}^2}(\mathbf{r}_{12}\cdot d\mathbf{c}_1)(\mathbf{r}_{12}\cdot d\mathbf{c}_2)\right] \qquad (3.7)$$

$$= \mathbf{u}_{12}\frac{\mu_o I_1 I_2}{4\pi r_{12}^2}\left[d\mathbf{c}_1 \cdot d\mathbf{c}_2 - \frac{3}{r_{12}^2}(\mathbf{r}_{12}\times d\mathbf{c}_1)\cdot(\mathbf{r}_{12}\times d\mathbf{c}_2)\right]. \qquad (3.8)$$

These expressions give the force exerted by the current element $d\mathbf{c}_2$ on the current element $d\mathbf{c}_1$ (Figure 3.2). The force is in the direction of the radial line joining the two current elements and, moreover, satisfies Newton's third law as $d^2\mathbf{F}_{12} = -d^2\mathbf{F}_{21}$. Ampère's force appears attractive for two parallel current elements orthogonal to the line joining them as seen from (3.7), while for axially located parallel current elements it is repulsive, which is more easily seen from (3.8). At a certain angle to the joining line there is no force at all between the parallel elements (Figure 3.3). This was criticized by Hermann Grassmann (1809–1877) as being unphysical. In 1845 he replaced (3.7) by the simpler-looking law:

$$d^2\mathbf{F}_{12} = \frac{\mu_o I_1 I_2}{4\pi r_{12}^3} d\mathbf{c}_1 \times (d\mathbf{c}_2 \times \mathbf{r}_{12}), \qquad (3.9)$$

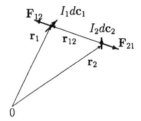

Figure 3.2. Ampère's force law for two current elements.

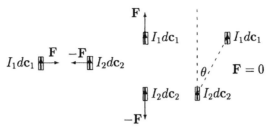

Figure 3.3. Ampère's law for two parallel current elements predicts both attractive and repulsive forces. At a certain angle the force disappears.

which is today given in most textbooks, e.g., [5,17,20], and often called Lorentz's or Biot-Savart's law. Equation (3.9) does not, however, satisfy Newton's third law for current elements. Equations (3.7) and (3.9) are generally considered equivalent because they give the same total force when integrated over two closed current loops [29]. However, the distribution of the force along the loops is different. The main difference between (3.7) and (3.9) comes when considering the force between elements on the same straight line, in which case (3.9) predicts no force at all in contrast to (3.7). Because high currents in a straight wire have been observed to cause rupture, breaking, and even explosion of the wire, it has been more recently claimed that (3.9) must be basically incorrect [3,10]. In his celebrated memoir of 1827 [2], Ampère arrived at (3.7) through four null-field experiments. In these (Figure 3.4) a force was compensated by another force whence the equality of the two forces gave

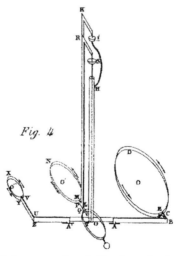

Figure 3.4. Ampère's null-force experiment. The forces by the loops on each side of the movable loop cancel one another when the sizes and distances of the two loops are in the same ratio and the current in all loops is the same. This leads to the $1/r^2$ force law.

information on the nature of the force. It has been claimed that the experiments were probably designed only to support the theory and had no original exploration value [6].

3.2.3   Ohm's Law

After the introduction of the Volta pile, it took quite some time before one started to understand what happened in the circuit connected to the poles of a chemical battery. One of the reasons was that the early batteries were highly unstable. Their internal resistance started to grow the moment the current started to flow. A second disturbing factor was the uncertainty about single-fluid and two-fluid theories of electricity. In the two-fluid case, did the two electric fluids flow from end to end or did they annihilate each other in the middle of the conductor?

Georg Simon Ohm (1787–1854) was not the first one to study the effect of the circuit, but he was the most successful. Replacing the chemical battery by the more stable thermal element, invented in 1822 by Thomas Seebeck (1770–1831), he performed experiments by observing the current in conductors of varying measures and materials through a magnetic needle. However, his book [25] of 1827 was not based on these experiments but, rather, written following the book of Joseph Fourier (1768–1830) on the flow of heat in materials. Fourier's book of 1822 [8] was a novel one in its mathematical handling of a physical problem through partial differential equations and boundary conditions. In contrast to Newton's action at a distance, it considered heat as an indestructible fluid and its propagation as a contiguous action from point to point in the medium. Ohm defined DC concepts corresponding to heat flow, temperature, and thermal conductivity as current, electroscopic force (potential), and electric conductance of the medium, respectively. After this, the vague terms "intensity" and "quantity" of electricity soon disappeared. He also realized the importance of the internal resistance of the battery which explained why increasing the number of turns of wire did not monotonously increase the magnetic force inside a coil. In electrostatics the conductivity of the material did not play any role, because the charge always stays on the surface of a conducting body. Ohm showed that the DC current was flowing through the whole cross section whence the nature of the material had a dominating effect on its magnitude. These points serve as a partial explanation as to why it took twenty-seven years after the invention of the Volta pile for the simple formula $V = IR$ to be discovered and why it took about a decade after that before it was widely accepted.

3.2.4   Neumann's Vector Potential

Faraday's 1831 induction experiment showed that there was an electromotive force (EMF) induced in a conducting loop; proportional to the rate of change of the magnetic force at the loop, but its mathematical formulation took some time.

In 1834 a simple rule was given by the Baltic German Emil Lenz (1804–1865). He stated that the direction of the EMF is so oriented that the current created by the EMF always opposes the change. If the change is by movement, there arises a force which tends to slow down the movement. However, this rule assumes positive resistance in the loop. If the resistance is negative, the rule works conversely and the induction would actually accelerate the change.

In 1845 Franz Neumann (1798–1895), in Königsberg, started from Lenz's rule and was first to formulate Faraday's induction law mathematically. He applied Ampère's law (3.7) to variable currents and showed that the interaction energy of two current loops can be expressed in the simple form:

$$W_{12} = \frac{\mu_o I_1 I_2}{4\pi} \oint_{c_1} \oint_{c_2} \frac{d\mathbf{c}_1 \cdot d\mathbf{c}_2}{r_{12}}. \tag{3.10}$$

This gave him a method for computing the mutual inductance L_{12} of two given loops C_1 and C_2 as

$$L_{12} = \frac{W_{12}}{I_1 I_2} = \frac{\mu_o}{4\pi} \oint_{c_1} \oint_{c_2} \frac{d\mathbf{c}_1 \cdot d\mathbf{c}_2}{|\mathbf{r}_1 - \mathbf{r}_2|}, \tag{3.11}$$

and the self-inductance of a single loop as

$$L_{11} = \frac{\mu_o}{4\pi} \oint_{c_1} \oint_{c_2} \frac{d\mathbf{c}_1 \cdot d\mathbf{c}_1'}{|\mathbf{r}_1 - \mathbf{r}_1'|}. \tag{3.12}$$

From the integral (3.10) he defined the interaction energy in terms of a new quantity, the vector potential function $\mathbf{A}(\mathbf{r})$, as

$$W_{12} = \oint_{c_1} \mathbf{A}_2(\mathbf{r}) \cdot I_1 d\mathbf{c}_1 = \oint_{c_2} \mathbf{A}_1(\mathbf{r}) \cdot I_2 d\mathbf{c}_2, \tag{3.13}$$

$$\mathbf{A}_i(\mathbf{r}) = \frac{\mu_o}{4\pi} \oint_{c_i} \frac{I_i d\mathbf{c}_i}{|\mathbf{r} - \mathbf{r}_i|}, \quad i = 1, 2. \tag{3.14}$$

Faraday's induction law for the EMF in loop 1 when the current I_2 is changed in loop 2 could now be expressed mathematically as:

$$\text{EMF}_1 = -L_{12} \partial_t I_2(t) = -\partial_t \oint_{c_1} \mathbf{A}_2(\mathbf{r}) \cdot d\mathbf{c}_1. \tag{3.15}$$

Because vector notations were introduced only in the 1880s, Neumann's original formulation was given in terms of lengthy component expressions.

3.2.5 Weber's Force Law

In 1846 Wilhelm Weber (1804–1891), then professor of physics in the Leipzig University, was able to combine Coulomb's and Ampère's laws in a single unified expression. To do this, he had to interpret the electric current elements

$I_1 d\mathbf{c}_1$, $I_2 d\mathbf{c}_2$ appearing in Ampère's law (3.7) in terms of point charges Q_1, Q_2 moving with the respective velocities $\mathbf{v}_1, \mathbf{v}_2$, as seen in Figure 3.5. Now a charge-density function of the form:

$$\rho_1(\mathbf{r},t) = Q_1 \delta(\mathbf{r} - \mathbf{r}_1 - \mathbf{v}_1 t) \tag{3.16}$$

corresponds to the current-density function;

$$\mathbf{J}_1(\mathbf{r},t) = \mathbf{v}_1 Q_1 \delta(\mathbf{r} - \mathbf{r}_1 - \mathbf{v}_1 t), \tag{3.17}$$

because they satisfy the continuity condition,

$$\nabla \cdot \mathbf{J}_1(\mathbf{r},t) = -\partial_t \rho_1(\mathbf{r},t). \tag{3.18}$$

Thus, replacing the differential current elements by finite ones as $I_1 d\mathbf{c}_1 = Q_1 \mathbf{v}_1$ and $I_2 d\mathbf{c}_2 = Q_2 \mathbf{v}_2$, Ampère's law (3.7) could be written as

$$\mathbf{F}_{12} = -\mathbf{u}_{12} \frac{Q_1 Q_2}{4\pi\varepsilon_o c^2 r_{12}^2} \left[2\mathbf{v}_1 \cdot \mathbf{v}_2 - \frac{3}{r_{12}^2}(\mathbf{r}_{12} \cdot \mathbf{v}_1)(\mathbf{r}_{12} \cdot \mathbf{v}_2) \right]. \tag{3.19}$$

Here we have introduced the velocity of light as $c^2 = 1/(\mu_0 \varepsilon_0)$ in the modern notation. Weber wished to express Ampère's force in terms of the distance between the charges:

$$r_{12}(t) = |\mathbf{r}_{12}| = \sqrt{\mathbf{r}_{12} \cdot \mathbf{r}_{12}}, \tag{3.20}$$

$$\mathbf{r}_{12} = (\mathbf{r}_1 - \mathbf{r}_2) + (\mathbf{v}_1 - \mathbf{v}_2)t, \tag{3.21}$$

and its time derivative (3.22), as well as its second derivative (3.23).

$$\partial_t r_{12} = \frac{1}{r_{12}} \mathbf{r}_{12} \cdot (\mathbf{v}_1 - \mathbf{v}_2), \tag{3.22}$$

$$\partial_t^2 r_{12} = \frac{2}{r_{12}} \mathbf{v}_1 \cdot \mathbf{v}_2 + \frac{2}{r_{12}^3}(\mathbf{r}_{12} \cdot \mathbf{v}_1)(\mathbf{r}_{12} \cdot \mathbf{v}_2)$$
$$+ \frac{1}{r_{12}}(\mathbf{v}_1 \cdot \mathbf{v}_1 + \mathbf{v}_2 \cdot \mathbf{v}_2) - \frac{1}{r_{12}^3}\left[(\mathbf{r}_{12} \cdot \mathbf{v}_1)^2 + (\mathbf{r}_{12} \cdot \mathbf{v}_2)^2 \right]. \tag{3.23}$$

The square of the former reads:

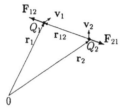

Figure 3.5. Weber's force law for two moving charges depends on the distance $\mathbf{r}_{12}$ and its time derivatives.

$$\left(\partial_t r_{12}\right)^2 = \frac{1}{r_{12}^2}\left[\left(\mathbf{r}_{12}\cdot\mathbf{v}_1\right)^2+\left(\mathbf{r}_{12}\cdot\mathbf{v}_2\right)^2-2\left(\mathbf{r}_{12}\cdot\mathbf{v}_1\right)\left(\mathbf{r}_{12}\cdot\mathbf{v}_2\right)\right]. \quad (3.24)$$

However, the two moving charges also affect one another through Coulomb's force. To obtain pure Amperian force, Weber used Fechner's hypothesis [16].

Weber's colleague Gustav Theodor Fechner (1801–1887) had studied forces between current conductors by describing the current as a symmetric flow of positive and negative charges of equal magnitude and opposite velocity [27]. Because at each point on the conductor the total charge was zero, there was no Coulomb force. (Later the two-fluid model was avoided by assuming that the current consists of point charges of single polarity, infinitesimal magnitude, and infinite density. In the limit, the moving charges produce an electric current with no Coulomb's force [17])

3.2.5.1 The Force Law. Considering a charge $Q_2' = -Q_2$ propagating at $\mathbf{r}_2$ with velocity $\mathbf{v}_2' = -\mathbf{v}_2$, its interaction with the previously defined charge Q_1 involves the distance function:

$$r_{12}' = |\mathbf{r}_{12}'|, \quad \mathbf{r}_{12}' = \left(\mathbf{r}_1-\mathbf{r}_2\right)+\left(\mathbf{v}_1+\mathbf{v}_2\right)t \ . \quad (3.25)$$

The differences of the previous functions at $t=0$, $\mathbf{r}_{12}' = \mathbf{r}_{12}$, now become

$$\partial_t^2 r_{12} - \partial_t^2 r_{12}' = -\frac{4}{r_{12}}\mathbf{v}_1\cdot\mathbf{v}_2 + \frac{4}{r_{12}^3}\left(\mathbf{r}_{12}\cdot\mathbf{v}_1\right)\left(\mathbf{r}_{12}\cdot\mathbf{v}_2\right), \quad (3.26)$$

$$\left(\partial_t r_{12}\right)^2 - \left(\partial_t r_{12}'\right)^2 = -\frac{4}{r_{12}^2}\left(\mathbf{r}_{12}\cdot\mathbf{v}_1\right)\left(\mathbf{r}_{12}\cdot\mathbf{v}_2\right). \quad (3.27)$$

Expressing the bracketed term in (3.19) in terms of these functions, Ampère's law can be written in the form:

$$\begin{aligned}
\mathbf{F}_{12} = &-\frac{1}{2}\mathbf{u}_{12}\frac{Q_1 Q_2}{4\pi\varepsilon_o c^2 r_{12}^2}\left(\frac{1}{2}\left(\partial_t r_{12}\right)^2 - r_{12}\partial_t^2 r_{12}\right) \\
&-\frac{1}{2}\mathbf{u}_{12}\frac{Q_1 Q_2'}{4\pi\varepsilon_o c^2 r_{12}^2}\left(\frac{1}{2}\left(\partial_t r_{12}'\right)^2 - r_{12}\partial_t^2 r_{12}'\right).
\end{aligned} \quad (3.28)$$

Now (3.28) can be interpreted as the force on the charge Q_1 at $\mathbf{r}_1$ moving with the velocity $\mathbf{v}_1$ as exerted by two charges at $\mathbf{r}_2$: First, by $Q_2/2$ moving with the velocity $\mathbf{v}_2$ and, second, by $Q_2'/2 = -Q_2/2$ moving with the velocity $\mathbf{v}_2' = -\mathbf{v}_2$. There is no Coulomb force because the total charge $Q_2 + Q_2'$ is zero. On basis of this, Weber interpreted the first term of (3.28) multiplied by a factor of 2 as the force exerted by Q_2 on Q_1 due to the motion of the two charges. Adding the Coulomb term (3.2), the Weber force law between two point charges in relative motion finally became [3,6,16,21]

$$\mathbf{F}_{12} = \mathbf{u}_{12} \frac{Q_1 Q_2}{4\pi\varepsilon_o r_{12}^2} \left(1 - \frac{1}{2c^2} (\partial_t r_{12})^2 + \frac{r_{12}}{c^2} \partial_t^2 r_{12} \right). \tag{3.29}$$

In its original form, (3.29) contained an unknown scalar C^2 which in modern notation equals $2c^2$. Weber realized that C is a quantity representing some velocity for which the Coulomb and Ampère forces on the moving charge are of the same magnitude. When in 1855 Weber and Kohlrausch made measurements, they found that C had the same order of magnitude as the velocity of light known from previous measurements.

3.2.5.2. Potential. Weber's force law (3.29) was adopted in Germany with enthusiasm as the unified theory explaining the electromagnetic phenomena. It satisfied Newton's third law for action and reaction, $\mathbf{F}_{12} = -\mathbf{F}_{21}$. Because it depended only on the distance of the two charges and its first and second time derivatives, it was independent of any reference frame, i.e., movement of the observers. (3.29) was criticized by Helmholtz for not satisfying the law of energy conservation. However, Weber could show that the force can be derived from the potential function

$$V_{12} = \frac{Q_1 Q_2}{4\pi\varepsilon_o r_{12}} \left(1 - \frac{1}{2c^2} (\partial_t r_{12})^2 \right), \tag{3.30}$$

which implies energy conservation. In fact, differentiating the potential gives

$$\partial_{r_{12}} V_{12} = -\frac{Q_1 Q_2}{4\pi\varepsilon_o r_{12}^2} \left(1 - \frac{1}{2c^2} (\partial_t r_{12})^2 \right) - \frac{Q_1 Q_2}{4\pi\varepsilon_o r_{12}} \times \frac{1}{c^2} (\partial_t r_{12}) \partial_{r_{12}} (\partial_t r_{12})$$

$$= -\frac{Q_1 Q_2}{4\pi\varepsilon_o r_{12}^2} \left(1 - \frac{1}{2c^2} (\partial_t r_{12})^2 + \frac{r_{12}}{c^2} \partial_t^2 r_{12} \right), \tag{3.31}$$

when inserting $\partial_{r_{12}} (\partial_t r_{12}) = 2\partial_t^2 r_{12} / \partial_t r_{12}$. Thus Weber's force (3.29) can be expressed as:

$$\mathbf{F}_{12} = -\mathbf{u}_{12} \partial_{r_{12}} V_{12} = -(\nabla r_{12}) \partial_{r_{12}} V_{12} = -\nabla V_{12}. \tag{3.32}$$

3.2.5.3 Neumann's Inductance. It turned out that Weber's force law could also reproduce Neumann's inductance formula (3.11) and Faraday's induction law (3.15), which gave more confidence to the theory.

Starting from the interaction potential (3.30) of two pairs of charges, Q_1, $Q_1' = -Q_1$ and $Q_2, Q_2' = -Q_2$ moving with the respective velocities $\mathbf{v}_1$, $\mathbf{v}_1' = -\mathbf{v}_1$ and $\mathbf{v}_2$, $\mathbf{v}_2' = -\mathbf{v}_2$ and using (3.24) for all four interaction terms $12, 12', 1'2, 1'2'$, the total potential can be written after cancellation of terms as:

$$V_{12} = \frac{Q_1 Q_2}{4\pi\varepsilon_o r_{12}} \frac{4(\mathbf{r}_{12} \cdot \mathbf{v}_1)(\mathbf{r}_{12} \cdot \mathbf{v}_2)}{c^2 r_{12}^2} = \frac{\mu_o (\mathbf{r}_{12} \cdot Q_1 \mathbf{v}_1)(\mathbf{r}_{12} \cdot Q_2 \mathbf{v}_2)}{\pi r_{12}^3}. \tag{3.33}$$

In terms of differential current elements

$$I_1 d\mathbf{c}_1 = Q_1 \mathbf{v}_1 + Q_1' \mathbf{v}_1' = 2Q_1 \mathbf{v}_1 \;, \quad I_2 d\mathbf{c}_2 = Q_2 \mathbf{v}_2 + Q_2' \mathbf{v}_2' = 2Q_2 \mathbf{v}_2 \;, \tag{3.34}$$

the expression becomes

$$dV_{12} = \frac{\mu_o I_1 I_2}{4\pi r_{12}^3}(\mathbf{r}_{12} \cdot d\mathbf{c}_1)(\mathbf{r}_{12} \cdot d\mathbf{c}_2) = \frac{\mu_o I_1 I_2}{4\pi} d\mathbf{c}_1 \cdot \frac{\mathbf{r}_{12}\mathbf{r}_{12}}{r_{12}^3} \cdot d\mathbf{c}_2 \;. \tag{3.35}$$

Here we can use the dyadic identity [19]

$$\frac{\mathbf{r}_{12}\mathbf{r}_{12}}{r_{12}^3} = \frac{\overline{\overline{I}}}{r_{12}} - \nabla\nabla r_{12}, \tag{3.36}$$

which is valid for ∇ operating on either $\mathbf{r}_1$ or $\mathbf{r}_2$. Integrating over two current loops C_1, C_2 and noting that a closed-path integral over a gradient vanishes, the interaction potential energy becomes

$$W_{12} = \frac{\mu_o I_1 I_2}{4\pi} \oint_{c_1} \oint_{c_2} \frac{d\mathbf{c}_1 \cdot d\mathbf{c}_2}{r_{12}} = L_{12} I_1 I_2 \;. \tag{3.37}$$

This gives directly Neumann's formula (3.11) for the mutual inductance L_{12} of two loops.

3.2.5.4 Faraday's Law.

Weber's law was derived to satisfy Coulomb's and Ampère's laws. In addition, Weber could show that it also included Faraday's induction law as formulated by Neumann. Actually, Faraday's law follows already from Ampère's law and conservation of energy, as was shown originally by Helmholtz in 1847 [13] through the following procedure [21].

Considering the loop 1 with resistance R_1 and generator with EMF U_1, the energy supplied by the generator during a differential time period dt is

$$dW_1 = U_1 I_1 dt \tag{3.38}$$

and the energy dissipated during the same period in Ohmic heat is

$$dW_1' = R_1 I_1^2 dt \;. \tag{3.39}$$

Now if I_2 is constant, the interaction energy W_{12} is constant, $dW_{12} = 0$ and, from energy conservation, we must have $dW_1 = dW_1'$, which defines the current as

$$I_1 = \frac{U_1}{R_1} \;. \tag{3.40}$$

However, if $I_2(t)$ changes with time, the interaction energy changes during the differential time period by

$$dW_{12} = L_{12}I_1\partial_t I_2(t)dt \ . \tag{3.41}$$

Expressing the energy conservation as

$$dW_1 = dW_1' + dW_{12} \tag{3.42}$$

leads to the condition

$$U_1 = R_1 I_1 + L_{12}\partial_t I_2(t) \ , \tag{3.43}$$

from which the current can be solved as

$$I_1 = \frac{1}{R_1}\left(U_1 - L_{12}\partial_t I_2(t)\right) . \tag{3.44}$$

This result can be interpreted so that the time-dependent current I_2 induces in the loop 1 an EMF defined by

$$\text{EMF}_1 = -L_{12}\partial_2 I_2(t) \ . \tag{3.45}$$

The minus sign here is due to adding the term (3.41) on the right-hand side of $dW_1 = dW_1'$ as in (3.42) according to the rule of Lenz. The final result coincides with (3.15) given by Neumann.

3.2.6   Electromagnetic Waves

Because they explained all known experiments, Neumann's and Weber's Newtonian theories dominated in continental electromagnetics until the late 1880s when the experiments by Heinrich Hertz (1857–1894) finally replaced them by Maxwellian electrodynamics. Before that there were a few attempts to apply the continental theories to light which was known to propagate with finite velocity. In 1858 Bernhard Riemann (1826–1866) assumed that the force between charges propagates at the velocity of light, which corresponds to a retarded scalar potential function. However, he withdrew his article from publication. A similar theory was published by Ludwig Lorenz (1829–1891) of Denmark in 1867 where he suggested that both scalar and vector potentials due to charge and current distributions can be expressed as the retarded integrals

$$\phi(\mathbf{r},t) = \int \frac{\rho\left(\mathbf{r}',t - |\mathbf{r}-\mathbf{r}'|/c\right)}{4\pi\varepsilon_o |\mathbf{r}-\mathbf{r}'|} dV'dt' \tag{3.46}$$

$$\mathbf{A}(\mathbf{r},t) = \int \frac{\mu_o \mathbf{J}\left(\mathbf{r}',t - |\mathbf{r}-\mathbf{r}'|/c\right)}{4\pi |\mathbf{r}-\mathbf{r}'|} dV'dt' \ . \tag{3.47}$$

However, Maxwell's theory with similar results had already been introduced two years earlier.

3.3   BRITISH ELECTROMAGNETICS

3.3.1   Faraday's Field Concept

The British branch of electromagnetics started from Faraday's induction experiment of 1831. He later interpreted it in terms of magnetic field lines as:

> *Whenever the number of lines of force passing through a*
> *closed circuit is altered, there is an electromotive force passing*
> *through the circuit, whose direction is such that it would itself*
> *produce lines of force passing through the circuit in opposite*
> *direction.* [7]

The field lines became visible with iron filings and Faraday considered them to be like rays of light (Figure 3.6). He gave them certain properties which showed themselves in the magnetic force. On one hand the force lines were like rubber strings with the tendency to shorten. On the other hand they created space around them by expelling each other laterally (Figure 3.7). These properties could describe the attractive and repulsive properties of the magnetic force. Eventually he considered tubes made by the force lines with cross sections inversely proportional to the field intensity. In this way Faraday developed a manageable physical insight even if he could not express these concepts mathematically. The idea of magnetic field lines found application in the design of electromagnetic generators which started the electrification of the globe towards the end of the 19th Century.

In 1837 Faraday studied the inductive capacity (permittivity) of different insulators and interpreted their reaction to electricity in terms of polarization of media. Similarity to magnetic polarization in iron suggested to him to define lines of electric force with corresponding properties. He described the polarization due to electric tension, where neutral molecules became electric dipoles. Electric field lines extended between electric charges of opposite polarity. In 1845 Faraday found that the plane of polarization of light was rotated

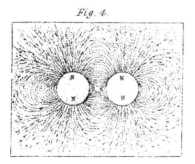
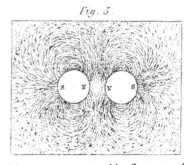

Figure 3.6. Faraday's concept of magnetic lines of force were suggested by figures made by iron filings in the vicinity of magnetized bodies. Because iron particles act like small magnets, there appears to be an attractive force along the magnetic line.

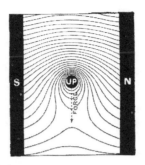

Figure 3.7. Faraday viewed the magnetic lines like rubber bands connectig opposite magnetic charges. The lines make tubes which expel each other laterally. Here the current flowing in a wire between the magnetic poles experiences a force due to the field lines.

when propagating along the magnetic field lines in a certain medium. The effect has since been called the Faraday rotation. This gave him a reason to speculate in 1846 about the electromagnetic nature of light, as some kind of oscillatory wave propagating along magnetic field lines.

3.3.2   Thomson

William Thomson (Kelvin) (1824–1907) had read Fourier's book [8] in 1840 just before entering the University of Cambridge at the age of 16. Inspired by it he published a paper "On the uniform motion of heat in homogeneous solid bodies and its connection with the mathematical theory of electricity", which used the mathematical analogy between stationary flow of heat and the electrostatic field. He showed that the lines of heat flow coincide with Faraday's electrostatic lines of force and derived the relation between the heat capacity and electric capacitance of conducting structures [26]. Thus, instead of considering the Coulomb force acting over a distance, it could be interpreted as flowing through the medium. In this case properties of the medium had an effect on the force. This started the mathematical representation of Faraday's field lines and tubes. In 1846 he also studied analogies between the linear displacement in an elastic solid and the electrostatic or magnetostatic force [30].

   To explain the Faraday rotation, in 1856 Thomson constructed a dynamical theory of magnetism based on assumed internal motion. He noted that the handedness of the Faraday rotation was reversed when the direction of propagation along the magnetic field was reversed while the handedness of the polarization rotation in chiral media like sugar solution was independent of direction of propagation. William Rankine (1820–1872) had suggested that heat constitutes of rotary motions associated with individual molecules, "molecular vortices". Thomson used this model when arguing that Faraday rotation could be explained by representing the magnetic line of force by a vortex in some intervening medium. Thus, magnetic force is not an interaction of static magnetic poles or stationary microscopic electric currents but a force between vortical motions [31]. This idea, sketched only briefly in his paper in 1856, would be consequently elaborated by Maxwell.

When analyzing time-dependent signals on a telegraphic cable in 1854, Thomson went too far with the heat analogy by adopting Fourier's diffusion equation. This corresponds to taking only the capacitance and resistance of the cable into account and it was later corrected by Heaviside. In 1855 Thomson investigated mathematically the discharge of a Leyden jar through a conducting wire and predicted that the discharge is a decaying oscillatory current. This fact, observed already using crude methods by Henry in 1842, was experimentally verified in 1857 by Feddersen.

3.3.3 Maxwell

James Clerk Maxwell (1831–1879) continued Thomson's application of analogies. Instead of heat, which was no longer considered as a substance, he based his analogy on the motion of an incompressible and massless fluid. Electric and magnetic equations were interpreted as relations between the velocity of the fluid in a porous medium and the pressure. The medium could be inhomogeneous and anisotropic. In an 1856 paper, namely the second part of his 1855 paper, *On Faraday's lines of force* [22], Maxwell distinguished between the "intensity" fields **E**, **H** and **A** and the "quantity" fields **D**, **B** and **J**. Between them he described electromagnetic "laws" corresponding to the following set of equations in modern vector language [14,20,23].

$$\nabla \times \mathbf{E} = -\partial_t \mathbf{B} \; , \tag{3.48}$$

$$\nabla \times \mathbf{H} = \mathbf{J} \; , \tag{3.49}$$

$$\nabla \cdot \mathbf{E} = \rho / \varepsilon_o \; , \tag{3.50}$$

$$\nabla \cdot \mathbf{B} = 0 \; , \tag{3.51}$$

$$\mathbf{B} = \nabla \times \mathbf{A} \; , \quad \mathbf{E} = -\partial_t \mathbf{A} \; . \tag{3.52}$$

3.3.3.1. Electromagnetic Clockwork. Each of the above equations could be interpreted by a suitable fluid-flow analogy. However, Maxwell wished to have a single model which would explain all known interactions between the electric fields, magnetic fields and electric currents. In a set of three papers "On physical lines of force" during 1861–1862 [23,27,28] Maxwell constructed such a physical model. For this Maxwell extended Thomson's idea of representing Faraday's magnetic field lines in terms of molecular vortices. They were rotating cells of fluid forming rotating tubes. The fluid was no longer assumed massless but obeyed the hydrodynamic laws of a real physical fluid. Because Faraday's magnetic field lines start and end in magnetic sources which can be replaced by Amperian current loops, rotary motion for magnetism appeared natural (Figure 3.8).

In Maxwell's representation the magnetic intensity force **H** corresponds to the angular velocity of the fluid and μ to the mass density of the fluid whence

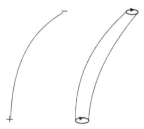

Figure 3.8. Faraday's magnetic line of force connecting positive and negative magnetic charges can be interpreted in terms of a rotating tube connecting equivalent current loops.

the quantity field $\mathbf{B} = \mu\mathbf{H}$ is related to the angular momentum. The energy density of the rotating fluid could now be written as $\mathbf{H}\cdot\mathbf{B}/2$. Rotating fluid has the tendency to shorten and broaden which can be verified by rotating a balloon filled with water. This corresponds to the properties of Faraday's force lines which tend to shorten and expel one another laterally.

To avoid friction between neighboring vortex tubes rotating in the same direction, Maxwell separated the tubes by ball bearings rotating in the opposite direction without slipping. To Maxwell these balls appeared as electric charges (Figure 3.9). In an insulating medium the balls could only rotate in fixed positions. In conductors, however, the balls could also move between the vortex cylinders. The driving force (pressure gradient) corresponded to the electric field **E**. Thus, a moving line of balls representing an electric line current creates a co-axial system of torus-shaped vortices around it whose angular velocity follows the correct 1/r law from the current line. A moving planar sheet of charges creates a plane current on each side of vortices with opposite angular velocities. In the more general case the movement of balls creates a difference in the

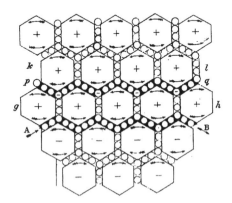

Figure 3.9. Maxwell's drawing depicting his mechanical model of electromagnetic fields. Direction of rotation of the vortex tubes with hexagonal cross section is shown by arrows. The line of balls A – B moving to the right represents an electric current. Because of the rigidity of the structure, the motion of all vortices is synchronous.

angular velocity of neighboring vortices which corresponds to the discontinuity caused by the current to the magnetic field. In this form Maxwell's system resembles a clockwork. A movement in any of its parts is simultaneously felt in all other parts without delay resulting in an electric current. Because of the rigidity of the structure, the motion of all vortices is synchronous.

Macroscopic phenomena could be explained by considering regions containing a large number of microscopic elements. In this form, Maxwell's model explained Ampère's law (3.49) and Faraday's law (3.48). The latter can be seen from the following: when trying to change the velocity of a vortex, because of the inertia of the surrounding vortices, there arises a force which opposes the change through the ball bearings. In this interpretation, Faraday's law corresponds to Newton's inertial law $\mathbf{F} = m\partial_t\mathbf{v}$ in the rotating system.

3.3.3.2 Electromagnetic Jelly. When Maxwell had explained mechanically Ampère's and Faraday's laws, he paused for about a year. It has been speculated that he had the intention to stop here, which would have been a pity because the greatest part was still to be discovered. The model was not yet perfect, because it did not explain electrostatics. With no magnetic field, the electric field in a non-conducting medium should create polarization in the medium according to Faraday. Maxwell solved this by assuming that the fluid cells are elastic. When the electric field exerts a force to the fixed charge balls, they tend to distort the fluid cells and the charges are displaced from their stationary locations by a distance proportional to the electric force, as seen in Figure 3.10. In this way the electric force is transferred through the medium because of the elasticity. The electric displacement is defined by $\mathbf{D} = \varepsilon\mathbf{E}$, where the constant "$\varepsilon$" depends on the elasticity of the fluid. Maxwell found that the energy density stored in the elastic displacement was of the form $\mathbf{E} \cdot \mathbf{D}/2$. When he computed the energy corresponding to two point charges, he obtained the classical result corresponding to Coulomb's force.

Because the time derivative of the electric displacement could be identified as a kind of electric current, Maxwell rewrote the equation (3.49) as:

$$\nabla \times \mathbf{H} = \partial_t\mathbf{D} + \mathbf{J} \quad , \tag{3.53}$$

which means that even in an insulating medium, one could have displacement currents. Since the structure now had elasticity, it was no longer rigid. Instead of clockwork, it acted more like jelly, in which a disturbance would propagate as a wave. Maxwell noticed this and found the expression for the velocity of the electromagnetic wave, which in modern notation is $1/\sqrt{\mu_0\varepsilon_0}$. The numerical value for this constant could be computed from the previous measurements by Weber and Kohlrausch and it was seen to coincide within one percent with the velocity of light known from measurements. Thus, Maxwell could conclude that,

We can scarcely avoid the inference that light consists in the transverse undulations of the same medium which is the cause of electric and magnetic phenomena.

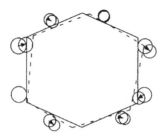

Figure 3.10. Assuming vortex tubes elastic so that fixed balls could be displaced by electric force made Maxwell's model elastic and capable of carrying mechanical waves. Displacement of electricity corresponds to current creating a magnetic field.

There was some luck with this conclusion, because actually the mechanical analogy was not meant for this kind of accuracy. In fact, it was found later that Maxwell had overlooked a factor of two when computing the transverse elasticity and even the light-velocity measurements of the time were not within an error of one percent [6].

3.3.3.3 Final Theory. Maxwell's original intention was to give a mechanical model for Faraday's force-field lines but it turned out to give much more. However, Maxwell himself realized that the model was only the starting point and discarded it. In his paper "A dynamical theory of the electromagnetic field" of 1865 [4,24,28], he replaced the vortex model with a dynamical justification of his field equations. For example, the displacement current was introduced by requiring that every current is closed and Ampère's integration law is valid for any closed path. In his 1873 book Treatise on Electricity and Magnetism [21] the full set of Maxwell equations were given in an expanded form with close to 20 scalar equations. The unknowns involved not only the field vectors but also potential quantities, as is seen from Figure 3.11. In the book these equations were not given as a compact set but, rather, dispersed in the commenting text over some ten pages in the book which did not help in their comprehension.

Maxwell did not do well in publicizing his theory. When giving his presidential lecture in 1870 in the British Association for the Advancement of Science, he referred to it as "another theory of electricity which I prefer" without even mentioning that it was his own. Freeman Dyson has pointed to this as a case when "modesty is not always a virtue" [20]. The field concept was new to physicists and the Maxwellian clockwork seemed like a step backwards from the clear-cut Newtonian astronomy to the Ptolemaic system. Moreover, fields obeyed partial differential equations, which were far more difficult to handle than the Newtonian force laws which had their high point in Weber's theory.

There were just a few contemporary physicists (sometimes called "the Maxwellians") who could understand the field concept and Maxwell's theory [15]. The acceptance was finally enforced by experimental verification of the displacement current and the existence of electromagnetic waves by Heinrich Hertz (1857–1894) in 1887–1888, almost a decade after Maxwell's death.

The Maxwell equations in the vector form mostly used today in electromagnetics were formulated by Oliver Heaviside (1850–1925) in 1886 [11]. He stripped the set of equations in Figure 3.11 from the potential quantities and called the simplified set "the duplex method." He also completed the formal symmetry by introducing magnetic sources and introduced the duality principle.

$$a = \frac{dH}{dy} - \frac{dG}{dz}$$
$$b = \frac{dF}{dz} - \frac{dH}{dx} \qquad \text{(A)} \qquad \mathbf{B} = \nabla \times \mathbf{A}$$
$$c = \frac{dG}{dx} - \frac{dF}{dy}$$

$$P = c\frac{dy}{dt} - b\frac{dz}{dt} - \frac{dF}{dt} - \frac{d\psi}{dx}$$
$$Q = a\frac{dz}{dt} - c\frac{dx}{dt} - \frac{dG}{dt} - \frac{d\psi}{dy} \qquad \text{(B)} \qquad \mathbf{E} = \mathbf{v} \times \mathbf{B} - \frac{\partial \mathbf{A}}{\partial t} - \nabla\phi$$
$$R = b\frac{dx}{dt} - a\frac{dy}{dt} - \frac{dH}{dt} - \frac{d\psi}{dz}$$

$$X = vc - wb$$
$$Y = wa - uc \qquad \text{(C)} \qquad \mathbf{F} = \mathbf{J} \times \mathbf{B}$$
$$Z = ub - va$$

$$a = \alpha + 4\pi A$$
$$b = \beta + 4\pi B \qquad \text{(D)} \qquad \mathbf{B} = \mu_o \mathbf{H} + \mathbf{M}$$
$$c = \gamma + 4\pi C$$

$$4\pi u = \frac{d\gamma}{dy} - \frac{d\beta}{dz}$$
$$4\pi v = \frac{d\alpha}{dz} - \frac{d\gamma}{dx} \qquad \text{(E)} \qquad \mathbf{J} = \nabla \times \mathbf{H}$$
$$4\pi w = \frac{d\beta}{dx} - \frac{d\alpha}{dy}$$

$$\mathfrak{D} = \frac{1}{4\pi} K \mathfrak{E} \qquad \text{(F)} \qquad \mathbf{D} = \epsilon \mathbf{E}$$

$$\mathfrak{K} = C \mathfrak{E} \qquad \text{(G)} \qquad \mathbf{J}_c = \sigma \mathbf{E}$$

$$\mathfrak{C} = \mathfrak{K} + \dot{\mathfrak{D}} \qquad \text{(H)} \qquad \mathbf{J} = \mathbf{J}_c + \frac{\partial \mathbf{D}}{\partial t}$$

$$u = p + \frac{df}{dt}$$
$$v = q + \frac{dg}{dt} \qquad \text{(H*)} \qquad \mathbf{J} = \mathbf{J}_c + \frac{\partial \mathbf{D}}{\partial t}$$
$$w = r + \frac{dh}{dt}$$

$$\mathfrak{C} = (C + \frac{1}{4\pi} K \frac{d}{dt}) \mathfrak{E} \qquad \text{(I)} \qquad \mathbf{J} = \sigma \mathbf{E} + \epsilon \frac{\partial \mathbf{E}}{\partial t}$$

$$u = CP + \frac{1}{4\pi} K \frac{dP}{dt}$$
$$v = CQ + \frac{1}{4\pi} K \frac{dQ}{dt} \qquad \text{(I*)} \qquad \mathbf{J} = \sigma \mathbf{E} + \epsilon \frac{\partial \mathbf{E}}{\partial t}$$
$$w = CR + \frac{1}{4\pi} K \frac{dR}{dt}$$

$$\rho = \frac{df}{dx} + \frac{dg}{dy} + \frac{dh}{dz} \qquad \text{(J)} \qquad \varrho = \nabla \cdot \mathbf{D}$$

$$\sigma = lf + mg + nh + l'f' + m'g' + n'h' \qquad \text{(K)} \qquad \varrho_s = \mathbf{n} \cdot (\mathbf{D}_1 - \mathbf{D}_2)$$

$$\mathfrak{B} = \mu \mathfrak{H} \qquad \text{(L)} \qquad \mathbf{B} = \mu \mathbf{H}$$

Figure 3.11. The final set of Maxwell equations (**A**) through (**L**) as labeled in [21], with their interpretation in modern Gibbsian vector notation.

$$\nabla \times \mathbf{E} = -\partial_t \mathbf{B} - \mathbf{J}_m, \quad \nabla \cdot \mathbf{B} = \rho_m, \tag{3.54}$$

$$\nabla \times \mathbf{H} = \partial_t \mathbf{D} = \mathbf{J}_e, \quad \nabla \cdot \mathbf{D} = \rho_e. \tag{3.55}$$

These equations, which were also simultaneously introduced in component form by Hertz, have since served as tools for analyzing a wide variety of electromagnetic problems.

3.4 CONCLUSION

In this chapter the evolution of the electromagnetic equations has been followed through the 19th century starting from Coulomb's force law and ending in Heaviside's version of the Maxwell equations. The evolution can be split into two distinct branches. The continental branch followed mathematical formalism based on the Newtonian action-at-a-distance philosophy while the British branch, opened by Faraday, followed the line-of-force idea and used physical analogies. Weber's force law was the high point of the continental branch. It gave the unified force law in a neat form which could explain all the known experiments consisting of static and low-frequency phenomena. The British branch ended in Maxwell's formulation which was a big mess of equations. It could also explain all the known phenomena. However, in addition, it could predict a new world of unknown phenomena, which gave an impetus to the electromagnetic research for the following century.

REFERENCES

[1]   F. U. T. Aepinus, *Tentamen Theoriae Electricitatis et Magnetismi*, St Petersburg, 1759. Translation, R.W. Home, P.J. Connor, *Aepinus's Essay on the Theory of Electricity and Magnetism*, Princeton, NJ: Princeton University Press, 1979.

[2]   A. M. Ampère, *Théorie mathématique des phénomènes électro-dynamiques uniquement déduite de l'expérience*, Paris: Chez Firmin Didot, 1827; reprint Paris: Jacques Gabay, 1990.

[3]   A. K. T. Assis, *Weber's Electrodynamics*, Dordrecht: Kluwer, 1994.

[4]   W. Berkson, *Fields of Force*, New York: Halsted Press, 1974.

[5]   D. K. Cheng, *Field and Wave Electromagnetics*, 2nd ed., Reading, MA: Prentice-Hall, 1989.

[6]   O. Darrigol, *Electrodynamics from Ampère to Einstein*, Oxford: Oxford University Press, 2000.

[7]   *The Encyclopaedia Britannica*, 9th ed., vol. VIII, New York: Samuel Hall, 1878, p.11.

[8]   J. Fourier, *Théorie de la chaleur*, Paris 1822; reprint Paris: Jacques Gabay, 1988.

[9]   W. Gilbert, *On the Magnet*, New York: Basic Books Inc., 1958.

[10]   P. Graneau, *Ampère-Neumann Electrodynamics of Metals*, Palm Harbor, FL: Hadronic Press, 1985.

[11]   O. Heaviside, *Electrical Papers*, New York: Chelsea 1970. Reprint of the first edition, London 1892; Vol. 1, p. 447 & Vol. 2, pp.172-175. The original article was published in *Phil. Mag.*, August 1886, p. 118.

[12]   J. L. Heilbron, *Electricity in the 17th and 18th Centuries*, Mineola, NY: Dover, 1999.

[13]   H. von Helmholtz, *Über die Erhaltung der Kraft*, Berlin: Verlag G. Reimer, 1847.

[14]   J. Hendry, *James Clerk Maxwell and the Theory of the Electromagnetic Field*, Bristol: Adam Hilger Ltd, 1986.

[15]   B. J. Hunt, *The Maxwellians*, Ithaca, NY: Cornell University Press, 1991.

[16]   E. Hoppe, *Geschichte der Physik*, Braunschweig: Vieweg und Sohn, 1926.

[17]   J. D. Jackson, *Classical Electrodynamics*, 3rd ed., New York: Wiley 1999, p.176.

[18]   C. Jungnickel and R. McCormmach, *Intellectual Mastery of Nature*, Volume I, Chicago: The University of Chicago Press, 1986, Chapter 6.

[19]   I. V. Lindell, *Methods for Electromagnetic Field Analysis*, 3rd ed., New York: Wiley and IEEE Press, 2002.

[20]   M. Longair, *Theoretical Concepts in Physics*, 2nd ed., Cambridge: Cambridge University Press, 2003, Chapter 5.

[21]   J. C. Maxwell, *Treatise on Electricity and Magnetism*, Vol. 2, 3rd ed., Oxford: Clarendon Press, 1904, pp. 480-493.

[22]   J. C. Maxwell, "On Faraday's lines of force", *Transactions of the Cambridge Philosophical Society*, Vol. 10, pp. 27-83, 1856.

[23]   J. C. Maxwell, "On physical lines of force", *Philosophical Magazine* (4th series). "Part I: The theory of molecular vortices applied to magnetic phenomena", Vol. 21, pp. 161-175, 1861; "Part II: The theory of molecular vortices applied to electric currents", Vol. 21, pp. 281-291, 338, 1861; "Part III: The theory of molecular vortices applied to static electricity", Vol. 23, pp. 12-24, 85, 1862.

[24]   J. C. Maxwell, "A dynamical theory of the Electromagnetic Field", *Philosophical Transactions*, Vol. 155, pp. 459-512, 1865. A shortened version in *Philosophical Magazine*, Vol. 29, pp. 152-157, 1865.

[25]   G. S. Ohm, *Galvanic Circuit Investigated Mathematically*, New York: Van Nostrand, 2002. Reprint of the 1891 translation of the 1827 German original.

[26]   A. Russell, "The eighth Kelvin lecture. Some aspects of Lord Kelvin's life and work", *Journal of the IEE*, Vol. 55, No. 261, December 1916, pp. 1-17.

[27]   D. M. Siegel, *Innovation in Maxwell's Electromagnetic Theory*, Cambridge: Cambridge University Press, 1991, Chapter 3.

[28]   T. K. Simpson, *Maxwell on the Electromagnetic Theory, A Guided Study*, New Brunswick, NJ: Rutgers University Press, 1997.

[29]   J. G. Ternan, "Equivalence of the Lorentz and Ampère force laws in magnetostatics", *J. Appl. Phys.*, Vol. 57, No. 5, March 1985 pp. 1743-1745.

[30]   W. Thomson (Kelvin), "On a mechanical representation of electric, magnetic and galvanic forces", *Cambridge and Dublin Mathematical Journal*, Vol. 2, pp. 61-64, 1847.

[31]   W. Thomson (Kelvin), *Reprint of Papers on Electricity and Magnetism*, London: Macmillan 1872, pp. 569-577.

[32]   E. Whittaker, *A History of the Theories of Aether and Electricity*, Volume 1, London: Thomas Nelson, 1951; reprint New York: Dover, 1989.

4

THE GENESIS OF MAXWELL'S EQUATIONS

OVIDIO M. BUCCI, *University of Naples Federico II*

4.1   INTRODUCTION

James Clerk Maxwell (Edinburgh, June 13, 1831 – Cambridge, November 5, 1879) is universally considered one of the greatest figures in the history of Science, and his name is usually associated to those of Newton and Einstein. This preeminent position is rightly attributed to Maxwell not so much for the number and the relevance of his scientific contributes, but for their nature. As Newton before him and Einstein after, he introduced in physics new ways of looking at phenomena, opening completely new conceptual (and practical) horizons and modifying the body of accepted theories and physical (and metaphysical) conceptions, namely what Kuhn [1] calls the "paradigms" of the "normal" science.

To appreciate the revolutionary (again, in Kuhn's sense) relevance of his work, it suffices to note that the constitutive elements of today's vision of a physical world made of particles interacting through fields, have their conceptual foundation in Maxwell's two main contributions: the kinetic theory of gases and the theory of the electromagnetic field.

It must be stressed that the development of both theories was not motivated by new experimental findings which did not fit in the existing paradigm, as it happened for the (restricted) Theory of Relativity and the Quantum Mechanics. In particular, in the case of electromagnetism, all the facts known in Maxwell's time had been satisfactorily interpreted within the Newtonian paradigm and incorporated in a theoretical framework of excellent predictive power, which was indeed exploited and further developed until Hertz's experimental verification of the most striking physical consequence of Maxwell's theory, i.e., the existence of electromagnetic waves.

The main motivation behind Maxwell's effort was philosophical even metaphysical, i.e., his adherence to a "world view", alternative to the dominant one introduced by Michael Faraday (Figure 2.7) in connection with his studies on electromagnetic induction and polarization.

The construction of a coherent and satisfactory theory based on this new alternative conception required ten years and an exceptional intellectual effort,

189

which led to the electromagnetic theory of light, and to the formulation of those equations we still adopt for the description of electromagnetic phenomena.

In this chapter, through an analysis of the three fundamental memoirs on Electromagnetism and of the extent scientific letters and manuscripts, we will try to elucidate the main stages of this enterprise in order to follow and clarify the evolution of Maxwell's thoughts, which ultimately led to the development of his theory of the electromagnetic field.

Maxwell was introduced to the study of magnetism by William Thomson (Lord Kelvin, Figure 4.1) when he was still an undergraduate student at Trinity College in Cambridge. However, his explicit interest in Electromagnetism started just after his successful graduation at the 1854 "Tripos" [1], when he was 23 (Figure 4.2), as it is attested by in a letter to Thomson dated February 20, 1854 [2]:

> But we [2] have a strong tendency to return to Physical Subjects and several of us wish to attack Electricity. Suppose a man to have a popular knowledge of electrical show experiments and a little antipathy to Murphy's Electricity [3], how ought he to proceed in reading & working so to get a little insight into that subject wh[ich] may be of use in further reading? If he wished to read Ampère Faraday &c how should they be arranged, and in what order might he read your articles in the Cambridge Journal?

Figure 4.1. William Thomson (1824–1907) in 1852.

---

[1] Formerly a bachelor of arts appointed to dispute, in a humorous or satirical style, with the candidates for degrees (so called from the three legged stool on which he sat), in Maxwell's time it denoted the final honours examination for the B.A. in mathematics. Maxwell confronted this grim week of hard tests in January 1854, in a vast, cold room called the "Senate House", and qualified second "wrangler" (i.e., second of the best).

[2] Maxwell and two fellows, members of the selected "Apostles Club", a traditional society of twelve undergraduates who met to exchange ideas by reading original essays which were thereupon discussed.

[3] A popular textbook of the day [3]

Figure 4.2. Maxwell in the Cambridge years holding a top for experimenting with color mixing.

In choosing electricity as a field of enquiry, he was selecting an area at the forefront of current research, and he was naturally drawn to the work of Michael Faraday, whose extraordinary series of experimental discoveries [4] formed (in Maxwell words) "the nucleus of everything electric since 1830", as well as to that of Franz Neumann (1798–1895), Gustav Kirchhoff (1824–1887) and Wilhelm Eduard Weber (Figure 4.3). The last had developed a comprehensive explanation of both electrodynamics and electromagnetic induction in the classic framework of a direct action at a distance between charges and currents (i.e., moving charges).

In the modern vector notation and units, which will be adopted henceforth, the Weber's expression for the force exerted *in vacuo* by a point charge, q_1, on another one, q_2, reads [5]:

$$F = q_1 q_2 \mathbf{r} / 4\pi\varepsilon_0 r^3 [1 - 1/2c^2 (dr/dt)^2 + 1/c^2 (rd^2r/dt^2)] \qquad (4.1)$$

being ε_0 the permittivity of vacuum, $\mathbf{r}$ the vector pointing from 1 to 2, r the corresponding distance, and c the ratio between the units of charge in the electromagnetic and electrostatic systems.

Maxwell made rapid progress in his studies. In a letter to Thomson, dated, November 13, 1854 [6], while qualifying himself as "an electrical freshman", he writes:

Then I tried to make out the theory of attractions of currents but tho' I could see how the effects could be determined I was not satisfied with the form of the theory which treats of

---

[4] It is a striking coincidence that Maxwell picked up Faraday's ideas in the very year in which he closed his investigations, publishing the third and last volume of the immense *Experimental Researches in Electricity* [4].

Figure 4.3. Wilhelm Eduard Weber (1804–1891).

elementary currents & their reciprocal actions, & I did not see how any general theory was to be formed from it. .... Now I have heard you speak of "magnetic lines of force" & Faradays seems to make great use of them but others seem to prefer the notion of attraction of elements of currents directly. Now I thought that as every current generated magnetic lines & was acted on in a manner determined by the lines thro[ough] wh:[ich] it passed that something might be done by considering "magnetic polarization" as a property of a "magnetic field" or space and developing the geometric ideas according to this view.

The explicit dislike for the continental approach is further expressed in a letter (May 5, 1855) to his father, John Clerk Maxwell [7]:

I am working away at electricity again, and have been working my way into the views of heavy German writers. It takes a long time to reduce to order all the notions one get from these men, but I hope to see my way through the subject and arrive at something intelligible in the way of a theory.

The attitude towards a description in terms of action at distance evidenced by these sentences contrasts with what we read in a manuscript written only four years before [8]:

There is however an another obvious method of treating the same subject [the interaction between material bodies] in which the action of the fixed bodies on the body considered, being always the same for the same position of the body is regarded not as directly dependent on the fixed bodies but as the immediate consequence of its particular position in space, that

> *space having been endued with certain properties by the action of the fixed bodies. When properties of this kind are attributed to portions of space occupied by matter or not, it must be carefully remembered that such properties imply nothing more than what is expressed in their definitions and that they are to be considered for the present as mere mathematical abstractions introduced to facilitate the study of certain phenomena.*

This clearly shows that since the beginning, Maxwell was strongly influenced by Faraday's conception that the transmission of forces is mediated by the action of contiguous particles of matter in the space between charged or magnetized bodies, i.e., through the action of lines of forces in space. This commitment to Faraday's ideas underlies all Maxwell's subsequent work and is central to the development of his field theory of electromagnetic phenomena.

Maxwell pursued the issues pointed out in the above letter to Thomson while mainly engaged in his second attempt to be elected Fellow at Trinity College, and in his work about color vision. In fact, in another letter to Thomson dated May 15, 1855 [9], he writes:

> *I am reading Weber's Elektrodynamische Maasbestimmungen which I have heard you speak of. I have been examining his mode of connecting electrostatics with electrodynamics, induction &c& I confess I like not at first* [again!].....*I am trying to construct two theories, mathematically identical, in one of which the elementary conceptions shall be about fluid particles attracting at a distance while in the other nothing (mathematical) is considered but various states of polarization, tension, etc., existing at various part of space. The result will resemble your analogy of the steady motion of heat ..., but applying it in a somewhat different way to a more general case to which the laws of heat do not apply.*

The analogy (between heat conduction and electrostatics) to which Maxwell makes reference, was proposed by Thomson in a paper of 1842 [10] and exploited by him to show that the Faraday's conceptions, which had been generally thought to be incompatible with the mathematical theory of electrostatics, could be indeed reconciled with it.

4.2   ON FARADAY'S LINES OF FORCE

The construction of the theory envisaged in the previously reported letter was carried out during the summer and autumn of the same year (see a letter to Thomson dated September 13th and a draft of the Autumn 1855 [11]), and led to the first of the three fundamental memoirs on electromagnetism: "On Faraday's Lines of Forces", which was presented to the Cambridge Philosophical Society in two parts, the 10th of December 1855, and the 11th of February 1856, and published in extenso the same year on the Transactions of the Society [12].

The first sentence of the paper is trenchant: "The present state of electrical science seems particularly unfavorable to speculation." Speculation is a suggestive word (from the Latin speculum, mirror), which reveals Maxwell's hope that a "reflection" on the various fragments of the "present state" of his science in the light of Faraday's ideas could allow to catch them in a single image, as a whole. Accordingly, at the very beginning of the first part, he states that the purpose of his work is:

> to show how, by a strict application of the ideas and methods of Faraday, the connection of the very different orders of phenomena which he has discovered can be clearly placed before the mathematical mind.

As forewarned in the above reported letter, the method he adapts to this end is that of the physical analogy, i.e., in his own words "that partial similarity between the laws of one science and those of another which makes each of them illustrate the other". The model exploited for the analogy is that of an imponderable and incompressible fluid moving through a resisting medium, which exerts on it a retarding force proportional to its velocity. The fluid can be supplied or swallowed by sources and sinks within the considered region of space, or from outside through its boundaries.

In the framework of the model, the Faraday's lines of force and tubes of flux correspond to lines and tubes of (steady) fluid motion, respectively, whose geometrical and dynamical properties are examined in detail in the first two sections of the first part of the memoir. All the results can be summarized in the couple of equations;

$$K\mathbf{v} = -\nabla p \ ; \ \nabla \cdot \mathbf{v} = S \tag{4.2}$$

wherein: $\mathbf{v}$ is the fluid velocity, p the pressure, S the source density, and K the friction coefficient (which Maxwell allows to vary in space and also to be anisotropic).

It is evident that by properly reinterpreting the quantities appearing in (4.2) according to Table 4.1: we immediately get the laws of magnetostatics (in absence of currents), electrostatics, and electric conduction. This is done by Maxwell in the third section, wherein he also starts to address the phenomena of electrodynamics and electromagnetic induction.

Table 4.1. Equivalence between Electrostatics and Magnetostatics

	K	$\mathbf{v}$	p	S
Magnetostatics	$1/\mu$	$\mathbf{B}$	Magnetic potential	0
Electrostatics	$1/\varepsilon$	$\mathbf{D}$	Electric potential	Charge density
Stationary currents	$1/\sigma$	$\mathbf{J}$	Electric potential	Source density

Concerning the mutual actions between currents, Maxwell stresses that, because we can only experiment with **closed** currents, the deduction of elementary laws of interaction between **current elements**, as Ampère did, necessarily involves some kind of a priori assumption. He writes:

> *We must recollect however that no experiments have been made on these elements of currents except under the form of closed current .... Hence if Ampère's formula applied to closed currents give true results, their truth is not proved for elements of currents unless we assume that the action between two such elements must be along the line which joins them. Although this assumption is most warrantable and philosophical in the present state of science, it will be more conducive to freedom of investigation if we endeavour to do without it, and to assume the laws of closed currents as the ultimate datum of experiment.*

Again, Faraday's ideas and theories, which are briefly outlined, should be exploited, as they allow reference only to the experimental laws on closed currents. With reference to electromagnetic induction, particularly appealing appears to Maxwell a somewhat vague concept introduced by Faraday: that of "electro-tonic state." The second part of the memoir is devoted to give a meaning to this concept, and to its exploitation for a comprehensive and unified description of the phenomena.

However, no physical analogy is now available, so that Maxwell writes at the end of the first part:

> *I can do no more than simply state the mathematical methods by which I believe that electrical phenomena can be best comprehended and reduced to calculation ... The idea of electro-tonic state, however, has not yet presented itself to my mind in such a form that its nature and properties may be clearly explained without reference to mere symbols, and therefore I propose in the following to use symbols freely, and to take for granted the ordinary mathematical operations.*

This was not an easy task, as it is testified, among others, by two letters to Thomson (February 14 and April 25, 1856) [13], as well as by the much higher mathematical level of the second part of the memoir. For the mathematical description of the electro-tonic state, he introduces what is called today the vector potential. Then by a full exploitation of the results of vectorial analysis available at his time, he can state the laws of electromagnetism and electromagnetic induction in explicit mathematical form, i.e.,:

$$\boldsymbol{B} = \nabla \times \boldsymbol{A} \; ; \; \nabla \cdot \boldsymbol{A} = 0 \tag{4.3}$$

$$\boldsymbol{J} = \nabla \times \boldsymbol{H} \tag{4.4}$$

$$\boldsymbol{E} = -\nabla \phi + \boldsymbol{E}_i \; ; \; \boldsymbol{E}_i = -\partial A / \partial t \, , \tag{4.5}$$

wherein the symbols have the usual meaning and $\boldsymbol{E}_i$ is the induced electric field.

Maxwell also provides the expression of the potential of a closed current in a magnetic field:

$$U = I \oint A_c \, dc \qquad (4.6)$$

from which all the dynamical actions can be derived. In (4.6) I is the current, A_c the tangential component of A, and the integral is obviously performed along the circuit.

The memoir closes with twelve examples, which were added by Maxwell to show the effectiveness of his approach, but are of no interest in our context.

Equations (4.3) (the first of which is obviously equivalent to $\nabla \cdot B = 0$, i.e., the fourth Maxwell equation) relate the electro-tonic state, described by the vector potential A, to the magnetic induction. Note explicitly that in order to make A unique, which is necessary if it must represent a physical property, he had to enforce what is called today the Coulomb gauge, i.e., $\nabla \cdot A = 0$.

Together with (4.3), the second of equations (4.5) expresses (for the first time) in local form the Faraday-Neumann law of the electromagnetic induction; (4.3) and (4.5) jointly are clearly equivalent to what we call now the first Maxwell equation.

Equation (4.4) states in local form the so (improperly) called Ampère law. It is noteworthy that Maxwell explicitly stresses that it can only be valid for closed currents, and says: "Our investigation are therefore for the present limited to closed currents; and we know little on the magnetic effects of any currents which are not closed."

Notwithstanding the achievement of his goal, Maxwell is well aware of the purely mathematical and somehow artificial character of his construction, and at the end of the memoir he writes:

> In these six laws [which summarize his results] I have endeavored to express the idea that I believe to be the mathematical foundation of the modes of thought indicated in the (Faraday's) Experimental Researches. I do not think that it contains even the shadow of a true physical theory; in fact, its chief merit as a temporary instrument of research is that it does not, even in appearance, account for anything.

To this, even excessive, understatement of the relevance of his work follows immediately (and a bit surprisingly) a great praise of the Weber's electrodynamics:

> There exists however a professedly physical theory of electro-dynamics, which is so elegant, so mathematical, and so entirely different from anything in this paper, that I must state its axioms, at the risk of repeating what ought to be well known. ... Here is then a really physical theory, satisfying the required conditions better perhaps than any yet invented, and put forth

by a philosopher whose experimental researches form an ample foundation for his mathematical investigations.

If it is so:

What is the use then of imaging an electronic state of which we have no distinctly physical conception, instead of a formula of attraction which we can readily understand?

Maxwell provides essentially two answers to this question. One is related to the dependence on velocity of the Weber's force[5]. As a matter of fact, just one year after its presentation, Hermann L. von Helmholtz (Figure 4.4) had published his famous and influential work which put on a firm theoretical basis the principle of the conservation of energy[6] [14]. Based on Helmholtz's results, Maxwell and others thought that Weber's electrodynamics did not comply with this principle. Although in 1848 Weber had shown that his force can be derived by a potential, it is only in 1869 and 1871 that he proved in detail that it satisfies the principle of the conservation of energy. After that Maxwell obviously corrected himself, but in the meantime the electromagnetic field theory had been fully developed.

Figure 4.4. H. L. von Helmholtz (1821–1894) in 1847.

---

[5] This is unavoidable in any theory relying on a direct action at distance between charges. On the other side, peculiar to Weber's theory is the assumption, first made in 1845 by the Leipzig professor Gustav Theodor Fechner, that electric currents consist of a streaming of opposite charges, equal in magnitude and traveling in opposite directions with equal velocities. This rather artificial, and ultimately false, assumption can be disposed of only introducing forces depending on the *absolute*, instead of *relative*, velocities, as made by R. J. E. Clausius in 1877 and 1880, just before the Michelson (1881) and Michelson and Morley (1887) experiments ruled out the existence of an absolute reference system.

[6] Helmholtz's paper was read to the Physical Society of Berlin on July 23, 1847, and was enthusiastically received by the younger physicists of the Society. However, the prejudices of the older generation, based largely on the dread of a revival of Hegel's natural philosophy, prevented its acceptance for the *Annalen der Physik*, so that it was eventually published as a separate treatise.

The other answer is more general and methodological. Maxwell says:

I would answer, that it is a good thing to have two ways of looking at a subject, and to admit that there are two ways of looking at it. Besides, I do not think that we have any right at present to understand the action of electricity, and I hold that the chief merit of temporary theory is, that it shall guide experiments, without impeding the progress of the true theory when it appears.

If read in the light of the previously reported letters and of the explicitly stated purpose of the memoir, these two sentences clearly unveil Maxwell's deep conviction that Weber's theory is unsatisfactory (hence "temporary") not so much because of its weak points but because of its being an **action at a distance** theory, while the way toward a "true theory" is that paved by Faraday, through his conception of an action mediated by the medium. However to develop a "professedly physical theory", mathematics is not enough: such a theory must also rely on sound physical, i.e., (for Maxwell's times) mechanical bases. This explains why Maxwell closes the first part of the memoir with a hope, which is also a program:

By a careful study of the laws of elastic solids and of the motions of viscous fluid fluids, I hope to discover a method of forming a mechanical conception of the electro-tonic state adapted to general reasoning.

4.3   ON PHYSICAL LINES OF FORCE

A relatively long time (according to Maxwell standards) had to pass before the hope expressed in the first memoir could be realized. In the meantime Maxwell became professor at Aberdeen University and then at the King's College in London, married Catherine Mary Dewar, daughter of the Rector of Aberdeen Marischal College, and published, among the others, five memoirs about the color vision, his outstanding memoir about the stability of Saturn's rings and the first of his epochal memoirs on the kinetic theory of gases.

However, he didn't stop reflecting on electromagnetism. In a letter to Cecil James Monro [15], one of the "Apostles", dated May 20, 1857, he writes:

This was a wet day & I have been grinding at many things and lately during this letter at a Vortical theory of magnetism & electricity which is very crude but has some merits, so I spin & spin.

The "vortical theory of magnetism & electricity" to which Maxwell makes reference is the theory of molecular vortices, proposed by Thomson [16] to explain the rotation of the plane of polarization of linearly polarized light by a magnetic field. Thomson supposed that this phenomenon, discovered by Faraday in 1845 [17], was caused by the rotation of molecular vortices in an ether, having their axis of rotation along the lines of forces of the magnetic field.

Maxwell's interest in this theory rapidly increases. In the letter to Faraday of November 9th of the same year [18], we read:

> *But there are questions relating to the connexion between magneto-electricity and certain mechanical effects which seem to me opening up quite a new road to the establishment of principles in electricity and a possible confirmation of the physical nature of magnetic lines of force. Professor W. Thomson seems to have some new lights on this subject.*

In a letter to Thomson dated 30 January 1858 [19] he outlines an experiment which could establish the effect of the rotating vortices on a freely rotating magnet. In the course of 1861 he actually realized the apparatus (Figure 4.5) and tried the experiment, though without success.

No wonder then, that the theory of molecular vortices is the cornerstone of Maxwell's second memoir, "On Physical Lines of Forces", which was published in the *Philosophical Magazine* in March, April and May 1861 (Parts I and II) and January and February 1862 (Parts III and IV) [20].

Again, the purpose of the work is clearly stated at the beginning:

> *My object in this paper is to clear the way for speculation in this direction* [i.e., the Faraday's point of view] *by investigating the mechanical results of certain states of tension and motions in a medium, and comparing these with the observed phenomena of magnetism and electricity. By pointing out the mechanical consequences of such hypothesis, I hope to be of some use to those who consider the phenomena as due to the action of a medium, but are in doubt as to the relation of this hypothesis to the experimental laws already established, which have generally been expressed in the language of other hypotheses* [action at distance].

After a brief summary of the content of his first memoir, Maxwell writes:

Figure 4.5. Apparatus to detect molecular vortices.

I propose now to examine the magnetic phenomena from a mechanical point of view, and to determine what tensions in, or motions of, a medium are capable of producing the mechanical phenomena observed. If, by the same hypothesis, we can connect the phenomena of magnetic attraction with electromagnetic phenomena and with those of induced currents, we shall have found a theory which, if not true, can only be proved to be erroneous by experiments which will greatly enlarge our knowledge of this part of physics.

Following Thomson's suggestions, the hypothetic medium consists of molecular vortices, with the axes directed along the magnetic lines of force, revolving with peripheral velocity proportional to the intensity of the magnetic field. The density of the medium is proportional to the magnetic permeability. By applying to this model the laws of continuum mechanics, in Part I (*The Theory of Molecular Vortices applied to Magnetic Phenomena*) Maxwell obtains the following expression for the force density in the medium:

$$f = H\nabla \cdot \mu H + 1/2\mu\nabla(H^2) + (\nabla \times H) \times (\mu H) - \nabla p . \qquad (4.7)$$

The first term is interpreted as the force exerted by the magnetic field on magnetic poles, "should we attribute magnetostatic interactions to imaginary magnetic matter" (Maxwell's words), whose charge density is accordingly given by:

$$p_m = \nabla \cdot \mu H = \nabla \cdot B , \qquad (4.8)$$

which is what we call the symmetrized fourth Maxwell equation.

The second term accounts for the fact, experimentally found by Faraday, those paramagnetic and diamagnetic bodies tend to move towards zones of higher and lower field intensity, respectively. Then, Maxwell notes that, according to Ampère's law (4.4), $\nabla \times H$ must be interpreted as the electric current density, so that the third term provides the density of force exerted by the magnetic field on currents:

$$f = J \times B . \qquad (4.9)$$

The last term in (4.7) is a standard pressure term, corresponding to the symmetrical part of the stress tensor, with no electromagnetic counterpart.

It must be stressed that in (4.9) electric current is introduced only by way of interpretation. Its presence is ensured by the existence of a non-irrotational magnetic field: we know that there must be a current, but in terms of the vortex model there is a missing link. Part II (*The Theory of Molecular Vortices applied to Electric Currents*) aims to provide this link.

After a brief summary of the results of Part I, Maxwell writes:

We have as yet give no answers to the questions, "How are these vortices set in rotations?" and "Why are they arranged according to the known laws of lines of forces about magnets and currents?" These questions are certainly of a higher order of difficulty than either of the former [considered in Part

I]. .... *We have, in fact, now come to enquire into the physical connection of these vortices with electric currents, while we are still in doubt as to the nature of electricity, whether it is one substance, two substances, or not a substance at all, or in what way it differs from matter, and how it is connected with it.*

The last sentence is particularly significant. As a matter of fact, it had been assumed, since the beginning of electrical science, that electricity was a substance of some sort - of one or two signs - whose flow in a wire gave rise to a current. The results obtained by Maxwell show that all phenomena could be accounted even if there were "no substance at all" resting on a charged conductor or flowing in a wire, but merely a state of affairs in the space around it. This is why Maxwell says that a satisfactory answer to the above questions: "...would lead us a long way towards that of a very important one, 'What is an electric current?' ".

To insert into his machine a new element corresponding in some way to electricity, Maxwell starts from an obvious difficulty of the model developed in the first part, the fact that contiguous vortices must move in opposite directions. To overcome this difficulty, he adopts a well-known mechanical arrangement: the interposition of a layer of massless round particles between contiguous vortices, playing the role of "idle wheels," as it is sketched in Figure 3.9, which appears in the memoir. These particles, which are in rolling contact with the vortices they separate, but do not rub against each other, play the role of electricity. Their motion of translation constitutes an electric current, while their rotation transmits the motion of the vortices from one part of the medium to another. The corresponding tangential stresses thus called into play constitute electromotive force.

By applying pure kinematical considerations to this model, Maxwell immediately derives the Ampère law (4.4). Then, he addresses the analysis of the dynamical behavior of his medium, starting with the determination of the (kinetic) energy density, w_c say, for which he gets:

$$w_c = 1/2\mu H^2 \tag{4.10}$$

Expression (4.10) is quite natural, in the light of the correspondence between vortex peripheral velocity and magnetic field intensity, and checks with the fact, first demonstrated by Thomson in 1853, that the energy of any magnetic system, whether consisting of magnets or of currents, can be evaluated by integrating (4.10) over all space.

Exploiting expression (4.10) and applying the laws of dynamics, Maxwell finds the relations between the motions of the vortices and the force they exert on the interposed layer of particles, i.e., the electromotive force due to the variations of the magnetic field. This leads to the law of electromagnetic induction; in the form we now call the first of Maxwell's equations:

$$\nabla \times \boldsymbol{E} = -\mu \partial \boldsymbol{H} / \partial t = -\partial \boldsymbol{B} / \partial t . \tag{4.11}$$

Finally, he shows that (4.11) is equivalent to (4.5), and generalizes this lastly to moving bodies, obtaining:

$$E = v \times (\mu H) - \nabla \phi - \partial A / \partial t \ , \tag{4.12}$$

wherein, v is the velocity of the body (with respect to the field).

Of course, Maxwell was well aware of the awkwardness of his mechanical model, as he clearly states at the end of the second part of the memoir. However, he had achieved his goal, as:

> *We have now shown in what way electro-magnetic phenomena may be imitated by an imaginary system of molecular vortices. Those who have been already inclined to adopt an hypothesis of this kind will find here the conditions which must be fulfilled in order to give it mathematical coherence, and a comparison, so far satisfactory, between its necessary results and known facts. Those who look in a different direction for the explanation of the facts, may be able to compare this theory with that of the existence of currents flowing freely through bodies, and with that which supposes electricity to act at distance with a force depending on its velocity, and therefore not subject to the law of conservation of energy.*

It is likely that Maxwell originally envisaged his paper as consisting of only these two parts. But during the summer of 1861, in his country house of Glenlair in Scotland (Figure 4.6), he developed his mechanical ether theory along new lines, which led to revolutionary and apparently unexpected results.

His excitement is clearly testified by all the extant (scientific) letters between summer 1861 and January 1862, just before the publication of the last two parts of the memoir. At the beginning of the first of these letters, dated 19 October 1861 [21] and addressed, perhaps not casually, to Faraday, we read:

> *The conception I have hit on has lead, when worked out mathematically, to some very interesting results, capable of testing my theory, and exhibiting numerical relations between optical, electric and electromagnetic phenomena, which I hope soon to verify more completely. ... I wish to get the numerical*

Figure 4.6. The Glenlair house of J. C. Maxwell in a photo of 1860.

value of the "electrical capacity" [permittivity] of various substances especially transparent ones ... Again I have not yet found any determination of the rotation of the plane of polarization by magnetism in which the absolute intensity of magnetism at the place of transparent body was given.

Then, he alludes to the interpretation of the electrostatic field in terms of distortion of an elastic medium that we will find in the third part of the memoir, and states his crucial result:

I have determined the elasticity of the medium in air, and assuming that it is the same with the luminiferous ether I have determined the velocity of propagation of transverse vibrations. The result is

193,088 miles per second

(deduced from electrical & magnetic experiments (performed by Kohlraush and Weber). Fizeau has determined the velocity of light

= 193,118 miles per second [7]

by direct experiments.

This coincidence is not merely numerical. I worked out the formulae in the country [Glenlair], before seeing Weber's number, which is in millimeters, and I think we have now strong reason to believe, whether my theory is a fact or not, that the luminifer and the electromagnetic medium are one. Supposing the luminous and electromagnetic phenomena to be similarly modified by the presence of gross matter, my theory says that the inductive capacity (static) is equal to the square of the index of refraction, divided by the coefficient of magnetic induction (air =1).

In its final part, the letter mentions some of the results on the gyromagnetic effect we will find in the fourth part, and his effort to detect experimentally the molecular vortices by means of the previously mentioned apparatus depicted in Figure 4.5.

It is clear from the above reported sentences that Maxwell was immediately aware of the relevance of his results as well as of the necessity of an experimental validation of their striking and unexpected consequences. As a matter of fact, when the day after he communicates his conclusion concerning the identity of luminiferous and electromagnetic medium to C. J. Monro [23], the latter immediately replies (October 23, 1861 [24]):

The coincidence between the observed velocity of light and your calculated velocity of a transverse vibration in your

---

[7] This is the value reported in: J.A. Galbraith and S. Haughton, *Manual of Astronomy*, London, 1855. Fizeau' original result [22] was 70,948 leagues per second, corresponding to 195,647 miles per second.

medium seems a brilliant result. But I must say I think a few
such results are wanted before you can get people to think that,
every time an electric current is produced, a little file of
particles is squeezed along between rows of wheels.

His quest for accurate values of the electrical permittivity is repeated in
two letters to Thomson, dated the 10th and 17th of December 1861 [25]. The first
of these letters contains also a quite complete summary of the contents of the
forthcoming parts.

To account for the phenomena of electrostatics, in the Part III (*The
Theory of Molecular Vortices applied to Static Electricity*) Maxwell extends his
model by providing the medium with elastic properties. Because the vortices'
rotation is no longer of interest, Maxwell speaks of "elastic cells" surrounded by
the layer of particles which play the role of electricity.

When these particles are displaced from their equilibrium positions,
they distort the cells and call into play a force arising from their elasticity, equal
and opposite to that which urges the particles away from their equilibrium
position. The state of the medium, in which the particles are displaced (and the
cells distorted) from their equilibrium position, is assumed to represent an
electrostatic field.

But what happens *during* the displacement? In the light of his model the
answer is apparently obvious, and Maxwell makes a statement which will be of
paramount relevance:

> *This displacement does not amount to a current because when*
> *it has attained a certain value it remains constant, but it is a*
> *commencement of a current, and its variations constitute*
> *currents, in the positive or negative direction, according as the*
> *displacement is increasing or diminishing. The amount of the*
> *displacement depends on the nature of the body, and on the*
> *electromotive force.*

Assuming a linear relationship between stress (i.e., electromotive force)
and displacement:

$$E = -E^2 h, \tag{4.13}$$

h being the displacement, and E^2 a constant, Maxwell explicitly determines the
dependence of E^2 on the elastic constants of the medium, under the assumption of
spherical particles.

Then, he performs the crucial step, i.e., the addition of the displacement
currents to the Ampère's law (4.4), thus obtaining:

$$J = \nabla \times H + \partial h / \partial t = \nabla \times H - (1/E^2)\partial E / \partial t, \tag{4.14}$$

which is essentially what today we call the second Maxwell equation.

However, note that the displacement current $\partial h / \partial t$ is *subtracted* from
and not *added* to the true current *J*. The correct result in terms of *E* is obtained
because of the minus sign in relation (4.13), which implies that in (4.13) *E* must
be identified with the elastic reaction exerted *by* the cells *on* the electric particles

and not vice versa, as it should be if, as previously asserted, the electromotive force is the cause of the displacement.

This means that while Maxwell is by that time sure that the displacement current must be taken fully into account, he is still not aware of the fact that the total current must be closed (divergenceless), as well as of the relevance of the step he performed. As a matter of fact, he exploits (4.14) only to derive [8], starting from the equation of continuity of electric charge, the Gauss law in the form:

$$\rho = 1/E^2 \, \nabla \cdot E, \qquad (4.15)$$

density, expressed in electromagnetic units.

Then, he determines the force acting between two charged bodies, and shows that, in order to comply with Coulomb law, E must be equal to the ratio between the measures of an electric charge in the electrostatic and electromagnetic systems. That ratio, which has the dimension of a velocity and is the same constant appearing in the Weber's force law (4.1), had been experimentally determined by Weber and Kohlrausch in 1856 [26], who found:

$$E = 310,740 \text{ km/s}.$$

After having established in this way the full correspondence between his mechanical model and the electromagnetic quantities, Maxwell proceeded to the determination of the velocity of propagation of transverse vibrations through the medium, which turns out to be equal (in air) to E.

Immediately he concludes:

> *The velocity of transverse undulations in our hypothetical medium, calculated from the electro-magnetic experiments of MM. Kohlrausch and Weber, agrees so exactly with the velocity of light calculated from the optical experiments of M. Fizeau, that we can scarcely avoid the inference that light consists in the transverse undulations of the same medium which is the cause of electric and magnetic phenomena.*

Maxwell had unexpectedly established the basis of the electromagnetic theory of light!

It must be stressed that the almost perfect coincidence between the constant c ($=E$) in (4.1) and the velocity of light *in vacuo* had been noted by Weber and Kohlrausch in their 1856 paper, but they did not consider this to be physically significant. Even more impressive is the fact that in 1857 Gustav Kirchhoff (Figure 2.14), applying Weber's electrodynamics to the study of the propagation of electrical signals along metallic wires [27], had shown that they propagate with *finite* velocity, which in the case of vanishingly small resistance is equal to the velocity of light [9]. Again, Kirchhoff did not develop the implications of this result. The contrast with Maxwell's attitude is a striking illustration of the paradigms' influence on the development of Science. If we look at

---

[8] To this end, he needs the minus sign in (5.13).

[9] Maxwell was quite certainly unaware of Kirchhoff's paper.

electromagnetic and optical phenomena under *different* paradigms, the Newtonian (action at distance between particles) and Cartesian (action by contact, through an interposed medium), respectively, a numerical coincidence between electromagnetic and optical properties appears fortuitous. On the other side, if we look at them under the *same* paradigm, as Maxwell did, the coincidence becomes physically relevant, and discloses a possible substantial unity, opening the way to a scientific revolution [10].

Part III closes with the derivation of the relationship between the index of refraction of a substance and its relative electric and magnetic constants:

$$n^2 = \varepsilon_r \mu_r \qquad (4.16)$$

mentioned in the above reported letter to Faraday, which could obviously provide a further check of the close connection between optical and electromagnetic phenomena.

In the fourth and last part of the memoir (*The Theory of Molecular Vortices applied to the Action of Magnetism on Polarized Light*), Maxwell applies his model to the analysis of the Faraday's effect, and he succeeds in determining a quantitative law for the rotation of the polarization plane, which was in agreement with all known experimental results. The law involves also the mean radius of the molecular vortices, which motivates the above mentioned efforts to detect their existence in an independent, non optical, way.

Since the beginning, Maxwell was fully persuaded of the correctness of his hypothesis on the nature of light. This is clearly shown by the above reported letters, as well as by his interest in a precise determination of the "electromagnetic" value of the light velocity, see the letters to Fleeming Jenkin [11] (August 27, 1863) and Thomson (September 27 and October 15, 1864) [28], as well as the manuscripts of September 1864 and February 1865 [29].

In particular, in the letter to Thomson dated October 15th, we find a sketch (Figure 4.7) of an experimental arrangement which establishes the principle of operation of the apparatus (Figure 4.8) he developed and used for the experiment reported in his 1868 paper on the subject [30]. In the same year, he also made his first attempts to explain the reflection and refraction of light in the framework of the theory see note of the October 1864 [31].

However, apart from the need of experimental corroborations, at least two points had to be addressed, before the theory could be considered satisfactory.

First of all, the equations for the electromagnetic field and the fundamental property of the total current needed to be clearly and explicitly noted.

Then, the properties of the electromagnetic (and optical) waves had to be derived from the equations and not, as in "Physical Lines of Force," exploiting

---

[10] Another outstanding example is provided by the coincidence of the inertial and gravitational mass, which led Einstein to the formulation of the Equivalence Principle. This principle is at the basis of the possibility of reducing gravitation to space-time geometry, hence of the General Theory of Relativity.

[11] An engineer, member of the British Association Committee, Thomson's agent in the great enterprise of laying the Atlantic cable.

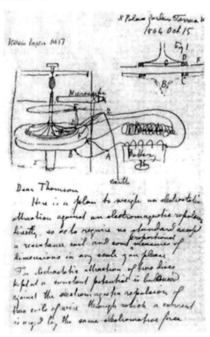

Figure 4.7. Sketch of an experimental arrangement to establish the ratio of electrostatic and electromagnetic units of electricity.

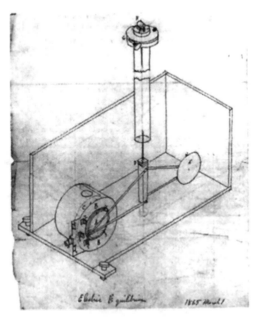

Figure 4.8. Maxwell's drawing (1865) of the torsion balance for the determination of the ratio of electrostatic and electromagnetic units of electricity.

a model so tricky to be considered "imaginary" by Maxwell himself.

This does not mean to give up a mechanical explanation of the phenomena, which would have been unthinkable in Maxwell's times, but to devise a field theory which could be still considered mechanical, but without all the scaffolding by aid of which it had been first erected. In other words, one should still assume the existence of an underlying medium, able to transmit the actions from one place to another by some kind of mechanical properties, but without specifying the physical nature of these properties and the structure of the medium.

4.4 A DYNAMICAL THEORY OF THE ELECTROMAGNETIC FIELD

The difficulty of the above mentioned goal is quite evident, and it required about three years to be achieved, during which Maxwell published just one minor paper of geometrical character.

The task was apparently completed in the summer, 1864; to which two fragmentary drafts date back [32]. One of these fragments contains some of the equations which will appear in the forthcoming third memoir, and still shows a sign of ambiguity in the relation connecting electric field and electrical displacement.

In a letter to Charles Hockin, dated 7 September 1864 [33], Maxwell writes: "I have also cleared the electromagnetic theory of light from all unwarrantable assumptions …"

On the 27th of October he presented the abstract of the memoir: *"A Dynamical Theory of the Electromagnetic Field"* to the Royal Society. This abstract was read on the 8th of December and published in the Transactions of the Society the following year [34].

The introduction, which summarizes in detail the motivations and the content of the memoir, opens with a praise of Weber's theory, after which Maxwell writes:

> *The mechanical difficulties, however, which are involved in the assumption of particles acting at a distance with forces which depend on their velocities are such as to prevent me from considering this theory as an ultimate one though it may have been, and may yet be useful in leading to the coordination of phenomena.*
>
> *I have therefore preferred to seek an explanation of the fact in another direction, by supposing them to be produced by actions which go on in the surrounding medium as well as in the excited bodies, and endeavoring to explain the action between distant bodies without assuming the existence of forces capable of acting directly at sensible distances.*
>
> *The theory I propose may therefore be called a theory of the* Electromagnetic Field, *because it has to do with the*

space in the neighbourhood of the electric or magnetic bodies, and it may be called Dynamical *Theory, because it assumes that in that space there is matter in motion, by which the observed electromagnetic phenomena are produced.*

The electromagnetic field is that part of space which contains and surrounds bodies in electric or magnetic conditions. It may be filled with any kind of matter, or we may endeavour to render it empty of all gross matter, as in the case of Geisslers tubes and other so-called vacua.

It is clear from these statements that Maxwell ascribes a physical, material, reality to the electromagnetic medium, and that in his view the electromagnetic phenomena are just the expression of the mechanical properties of this medium. This is by no means surprising: on the contrary, it is in complete agreement with the then widely accepted view that the propagation of light and radiant heat consists of undulations of "omnipervasive luminiferous ether". In fact, he writes:

We may therefore receive, as a datum from a branch of science independent of that with which we have to deal, the existence of a pervading medium, of small but real density, capable of being set in motion, and of transmitting motion from one part to another with great, but not infinite velocity. Hence the parts at this medium must be so connected that the motion of one part depends in some way on the motion of the rest; and at the same time these connections must be capable of a certain kind of elastic yielding, since the communication of motion is not instantaneous, but occupies time.

The medium is therefore capable of receiving and storing up two kinds of energy, namely, the "actual" energy depending on the motions of its parts, and "potential" energy, consisting of the work which the medium will do in recovering from displacement in virtue of its elasticity.

The propagation of undulations consists in the continual transformation of one of these forms of energy into the other alternately, and at any instant the amount of energy in the whole medium is equally divided, so that half is energy of motion, and halt is elastic resilience.

A medium having such a constitution may be capable at other kinds of motion and displacement than those which produce the phenomena of light and heat, and some of these may be of such a kind that they may evidence to our senses by the phenomena they produce.

However, this time the main question is not, "how is it made?", but "how does it work?" Only very general assumptions are made concerning the properties of the medium, mainly the capacity of receiving and storing both "actual" (i.e., kinetic) and "potential" energy, some kind of elasticity, and the fact that it must be subject to the general laws of dynamics. The question is answered in the following two parts of the memoir (*On Electromagnetic Induction* and *General Equations of the Electromagnetic Field*).

Starting from the laws of electromagnetic induction, expressed in the language of Lagrangian dynamics [12], Maxwell finds the generalized co-ordinates and momenta of the mechanical system in terms of electromagnetic quantities, and then its energy (i.e., its Hamiltonian). By requiring that to get the total motion of electricity the displacement currents must be added to the true one, he obtains the following set of general equations of the electromagnetic field (see Table 4.2). And Maxwell's equations are born! Maxwell however originally wrote these equations in the scalar form.

The equations are then applied to deduce the mechanical actions of the field (Part IV), to the electric condensers (Part. V), and to the calculation of the coefficients of mutual induction (Part VII).

Part VI is devoted to the electromagnetic theory of light. The purpose is clearly stated in the first sentence:

> At the commencement of this paper we made use of the optical hypothesis of an elastic medium through which the vibrations of light are propagated, in order to show that we have warrantable grounds for seeking, in the same medium, the cause of other phenomena as well as those of light. We then examined electromagnetic phenomena, seeking for their explanation in the properties of the field which surrounds the electrified or magnetic bodies. In this way we arrived at certain equations expressing certain properties of the electromagnetic field. We now proceed to investigate whether these properties of that which constitutes the electromagnetic field, deduced from electromagnetic phenomena alone, are

Table 4.2 Maxwell's Set of General Equations of the Electromagnetic Field

A	$C = K + \partial D / \partial t$	eq. of total currents
B	$\mu H = \nabla \times A$	eq. of magnetic force
C	$\nabla \times H = C$	eq. of currents
D	$E = v \times B - \partial A / \partial t - \nabla \Phi$	eq. of electromotive force
E	$E = kD$	eq. of electric elasticity
F	$E = \rho K$	eq. of electric resistance
G	$\nabla \cdot D = e$	eq. of free electricity
H	$\nabla \cdot K = -\partial e / \partial t$	eq. of continuity

[12] This approach was a relatively new historical development in Maxwell's time. Thomson and Peter Guthrie Tait set out a virtual manifesto of the new generalized mechanics in their ambitious Treatise on Natural Philosophy [35]. Tait, first wrangler at Maxwell's Tripos, was a good mathematician and a lifelong friend of Maxwell.

sufficient to explain the propagation of light through the same substance.

Then, Maxwell proceeds to deduce the properties of electromagnetic waves from the field equations (Figure 4.9), even if in somewhat involved way, and their propagation in isotropic (possibly lossy) and anisotropic media is examined in detail and compared with that of optical waves.

This allows him to conclude:

Hence electromagnetic science leads to exactly the same conclusions as optical science with respect to the direction of the disturbances which can propagate through the field; both affirm the propagation of transverse vibrations and both give

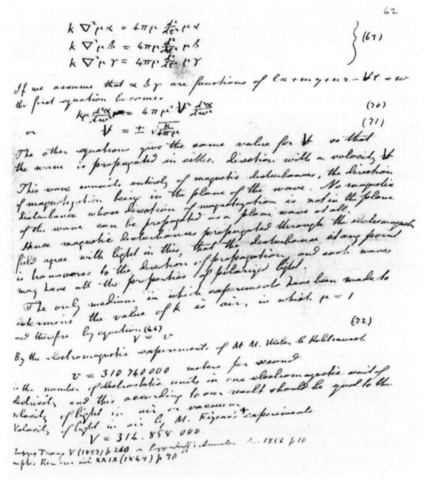

Figure 4.9. Maxwell's manuscript on electromagnetic theory of light.

the same velocity of propagation.

Therefore:

> *The agreement of the results seems to show that light and magnetism are affections of the same substance, and that light is an electromagnetic disturbance propagated through the field according to electromagnetic laws.*

The establishment of the electromagnetic theory of light is accomplished. However, nothing is said (or will be ever said by Maxwell) about the reflection and refraction of electromagnetic waves. This is not casual, but reflects a serious drawback of the Maxwell approach, i.e., the difficulty, or even the impossibility of deducing the correct conditions at the interface starting from purely mechanical considerations.

Another peculiar aspect of the Maxwell approach concerns the nature of electric charge and current. In accordance with his emphasis on the role of the medium, Maxwell (and the British Maxwellians after him) considered charges and currents not as the *sources* of the field, but, vice versa, as a *product* of the field itself. In other words, the ether is the only fundamental physical (i.e., mechanical) entity, and the description of the phenomena must be obtained by a proper characterization of its dynamical properties, i.e., its Hamiltonian.

This attitude was doomed to failure, because matter possesses its own degrees of freedom, so that it constitutes a dynamical system *independent* of (even if *coupled* to) the electromagnetic field. Moreover, it explains the inability of Maxwell and the Maxwellians to consider the question of the generation of electromagnetic waves distinct from the light.

It is well known that this crucial validation of Maxwell theory was performed by Hertz only in October 1886, almost twenty five years after his first announcement, whereas the definitive abandonment of the ether theories had to wait the epochal paper of Einstein in 1905. In the meantime, the atomic theory of matter had been developed and the quantum revolution had started. But Maxwell's equations stood unchanged and today they are still those created by his genius 145 years ago.

REFERENCES

[1]   T. S. Kuhn, *The Structure of Scientific Revolutions*, Chicago: The University of Chicago Press, 1962.

[2]   P. M. Harmon (Ed.), *The scientific letters and papers of James Clerk Maxwell*, Cambridge: Cambridge University Press, 1990, Vol. 1, pp.237-238.

[3]   R. Murphy, *Elementary Principles of the Theories of Electricity, Heat and Molecular Actions. Part. I. On Electricity*, Cambridge, 1833.

[4]   M. Faraday, *Experimental Researches in Electricity*, London, Taylor & Francis, 3 vols. 1839, 1844, 1855.

[5]   W. Weber, "Elektrodynamische Maassbestimmungen, uber ein

allgemeines Grundgesetz der Elektrischen Wirkung", *Leipzig Abhandl.*, 1846, pp. 211-378.

[6]   P. M. Harmon (Ed.), op. cit., Vol. 1, pp. 254-263.

[7]   Idem, p. 294.

[8]   Ibidem, pp. 210-211.

[9]   Ibidem, pp. 305-313.

[10]   W. Thomson, "On the uniform motion of heat in homogeneous solid bodies, and its connection with mathematical theory of electricity", *Camb. Math. J.*, 1842, Vol. 3, pp. 71-84.

[11]   P. M. Harmon (Ed.), op. cit, Vol. 1, pp.319-324 and 337-352.

[12]   Trans. Camb. Phil. Soc., X, 1856, pp. 27-83; W. D. Niven (Ed.), *The Scientific Papers of James Clerk Maxwell*, Cambridge, 1890, Vol. 1, pp. 155-229.

[13]   P. M. Harmon (Ed.), op. cit, Vol. 1, pp. 387-391 and 406-409

[14]   H. L. von Helmholtz, *Uber die Erhaltung der Kraft*, Berlin, G. A. Reimer, 1847.

[15]   P. M. Harmon (Ed.), op. cit, Vol. 1, pp. 505-507.

[16]   W. Thomson, "Dynamical illustrations of the magnetic and helicoidal rotatory effects of transparent bodies on polarized light", *Proc. Roy. Soc.*, Vol. 8, pp. 150-158, 1856.

[17]   M. Faraday, "Experimental Research in electricity.-Nineteenth series. On the magnetization of light and the illumination of magnetic lines of force", *Phil. Trans.*, Vol. 136, 1846, pp. 1-20.

[18]   P.M. Harmon (Ed.), op. cit, Vol. 1, pp. 548-552.

[19]   Idem, pp. 578-581.

[20]   *Phil. Mag.*, XXI, pp. 161-175, 281-291, 338-348, 1861 (Parts I and II); *Phil. Mag.*, XXIII, pp. 12-25, 85-95, 1862 (Parts III and IV).

[21]   P. M. Harmon (Ed.), op. cit, Vol. 1, pp. 683-688.

[22]   H. L. Fizeau, "Sur an expérience relative à la vitesse de la propagation de la lumière", *Compte Rendus*, Vol. 29, 1849, pp. 90-92.

[23]   P. M. Harmon (Ed.), op. cit, Vol. 1, pp. 690-691.

[24]   L. Campbell and W. Garnett, *The Life of James Clerk Maxwell. With a Selection from his Correspondence and Occasional Writings and a Sketch of his Contributions to Science,* London, p. 329, 1882.

[25]   P. M. Harmon (Ed.), op. cit, Vol. 1, pp. 692-698 and 699-702.

[26]   R. Koholraush and W. Weber, "Elektrodynamische Maassbestimmungen insbesondere Zuruckfuhurung der Stromintensitatsmessungen auf mechanishes Maass", *Abhandl. Der Konig. Sachsischen Gesell. der Wissenschaften*, Math.-Phys. Klasse, Vol. 3, 1857, pp. 219-292.

[27]   G. Kirchhoff, "Uber die Bewegung der Electricitat in Drahten", *Ann. Phys.*, Vol. 100, 1857, pp. 193-217.

[28]   P. M. Harmon (Ed.), op. cit, Vol. 2, pp. 110-111, 172-175 and 176-181.

[29]   Idem, pp. 165-171 and 204-206.

[30]   J. C. Maxwell, "On a Method of Making a Direct Comparison of Electrostatic with Electromagnetic Force; with a Note on the Electromagnetic Theory of Light", *Phil. Trans.*, Vol. 158, 1868, pp. 643-657.

[31]   P. M. Harmon (Ed.), op. cit, Vol. 2, pp. 182-185.
[32]   Idem, pp. 158-159 and 160-163.
[33]   Ibidem, p.164.
[34]   J. C. Maxwell, "A Dynamical Theory of the Electromagnetic Field", *Phil. Trans. Roy. Soc.*, Vol. 155, 1865, pp. 459-512.
[35]   W. Thomson and P. G. Tait, *Treatise on Natural Philosophy*, Oxford, Clarendon Press, 1867.

5

MAXWELL, HERTZ, THE MAXWELLIANS AND THE EARLY HISTORY OF ELECTROMAGNETIC WAVES

DIPAK L. SENGUPTA, *University of Michigan, Ann Arbor, MI*
TAPAN K. SARKAR, *Syracuse University, Syracuse, NY*

5.1   INTRODUCTION

In 1864 Maxwell conjectured from his famous equations that light is a transverse electromagnetic wave. Maxwell's conjecture does not imply that he believed light could be generated electromagnetically. In fact, he was silent about electromagnetic waves and their generation and detection. It took almost a quarter of a century before Hertz discovered electromagnetic waves and his brilliant experiments confirmed Maxwell's theory. Maxwell's ideas and equations were expanded, modified and made understandable by the efforts of Hertz, Fitzgerald, Lodge and Heaviside, the last three being referred to as the Maxwellians. A cursory overview is provided of the early history of electromagnetic waves up to the death of Hertz in 1894. The work of Hertz and the Maxwellians are reviewed briefly in the context of electromagnetic waves. It is found that historical facts do not support the views proposed by some in the past that Hertz's epoch making findings and contributions were "significantly influenced by the Maxwellians".

In the year 1864 James Clerk Maxwell (1831–1879) proposed his Dynamical Theory of the Electromagnetic Field [1] where he observed theoretically that electromagnetic disturbance travels in free space with the velocity of light. He then conjectured that light is a transverse electromagnetic wave. Although the idea of electromagnetic waves was hidden in the set of equations proposed by Maxwell, he had in fact said virtually nothing about electromagnetic waves other than light nor did he propose any idea to generate such waves electromagnetically. It has been stated [2] that: *There is even some reason to think that he (Maxwell) regarded the electrical production of such waves as impossibility.* Heinrich Hertz (1857–1894) discovered electromagnetic waves around the year 1888 [3]; the results of his epoch making experiments and his related theoretical work confirmed Maxwell's prediction and helped the

215

general acceptance of Maxwell's electromagnetic theory. However, it is not commonly appreciated that *Maxwell's theory that Hertz's brilliant experiments confirmed was not quite the same as the one Maxwell left at his death in the year 1879* [2]. It is interesting to note how the relevance of electromagnetic waves to Maxwell and his theory prior to Hertz's experiments and findings are described in [2]:

"..... *Thus Maxwell missed what is now regarded as the most exciting implication of his theory, and one with enormous practical consequences. That relatively long electromagnetic waves or perhaps light itself, could be generated in the laboratory with ordinary electrical apparatus was unsuspected through most of the 1870s.*"

Maxwell's ideas and equations were expanded, modified and made understandable after his death mainly by the efforts of Heinrich Hertz, George Francis FitzGerald (1851–11901), Oliver Lodge (1851–1940) and Oliver Heaviside (1850–1925) of which the last three were christened as "The Maxwellians" by Heaviside [2,4].

The history of electromagnetic waves up to the year 1894 is briefly reviewed in this chapter. A short discussion of Maxwell's original equations and some brief comments on the work of Hertz and the Maxwellians are given in the context of electromagnetic waves. It is found that historical facts do not support the views expressed by some in the past [4] that Hertz's epoch making findings and other contributions in electromagnetics were "significantly influenced by the Maxwellians".

5.2   SPECULATIONS OF ELECTROMAGNETIC PROPAGATION BEFORE MAXWELL

There were natural philosophers and scientists before Maxwell who speculated [5] on the manner in which electric and magnetic influences or effects are transmitted through space. The prince of mathematics Karl Friedrich Gauss (1777–1855) in 1855 tossed with the idea that electric actions propagate between the charges with finite velocity but he resolved not to publish his researches because he could not design a mechanism to achieve that transmission. More than one attempt to realize Gauss's aspiration were made by his pupil Riemann who in 1853 proposed [5] to replace Poisson's equation for the electrostatic potential by a wave equation according to which the changes in potential due to changing electricity would propagate outward from the charges with the velocity of light. Although this is in agreement with the view, which is now accepted as correct, Riemann's hypothesis was too trivial to serve as the basis of a complete theory. Two papers appeared in Poggendorf's *Annalen* for 1867. The first by Bernhard Riemann who in 1858 showed that by appropriately modifying Laplace's equation, one can obtain the wave equation, but was withdrawn before publication. The second paper by M. Lorenz shows that on Weber's theory periodic electric disturbances could propagate with a velocity that of light.

It is now known [6] that in a deposition with The Royal Society (London) entitled "The Original Views" Michael Faraday (1791–1867) tossed with the idea that electric and magnetic effects "are progressive and require finite time for their transmission". Faraday did not find time to provide experimental evidence to support his views and hence wished the deposition submitted in 1832 to remain unopened for at least 100 years. It should be noted that Maxwell in his 1864 paper [1] comments on Faraday's thought in this regard in the following manner: *The conception of the propagation of transverse magnetic disturbances to the exclusion of normal ones is distinctly set forth (Philosophical Magazine, III, p. 447) in his "Thoughts on Ray Vibrations". The electromagnetic theory of light, as proposed by him, is the same in substance as that which I have begun to develop in this paper, except that in 1846 there were no data to calculate the velocity of propagation.*

5.3   MAXWELL'S ELECTROMAGNETIC THEORY OF LIGHT

In his treatise *Electricity and Magnetism*, Maxwell developed a mechanically equivalent model of the electromagnetic field. *On the Dynamical Theory of the Electromagnetic Field* in 1864, Maxwell reverses his mode of treating electrical phenomena adopted by first arriving at the laws of induction and then deducing the mechanical attraction and repulsions. In his 1864 paper read at the Royal Society (London) Maxwell introduced 20 equations involving 20 variables [1]. These equations together expressed mathematically virtually all that was known about electricity and magnetism then. Through these equations Maxwell essentially summarized the work of Hans C. Öersted (1777–1851), Karl F. Gauss (1777–1855, Andre M. Ampère (1775–1836), Michael Faraday (1791–1867) and others, and added his own radical concept of "Displacement Current" to complete the theory.

To place Hertz's contributions in proper perspective in the context of Maxwell's proposed Dynamical theory of Electromagnetic Field and his conjecture therein regarding the nature of light [1], and also for historical reasons, it is appropriate to review the original 20 equations introduced by Maxwell and how he arrived at the crucial conjecture mentioned earlier. For this purpose Maxwell's original variables and equations are recast under modern notation. Table 5.1 shows the names and symbols used by Maxwell for the variables along with their identification by modern vector/scalar notation.

Observe that the set of three quantities appearing in each of the first six entries in the center column of Table 5.1 represent the three rectangular (x, y, z) components, respectively, of the corresponding vector quantity given in the right most column. Maxwell also indirectly used another variable (not shown in Table 5.1) named "the magnetic induction" whose three components in an isotropic medium being μ_α, μ_β and μ_γ with μ being the "coefficient of magnetic induction". We now call it the magnetic flux density vector $\boldsymbol{B} = \mu\boldsymbol{H}$, μ being the permeability of the medium.

With the variables given in Table 5.1 and for an isotropic medium Maxwell introduced 20 equations in component forms numbered (A) - (H) which formed the basis for his proposed Dynamical Theory of Electromagnetic Field

[1]. In our modified notation these equations maybe represented by the following:

$$J_T = J + \frac{\partial D}{\partial t} \tag{A}$$

$$\mu H = B = \nabla \times A \tag{B}$$

$$\nabla \times H = 4\pi J_T = 4\pi \left[J + \frac{\partial D}{\partial t} \right] \tag{C}$$

$$E = \begin{array}{c} \mu(v \times H) \\ or \\ (v \times B) \end{array} \quad -\frac{\partial A}{\partial t} - \nabla \psi. \tag{D}$$

[*Note*: Maxwell called eq. (D) as the equations of electromotive force in a conductor moving with velocity v in an isotropic medium].

TABLE 5.1. Twenty Variables Originally Introduced by Maxwell ([1] page 71)

Variable Name Used by Maxwell (Equivalent Modern Name)	Symbol Used by Maxwell	Modern Equivalent Vector/Scalar
Electromagnetic Momentum (Magnetic Vector Potential)	F, G, H	A
Magnetic Force (Magnetic Field Intensity)	α, β, γ	H
Electromotive Force (Electric Field Intensity)	P, Q, R	E
Current Due to True Conduction (Conduction Current Density)	p, q, r	J
Electric Displacement (Electric Flux Density)	f, g, h	D
Total Current Including Variation of Displacement (Conduction plus Displacement Current Density)	$\begin{cases} p^1 = p + \dfrac{df}{dt} \\ q^1 = q + \dfrac{dg}{dt} \\ r^1 = r + \dfrac{dh}{dt} \end{cases}$	J_T
Quantity of Free Electricity (Volume Density of Electric Charge)	e	ρ
Electric Potential (Electric Scalar Potential)	ψ	ψ

$$E = k\boldsymbol{D}, \tag{E}$$

where k is the "coefficient of electric elasticity" called by Maxwell. [Note: Compare Eq. (E) with the modern Eq. $\boldsymbol{D} = \varepsilon\boldsymbol{E}$, ε being the permittivity of the medium].

$$E = \rho'\boldsymbol{J}, \tag{F}$$

ρ' being the "specific resistance" or resistivity of the material. [Note: (i) Maxwell used the symbol ρ instead of ρ'. We are using ρ' so as not to conflict with our notation ρ for volume charge density]. (ii) Compare Eq. (F) with the modern equation $\boldsymbol{J} = \sigma\boldsymbol{E}$, with $\sigma = \dfrac{1}{\rho'}$].

$$\nabla \cdot \boldsymbol{D} = \rho \tag{G}$$

$$\nabla \cdot \boldsymbol{J} + \frac{\partial \rho}{\partial t} = 0 \tag{H}$$

It is clear that there are 20 equations in Eqs. (A) - (H). It should be noted that Maxwell used Gaussian system of units in which the electric and magnetic quantities are expressed in cgs electrostatic and cgs electromagnetic units, respectively (i.e., *E.S.U.* and *E.M.U.*, respectively). The appearance of the factor 4π in Eq. (C) is due to the use of this unit. It is important to retain the original units used by Maxwell, which is essential to appreciate how Maxwell arrived at his famous conjecture.

There are two theoretical set of units [20]. In the electrostatic system (*E.S.U.*), the fundamental unit is the unit of charge, this being defined as a charge such that two such charges at unit distance apart in air exert unit force upon one another. From the unit of charge we can get other units for the electric force, of electric potential, of electric current and so on. In the electromagnetic system (*E.M.U.*) the fundamental unit is the magnetic pole, this being defined to be such that two such poles at unit distance apart in air exert unit force upon one another. So also from the unit magnetic pole can be derived other units of magnetic force, of magnetic potential, of strength of a magnetic shell and so on in which we measure quantities which occur in magnetic phenomenon [20]. After Maxwell, the *E.M.U.* and *E.S.U.* systems described above were not adopted in their pure forms. Instead a composite system was devised called *Gaussian units*. This is a mixture of the *E.M.U.* and *E.S.U.* systems. From the former it takes the 'magnetic units' of field strength, flux, flux density, and magnetization and so on. From the latter it takes the 'electric units' charge, current, permittivity and so on.

The use of the parameter k in Eq. (E) needs some explanation. From the considerations of mechanical forces experienced by electric and magnetic charges, and the relationship between *E.M.U.* and *E.S.U.* units, Maxwell showed that ([1], page 569)

$$k = 4\pi \frac{C^2}{K}, \tag{5.1}$$

where, K is the "inductive or specific inductive capacity" (or, the dielectric constant) of the medium, and

$$C = (one\ emu\ of\ electric\ charge) / (one\ esu\ of\ electric\ charge)$$

in Gaussian units, for air (free space) $\mu = 1$ and $K = 1$.

If electric phenomena were entirely disassociated from magnetic phenomena, then two entirely different set of units would be necessary and there would be no connection between them. But the discovery of the connection between electric currents and magnetic forces enables us at once to form a connection between these two sets of the units. That is exactly how Maxwell proceeded. At this point, it is important to note that Maxwell was the originator of dimensional analysis. He first wrote the dimensions of all the quantities in the two systems and observed the expressions of the same quantity in these systems. Maxwell then wrote the two expressions in the E.S.U and E.M.U units for the current to obtain the relationship between the units of the two systems. He observed that

$$\frac{E.S.U.\ of\ current}{E.M.U.\ of\ current} = \frac{M^{1/2}L^{3/2}T^{-2}K^{1/2}}{M^{1/2}L^{1/2}T^{-1}\mu^{-1/2}} = \frac{L/T}{\dfrac{1}{\sqrt{K\mu}}}$$

$$= \frac{unit\ velocity}{velocity} = pure\ number\ \left(say\ \frac{1}{c}\right)$$

this ratio is equal to a unit velocity divided by velocity, and this ratio of the two units must be a pure number [21]. Weber and Kohlrausch was the first to compute this ratio by measuring the capacity of a condenser electrostatically by comparison with the capacity of a sphere of known radius, and electromagnetically by passing the discharge from the condenser through a galvanometer. Their measured data revealed that this ratio is approximately equal to 3.0001×10^{10}[20]. For the velocity of propagation of light in air the following experimental value was noted 2.991×10^{10}. Thus the two quantities agree to within a difference which is easily within the limits of the experimental error. This led Maxwell to suggest that the phenomenon of electromagnetic propagation was in effect, identical with the propagation of light.

Maxwell assigned strong physical significance to the vector and scalar potentials A and ψ both of which played dominant roles in his formulation. He also assumed a hypothetical mechanical medium to justify the existence of displacement current in free space; this assumption produced strong opposition to Maxwell's theory from many scientists of his time. It is well known now that Maxwell's equations, as we know them now, do not contain any potential variable neither does his electromagnetic theory require any assumption of an artificial medium [as shown by Hertz and Heaviside] to sustain his "Displacement Current" in free space. The original interpretation given to the "Displacement Current" by Maxwell is no longer used; however, we still retain the term in honor of Maxwell. Although modern Maxwell's equations appear in a modified form, the equations originally introduced by Maxwell in 1864 formed the foundation of electromagnetic theory, which together may very well be refereed to as Maxwell's electromagnetic theory.

Maxwell now assumed that a plane wave is propagating through the field with velocity V in a direction given by the unit vector $\hat{w}$ whose direction cosines in the x, y, z directions are l, m, n, respectively. Then all electromagnetic functions will be function of

$$w = lx + my + nz - Vt. \tag{5.2}$$

By using the magnetic force equation (B) Maxwell showed that

$$\mu \boldsymbol{H} \cdot \hat{w} = 0, \quad \text{i.e.,} \quad \mu \boldsymbol{H} \perp \hat{w}, \tag{5.3}$$

which implies that the "direction of magnetization" must be in the plane of the wave (i.e., in the wavefront). Assuming an insulating and stationary isotropic medium ($\boldsymbol{J} \equiv 0, V \equiv 0$) he obtained the following from Eqs. (B), (C), and (D):

$$k[\nabla(\nabla \cdot \boldsymbol{A}) - \nabla^2 \boldsymbol{A}] + 4\pi\mu \left[\frac{\partial^2 \boldsymbol{A}}{\partial t^2} + \nabla\left(\frac{\partial \psi}{\partial t} \right) \right] = 0, \tag{5.4}$$

where

$$\nabla = \hat{x}\frac{\partial}{\partial x} + \hat{y}\frac{\partial}{\partial y} + \hat{z}\frac{\partial}{\partial z}.$$

The three rectangular components of Eq. (5.4) are the same as those given by Maxwell as Eq. (68) in ([1] – p. 578). Maxwell then eliminated A and ψ from each of the 3 equations in Eq. (5.4) and obtained 3 similar equations for the three rectangular components of $\boldsymbol{H}$. Since this is not obvious, we shall briefly outline the procedures involved. After taking the difference of y- and z- derivatives of the z- and y- components, respectively, of Eq. (5.4) and making use of Eq. (B), it can be shown that the x-component of $\boldsymbol{H}$ satisfies the following equation:

$$k \nabla^2 \mu H_x - 4\pi\mu \frac{\partial^2 (\mu H_x)}{\partial t^2} = 0. \tag{5.5}$$

Equations similar to Eq. (5.5) are obtained for H_y and H_z, and the three equations were shown as a set of equations numbered {Eq. (69)} in [[1] - p. 579].

Assuming that H_x, H_y, H_z are functions of w (given by Eq. (5.2)), it can be shown that Eq. (5.5) yields,

$$k\mu \frac{d^2 H_x}{d w^2} = 4\pi\mu^2 V^2 \frac{2d^2 H_x}{d w^2}, \tag{5.6}$$

which implies

$$V = \pm \sqrt{\frac{k}{4\pi\mu}}. \tag{5.7}$$

Similarly, the other equations for H_y, H_z yields the same value of V, so that the wave propagates in either direction $\pm \hat{w}$ with velocity V.

At this stage we quote Maxwell ([1] - p. 579): *This wave consists entirely of magnetic disturbances, the direction of magnetization being in the plane of the wave. No magnetic disturbance whose direction of magnetization is not in the plane of the wave can be propagated as a plane wave at all.*

Hence magnetic disturbances propagated through the electromagnetic field agree with light in this, that the disturbance at any point is transverse to the direction of propagation, and such waves may have all the properties of polarized light.

Although Maxwell considered only magnetic disturbance in [1], he later showed that [11] $E \perp w$ $(i.e., E \circ w = 0)$ and also $E \perp H$. Thus both electric and magnetic disturbance lie in the plane of the wave and they are mutually orthogonal and together propagate as a plane electromagnetic wave. We quote Maxwell again [7]: *The mathematical form of the disturbance therefore agrees with that of the disturbance which constitutes light, being transverse in the direction of propagation.*

With k given by (5.1), Eq. (5.7) yields:

$$V = \frac{C}{\sqrt{K\mu}}. \tag{5.8}$$

Maxwell considered air (free space) for which $K = 1$ and $\mu = 1$ (in Gaussian units), thus:

$V = C=$ *(one unit of electric charge in emu) / (one unit of electric charge in esu)*

Purely electromagnetic measurements [8] yielded:

$C = 314, 740,000$ *meters/sec* $= c =$ *velocity of light in free space.*

Maxwell now makes his most significant conjecture: ([1], p. 580) *The agreement of the results seems to show that light and magnetism are affections of the same substance, and that light is an electromagnetic disturbance propagated through the field according to electromagnetic laws.*

Thus, the most important consequence of Maxwell's proposed equations was to establish the possibility of electromagnetic wave propagating with the velocity, which could be calculated from the results of purely electrical measurements. Indeed as mentioned earlier, electromagnetic measurements [8] indicated that the velocity equals to that of light in free space. This led Maxwell to his famous conjecture that light is a transverse electromagnetic wave, a conjecture later verified by Hertz. However, it is important to note that Maxwell did not comment anything about the generation of light waves and/or electromagnetic waves of lower frequencies by electromagnetic means. There is no indication left behind by him that he believed such is even possible. Maxwell did not live to see his prediction confirmed experimentally and his electromagnetic theory fully accepted. The former was confirmed by Hertz's brilliant experiments; his theory received universal acceptance and his original equations in a modified form became the language of electromagnetic waves and electromagnetics mainly due to the efforts of Hertz and Heaviside.

5.4 ACCEPTANCE OF MAXWELL'S THEORY

It is well documented how the investigations of Hertz and the Maxwellians during the years 1879 (the year Maxwell died) to 1894 (the year Hertz died) finally led to the acceptance of Maxwell's theory by the scientific community. The contributions of the Maxwellians are described in [2,4], and that of Heaviside in particular [2,9] and also in Chapter 6. Hertz's original contributions can be found in [6,10,11] and also in section 11.2.3; descriptions of Hertz's experimental arrangement are given in [12,13] and of Hertz's biography and work in [14]. Since this has been discussed in detail in the literature, in the following we give only short comments on some appropriate items in the context of the topic of our interest.

5.4.1 Maxwell's Equations

Maxwell's original equations were modified and later expressed in the form we now know as "Maxwell's Equations" independently by Hertz and Heaviside. Their work discarded the requirement of a medium for the existence of "Displacement Current" in free space and they also eliminated the vector and scalar potentials from the fundamental equations. Thus, Hertz and Heaviside, independently, expressed Maxwell's equations involving only the four field vectors E, B, D, H. Although priority is given to Heaviside for the vector form of Maxwell's equations, it is important to note that Hertz's 1884 paper [15] provided the rectangular form of Maxwell's equations which also appeared in his later paper of 1890 [16]. It is to be noted that the coordinate forms of the equations given in [15] were first obtained by Hertz.

It is appropriate to mention here that the importance of Hertz's theoretical work [15] and its significance appear not to have been fully recognized [17]. In this paper, Hertz started from the older action-at-a distance theories of electromagnetism and proceeded to obtain Maxwell's equations in an alternate way that avoided both the mechanical models Maxwell had originally used and his formulation of displacement current; in fact, this paper formed the basis for all of Hertz's future theoretical and experimental contributions to electromagnetism. D'Agostino [18] was the first to point out the importance of this paper in the development of Hertz's ideas. New insights into Hertz's theory of electromagnetism are discussed in [17] where the entire work was recast in modern notation for ease of understanding. The authors of [17] then concluded, *"...It is remarkable that an alternate method was available to derive Maxwell's equations based on quite a different approach. The physical insight of Hertz's work seems not to have been well appreciated in the past. The contents of Hertz's theory probably had a great impact on his design of experiments years later.* D'Agostino points out [18] that in his 1884 paper Hertz developed a theory of free propagation of electromagnetic forces which was inspired by purely electromagnetic, not optical, phenomena. The fundamental contribution of Hertz's development was a theory of source-field relation, unknown to Maxwell.

In contrast to the 1884 paper, in his 1890 paper [16] Hertz postulated Maxwell's equations rather than deriving them alternatively. The equations written in component forms, rather than in vector form as done by Heaviside [9], brought unparalleled clarity to Maxwell's theory. After reading this 1890 paper by Hertz, Arnold Sommerfield had this to say: *It was as though scales fell from my eyes when I read Hertz's great paper* [19]. The paper entitled "Fundamental Equations of Electrodynamics for Bodies at Rest" served as a model for Sommerfeld's lecture on electrodynamics since his student days [19].

5.4.2   Electromagnetic Waves

A few months before the death of Maxwell, Lodge began to look into the possibility of producing electromagnetic waves [2]. He recorded several ideas in his Laboratory Notebook regarding the possibility of generating light electromagnetically. The first unambiguous description of how to generate electromagnetic waves other than light were given by FitzGerald and Lodge between 1879 and 1883 [2]; however, they did not have any idea as to how to detect them. Thus, although the Maxwellians utilized Maxwell's theory to show the possibility of generating electromagnetic waves, they grappled with the idea of actually producing and detecting such waves in practice without any success. This was to be accomplished by Hertz.

Maxwell's predictions and theory were confirmed by a set of brilliant experiments conceived and performed by Hertz who generated, radiated (transmitted) and received (detected) electromagnetic waves of frequencies lower than light. His initial experiment started in 1887 and the decisive paper on the finite velocity of electromagnetic waves in air was published in 1888 [3]. English translation of Hertz's original publications [10] on experimental and theoretical investigation of electric waves is still a decisive source of the history of electromagnetic waves and Maxwell's theory. Description of Hertz's experimental setup and his epoch making findings are given in [12].

It is important to note that Hertz and the Maxwellians were not aware of each other's work until Hertz published his 1888 work. The Maxwellians appreciated Hertz's brilliant work and its implication. By early 1889 FitzGerald and his assistant Trouton repeated most of Hertz's main experiments [2,4]; Lodge and his group in Liverpool repeated Hertz's results in 1890 [4] and *The Maxwellians quickly gave Hertz's experiments the widest possible publicity and labeled them from the first as a decisive new confirmation of Maxwell's theory* [2]. FitzGerald was instrumental in awarding Hertz the Rumford Medal of The Royal Society.

After the 1888 results, Hertz continued his work at higher frequencies and his later papers conclusively proved the optical properties (reflection, polarization etc.) electromagnetic waves and thereby provided unimpeachable confirmation of Maxwell's theory and predictions.

It is appropriate to mention here that there were other people who before Hertz observed electromagnetic waves; however, they could not relate their observations to Maxwell's theory. Susskind [14] describes that during 1875-1882 Thomas Alva Edison, Elihu Thomson, Amos Dolbear and David Edwards Hughes observed some form of electromagnetic waves. However, none of them were well versed in Maxwell's theory or electromagnetics and could not correlate such observations with electromagnetic waves and thereby missed the deep significance of their observations. Hughes [6] did detect the standing waves with nodes and antinodes at fixed distances produced by interference between incident and reflected waves, but he did not realize that until much later. Hertz, a trained physicist, believer in Maxwell's theory observed electromagnetic waves and related his findings to Maxwell's theory and thereby became the discoverer of electromagnetic wave.

5.5   HERTZ AND THE MAXWELLIANS

The Maxwellians, a book by Hunt [2], describes historically and chronologically the investigations of FitzGerald, Lodge and Heaviside, and of Hertz which helped to the final acceptance of Maxwell's theory and predictions by the scientific community.

Another book entitled *Hertz and the Maxwellians* by O'Hara and Pritcha [4] discusses selected works of Hertz and it documents for the first time in one place the correspondence between Hertz and the Maxwellians and English translations of the correspondence. In addition, it contains much valuable information about Hertz's personal life. O'Hara and Pritcha begin their book with the following statements [4]...

> *whereas the beginning of Hertz's career coincided with the death of Maxwell and no direct contact between them would have been possible, and association, mainly in the form of correspondence, between Hertz and the disciples of Maxwell in Britain and Ireland did come about and had a significant influence on his thought and researches. One of the aims of this study is to illuminate their association and to consider the extent of its influence of Hertz's work....*
>
> *.......It will be argued that Hertz's conversion to Maxwell's theory was a gradual process which was influenced by his association with the 'Maxwellians': his experimental work too was to benefit from these contacts.*

On the basis of our study it is argued that the basic premise of the book [36] quoted above is flawed. Since the first letter appearing in [4] is dated 1888 after the publication of Hertz's seminal paper [3] and since there is no evidence that there was any contact between Hertz and the Maxwellians before 1888, it is not possible "that the Maxwellians significantly influenced Hertz's researches in electromagnetics". Due to the late date at which correspondence started between Hertz and them, the most that could have influenced his work is that represented by the last two papers on electromagnetics included in the collected works, Electric Waves [10].

It is appropriate here to quote Hertz's own words to comment on this [10]...

> I may be permitted to record the good work done by two English colleagues who at the same time as myself were striving towards the same end. In the same year in which I carried out the above research, Professor Oliver Lodge in Liverpool investigated the theory of the lightning inductor and in connection with this carried out a series of experiments on the discharge of conductors which led him on the observations of oscillations and waves in wires. In as much as he entirely accepted Maxwell's views, and eagerly strove to verify them, there can scarcely be any doubt that if I had not anticipated him he would also have succeeded in observing waves in air, and had some years before endeavored to predict with the aid of theory the possibility of such waves, and to discover the conditions for producing them. My own experiments were not influenced by the researches of these physicists, for I only knew them subsequently. Nor indeed do I believe that it would have been possible to arrive at a knowledge of these phenomena by the aid of theory alone. For their appearance upon the scene of our experiments depends not only upon their theoretical possibility, but also upon a special and surprising property of the electric spark which could not be foreseen by any theory.

There is no evidence of interaction of any kind between Hertz and the Maxwellians during or before the preparation of Hertz's 1884 paper [15] on the modified form of Maxwell's equations. It is interesting to note that in his 1890 paper on Maxwell's equations [16] Hertz made the following comments [9] on Heaviside's work on the similar topic: ...

> Again the incompleteness of form referred to renders it more difficult to apply Maxwell's theory to special cases. In connection with such appreciation I have been led to endeavor for some time past to sift Maxwell's formulae and separate their essential significance from the particular form in which they first happened to appear. The results at which I have arrived are set forth in the present paper. Mr. Oliver Heaviside has been working in the same direction ever since 1885. From

*Maxwell's equations he moves the same symbols as myself;
and the simplest form which these equations (These equations
will be found in the Phil. Mag. for February 1888. Reference
is there made to earlier papers in The Electrician for 1885, but
this source was not accessible to me) thereby attain is
essentially the same as that at which I arrive. In this respect,
then, Mr. Heaviside has the priority.*

The four equations in vector notation containing the four electromagnetic field vectors first published by Heaviside, are now commonly known as Maxwell's equations. However, Einstein and Heaviside referred to them as the Maxwell-Hertz's and Maxwell-Heaviside and Hertz equations, respectively [2].

As described in [4], contact between Hertz and the Maxwellians started and deepened, after Hertz's 1888 paper was published. It is possible that contact with Heaviside may very well have modified Hertz's 1890 paper. However his 1884 paper where Hertz introduced the coordinate form of modified Maxwell's equations was worked out before the above mentioned contact with Heaviside.

It is beyond any doubt that Maxwell's original theory and thinking had profound influence on Hertz. But Hertz's epoch making discovery of electromagnetic waves, his researches in electromagnetics and the manner in which he formulated Maxwell's equations, in fact, the researches by which Hertz validated and confirmed Maxwell's theory and predictions were certainly not significantly influenced by the Maxwellians, as implied in [4].

5.6   CONCLUSION

Maxwell's original equations forming the foundation of electromagnetic theory and his famous conjecture of light as electromagnetic wave have been briefly discussed in the context of electromagnetic waves in general. The early history of electromagnetic waves and the part played by Hertz and the Maxwellians (i.e., FitzGerald, Lodge and Heaviside) towards the confirmation and acceptance of Maxwell's theory have been presented. It is found that the Maxwellians had minimum or no influence on Hertz's discovery of electromagnetic waves and on his other accomplishments in electromagnetics.

REFERENCES

[1]   J. C. Maxwell, "A Dynamical Theory of the Electromagnetic Field", *Phil. Trans.*, Vol. 166, pp. 459-512, 1865. (Reprinted in *The Scientific Papers of James Clerk Maxwell*, Dover, New York, 1952: Vol. 1, pp. 528-597).

[2]   B. J. Hunt, *The Maxwellians*, Ithaca and London: Cornell Press University Press, 1991.

[3]   H. Hertz, "On the Finite Velocity of Propagation of Electromagnetic Action" (*Sitzungsber. D. Berl. Akad. D. Wiss.*, Feb. 2, 1888. *Wiedemann's Ann.*, Vol. 34, pp. 551), in *Electric Waves*, H. Hertz translated by D. E.

Jones, New York, Dover Publications Inc., 1962.

[4]   J. G. O'Hara and W. Pritcha, *Hertz and the Maxwellians*, London: Peter Peregrinus Ltd., 1987.

[5]   E. Whittaker, *A History of the Theories of Aether and Electricity.* London and New York: Thomas Nelson and Sons Ltd., 1962.

[6]   G. R. M. Garratt, *The Early History of Radio from Faraday to Marconi*, London : The Institution of Electrical Engineers, 1995.

[7]   J. C. Maxwell, *A Treatise on Electricity and Magnetism*, Vol. II, Third Edition, Academic Reprints, Stanford, CA, 1953.

[8]   R. Kohlrausch, *Leipzig Transactions*, Vol. 5., 1957, p. 260 or W. Weber, *Poggendorff's Annalen*, Aug. 1856, p. 10.

[9]   P. J. Nahin, *Oliver Heaviside: Sage in Solitude*, New York: IEEE Inc., 1988.

[10]   H. Hertz, *Electric Waves* (Authorized English Translation by D. E. Jones), New York: Dover Publications, Inc., 1962.

[11]   J. F. Mulligan, *Heinrich Rudolph Hertz* (1857–1894), New York: Garland Publishing, Inc., 1994.

[12]   J. H. Bryant, *Heinrich Hertz: The Beginning of Microwaves*, IEEE Service Center, Piscataway, NJ, 1988.

[13]   R. Gerhard-Multhaupt et. al., *Heinrich Hertz*, Heinrich Hertz Institute, Berlin, 1988.

[14]   C. Susskind, *Heinrich Hertz*, San Francisco Press Inc., 1995.

[15]   H. Hertz, "On the Relations between Maxwell's Fundamental Equations of the Opposing Electromagnetics" (in German), *Wiedemann's Annalen*, Vol. 23, pp. 84-103, 1884. English translation is given in [16], p. 127-145.

[16]   H. Hertz, "On the Fundamental Equations of Electromagnetics for Bodies at Rest", in [10], pp. 195-240.

[17]   C-T. Tai and J. H. Bryant, "New Insights into Hertz's Theory of Electromagnetism", *Radio Science*, Vol. 29, No. 4, July-August 1994, pp. 685-690.

[18]   Salvo D'Agostino, "Hertz's Researches on Electromagnetic Waves", *Hist. Stud. Phys. Sci.*, Vol. 6, 1975, pp. 261-323.

[19]   A Sommerfeld, *Electrodynamics*, New York: Academic Press, 1964, p.2.

[20]   J. H. Jeans, *The Mathematical Theory of Electricity and Magnetism*, Cambridge: Cambridge University Press, 1908.

[21]   S. S. Attwood, *Electric and Magnetic Fields*, New York: John Wiley & Sons, 1932.

6

OLIVER HEAVISIDE

HUGH GRIFFITHS, *University College, London*

6.1   INTRODUCTION

Oliver Heaviside is unusual – perhaps even unique – among the scientists and engineers described in this book, in that he had no formal education beyond leaving school at the age of sixteen. He was somewhat deaf and lived a rather solitary life, and few people could ever claim to have known him at a personal level. In spite of these factors his contribution to several aspects of radio engineering has been immense, and his name is attached to at least two phenomena – the Heaviside step function and the Heaviside layer – in common use today.

This chapter sets out to explain something of Heaviside's life, his unusual character, and the nature and significance of his contributions.

6.2   HEAVISIDE'S LIFE

Oliver Heaviside was born on May 18, 1850, at 55 King Street [1] in Camden Town in north London (Figure 6.1). He was the youngest of four brothers. His father, Thomas Heaviside, was a wood engraver originally from Stockton-on-Tees in north-east England, and had come to London in 1849. Camden Town in the middle of the 19th century was typical of the developing Victorian metropolitan life, with its crowded, smoke-polluted environment.

One survey [2] identifies King Street as being near an area described as "lowest class; vicious, semi-criminal". Camden is now very fashionable, with property fetching outrageous prices.

In a letter written in 1897 to the Irish physicist, George FitzGerald, he recalled:

> *I was born and lived 13 years in a very mean street in London,*
> *with the beer shop and baker and grocer and coffee shop right*

---

[1] The house no longer exists; also King Street has been renamed Plender Street.

[2] C. Booth, *Life and Labour of the People in London*, Vol. 5 (first series: poverty), 1902: Descriptive map of London Poverty, 1889. The area of the map immediately to the south of Camden Town is reproduced in Professor Steve Jones's book *In the Blood: God, Genes and Destiny* (Harper Collins, 1996).

Figure 6.1. Heaviside's birthplace: 55 King Street, Camden Town.

opposite, and the ragged school just around the corner.
Though born and raised up in it, I never took to it, and was
very miserable there, all the more so because I was so
exceedingly deaf that I couldn't go and make friends with the
boys and play about and enjoy myself. And I got to hate the
way of tradespeople, having to fetch the things, and seeing all
their tricks. The sight of the boozing in the pub made me a
teetotaler for life.

The Parr's Head public house on Plender Street was built in 1861 (Figure 6.2), and may well have been the pub to which Heaviside was referring. Subsequent to this the family moved to 117 Camden Street (Figure 6.3). This house is still standing.

The sister of Oliver's mother had married Sir Charles Wheatstone (she had actually been Wheatstone's cook, so this must have been an unusual marriage, particularly in Victorian times). Wheatstone took a strong interest in his nephews, and got them to learn French, German and Danish, and no doubt sowed the seeds of Heaviside's interest in telegraphy.

Oliver suffered from Scarlet Fever in his youth, and as noted in his letter to FitzGerald, this left him partially deaf. It is very likely that this was a significant factor in his solitary lifestyle.

He did reasonably well at school, and achieved just under 44% in the school leaving exam at the age of 15, and gaining first place in Natural Sciences (but only achieving 15% in Euclidean geometry). He later had some strong criticisms of the teaching methods of the time:

Euclid is the worst. It is shocking that young people should be
addling their brains over mere logical subtleties, trying to

understand the proof of one logical fact in terms of something equally ... obvious, and conceiving a profound dislike for mathematics, when they might be learning geometry, a most important fundamental subject. I hold the view that it is essentially an experimental science, like any other, and should be taught observationally, descriptively and experimentally...

and

I feel quite certain that I am right in this question of the teaching of geometry, having gone through it at school, where I made the closest of observations on the effect of Euclid on the rest of them. It was a sad farce, though conducted by a conscientious, hard-working teacher. Two or three followed, and were made temporarily into conceited logic-choppers, contradicting their parents....

He left school at sixteen, and pursued his studies at home. Then, at the age of eighteen he took a job (the only paid employment he ever had) with the Great Northern Telegraph Company, working in Newcastle and in Denmark. There seems little doubt that Wheatstone was instrumental in getting him this job. During this time he had started to study Maxwell's work, and began to publish articles in the *Philosophical Magazine* and elsewhere, on various aspects of circuit and telegraph theory. He was greatly influenced by reading Maxwell's work, and later recorded:

Figure 6.2. The Parr's Head public house in Plender Street (formerly King Street), built in 1861, when Oliver would have been 11.

Figure 6.3. 117 Camden Street – as it is today – where Heaviside's family lived till 1876.

I remember my first look at the great treatise of Maxwell's when I was a young man. Up to that time there was not a single comprehensive theory, just a few scraps; I was struggling to understand electricity in the midst of a great obscurity. When I saw on the table in the library the work that had just been published (1873), I browsed through it and I was astonished! I read the preface and the last chapter, and several bits here and there; I saw that it was great, greater and greatest, with prodigious possibilities in its power.

and

I was determined to master the book and set to work. I was very ignorant. I had no knowledge of mathematical analysis (having learned only school algebra and trigonometry, which I had largely forgotten), and thus my work was laid out for me. It took me several years before I could understand as much as I possibly could. Then I set Maxwell aside and followed my own course. And I progressed much more quickly.

In his Electrical Papers he recorded:

It will be understood that I preach the gospel according to my interpretation of Maxwell.

After six years, in 1874, he left this job and returned to live with his parents. Two years later the family moved to 3 St. Augustine's Road [3] (Figure 6.4), a few hundred yards from Heaviside's birthplace. The house was owned by the Midland Railway Company.

He apparently liked to work late at night, and liked to stoke up his fire so that the temperature was high – "hotter than hell" as one observer described it.

In 1889 (when Oliver would have been thirty-nine) the family moved to Paignton, near Torquay, in south-west England. Oliver's brother Charles was a partner in a music business there (apparently Wheatstone was also interested in music, and even invented something called the "English concertina").

In the summer of 1908, after his parents had died, he moved to a house called "Homefield" in Torquay, and was looked after by Miss Mary Way, the sister of his brother's wife. In 1916 she left, so from 1916 till his death in 1925 (at the age of seventy-four) he lived entirely alone.

Recognition did ultimately come to Heaviside (Figure 2.20). He was elected Fellow of the Royal Society in 1891. He was sent a formal notice asking him to come to London to be admitted, but wanted nothing of it, and even wrote a little poem in his notebook (Figure 6.5):

---

[3] This house also no longer exists – there are now just some lock-up garages on the site.

Figure 6.4. 3 St. Augustine's Road, Camden Town.

Figure 6.5. A page from Heaviside's notebook.

Yet one thing More
*   Before*
Thou perfect Be.
Pay us three poun'
Come up to Town
And then admitted Be.
But if you Won't
Be Fellow, then Don't.

He was granted a Civil List pension of £120 per annum (approximately $15,000 in today's currency) in 1896 "in consideration of his work in connection with the theory of electricity". It is likely that Lord Rayleigh was instrumental in arranging this. The pension was supplemented by an additional £100 in 1914 "in recognition of the importance of his researches in the theory of high-speed telegraphy and long distance telephony". In 1905 the University of Göttingen, in Germany, conferred on him the degree of Ph.D. *honoris causa*. The citation, freely translated from Latin, reads as follows:

That Eminent Man
Oliver Heaviside
An Englishman by Nation, dwelling
at Newton Abbot
Learned in the Artifices of Analysis
Investigator of the Corpuscles which are
Wont to be called Electrons
Persevering, Fertile, Happy, though
given to a Solitary Life
Nevertheless among the Propagators
of the Maxwellian Science
Easily the First

From about 1918 he called himself The Worm, and signed himself Oliver Heaviside, W.O.R.M. His closest friend (if he had one) during this period was Dr George Searle, of the University of Cambridge.

Searle was a mathematician and experimental physicist (Figure 6.6). As a young man he had worked with J. J. Thompson on determining the velocity of electromagnetic waves. He died in 1954, and his obituaries show that his teaching at Cambridge was very highly regarded. He was 'a man of striking appearance, arresting voice and forceful personality'. Dr Arnold Lynch suggests that the character of M. H. L. Gay in C. P. Snow's *Strangers and Brothers* series of novels may be based on Searle [4].

Searle and his wife visited Heaviside on numerous occasions, and Searle was uniquely able to write what is without doubt the most faithful description of

---

[4] A. C. Lynch, 'The sources for a biography of Oliver Heaviside', *History of Technology*, Vol. 13, pp. 145-149, 1991.

Figure 6.6. G. F. C. Searle.

Heaviside's character. This account, in the form of a 50,000 word draft, was found by Ivor Catt, and published by him in 1987. Lynch notes that it looks incomplete, but that it is the whole of what Catt had obtained.

Searle described Heaviside as 'impish', and it is true that he had quite a sense of humor. He sometimes liked to use his own versions of Latin in his writings – thus he wrote to Searle:

> *Dr S. Te igitur. V.P. Ego te remittare £10 (decem pundi Anglorum) in returno Ionorum.*

and Mary Way, his housekeeper up until 1916, was described as *"mulier bestissima"* – the very best woman.

Searle was quite dismissive of Heaviside's use of Latin – and of Greek – but it is clear that he was just playing games with language. Another example of this is that Heaviside sometimes used an (imperfect) anagram of his own name – "O he is a very devil".

Mary Way had a lot to put up with; he referred to her as 'the baby', and if she went out and did not return by the time Heaviside expected, she might find him in the garden with a lighted candle, "looking for her dead body".

From 1916 till his death Heaviside lived alone. His health was sometimes poor, and he clearly found looking after himself quite difficult. He wrote in a letter to Searle:

> *I made some jam the other day out of some apples the boys had not stolen and some blackberries which I could not eat. But I am not fit for a cook; I forget. Then it all goes to cinder, to be discovered hours later. Or, if I boil an egg, I am startled by a loud report; either I did not put any water in or else it has all boiled away.*

In 1921 the Institution of Electrical Engineers (IEE) instituted its Faraday Medal, and selected Heaviside as its first recipient. He was asked what form he thought the medal ought to take, and replied 'about three inches in diameter, one inch thick, and made of solid gold'. Since it was out of the question for him to travel to London, the then President of the IEE, J. S. Highfield went to Torquay to present him with it.

> *It was my duty as President to present the Medal, and the duty was an interesting and pathetic one. Heaviside lived entirely alone in a pleasant house in Torquay – a house decaying from long neglect. I called first by appointment and found him waiting in the weed-covered drive in an old dressing-gown, armed with a broom, trying vainly to sweep up the fallen leaves.*
>
> *I saw him again in the August of 1922 .... He seemed greatly improved in health, and there is no doubt that the interest taken and shown in him by his fellow workers of the Institution had really cheered him; in fact he seemed quite happy.*
>
> *(Highfield)*

Heaviside died on 3 February 1925, and his body is buried in Paignton Cemetery. Although there are no blue plaques in Camden commemorating his birth or where he lived, the IEE erected one in Torquay (Figure 6.7).

All of the above paints a picture of a solitary man of eccentric behavior, awkward in company and with very few friends. No doubt much of this was due to his deafness from a young age. His eccentricity may perhaps be summarized by his paragraph which introduces Volume 3 of *Electromagnetic Theory*:

Figure 6.7. The plaque erected by the IEE to commemorate Heaviside.

The following story is true. There was a little boy, and his father said, 'do try to be like other people. Don't frown.' And he tried and tried, but could not. So his father beat him with a strap, and then he was eaten up by lions.

6.3   HEAVISIDE'S CONTRIBUTIONS

Heaviside's published output was quite prodigious, and he published in *The Electrician* and the *Philosophical Magazine* and elsewhere, as well as collected papers in two books: *Electrical Papers* (in two volumes) and *Electromagnetic Theory* (in three volumes).

6.3.1   Transmission Lines

Heaviside's initial work was on telegraphy – what we now call transmission lines, in the form of submarine cables. The existing theory on the subject was due to Kelvin, and gave the voltage $v(x, t)$ on a cable, in terms of the resistance R and capacitance C per unit length, thus:

$$RC\frac{\partial v}{\partial t} = \frac{\partial^2 v}{\partial x^2}.$$

This is exactly similar in form to Fourier's equation for thermal diffusion, and indeed it must have seemed natural to think of the problem in terms of "electric diffusion".

However, this failed to account for a number of observations. In particular, it had been noticed that there was a maximum signaling speed above which the Morse characters smeared into each other – intersymbol interference or dispersion in modern parlance – and that this could even be asymmetrical, in other words that a particular cable could carry information more rapidly in one direction than the other. In three papers in the *Philosophical Magazine*, the first of which was published in 1876, he presented a full theory. He approached the problem in terms of the propagation of waves along the transmission line, and realized that it was necessary to take into account the inductance of the line. From this starting point he derived the differential equation (the now widely known telegrapher's equation)

$$\frac{1}{LC}\frac{\partial^2 v}{\partial x^2} = \frac{R}{L}\frac{\partial v}{\partial t} + \frac{\partial^2 v}{\partial t^2},$$

in which L is the inductance per unit length.

He derived and presented solutions to this equation for various configurations and initial conditions, which accounted for the observed phenomena including the asymmetrical effects. He also published in a paper entitled "Electromagnetic Induction and its Propagation", in *The Electrician* in

June 1887, the conditions for the distortionless (i.e. dispersionless) transmission line, and showing that the self-inductance of the line was essential – indeed that periodic inductive loading could be used to realize the condition for dispersionless transmission and achieve greater rate of transmission of information. This can be seen by requiring the second term in the telegrapher's equation to be negligible, i.e.

$$\frac{R}{L}\frac{\partial v}{\partial t} \to 0,$$

which is satisfied when L is large. Under these conditions the equation reduces to:

$$\frac{1}{LC}\frac{\partial^2 v}{\partial x^2} = \frac{\partial^2 v}{\partial t^2},$$

whose solution is a wave propagating (in either direction) with velocity $1/\sqrt{LC}$, independent of frequency.

Heaviside even proposed, in a letter to B. A. Behrend[5], that the process of inductive loading should be known as *heavification*, which he felt was expressive and gave just credit to its inventor – but this was never adopted, of course.

The papers were not, however, written in a very accessible style, and tended to leave out stages in derivations that would have made them much easier to follow. Heaviside prefaced the first Volume of *Electromagnetic Theory* with:

> *Fault has been found with these articles that they are hard to read. They were, perhaps, harder to write.*

Heaviside had quite a major disagreement with the establishment, in the form of William Preece (later Sir William Preece), who was Engineer-in-Chief of the Post Office, a Fellow of the Royal Society and a President of the IEE. Preece took public pleasure in announcing his disdain for mathematicians in general and Heaviside in particular:

> *Mr Heaviside ... asserts (as I think somewhat impertinently) that 'it was very much his own fault' if Maxwell is 'still fully to be appreciated' - which appears to be Heavisidean English for 'not yet quite understood'. Well, if Maxwell's expositions are anything approaching Heaviside's in obscurity, no wonder.*
>
> (Preece)

He went on to criticize Heaviside's assertion that induction in transmission lines was essential:

---

[5] B. A. Behrend, 'The work of Oliver Heaviside', *The Electrical Journal*, January/February 1928; also published at the end of Vol. 1 of the third edition of *Electromagnetic Theory* (O. Heaviside, pp. 469-504, Chelsea Publishing Company, New York, 1971.

> *... self induction in various forms has proved a bête noire which required all our knowledge and all our skill, not only to master, but to comprehend; for the effects of self-induction are invariably ill effects*

and

> *Possibly the magnetic susceptibility of the iron is the cause of this. The magnetisation of the iron acts as a kind of drag on the currents. It is well known that telephones always work better on copper than iron wires, doubtless for the same reason.*

(Preece)

Later, in the United States, Michael Idvorsky Pupin of Columbia University and George Campbell of the Bell Company clashed when they tried to patent the concept of inductive loading, and although Pupin's original patent application was denied due to Heaviside's prior work, Pupin was ultimately able to patent the idea [6]. Ironically, Preece in 1908 then claimed that he had known all along that self-induction could be beneficial.

Another of Heaviside's major contributions to transmission line theory is contained in a patent in his name dated 6 April 1880, and entitled "Improvements in Electrical Conductors, and in the Manner of Using Conductors for Telephonic and Telegraphic Purposes". This patent was unearthed by Kraeuter and described in his book *British Radio and Television Pioneers: A Patent Bibliography*, and is mentioned also in the preface to the second edition of Nahin's book. The key paragraphs are as follows:

> *When a number of wires run parallel to one another, either suspended or otherwise, any change in the current flowing in one wire causes currents in all the rest by induction, and the effect may be so great as to seriously interfere with the working of telephonic circuits, and to a less degree of ordinary telegraphic circuits also.*
>
> *It is a common practice to complete the circuit by means of a second wire instead of the earth and it is well known that the inductive interference is thereby reduced in magnitude, the induced electro-motive forces in one wire canceling those in the other to a certain extent. The nearer the wires are brought together the less does the inductive interference become, but it cannot be altogether eliminated in this way, because the axes of the two wires cannot be made to coincide so that they shall be both at the same distance from any disturbing wire.*

---

[6] The story is told in detail by Norbert Wiener in his book *Invention: the Care and Feeding of Ideas* (Cambridge, MA: MIT Press, 1993).

> *My improvements have for object to obtain perfect
> protection, and to render a circuit completely independent
> under all circumstances of external influence. For this purpose
> I use two insulated conductors for the circuit, and place one of
> them inside the other; thus one conductor may be a wire, and
> the other a tube or sheath, thus forming a compound conductor
> consisting of a central wire surrounded by an insulated
> covering, which is in turn surrounded by a conducting tube or
> sheath, which must also be insulated. When the tube and inner
> wire are electrically connected at both ends of the line, as
> through apparatus in the usual manner, the circuit as thus
> described is completely independent of other circuits, and any
> number of such circuits, each containing an insulated tube and
> inner wire, may be laid side by side and worked without any
> mutual inductive interference from other wires worked in
> ordinary manners.*

This is interesting, because it describes nothing less than the ubiquitous coaxial cable. Such cables had been in use for many years previously, in transatlantic submarine telegraphy, but the idea had apparently not been patented. What is also curious is that Heaviside evidently did not make much (if any) money from this patent. It seems most likely that he sold the rights for a nominal sum.

It is frequently supposed that Heaviside was purely a theoretician, but that is not in fact the case. Appleyard records that he conducted experiments at 3 St. Augustine's Road with apparatus consisting of battery, carbon contact-blocks, a watch and a galvanometer. He also quotes a passage from Heaviside's notebook at that time:

> *Father smells acid in the room. Two or three evenings, I said,
> at hazard, it was the electricity. Query, ozone generated by
> sparking, or nothing to do with it. Father says it is just like the
> battery he made when he was a boy, and that it is my battery. I
> didn't say it wasn't. What is the best arrangement to get the
> greatest variation of resistance in the circuit? I find that the
> internal and external resistances must be equal.*

The final sentence seems to be a statement of the Maximum Power Transfer Theorem.

Dr Arnold Lynch notes that another example of Heaviside's experimental work is that he developed an experimental method for measuring inductance using an improved form of the ballistic bridge [7]. He explored various circuits with inductors in two of the four arms or in one of the familiar arms and

---

[7] O. Heaviside, 'On the use of a bridge as an inductive balance', *The Electrician*, Vol. 16, pp. 489-491, 1885.

in the source or the detector. However, he failed to notice that if the mutual inductance between two of the four arms was big enough, the result was an excellent bridge; this remained undiscovered for forty years before Blumlein invented it.

6.3.2   Maxwell's Equations.

While Maxwell's Treatise is very clear on the nature and importance of the Displacement Current, it does not present Maxwell's equations in the form in which we now recognize them [8]. The equations (of which there are many more than four) are written in terms of several different variables: fields and potentials, as well as current density, and they are in Cartesian form, with each component stated separately. Heaviside was convinced that the formulation should be in terms of fields (E and H) rather than potentials, and in the general vector notation which makes the symmetry and elegance of the equations so clear. He stated [9]

> ... *in particular the divergence of the displacement measures the density of electrification. Similarly, the divergence of the inductance measures the "magnetification," if there is any to measure, which is a very doubtful matter indeed. There is no evidence that the flux induction has any divergence...*

> ... *Divergence is represented by div, thus:* —

$$\mathrm{div}\mathbf{D} = \rho$$
$$\mathrm{div}\mathbf{B} = \sigma$$

> *if ρ and σ are the electrification and magnetification densities respectively."*

and

> ... *so that we have, in terms of the forces* **E** *and* **H**

$$\mathrm{curl}\left(\mathbf{H} - \mathbf{h}\right) = \mathbf{J}$$
$$-\mathrm{curl}\left(\mathbf{E} - \mathbf{e}\right) = \mathbf{G}$$

$\big[$ **J** had already been defined to include the displacement current, $\mathbf{J} + \dfrac{\partial \mathbf{D}}{\partial t}$ in modern notation, and **G** as the 'magnetic current', $\dfrac{\partial \mathbf{B}}{\partial t}$ $\big]$.

> ... *under e we include*

---

[8] J. C. Maxwell, "General equations of the electromagnetic field': Chapter IX in *A Treatise on Electricity and Magnetism*, Vol. 2, 1873; 3rd edition, Dover: New York, 1954.

[9] *Electromagnetic Theory*, Vol. I, p. 38, p. 50.

    (1) Voltaic force.

    (2) Thermo-electric force.

    (3) The force of intrinsic electrisation.

    (4) Motional electric force.

    (5) Perhaps due to various secondary causes, especially in connection with strains.

And under **h** *we include*

    (1) The force of intrinsic magnetization.

    (2) Motional magnetic force.

    (3) Perhaps due to secondary causes. "

It is interesting to note that Heaviside was always trying to state the equations in the most general form, but if the improbable contributions are neglected, and if modern symbols are used, the four equations reduce to the familiar

$$\mathrm{div}\mathbf{D} = \rho \qquad\qquad \mathrm{div}\mathbf{B} = 0$$

$$\mathrm{curl}\mathbf{H} = \mathbf{J} + \frac{\partial \mathbf{D}}{\partial t} \qquad\qquad \mathrm{curl}\mathbf{E} = -\frac{\partial \mathbf{B}}{\partial t}.$$

    Hertz, independently, was working on the same problem and arrived at the same conclusions, but Hertz acknowledged Heaviside's priority:

> *I have been led to endeavour for some time past to sift Maxwell's formulae and to separate their essential significance from the particular form in which they first happened to appear. The results at which I have arrived are set forth in the present paper. Mr Oliver Heaviside has been working in the same direction ever since 1885. From Maxwell's equations he removes the same symbols [the potentials] as myself; and the simplest form which these equations thereby attain is essentially the same as that at which I arrive. In this respect, then, Mr Heaviside has the priority ....*

(Hertz)

6.3.3 Operational Calculus

One of Heaviside's most original contributions was in operational calculus and its application to circuit theory. The concepts of operational calculus had been around for nearly two centuries, due to mathematicians such as Bernoulli, Lagrange, Laplace, Cauchy and Boole [10], and were developed with full mathematical rigor. Heaviside was aware of at least some of this background,

---

[10] E. Koppelman, 'The calculus of operations and the rise of abstract algebra', *Archive for History of Exact Sciences*, Vol. 8, pp. 155-242, 1971/72; quoted by Paul Nahin in *Oliver Heaviside: Sage in Solitude.*

and indeed possessed a copy of Boole's book on differential equations and operators.

Heaviside developed a system of operational calculus and used it to solve problems in circuit theory. In this system the symbol p was used to denote differentiation with respect to time, and hence p^{-1} to integrate. He was able to manipulate these operators algebraically, and was quite happy to use things such as $p^{1/2}$ or expand them as a series. He also introduced the step function which bears his name, and to which he gave the symbol **1**. Thus

$$\frac{1}{p} = \int_0^t du$$

$$\frac{1}{p}\mathbf{1} = \int_0^t 1 dt = \begin{cases} t & t \geq 0 \\ 0 & t \leq 0 \end{cases}$$

$$\frac{1}{p^n}\mathbf{1} = \begin{cases} \dfrac{t^n}{n!} & t \geq 0 \\ 0 & t \leq 0 \end{cases}$$

and even

$$p^{\frac{1}{2}}\mathbf{1} = \frac{1}{\sqrt{\pi t}} \quad .$$

Appleyard summarizes Heaviside's work as a:

> ... *remarkable article, written in 1887 but not published until 1892, on telegraph circuits, and his communications to the Philosophical Magazine of December 1887 on resistance and conductance operators, the whole comprising, in about 200 pages, the foundation of modern theory of telephonic and telegraphic transmission, united with dynamics in the conception of forces and stresses. In effect, Heaviside's Expansion Theorem enables an explicit expression for the currents as functions of time to be derived, for any network, from the differential equations, by means of intermediate operational equations, under the conditions that (i) the currents are initially zero, and (ii) given potential differences are applied at various given points in the network. Or conversely, his theorem enables the potential differences to be derived from the currents. His method consists in prescribing rules for obtaining the operational equations, and rules for translating the solution of the operational equations into solutions of the original differential equations.*

These methods were extremely powerful, but the mathematical establishment was rather suspicious. Later on, when Heaviside had submitted a paper to the Royal Society (two previous ones having been published), the referee stated that Heaviside was 'ignorant of the modern developments of the theory of linear differential equations' and that he was 'trying to find a royal road to results which have already been established by exact reasoning'. Lord Rayleigh, then Secretary of the Royal Society, wrote to Heaviside saying:

> *I am desired to return to you the thanks of the Royal Society for your paper 'On Operators in Physical Mathematics, Part III', and to inform you that the Committee of Papers, not thinking it expedient to publish it at present, have directed your manuscript to be deposited in the Archives of the Society.*
>
> (Lord Rayleigh)

Heaviside was rather annoyed at this, and wrote to one of his correspondents, FitzGerald:

> *As regards my paper, I think I had better drop it. I don't care to write for any medium where I am not welcome. The way the R.S. behaves is extraordinary. They have lots of money to pay the cost of publication, and they deliberately refuse to use their opportunities. It is such an impractical Society.*

FitzGerald replied:

> *I do not think it would be at all dignified or wise to take any extreme step because a referee differed from one as to the value of one's own papers. The best attitude is one of pity and regret that others should be so blind.*
>
> (FitzGerald)

Heaviside's approach to circuit analysis enjoyed some success in the early 20th century, and a number of textbooks were published on his methods. However, in 1937 the German mathematician Gustav Doestch published a book[11] showing how the Laplace transform could be applied to circuit analysis, and these methods have been since been almost universally adopted and taught.

6.3.4   The Heaviside Layer

Marconi's experiments had led to the first transatlantic radio communication in December of 1901, using equipment designed and built by John Ambrose Fleming (see Chapter 10). There was (and still is) some debate about the actual mechanism of propagation. In June 1902 Heaviside published the following, in an invited contribution to the 10th edition of the *Encyclopaedia Britannica*:

> *Seawater ... has quite enough conductivity to make it behave as a conductor for Hertzian waves and the same is true in a*

---

[11] G. Doetsch, *Theorie und Anwendung der Laplace-Transformation*, Springer-Verlag, 1937; republished by Dover, NY, 1943.

more imperfect manner of the Earth. The irregularities make confusion, no doubt, but the main waves are pulled round by the curvature of the Earth, and do not jump off.

There is another consideration. There may possibly be a sufficiently conducting layer in the upper air. If so, the waves will, so to speak, catch on to it more or less. Then the guidance will be by the sea on one side and the upper layer on the other.

In fact Kennelly, in America, had made much the same suggestion more or less simultaneously and there is evidence that both were anticipated by a physicist, Balfour Stewart. Nevertheless, the name "Kennelly-Heaviside Layer" or even just "Heaviside Layer" is pretty much universally recognized.

Those who have seen the Musical 'Cats', which is based on a set of poems by T.S. Eliot, may remember that the big final number consists of Grizabella, the old cat, being taken up to the Heaviside Layer to be reborn. We can imagine that would have appealed to Heaviside's sense of humor.

6.4 CONCLUSIONS

It can be appreciated from all of the above that Heaviside was a complex and unusual character. Despite his solitary lifestyle, he did correspond quite extensively with other contemporary scientists and mathematicians, and he published a great deal, so there is plenty of material about him to study. His work was ultimately recognized, and his legacy to us today is certainly significant, not the least by way of the Heaviside step function and the Heaviside Layer.

It is likely that his insight and his unusual and creative approaches to solving problems were due to his *lack* of a university education; there is perhaps a lesson for those of us who teach university courses, and we may ask, if another Heaviside were to appear, how the scientific establishment would recognize and treat him or her.

6.5 ACKNOWLDGMENTS

I am most grateful to the many people from whom I have learned about Heaviside and his life and work, both from their writings and from discussions with them. I would particularly like to mention Dr. Arnold Lynch and Professor Alex Cullen, both of University College London, and Professor John Roulston, lately from BAE Systems, Edinburgh. I have drawn heavily on Paul Nahin's excellent and painstakingly researched book *Oliver Heaviside: Sage in Solitude* and I am most grateful to him for his comments. Special thanks are due to Richard Forward and Anne Locker of the IEE archives; I am grateful to the IEE for permission to reproduce Figures 6.1, 6.4, 6.5 and 6.7, and to the Cavendish Laboratory, University of Cambridge, for permission to reproduce Figure 6.6.

While I have taken great pleasure in unearthing each piece of information about Heaviside and his work, responsibility for any omissions or errors must remain with me.

REFERENCES

[1] R. Appleyard, *Pioneers of Electrical Communication*, New York: MacMillan Company, 1930; Chapter 8, Oliver Heaviside.

[2] B. A. Behrend, "The work of Oliver Heaviside", *Electric Journal*, January/February 1928; also published at the end of Vol. I of the third edition of *Electromagnetic Theory* (O. Heaviside), pp. 469–504, New York: Chelsea Publishing Company, 1971.

[3] F. E. Hackett, "FitzGerald as revealed by his letters to Heaviside", *Scientific Proceedings of the Royal Dublin Society*, Vol. 26, No. 1, pp. 3-7, 1952/54.

[4] O. Heaviside, *Electromagnetic Theory* Vols. 1-3.

[5] H. J. Josephs, *Heaviside's Electric Circuit Theory*, Methuen's Monograms on Physical Subjects, London, 1946.

[6] H. J. Josephs, unpublished biography of Heaviside, IEE Archives.

[7] D. W. Kraeuter, *British Radio and Television Pioneers: A Patent Bibliography*, Lanham, MaryLand: Scarecrow Press, 1993.

[8] Per A. Kullstam, "Heaviside's operational calculus: Oliver's revenge", *IEEE Trans. Education*, Vol. 34, No. 2, pp. 155–156, May 1991.

[9] Per A. Kullstam, "Heaviside's operational calculus applied to electrical circuit problems", *IEEE Trans. Education*, Vol. 35, No. 4, pp. 266-277, November 1992.

[10] A. C. Lynch, "The sources for a biography of Oliver Heaviside", *History of Technology*, Vol. 13, pp. 145-149, 1991.

[11] A. C. Lynch, review of *Oliver Heaviside – The Man* (I. Catt ed.), *Electronics and Power*, Vol. 33, p. 469, July 1987.

[12] P. Nahin, *Oliver Heaviside - Sage in Solitude*, IEEE Press, 1988; second edition, Baltimore: Johns Hopkins University Press, 2001.

[13] J. G. O'Hara, and W. Pricha, *Hertz and the Maxwellians*, Peter Peregrinus, Stevenage, 1987.

[14] G. F. C. Searle, *Oliver Heaviside - The Man* (I. Catt ed.), C.A.M. Publishing, St. Albans, 1987.

[15] E. T. Whittaker, "Oliver Heaviside", *Bulletin of the Calcutta Mathematical Society*, Vol. 20, pp. 199-220, 1928; also published in the preface to the third edition of *Electromagnetic Theory* (O. Heaviside), Chelsea Publishing Company, New York, 1971, and in Moore, D.H., *Heaviside Operational Calculus*, New York, 1971.

[16] N. Wiener, *Invention: the Care and Feeding of Ideas*, Cambridge, MA: MIT Press, 1993.

[17] I. Yavetz, *From Obscurity to Enigma: The Work of Oliver Heaviside, 1872-1889*, 1995.

[18] *The Heaviside Centenary Volume*, Institution of Electrical Engineers, 1950.

7

WIRELESS BEFORE MARCONI

I.V. LINDELL, *Helsinki University of Technology, Finland*

7.1 INTRODUCTION

In this short overview we consider steps leading to the introduction of the practical wireless telegraph by Marconi. The principle of operation in these pre-Marconi experiments can be divided in three groups: conduction telegraph, induction telegraph and electromagnetic telegraph. The main reference for the present text is the book *A History of Wireless Telegraphy 1838 – 1899* by Fahie from 1900 [8]. Another major reference by Thörnblad from 1911 [21] (in Swedish) gives approximately the same material with an emphasis on later developments of the electromagnetic wireless. At the end of this chapter, other more specialized; references on the topic can be found.

7.2 CONDUCTION TELEGRAPH

Telegraphy based on current distribution in the ground was developed mainly in the hope of obtaining communication over watery regions like the sea and rivers where overhead lines could not be used. Sea cables turned out to be expensive, unreliable and, of course, useless for moving ships.

7.2.1 Early Ideas

The possibility of transmitting the electric force over wires was discovered by Stephen Gray (1666–1736), in London, in 1729 when making electrostatic experiments as shown in Figure 7.1. The series of experiments was carefully recorded in the Proceedings of the Royal Society, which is why we now know that the transmission line was invented at 10:00 AM on July 2, 1729. The idea of transmitting information over wires using the electrostatic force was first suggested in the 1750's and first known experiments were made in the 1770s. The first suggestion of using electricity without wires for telegraphic purposes seems to come from Francisco Salvá i Campillo (1751–1828), a Spanish physician [16]. After suggesting telegraphy over the sea using an insulated wire and water as the return conductor, before the Academy of Sciences in 1795 under the title "On the application of electricity to telegraphy", he continued with a

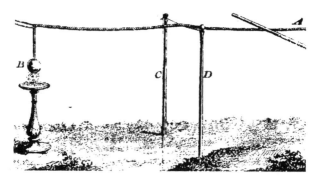

Figure 7.1. The first transmission line by Gray conveyed the electrostatic force from a rubbed glass tube to an ivory ball over common packthread hanging on isolating strings.

remark that the wire could also be totally disposed of. His idea was based on the analogy of a charged capacitor whose plates contain charge of opposite polarity. By connecting one of the plates to water a spark would indicate that the other plate has been connected to water. Thus, charging two distant land areas with opposite charges, one can test whether the other one has been connected to sea or not, which can be used to send one bit of information on an agreed topic over the sea.

After Volta's pile was introduced, Samuel von Sömmering (1755–1830) from Munich developed a working galvanic telegraph with a wire for each letter of the alphabet, which covered the distance of 12 kilometers in 1812 [12]. He also tried to replace two wires by troughs filled with water but noted that, although with two separate troughs the connection worked fine, for both electrodes in the same trough the communication discontinued. Had the trough been wider he could have produced the first wireless conduction telegraph. Professor Karl Steinheil (1801–1870), from Göttingen University (Figure 7.2)

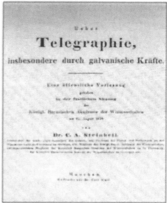

Figure 7.2. Steinheil was the first to make experiments with wireless electric telegraph making use of the conduction currents in the ground. On the right a printed version of his public talk on the topic in 1838.

made public the idea of using ground as a connecting medium. Karl Friedrich Gauss (1777–1855), who had a working two-wire telegraph line in Göttingen since 1833, suggested that Steinheil try to substitute the tracks of a newly built railroad between Nuremberg and Fürth for telegraphic wires. Experiments in 1838 showed that because of poor insulation this turned out to be highly impractical. However, this gave him the idea of using ground as the return conductor with a single overhead wire. The idea would save 50 percent of wire expenses in subsequent telegraph lines. Steinheil was aware that the density of current from an isolated ground electrode decays as the inverse square of the distance and part of the current could be extracted with two submerged electrodes. When making experiments he found that signals could be received at a distance of 20 meters without wires, which showed the possibility of a conduction telegraph. He predicted that with a more sensitive current detector and larger ground electrodes greater distances could be covered.

7.2.2   Morse's Wireless

Samuel Morse (1791–1872) is best known for his system of galvanic telegraph developed in the 1840's, which surpassed all the other contemporary systems by its speed and practicality. In 1842–1844 he also made the first known experiments with wireless telegraph over water as shown in Figure 7.3. The idea of trying to use water as the conductor occurred to him when an anchor had cut an insulated cable. He laid a pair of electrodes on each side of a channel of water 80 feet wide, and conducted experiments by changing the number of galvanic elements in the transmitter, distance of the pair of electrodes on each side of the channel, and size of the electrode plates. Observing the galvanometer readings, a conclusion was made that for best communication the distance between the electrodes on each side of the channel should be three times the width of the channel. After this, experiments were made over a river for a distance of a mile.

7.2.3   British and French Experiments

Since the British Isles are surrounded by the sea, many experiments with the

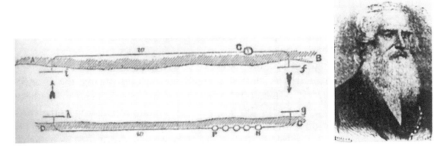

Figure 7.3. Morse's experiment with wireless telegraph shows two pairs of electrodes submerged in a channel with galvanometer G as the detector. Best reception required separation of electrodes three times the width of the channel.

wireless sea telegraph were made there during the 1840s and 1850s. A Scotsman James Bowman Lindsay (1799–1862), a teacher of mathematics in Dundee, pondered about the possibility of making a wireless telegraph connection between England and America. In an interesting letter to *Northern Warder* in 1845 he also suggested building a world-wide web of sea cables connecting a number of stations on different continents, and gave an estimate for the expenses of such a web. His idea was to give each station a fixed time slot for transmission during which all the other stations remain silent. The slot could be divided into smaller slots, one for transmission to each of the other stations. Because synchronizing the clocks around the world within an accuracy of seconds was possible through the telegraph, transmission from any station to any other station could take place at a certain hour during the 24 hour period.

In 1853 Lindsay gave a public lecture on "Telegraphic Communication," a report of which subsequently appeared in *Dundee Advertiser*. At this time, marine cable connections had been successfully laid out from England to Ireland and France but the first cable over the Atlantic was yet to come in 1858. In his patent application from 1854, Lindsay emphasized that the distance between the electrodes must be larger than the distance of communication, otherwise the batteries must be made very strong. In his view such a wireless connection to Ireland and France was practicable. On the other hand, because of the size of the British Isles, a connection to America would require a battery with zinc plates of 130 square feet and electrode plates of 3000 square feet for a signal observable to a galvanometer with a coil weighing 200 pounds. Such a huge coil was used by Morse on the Washington-Baltimore line in 1844 where the telegraph first created interest in the American public. After making successful experiments over a river for a distance of a kilometer, he learned to his disappointment that similar work was done by Morse a decade before. Nevertheless, Lindsay continued with experiments at various places, with largest distance of nearly two miles, and tried without success to convince the public on the advantage of the wireless telegraph.

In 1849 J. W. Wilkins from England suggested a wireless telegraph between England and France. His system applied overhead wires parallel to the Channel with electrodes at their ends submerged in the water. Thus, also, magnetic induction should have played some minor part in the connection. During 1852 till 1872, brothers Edward and Henry Highton made experiments with two bare metal wires laid in the water with transmitter and receiver connected at their ends. Similar attempts by other British experimenters have been described in Fahie's book.

Wireless telegraph became suddenly of utmost interest to the citizens of Paris after the German troops started a siege of the city during the Franco-Prussian war 1870–1871. After considering different communication methods over the enemy lines such as acoustic telegraph by sound propagating in the earth or overhead wire held up with balloons; a suggestion to use the river Seine as the medium for conductive wireless was made by the instrument maker Jean G. Bourbouze (1826–1889). His idea was to submerge transmitting electrodes at one point in the river and receiving electrodes at another point behind the enemy lines. After preliminary experiments, in December 1870 an agent was sent in a balloon to Champagne, from whence he continued to Le Havre to order the necessary equipment from England. After about a month, the agent returned to the bank of Seine with the apparatus only to find that the river was frozen. A test of this method to communicate with Paris could not be made because, when after ten days of waiting and the ice finally melted, Paris had capitulated just the day before.

7.2.4   Loomis's Wireless Telegraph

In 1872 a strange wireless telegraph using upper layers of the atmosphere as a communication medium was patented by an American dentist Mahlon Loomis (1826–1886). It was known since Franklin's kite experiments that electricity could be extracted from the atmosphere even on fair weather. It was also known that air in low pressure had higher conductivity as shown by phenomena in Geissler's tubes. Since the pressure became smaller at higher altitudes, which was proved in the many scientific balloon trips, it could be guessed that one could communicate through higher layers much like through the ground. Now Loomis had the idea to connect these two facts. Two kites with conducting wires were sent to the same height. A simple switch in the transmitting kite connecting the wire to a grounded electrode would create a disturbance in the available atmospheric electricity and create a current which should be observable in the electrometer connected to the grounded wire of another kite. No other source of electricity was needed.

Loomis claimed to have made in 1866 successful experiments between two Blue Ridge mountaintops in West Virginia to the distance of 14 miles, as seen in Figure 2.17. The kite wires were 600 feet in length and the detector was either a galvanometer [5] or an electrometer [8]. During an agreed period of time the transmitter switch was operated and the meter in the receiving end was observed. As the result of the experiment it was only stated that "messages were sent and received". In 1869 the U.S. congress decided to give $50,000 for the development of the "aerial telegraph", and although signed by President Grant, it eventually came to nothing. Loomis was not able to receive any financial aid even when he described the prospect of making a wireless connection between the Rocky Mountains and the Alps.

7.2.5   New Detector

Inventing the telephone in 1876 brought a new detector to the experimenters, the Bell receiver. While no movement could be seen in the most sensitive mirror galvanometer, a faint signal could be heard as a clear noise in the telephone receiver.

Bell's telephone originally used the same electromagnetic device as a receiver and a transmitter and it had to be moved between mouth and ear. Faint currents, difficult to hear without shouting, were induced in the moving coil by a permanent magnet without an exterior current source. When Edison sent signals over telegraph lines using his own microphone and a chemical battery, the much stronger currents were often overheard by the subscribers of Bell telephone lines without galvanic contact between the circuits. Emergence of this highly sensitive detector also had an impact on studying wireless telegraphy.

Professor John Trowbridge (1843–1923) from Harvard University was first to apply the telephone receiver to wireless communication between ships and stations on the shore. His interest started when he realized that time signals from the Harvard observatory were heard in all Boston telephones at a distance of 6 kilometers, which he found was due to a leaking current. Connecting a telephone receiver between two points in the ground, the signal was heard except when the points were on the same equi-potential curve. To make a large enough current needed for the ship communication, a less expensive source than a chemical battery, the dynamo powered by the steam engine, was invented in 1867.

After Edison's incandescent lamp came to the market in 1879 a dynamo was also needed in the ships for lighting. In the 1880s, Trowbridge developed conduction telegraphy by setting one of the electrodes at the front of the ship and the other one at the end of an insulated cable towed by the ship. When the distance between the two electrodes was half a mile, one could communicate with another ship to the distance of half a mile, which was a real step forward in foggy weather when the signal flags can't be seen. Another idea was to study the potential field of the other ship by a magnetic needle to find out the location of the ship. Similar experiments for wireless telegraphy for ships were also made by Alexander Graham Bell (1847–1922) who used a chopped DC current from a chemical battery for a source and telephone receiver for a detector.

7.2.6   Last Steps

Professor Erich Rathenau from Berlin finally developed the conduction wireless telegraph to its peak by extending its range to the distance of about 5 kilometers over water. Learning about the British experiments, and at the request of the Berlin Electrical Society, Rathenau made extensive tests in 1894 in Lake Wannsee close to the imperial palace at Potsdam. The transmitter electrodes were submerged at the distance of 500 meters along the shore and two boats with a receiving electrode on each were at a variable distance and connected by a cable, Figure 7.4. To increase the sensitivity, he used a telephone receiver with an iron tongue instead of the common disk. The receiver was tuned to the carrier frequency of the Morse signal obtained from a buzzer. Varying the distances of the receiving electrodes and their position with respect to the direction of the transmitter, optimal parameter values were observed. 150 Hz was found the best carrier frequency for this purpose. Rathenau pointed out that transmission of multiple signals could be made possible using different carrier frequencies. Compared to earlier results, five kilometers does not sound a major step forward. However, previously, the distance between transmitting electrodes was required to be larger than the transmitting distance, while for the new tuned system it

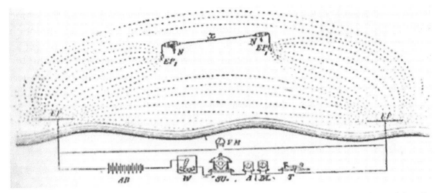

Figure 7.4. Rathenau's experiment on Lake Wannsee used tuned metal reed in the receiver.

could be only one tenth of that. In 1896 these experiments were continued by Strecker who achieved wireless reception at a distance of 17 kilometers. However, Marconi's experiments, starting the same year, would soon make the conductive wireless obsolete.

7.3 INDUCTION TELEGRAPH

Induction telegraph is based on either electric or magnetic induction between transmitting and receiving circuits. Since it utilizes quasi-static fields, range of transmission is comparable to that of the resistive telegraph.

Magnetic induction was discovered by Michael Faraday (1791–1867) and Joseph Henry (1797–1878) at about the same time although the credit went to the former. The latter was probably first to suggest application of magnetic induction to wireless telegraphy (Figure 7.5). A more detailed plan was given by Trowbridge in an article of 1891. He predicted that its main application could be

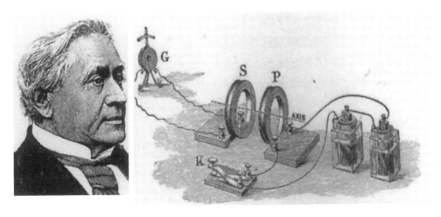

Figure 7.5. Joseph Henry suggested telegraphy based on magnetic induction.

between ships in foggy weather for relatively close distances. Because the sound from a horn tends to reflect from different air masses in a thick fog, direction finding tends to become erratic with a danger of collision. The magnetic induction can be used when the current is interrupted by a buzzer. Trowbridge noted that for a regular-sized battery the induced current is detectable only when the coils are at a distance of about their own size. For example, a transmission distance of half a mile would require a coil of the size 800 feet with ten turns. By rotating the coil, direction finding in a fog could be possible. Otherwise, he could not see a more general application of magnetic or electric induction as the principle of wireless telegraph.

7.3.1   Dolbear's Wireless Telephone

Professor Amos E. Dolbear (1837–1910), Figure 7.6 (right), from Tuft's College, Boston, invented the electrostatic telephone receiver in 1880. It consisted of a flexible capacitor plate, which created sound when time-varying voltage was applied. In contrast to Bell's receiver, the impedance was high and thus, it could be used to detect very small currents if the voltage was high enough. When using his receiver in the telegraph line, he observed that reception was possible even when the line was broken. He made experiments with different lengths of the broken section and noticed that metal obstacles like a tin roof or a kite attached to the broken line helped in the detection of the signal. This gave him the reason to experiment with wireless telegraphy in the 1880s [7]. He patented a device in 1883 and gave a description of it in *Scientific American* in 1886. The scheme resembles somewhat that of Marconi's radio equipment ten years later because both have an antenna and a ground electrode, Figure 7.6 (left). However, Dolbear's explanation of its operation was based on the excitation and detection of ground currents, although quasi-static capacitive induction must have played a role in bridging the gap because of the small frequencies involved. In 1883 using

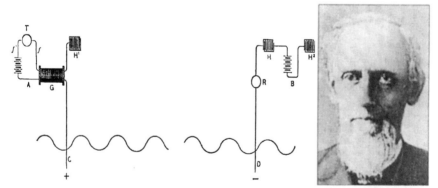

Figure 7.6. Dolbear's wireless telegraph resembles that of Marconi with antennas and ground electrodes. However the frequency of operation in the first experiments was too small for the electromagnetic waves. Dolbear interpreted the operation through currents in the ground.

a kite at 100 meters he obtained a communication to the distance of 1/4 mile. Later, attaching a spark gap to the transmitting side, the electromagnetic waves started to play the major role. The communication distance was finally extended to 12 miles, but before that Marconi had already entered the scene.

7.3.2   Edison's Wireless Telegraph

In 1885 Thomas A. Edison (1847–1931) filed a patent application "Means for transmitting signals electrically", which aimed at communication between ship and ground station or two ships. The figure in the patent shows antennas on ship masts and towers in ground stations, Figure 7.7. For beyond-the-horizon communication, antennas would be held by balloons. Applying electromagnetic waves, the invention would have preceded Marconi by a decade and even Hertz's experiments by a year or two. However, the system of operation was based on electrostatic induction and, although the patent was finally granted in 1891, it was never tested. A more practical invention was the patent "Railway telegraphy" from 1886, Figure 7.8. Telegraph communication from and to a moving train had already been suggested in Britain in the 1830's when train traffic and telegraphy were both novelties. There was an early need for a railway telegraph to avoid collisions of trains moving on single tracks. Edison wished to

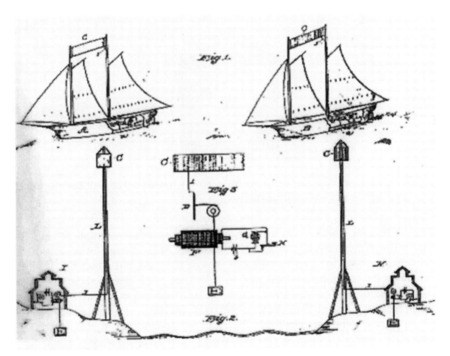

Figure 7.7. Edison's patent for wireless telegraphy between ships and ground stations used chopped current and telephone receivers. The antennas relied on quasi-static capacitive coupling.

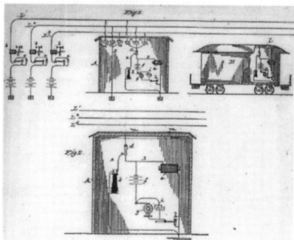

Figure 7.8. Edison's wireless telegraph from moving trains worked well but was ahead of its time and created no economic success.

make a system, which could be connected to the general telegraph network so that businessmen could work while traveling. His design was based on electrostatic induction between a metal roof of the train and a number of wires parallel to the tracks hanging on poles at a distance of 25 feet from the train. The telegraph stations along the track had a similar capacitive coupling to the wires. When receiving, the plate of the train was grounded through a telephone receiver to the tracks. When transmitting, the plate was connected to the Morse key through a battery chopped at 600 Hz. With this system, passengers could send and receive wireless messages with the outside world. In a test in 1887 a message was actually sent to London from a moving train. However, contrary to Edison's anticipations, businessmen wouldn't want to work while traveling, which soon made the invention unprofitable [2]. This somewhat resembles the fate of the picture telegraph by Giovanni Caselli (1815–1891): the invention was good but in the 1850's it did not respond to the needs of the public.

7.3.3 Stevenson and Preece

Wireless telegraph applying magnetic induction as suggested by Trowbridge was finally put into practice in Great Britain where the need for communication over water was imminent. In 1892 C. A. Stevenson made experiments with the induction between two coils, trying to find their optimum number of turns, resistance, relative position, and geometry. His aim was to build a connection between the island Muckle Flugga and mainland Scotland whose distance was 800 yards. For that purpose two coils of the diameter 200 yards and nine turns were erected and the transmission coil was fed with a current of one ampere.

Similar experiments with parallel wires on both sides of the water were made by Sir William H. Preece (1834–1913) over the Bristol Channel. In 1892 he had connections of 3.3 and 5.3 miles to two islands. These were used later when he assessed the performance of Marconi's wireless system. In 1894 Preece constructed a 4 mile wireless inductive telegraph over Kilbrannan Sound using parallel wires of 6 miles in length at the height of 500 feet. The array acted like a one-to-one transformer made of two rectangular loops, Figure 7.9. However, replacing the return conductors by ground electrodes at both ends, he found that the connection was clearly improved. Preece concluded that the effective areas of the loops were increased because of the distributed ground currents so that the connecting magnetic flux was almost doubled. In 1896 Preece met Marconi and found his method much better because "with Hertzian waves, conductors of very moderate length are needed."

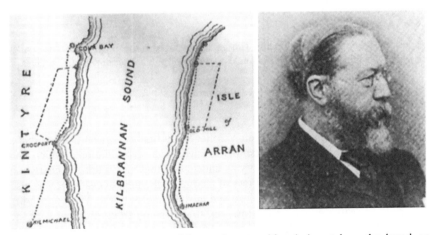

Figure 7.9. Preece made successful experiments with wireless telegraphy based on magnetic induction between loops of wire shown by dashed lines in this example.

7.4   ELECTROMAGNETIC TELEGRAPH

With better luck, the name of the inventor of radio would have been Luigi Galvani (1737–1798) who accidentally found, in 1789, that a spark from an electrostatic machine created convulsions in a frog leg when its nerve was simultaneously touched by a metal scalpel as seen in Figure 7.10. It was a lucky coincidence that Galvani was a medical doctor with an interest in frogs because the frog leg happened to be utterly sensitive to tiny electric currents. On the other hand, it was also a poor coincidence because from then on he concentrated his interest in animal electricity and did not continue to study the relation between the spark and the convulsions. The outcome of Galvani's work was the chemical source of electricity which was known as the Volta pile because it was Alessandro Volta (1745–1829) who solved the problems in Galvani's later experiments. After Marconi's success in wireless, Galvani's first experiment was remembered and repeated on a somewhat larger scale by a Frenchman Lefeuvre, early in the 1900s, with the help of a spark transmitter. The transmitter was situated in the Eiffel tower in Paris and the frog leg was connected to a receiver 319 kilometers away in Rennes. When the transmitter sent a Morse signal, a needle attached to the leg made the corresponding mark on a rotating cylinder covered by soot [14].

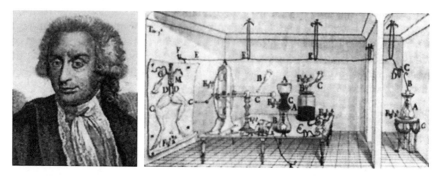

Figure 7.10. Continued experiments with his observation on the effect of electric sparks on frog legs could have made Galvani an early inventor of the electromagnetic wireless.

7.4.1   Henry

Joseph Henry a professor of Princeton College was first to suggest in 1842 that a spark from a Leyden jar (a capacitor) created electric oscillations, which can propagate in space. At this time, it was understood that in a spark the charge on the positive side moved to fill the loss of charge on the negative side. It was also known that if a steel needle was close and at right angles to the wire connecting the sides, it became magnetized. However, there was a certain problem with the magnetization of the needle found by a Frenchman Felix Savary in 1827. The direction of magnetization seemed to be erratic and showed no simple law as a function of amount of charge or distance from the wire, in particular, when the needle was thin. After making careful experiments, Henry came to the conclusion that the charge/discharge is actually oscillatory with decaying amplitude. The direction of magnetization is reversed on each oscillation until the amplitude becomes smaller than the level of saturation. Obviously, the saturation effect depended on the distance from the spark. Henry started to study the distance effect of the magnetic induction and enclosed the needle in a small coil connected to a loop of wire. When a spark of one inch was made in a room upstairs through a loop of wire, a similar loop in the cellar 30 feet below could magnetize the needle. Eventually, he could magnetize a needle at 220 feet away from the spark. Henry called this effect an extraordinary case of induction. He also observed magnetization created by lightning, which was definitely an electromagnetic wave effect [8,20].

7.4.2 Edison's Etheric Force

In 1875 Edison came very close to inventing the electromagnetic wireless. He noticed that a spark in the telegraphic circuit containing inductance created sparks between two distant conducting objects without any galvanic connection to the circuit except the ground. Observing these sparks, he could actually send messages without wires! When studying the received sparks he was puzzled because they could not be detected by a galvanometer, had no polarity and gave no chemical reactions like sparks from a Volta pile (Figure 7.11) Edison's conclusion was that this was due to a new physical force: "a radiant force, somewhere between light and heat on the one hand and magnetism and electricity on the other; in short, something new to science."

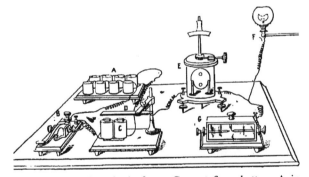

Figure 7.11. Edison's setup for studying the etheric force. Current from battery **A** is chopped by key **B** whence the inductance **C** gives a spark **D**. Galvanometer **E** does not react but there is a visible spark in the spark gap **G** grounded in the gas mains **F**.

He started to call it the "etheric force". For its detection, he constructed a spark-gap detector "the etheroscope," a dark box with two pointed conductors whose separation could be changed. Progress could be read from the weekly pages of *Scientific American* at the end of 1875 and in early spring 1876 [13,17]. The research ended when Edwin James Houston (1847–1914) and Elihu

Thomson (1853–1937) proved the same year that the sparks were actually oscillatory high-frequency electric currents, which could not be detected by DC equipment. Unfortunately, Edison dropped this phenomenon in favor of other more profitable research. With some more work he could have easily arrived at some form of electromagnetic wireless telegraph.

7.4.3   Maxwell and Hertz

When, in 1864 James Clark Maxwell (1831–1879) (Figure 2.16) had completed the set of electromagnetic equations with an additional term, he was able to predict the existence of electromagnetic waves. Experimental verification took a long time because there were no devices to generate and detect such waves for high enough frequencies. Heinrich Hertz (1857–1894) (Figure 2.23) was finally able to construct equipment needed to make the definite tests at Karlsruhe University in 1887. He proved that electromagnetic waves were similar to light except for the wavelength. He also showed that to effectively generate and detect these waves the size of the antennas must be that of the wavelength. Like Edison, Hertz used a small spark gap as a detector where the spark could be seen with a magnifying glass or, more often, it could be heard as a crackling noise. This kind of detector was unsuitable for telegraphic use. Hertz's interest was in physics and he was followed by a group of physicists interested in studying the "electromagnetic light".

7.4.4   Hughes

David Edward Hughes (1831–1900) (Figure 7.12, right) made experiments similar to Edison's but he, as well, failed to be the inventor of radio. Hughes was the inventor of letter-printing telegraph (1855) and a sensitive microphone (1878) which was used with the early telephone. The microphone was based on a loose contact made by a piece of carbon whose resistance varied with the pressure

Figure 7.12. The microphone by Hughes was based on a loose contact of a pointed carbon bar which made it sensitive to catch the faintest vibrations. It also served as a nonlinear detector in his wireless experiments.

wave created by voice. In 1879 right after the death of Maxwell and some 7 years before Hertz's experiments, when working with an impedance bridge he had invented, he noticed that a poor contact in the bridge circuit created noise in a telephone circuit containing the carbon microphone and with no galvanic contact to the bridge circuit (Figure 7.12, left). Interested in the phenomenon he found that the noise was caused by any sparks. Let us quote his letter from 1899 to Fahie:

> *Further researches proved that an interrupted current in any coil gave out at each interruption such intense extra currents that the whole atmosphere of the room (or in several rooms distant) would have a momentary invisible charge, which became evident if a microphonic joint was used as a receiver with a telephone. This led me to experiment upon the best form of a receiver for these invisible electric waves, which evidently permeated great distances, and through all apparent obstacles, such as walls &c. I found that all microphonic contacts or joints were extremely sensitive.* [8, p.290]

When writing this letter Hughes knew about the coherer and mentioned that it was based on the same phenomenon as his microphone. It became known to be more sensitive as a detector than Hertz's spark gap. In a further experiment trying to find out the range of the effect, he moved out from his home and found that the noise in the telephonic circuit radiated to a distance of 500 yards. In a demonstration to some Fellows of the Royal Society, Professor Stokes could not accept the explanation of electricity being conducted though the air but considered it to be due to simple magnetic induction. This seemed to be the only explanation before Hertz's experiments. Hughes, a professor of music, rejected an invitation to give a talk on his experiments to the Royal Society, perhaps because of his incomplete knowledge on scientific matters, and decided to do more experiments before publication. However, after Hertz's experiments, Hughes realized that it was too late to publish anyway. His handwritten notes on the experiments from the years 1878–1886 are preserved in the British Library as MSS 40161/3 [9].

7.4.5   The Coherer

After it was known that electromagnetic waves could be generated by electric sparks, the main problem was to find an effective way for their detection. The most practical detector in the early wireless telegraph was the coherer, a nonlinear circuit element, whose operation was based on the change of contact resistance of metal or carbon particles in an electric field. The first to deal with such a device was the English engineer Alfred Varley (1832–1921) who developed a protection against lightning surges in telegraphic lines. In 1866 he discovered that dust of conducting material, like coal or lead, was not conducting for small voltages but becomes so for larger voltages. He developed a lightning bridge (Figure 7.13, top left), which protected the sensitive telegraphic equipment by short-circuiting it for large voltage waves. The bridge consisted of a hollow piece of wood with two pointed metal conductors, at a distance of one-eighteenth of an inch, surrounded by powder of carbon mixed with some non-conducting material. With pure carbon the device would have been too sensitive.

Another inventor for the coherer was Temistocle Calzecchi-Onesti (1853–1922), an Italian professor, who found in 1884–1885 that small copper filings between two plates of brass (**D**, **E** in Figure 7.13, bottom) had conducting and non-conducting states. The latter state could be transformed to the former by applying a voltage which was high enough. The high voltage was created in the inductance **D** who's current from source **A** was suddenly changed by a switch **F**. Studying different materials between the plates it was found that some of them were sensitive enough to obtain the conducting state from a nearby spark.

Professor Edouard Branly (1844–1940) (Figure 7.13, photo), from Paris, made very thorough experiments similar to those of Calzecchi-Onesti in 1890 and published them in 1891. Through a set of quantitative measurements he was able to observe resistance changes of the ratio 1:1000. The publication made the device finally known. It was called the coherer by Lodge and proved superior to Hertz's spark detector in sensitivity [4, 15].

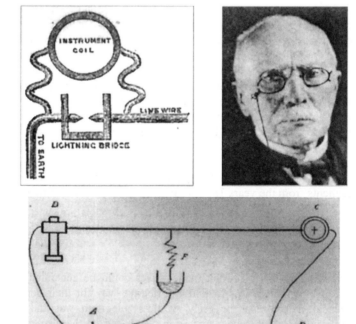

Figure 7.13. Varley's lightening bridge (top left) and Calzecchi-Onesti's experiment with copper filings between two plates (bottom) preceded Branly's (photo) thorough study of the coherer.

7.4.6 Tesla

Nikola Tesla (1856–1943) (Figure 2.24) was concerned with global views of telecommunication. Just like Loomis a few decades earlier, Tesla was convinced of the existence of a conducting layer in the atmosphere and he made an effort to make use of it for transmitting information and electric energy over long distances (Figure 7.14). Tesla started to consider the whole world as a resonator, whose basic resonance frequency he estimated as 6 Hz. Using rotating generators of his own design for carrier frequencies between the basic resonance and 20 kHz, with high towering antennas and ground electrodes, he planned to create a "stationary wave" propagating from his transmitter to any point on the globe. In 1893 he filed a patent on such a system which included the ideas of continuous-wave transmission with tuning and multiple phased radiators to increase and direct the radiated power. However, after starting to build an expensive world-wide transmitter, he lost the financial support because of Marconi's success with less costly equipment and the grand plan was discontinued. His patent was later used in an American court against Marconi's patent when determining the originator of the electromagnetic wireless telegraph.

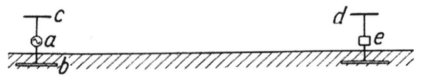

Figure 7.14. Tesla's patent for global wireless telegraphy involved continuous wave and tuning.

7.4.7 Lodge and FitzGerald

Sir Oliver Lodge (1851–1940) (Figure 7.15, left), professor in the University College at Liverpool, was one of the few who knew well Maxwell's theory. He lived a long life which included attending Maxwell's lectures and witnessing the development of radio, radar, and television. Inspired by Hertz's experiments he worked during 1888–1894 to clarify the connection between electromagnetic waves and optics. He was especially interested in finding a detector for the electromagnetic waves, which would correspond to the eye in optics and ended in using Branly's coherer. Because of his physics background Lodge was not primarily interested in the development of wireless telegraphy. However, he saw this as a possibility, but before Marconi's success, was rather pessimistic concerning its range of operation. George Francis FitzGerald (1851–1901) (Figure 7.15, right), professor at Trinity College, Dublin, had published in 1883 a formula for the power radiated by a small loop antenna, which showed it as proportional to the fourth power of the frequency [19]. It was easy to forget that this was only valid for a small antenna and imagine that efficient wireless connection required high frequencies. Because of FitzGerald's formula, Lodge

Figure 7.15. Lodge (left) and Fitzgerald (right) were Maxwellian physicists whose approach to wireless telegraphy was guided by the optical analogue.

also gave the careless estimation that the maximum practical communication distance with electromagnetic waves would be limited to half a mile. After Hertz' death in 1894 Lodge gave a public demonstration showing that signals could be transmitted through the walls to a distance of 60 yards in Oxford University. In the experiment he was first to use Branly's coherer as a detector instead of Hertz's spark gap. Since the receiver was a mirror galvanometer instead of a Morse inker, it has been argued that this was not a real demonstration of the wireless telegraph and thus, Lodge cannot be considered as the inventor of the radio [1, 10].

7.4.8 The Visionaries

Before Marconi there were several visionaries who could see electromagnetic waves being used in telecommunication. Simple use as a beacon was suggested by Richard Threlfall (1861–1932) in a meeting in Sydney in 1890. A Hertzian transmitter in a lighthouse would act like a torchlight, which would be visible also through a thick fog when a Geissler tube is used as a detector. A similar idea, ship to shore communication on foggy days, was also suggested by Alexander Pelham Trotter (1857–1947) in an 1891 article in *The Electrician*. However, full possibilities of Hertzian waves were predicted by William Crookes (1832–1919) (Figure 7.16, left), in an article published by *Fortnightly Review* in 1892.

    In the article he pointed out that Hertzian waves made possible telegraphic communication without wires, poles, and cables in spite of fog or brick walls. He could also see the advantage of narrowband tuning and directed beams for multiple simultaneous transmission and secrecy. It is not clear what the real impact of this article was on the further development of electromagnetic wireless. It has been noted that there were very few references to the article before Marconi's success. This may be due to his other, less realistic, visions

Figure 7.16. Crookes (left) wrote an article foreseeing the possibilities of Hertzian waves for wireless telegraphy while Marconi's (right) success was based on his engineering approach.

concerning the use of Hertzian waves for improving harvests, killing parasites, purifying sewage, eliminating diseases, and controlling weather. However, after 1897 when Marconi's wireless had proved its viability, references to the article started to emerge, most probably to weaken Marconi's claim for priority [10,11].

7.4.9   Finally, Marconi

In contrast to Lodge and other Maxwellians who were experimenting with the "electromagnetic light", Guglielmo Marconi (1874–1937) (Figure 7.16, right) did not have the ballast of physics background. Rather, he considered wireless as a possibility, much like a telegraph engineer, with an invisible line replacing the overhead wire. Thus, originally, frequency did not play any specific role for him and using earth as the return conductor gave him the earth electrode as a natural part of the setup. For him, the main problems were the size and form of the antenna and the sensitivity of the detector. It was only later that frequency came to his plans, in terms of tuning, but this falls beyond the scope of the present review. Since wireless took the path he originated with his first experiments in 1896, Marconi has been generally considered as the inventor of radio [3,18].

REFERENCES

[1]   H. G. J. Aitken, *Syntony and Spark – the Origins of Radio*, Princeton, NJ: Princeton University Press, 1976, p.123.

[2]   N. Baldwin, *Edison, Inventing the Century*, New York: Hyperion, 1995.

[3]   R. Barrett, "Popov versus Marconi: the century of radio," *GEC Review*, Vol. 12, No. 2, pp. 107-116, 1997.

[4]   C. Blondel, "Edouard Branly, dalla parte della scienza," *Cento anni di radio, le radici dell'invenzione* (ed., A. Guaglini, G. Pancaldi), Torino: Edizione Seat, 1995, pp. 303-354.

[5]   S. P. Bordeau, *Volts to Hertz...the rise of electricity*, Minneapolis: Burgess Publishing Company, 1982.

[6]   D. E. Dunlap, *Radio's 100 Men of Science*, New York: Harper & Brothers, 1944.

[7]   J. Erskine-Murray, *A Handbook of Wireless Telegraphy, its Theory and Practice*, London: Crosby Lockwood and Son, 1907.

[8]   J. J. Fahie, *A History of Wireless Telegraphy 1838-1899, including some bare-wire proposals for subaqueous telegraphs*, New York: Dodd, Mead and Co., 1900.

[9]   G. R. M. Garratt, *The Early History of Radio from Faraday to Marconi*, London: IEE History of Technology Series 20, 1994.

[10]   S. Hong, "Marconi and the Maxwellians: The Origins of Wireless Telegraph Revisited," *Technology and Culture*, Vol. 35, No. 4, pp. 717-749, 1994.

[11]   S. Hong, *Wireless*, Cambridge, MA: MIT Press, 2001.

[12]   Th. Karrass, *Geschichte der Telegraphie*, Braunschweig: Vieweg und Sohn, 1909.

[13]   I. V. Lindell, *Sähkötekniikan Historia* (History of Electrical Engineering, in Finnish), Helsinki: Otatieto, 1994.

[14]   E. Nesper, *Der Radio-Amateur (Radio-Telephonie)*, 6th ed., Berlin: Springer, 1925.

[15]   T. W. Pegram, R. B. Molyneus-Berry, and A. G. P. Boswell, "The coherer era, the original Marconi system of wireless telegraphy," *GEC Review*, Vol. 12, No. 2, pp. 83-93, 1997.

[16]   J. Romeu and A. Elias, "Early proposals of wireless telegraphy in Spain: Francisco Salvá Campillo (1751-1828)," *IEEE Antennas and Propagation Symposium*, Boston 2001.

[17]   *Scientific American*, Vol. 33, 1875. Articles "The discovery of another form of electricity" pp. 400-401, "The new phase of electric force", p. 401. In Vol. 34, 1876, "Etheric force and weak electric sparks", p.2, "Mr. Edison's electric discovery", p. 17, "One of Mr. Edison's curious experiments", p. 33, and subsequent articles on the same topic.

[18]   R. W. Simons, "Guglielmo Marconi and early systems of wireless communication," *GEC Review*, Vol. 11, No. 1, pp. 37-55, 1996.

[19]   S. S. Swords and B. K. P. Scaife, "George Francis FitzGerald 1851-1901," *Radio Science Bulletin*, No. 278, pp. 5-13, September 1996.

[20]   S. P. Thompson, *Elementary Lessons in Electricity and Magnetism*, 7th ed., New York: Macmillan Company, 1996.

[21]   T. Thörnblad, *Trådlös telegrafi* (Wireless telegraphy, in Swedish), Stockholm: P.A. Nordstedt & Söners Förlag, 1911.

8

NIKOLA TESLA AND HIS CONTRIBUTIONS TO RADIO DEVELOPMENT

ALEKSANDAR MARINCIC, *University of Belgrade, Serbia*

8.1   INTRODUCTION

Nikola Tesla (Figure 2.24) belongs to that rare set of human beings who devoted all his long life in search of the secrets of nature whose disclosure would help mankind to cope with the everlasting fight with unmerciful powers of Nature. There were so many things that Tesla contributed to and a few are listed here: he succeeded in inventing polyphase alternating current system of generation, transmission and utilization which we still use today, he succeeded in generation of high frequency currents and opened numerous fields and their uses in radio, for light production, industry, medicine, etc. In later life, he also fulfilled his childhood dream of making bladeless turbine and pumps. With his writing and speeches he impressed many people who knew something about his life and both his finished and unfinished research into the secrets of electricity.

Nikola Tesla was born in Smiljan, Lika in 1856 as the fourth child of Milutin and Djuka Tesla. His father was a well-educated priest of the Serbian Orthodox Church. Nikola's mother was also intelligent and talented and he often said that his mother influenced his life as an inventor. His technical education was limited to two years polytechnic studies at Gratz, Styria, where he devoted himself to mathematics, physics and mechanical engineering. From Gratz he went to Prague with the object of completing his scientific education and on philosophical studies at the University. From Prague he went to Budapest to work in a new telephone company. It was there that in 1882 he invented his induction motor and an alternating-current system of power transmission. Seeking better opportunities to find people who were interested in his invention, he accepted a position of an electrical engineer for a French Company in Paris, where he remained for two years. Another important step in his life was acceptance of the position of a designer to build direct current dynamos and motors for the Edison Company in New York, where he arrived in 1884 in hope of finding the "the land of golden promise". Edison was not interested in Tesla's alternating currents system and Tesla soon left Edison, after a bitter struggle. In

1885, the Tesla Electric Light Company was formed but Tesla had to work on electric arc light. It was not until he formed the new Tesla Electric Company, which enabled him to realize his inventions and develop working models of motors, generators and transformers. During the years 1887 and 1888 Tesla applied and was granted more than 30 patents for his inventions. In 1888, the American Institute of Electrical Engineers invited Tesla to give a lecture on his work on the alternating-current system. After that lecture Tesla became famous. In 1889 George Westinghouse approached Tesla and soon they completed an agreement for transferring exclusive license of Tesla's polyphase current patents to the Westinghouse Company. In 1895, the Westinghouse Company at Niagara Falls built on the world's largest scale the first Tesla alternating-current system.

8.2   INVENTION OF THE TESLA COIL

Nikola Tesla's polyphase low frequency alternating-current system essentially solved the problem of generating, transmitting and utilization of electrical power. For the operation of his induction and other types of AC motors he advocated the use of frequencies lower than 133 Hz used in the single-phase Westinghouse's lighting system. Towards the end of nineties, Tesla started experiments with alternating current machines "capable of giving more than two million reversals of current per minute". In his famous lecture on alternating currents of very high frequency [1] he described many experiments with two types of generators [2,3]. One of the machines had "*a disk-shaped armature that anticipated later General Electric design and reached a frequency of 15 kHz, the highest frequency attained by any alternator before 1900*" [4]. In the same lecture Tesla described a new method of producing light with a single electrode bulb and a novel apparatus for producing high frequency and high voltage (Figure 8.1) to be known later as "the Tesla coil" [5]. In a book [6], W. H. Eccles remarked about this coil that "*it was invented not for wireless but for making vacuum lamps glow without external electrodes, and it later played a principal part in other hands in the operation of big spark station*". In his lecture, Tesla also proposed, as "the ideal way of lighting a hall or room", a system composed of high frequency alternator connected in series with a condenser in the primary of a high frequency transformer. The secondary of the transformer was connected to an insulated metal plate suspended from the ceiling to the ground, "their sizes being carefully

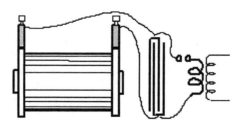

Figure 8.1. First drawing of the Tesla Coil driven by Ruhmkorff coil presented in his 1891 lecture.

determined". According to his explanation, an illuminating device could be moved and put anywhere, even beyond the plates. We are mentioning this because it seems to be the first use of generator driving a transformer with resonant primary and secondary circuits, and, in a way, indicates that he already started to think about wireless energy transmission. Here he mentioned electrostatic field effects and electromagnetic inductive effects, expecting that the latter may be more suitable as their strengths decay simply with the distance, as compared to the electrostatic effects where the field strength "diminish nearly with the cube of the distance from the coil". He also mentioned that by making use of resonance we might obtain the required electromotive force at a distance.

In 1891 Tesla published more than ten papers, most of them in connection with his research in the field of high frequency currents, in addition to ten patent applications, accepted later in the United States and abroad. It is surprising that after so many activities he, at the beginning of 1892, visited London and Paris talking about his future experiments with alternating currents of high potential and high frequency [7]. He disclosed his new achievements in obtaining better operation of his high frequency spark generator by producing rapid succession of sparks, either by employment of a magnet, simple or multiple air gaps or by various designs of mechanical interrupters. Many of these inventions were later 'reinvented' by others without referring to Tesla. On wireless energy transmission, Tesla made the following remark in the demonstration of high frequency driven motors by a single wire:

> It is quite possible, that such "no wire" motors, as they might be called, could be operated by conduction through the rarefied air at considerable distances. Alternate currents, especially of high frequencies, pass with astonishing freedom through even slightly rarefied gases. The upper strata of air are rarefied. To reach a number of miles into space requires the overcoming of difficulties of a merely mechanical nature. There is no doubt that with the enormous potentials obtainable by the use of high frequencies and oil insulation luminous discharges might be passed through many miles of rarefied air, and that, by thus directing the energy of many hundreds or thousands of horse-power, motors or lamps might be operated at considerable distances from stationary sources.

As the main subject of the lecture was light production he did not continue to talk about wireless transmission but turned to description of many types of single electrode bulbs. He was impressed by carborundum – material produced to replace ordinary diamond powder for polishing precious stones. Carborundum, he stated, "withstands excessively high degrees of heat, and it is barely deteriorated by molecular bombardment, and it does not blacken the glass globe as ordinary carbon does". He produced many bulbs with buttons of carborundum and believed that they can produce even 20 times more light by means of currents of very high frequencies as compared with the light produced by the present incandescent lamps with the same expenditure of energy. At the end of the lecture Tesla again returned to transmission without wires and wondered why with the existing knowledge and experience gained, no attempt is being made to disturb the electrostatic or magnetic condition of the earth, to 'transmit, if nothing else, intelligence'.

Even at this time, it was obvious that Nikola Tesla did not belong to the group of great theoreticians who develop new frontier of science, or to the class of great practitioners who invent many useful things for our lives. He belonged to a group of pioneers who opened up new fields of technology. He developed his own theories based on experimental observations about a number of electrical phenomena that helped him to invent his polyphase system, and later a new concept of high frequency currents and their uses in lighting, radio, medicine and industry. From the very first appearance in the AC field, he concentrated on efficient transmission phenomena based on step-up and step-down transformers. Others researchers developed low frequency single-phase AC transformers, but he invented a polyphase high power, oil cooled, iron-copper core transformer to suit his new system of polyphase currents. He was exploring completely new territories when he entered into the field of high frequency AC. In this case, he discovered quickly that the iron core was a disadvantage and he disposed of the iron and produced an air-core transformer, known since as the Tesla coil. Usually, this transformer was a part of a high frequency generator used to produce a high voltage in the secondary. Tesla discovered that the length of the wire in the secondary should be about quarter of a wavelength at the operating frequency! The operating frequency of this generator is determined mainly by the resonant circuit in the primary and the secondary circuit is composed of the inductance and self-capacitance of the secondary coil. It is interesting to observe that the secondary circuit of the Tesla coil was very close to the self resonant condition. Tesla never made full theoretical analysis of his transformer, but in tuning for the induced maximum secondary voltage he certainly used "cut and try method". The coupling between the primary and the secondary also played an important role and to these days this transformer is still studied with more or less by an approximate theory to study its operation.

The drawing shown in Figure 8.2 was found as a slide among Tesla's belongings now in Nikola Tesla Museum in Belgrade. According to Tesla's caption, these diagrams are "illustrating various ways of using high frequency alternator in the first experiment at Grand Street Laboratory 1891-1892". Only some of these diagrams have been published by Tesla in his lectures [1,7,8], so this is an important document throwing new light on an exceptionally fertile but relatively little known Tesla's work [9]. Circuits like the one shown in that figure are to be found later in his famous patents filed in 1897 on his apparatus and on the system for wireless transmission of power [10,11] and in a patent on a high frequency electrical transformer [12].

In February 1893 Tesla delivered a third lecture on high-frequency currents before the Franklin Institute in Philadelphia [8], and repeated it in March at the National Electric Light Association in St. Louis. In the first part of the lecture he described a '*method of conversion of low frequency AC or DC current into high-frequency currents, all based on using spark gap discharge*'. Just as in

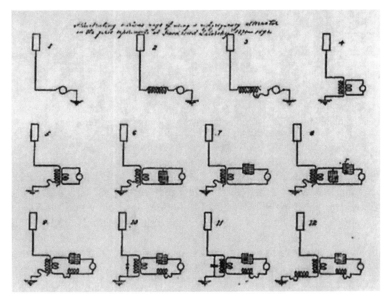

Figure 8.2. Various connections of high frequency transformers used by Tesla in 1891–1892 (Tesla's own slide, now at the Nikola Tesla Museum, Belgrade).

the previous two lectures he performed again many experiments with high frequency currents and various resonant circuits, illustrating that the current can pass through an open circuit consisting of a coil connected to a terminal of a generator and an insulated plate. Today specialists can easily build circuits used in these experiments but in 1893 they were fascinating and novel. The most significant part of this lecture refers to a system for "*transmitting intelligence or perhaps power, to any distance through the earth or intervening medium*". What Tesla described in this lecture should be taken to be the foundation of radio engineering, since it embodied the following principles and ideas of fundamental importance, namely:

- the principle of adjusting for resonance to get the maximum sensitivity in a selective reception,
- inductive link between the driver and the tank circuit,
- an antenna circuit in which the antenna appears as a capacitive load [13].

Tesla also correctly noted the importance of the choice of frequencies and the advantages of a continuous carrier for transmitting signals over great distances [6]. Figure 8.3 illustrates his scheme that he presented in his lecture.

Full explanation of Tesla's radio system is contained in the four-tuned circuit he patented through the two patents submitted in September 1897 [10,11]. At that time, the only known form of electromagnetic wave transmission was accomplished by Heinrich Hertz, who in 1888 generated using spark gaps and a dipole antenna as shown in Figure 8.4 free-space waves [14]. During the year or two following Hertz's publications of his discoveries, many physicists repeated

Figure 8.3. Tesla's proposition from 1893: S – high frequency generator: B-ground plate; P-elevated plate (antenna). Receiver is tuned to the transmitter frequency.

Hertz's experiments with electromagnetic waves and knew that they could control their wave-lengths by altering the dimensions of the conductors attached to the spark gap. Instead of using Hertz's 'ring resonator with a spark gap' as a receiver, Oliver Lodge, Guglielmo Marconi, Aleksander Popov and others had introduced a more sensitive device, invented by Edouard Branly in 1890 and called the coherer by Lodge [6]. Marconi introduced a tall vertical antenna, one end of which was connected to the ground as shown in Figure 8.5, a combination of Hertz's transmitter and Tesla's connection that enabled generation of guided waves around the earth. Although Tesla was the first to generate a ground-guided wave, which soon became the only one used in the early development of radio wave transmission technology, it is hardly mentioned in the history of radio and never in the printed documents of Marconi. For example, in 'A century of wireless', published by the European Broadcast Union in 1995, there is no mention of Nikola Tesla in the review: "Six great pioneers of radio: Michael Faraday, James Clerk Maxwell, Heinrich Hertz, Oliver Lodge, Aleksandr Stepanovich Popov and Guglielmo Marconi" [15].

In developing his 1893 system, 'using a single or no wire for electrical energy transmission', Tesla was slowed down because of many other side activities, such as the inauguration of the polyphase system built by George Westinghouse at the World Fair in Chicago, which opened to the public from May 1, 1893 till October 30, 1893. At this fair, Tesla had his own stand showing his inventions in the area of low and high frequency currents. In 1893 Tesla applied for six patents in the United States and two in Germany. In 1894 he applied for 8 patents abroad. On March 13, 1895 Tesla's laboratory in South Fifth Avenue was burned, and that stopped his research in the field of high frequency currents for sometime. Throughout 1895 he did not submit any patent

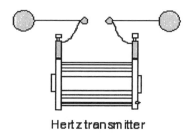

Hertz transmitter

Figure 8.4. Basic Hertz's transmitter from 1887.

or wrote any article. In his biography Tesla described this event as:

> *This calamity sent me back in many ways and most of that year*
> *had to be devoted to planning and reconstruction. However, as*
> *soon as circumstances permitted, I returned to the task.*
> *Although I knew that higher electro-motive forces were*
> *attainable with apparatus of larger dimensions, I had an*
> *instinctive perception that the object could be accomplished by*
> *the proper design of a comparatively small and compact*
> *transformer* [16].

In 1896, he indeed recovered and applied for seven patents in the United States and eight patents abroad. He also published eleven papers in various

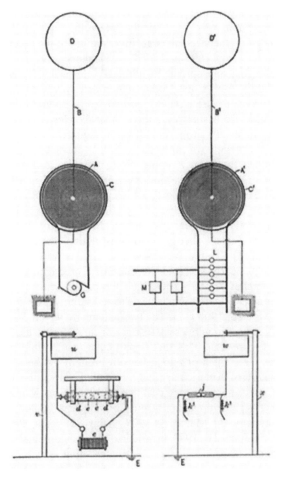

Figure 8.5. Tesla's system of four tuned circuits (US Patents [10,11]) and Marconi's system as shown in his US patent 11,913 June 4,1901. Applications to all these patents are filed in 1897.

periodicals, and the majority of these papers were devoted to his X-ray research. At this time he completed and patented design of many types of apparatus for producing electrical currents using high frequency-spark generators and circuit controllers. In his new laboratory at Houston Street, New York, he generated 4 million volts and the discharges extended through a distance of 16 feet.

A year after Tesla's lecture in 1893, Oliver Lodge transmitted Hertzian waves over a distance to a receiver using a symmetrical dipole antenna connected to a coherer. Popov demonstrated in 1895 a somewhat improved coherer with a vertical wire as an antenna and used the earth connection instead of a symmetrical dipole as done by Hertz and Lodge. In 1896, Marconi came to England with his improved apparatus and continued experiments with the transmitter shown in Figure 8.5 and with a coherer placed between the receiving antenna and ground. The coherer in each apparatus required to be tapped (de-cohered) after receiving a pulse that made it conducting. In all these receivers no resonance phenomenon was used at the receiving side and the operating frequency depended on the size and capacitive loading of the antenna.

During the years 1896 and 1897 researchers in all parts of the world discussed wavelengths appropriate to wireless telegraphy. It was known that Lodge's and Righi's oscillators produced damped waves of few centimeters, Hertz generated waves of a few meters, and waves of several kilometers long by Tesla. Hertz deduced from Maxwell's theory that the electric force produced at a large distance is inversely proportional to the square of the wavelength and proportional to the length of antenna and the current through it. By this reasoning, the higher the frequency, the stronger field at a distance is expected. In explaining the operation of his radio system announced in 1893, Tesla came with his original theory, explaining that his system is not producing significant free-space radiation but that it makes use of conduction current by disturbing the electrostatic charge along the earth's surface. In the early days of radio some writers, misled by Tesla's theory, made distinction between the Tesla and Marconi systems [17]! Today we know that this is wrong and that Tesla's experiments were just early attempts to use the ELF propagation mode [18].

On September 2, 1897 Tesla filed the patent application No. 650,343, subsequently granted as patent No. 645,576 of March 20,1900 [10] and patent No. 649,621 of May 15, 1900 [11]. The two patents by which Tesla protected his system and apparatus for wireless transmission are known as "system of four tuned circuits" (Figure 8.6). This fact is particularly important in the history of radio. They were the subject of a long lawsuit brought by the Marconi Wireless Telegraph Company of America against the United States of America, alleging that they have used wireless devices that infringed on Marconi patent No. 763,772 of June 28, 1904, on an application filed Nov. 10, 1900, and assigned to the Marconi company on March 6, 1905. After 25 years, the United States Supreme Court on June 21, 1943 invalidated the fundamental American radio patent of Marconi No.763,772 as containing nothing which was not already contained in patents granted to Lodge, Tesla, and Stone. Marconi was granted eight other US patents for wireless apparatus on applications filed between the filing dates of Nos. 586,193 and 763,772 [19]. The Supreme Court cited Tesla's system in its deliberations:

> *The Tesla patent No.645,576, applied for September 2, 1897 and allowed March 20, 1900, disclosed a four circuit system, having two circuits each at transmitter and receiver, and recommended that all four circuits be tuned to the same frequency. Tesla's apparatus was devised primarily for transmission of energy of any form of energy-consuming device by using the rarefied atmosphere at high elevations as a conductor when subjected to the electrical pressure of a very high voltage. But he also recognized that his apparatus could, without change, be used for wireless communication, which is dependent upon the transmission of electrical energy. His specifications declare:* "The apparatus which I have shown will obviously have many other valuable uses – as, for instance, when it is desired to transmit intelligible messages to great distances..." *Tesla's specifications disclosed an arrangement of four circuits, an open antenna circuit coupled, through a transformer, to a closed charging circuit at the transmitter, and an open antenna circuit at the receiver similarly coupled to a closed detector circuit.*

Tesla proposed to tune all the circuits to the same frequency and select the length of the wires in the secondary of the transmitter and the primary of the

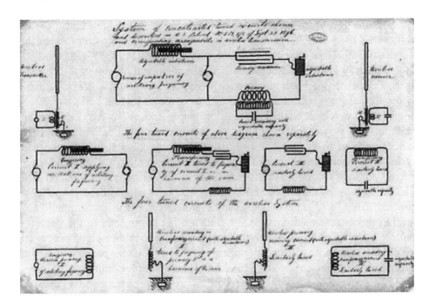

Figure 8.6. Tesla's drawing of four tuned circuits. The diagrams in Tesla's patents are as shown in upper part of Figure 8.5. (copyright Nikola Tesla Museum, Belgrade).

receiver so that the highest potentials are located at the antenna terminals. As, the input impedances of Tesla's antennas are capacitive such an adjustment would guarantee that both antenna circuits are tuned to the same frequency.

Along with his work on the improvement of his high frequency oscillators Tesla was continuously exploring application of these currents. His work on X-ray generating apparatus is reported in a series of articles in 1896 and 1897. On April, 6 1897, he also gave a lecture to the New York Academy of Science [20]. In a lecture to the American Electro-therapeutic Association in Buffalo he described the use of high frequency oscillator for therapeutic and other purposes. As an introduction to this lecture Tesla said:

> *Some theoretical possibilities offered by currents of very high frequency and observation which I casually made while pursuing experiments with alternating currents, as well as the stimulating influence of the work of Hertz and of views boldly put forth by Oliver Lodge, determined me some time during 1889 to enter a systematic investigation of high frequency phenomena, and the results soon reached were such as to justify further efforts towards providing the laboratory with efficient means for carrying on the research in this particular field, which has proved itself so fruitful since. [21]*

8.3   RADIO CONTROLLED VEHICLE

Nikola Tesla's invention of remote control by radio waves appeared in the early phase of radio development. It followed his work on "Tesla coil and oscillation transformer (1889−1892), researches and experiments with currents of high frequency (1889−1898) and Tesla Wireless System (1891−1893)" [22].

On July 1, 1898 Tesla filed a US patent application and in just over four months, the patent on Method of and Apparatus for Controlling Mechanism of Moving Vessels was granted to him [23] as seen in Figure 2.31. In 1898, he demonstrated wireless control of model ships and predicted the imminent completion of a system that could transmit both power and intelligence over long distances without wires [4]. The principle he was developing was applicable to "any kind of machine that moves on land or in the water or in the air", and to show this to an audience he constructed a boat. A storage battery placed within furnished the power, the speed and the direction of rotation of the propeller was controlled from a distance. The rudder was controlled by another motor executing the directive commands. To transmit his commands to the boat Tesla considered use of radio waves – thereby not using a line-of-sight propagating path of the Hertzian waves, or any optical ray-like rectilinear propagation, which would require that the controlled vessel be in the operator continuously. In order to avoid 'the line of sight' control, he used waves that propagated in all directions through space, and used such circuits within the boat, which were exactly tuned to the electrical vibrations that were sent from the distant transmitter. Following his usual method of presenting new results, the patent on remote control is full of details about his wireless means of transmitting the command signals to the moving object. He preferred to use high frequency currents that generated electromagnetic waves capable of reaching the moving object, although it is not in the line of sight path from the operator position. He also suggested the use of tuned circuits at the receiver to improve sensitivity and selectivity. His transmitting antenna was an elevated conductor connected to one terminal of the Tesla's coil whose other end was connected to an earthed metal body, or the two terminals can be connected to two remote points with grounded metal plates. The first case will require the use of a monopole antenna and the second case will require a loop antenna formed by the current spreading between the plates through the ground. He noticed that if the loop antenna is used one has to consider appropriate orientations of the transmitting and receiving circuits which shows that he was aware of an antenna radiation pattern. At that time Tesla had made many experiments with various antennas but never patented the monopole or the loop antenna as a separate device.

Tesla performed some work on remote control in his laboratory on 35 South Fifth Avenue. When this laboratory burned down in March 1895, it was a terrible blow to him. Many experiments were stopped until the end of 1895, when he opened a new laboratory at 46 East Houston Street. Regarding this laboratory he created, he had this to say:

> *Striking demonstrations, in many instances actually transmitting the whole motive energy to the devices instead of simply controlling the same from distance. In '97 I began the construction of a complete Automaton in the form of a boat, which is described in my original specification #613,809... This application was written during that year but the filing was delayed until July of the following year, long before which date the machine had been often exhibited to visitors who never seized to wonder at the performances... In that year I also constructed a larger boat, which I exhibited, among other things, in Chicago during a lecture before the Commercial Club. In this lecture I treated the whole field broadly, not limiting myself to mechanisms controlled from distance but to machine possessed of their own intelligence. Since that time I have advanced greatly in the evolution of the invention and think that the time is not distant when I shall show an automaton which, left to itself, will act as though possessed of reason and without any willful control from the outside. Whatever be the practical possibilities of such an achievement, it will mark the beginning of a new epoch in mechanics. [25]*

In the following 14 months, after submitting his US patent on the apparatus and system for controlling moving objects, Tesla submitted 10 patents on this topic in other countries, but one had to wait many years before remote control became a reality!

It is also interesting that Tesla's patent on remote control was listed in "Historical Perspectives of Microwave Technology" in the *Special Centennial Issue of the IEEE Microwave Theory and Techniques Society* as the first among selected patent abstracts, which date from 1898 to 1970. In the same issue Tesla's contributions were mentioned in several articles dealing with the development of Microwave Communication, with the History of Biological Effects and Medical Application of Microwave Energy, and especially in connection with the History of Power Transmission by Radio Waves.

8.4   COLORADO SPRINGS LABORATORY

Continuing his work on high frequency and very high voltages Tesla in 1899 moved to a new laboratory in Colorado Springs (Figure 8.7). Tesla's arrival in Colorado Springs was reported in the press. According to the Philadelphia *Engineering Mechanics*, Tesla arrived on May 18, 1899, with the intention of carrying out intensive research in wireless telegraphy and studying properties of the upper atmosphere. In his article written upon his return to New York [22], he writes that he came to Colorado Springs with the following goals:

- to develop a transmitter of high power;
- to perfect means for individualizing and isolating the energy transmitted;
- to ascertain the laws of propagation of currents through the earth and the atmosphere.

Figure 8.7. Colorado Springs experimental station (copyright Nikola Tesla Museum, Belgrade).

Tesla spent about eight months in Colorado Springs. A snapshot of his work and the results obtained during this period can be seen from the articles in *American Inventor* and *Western Electrician*. For example, it is stated that Tesla intended to carry out wireless transmission of signals to Paris in 1900. An article in November 1899 reports that he was making rapid progress with his system for wireless transmission of signals and that there was no way of interfering with messages sent by it. Tesla returned to New York on Jan. 11, 1900.

The diary [9] which Tesla kept at that time gives a detailed day-by-day description of the research in the period June 1, 1899 to Jan. 7, 1900. Unlike many other records in the archives of Nikola Tesla Museum in Belgrade, the Colorado Springs diary is continuous and orderly. Since it was not intended for publications, Tesla probably kept it as a way of recording his research results. It could perhaps also have been a safety measure in case the laboratory had been destroyed, an eventuality by no means unlikely considering the dangerous experiments he was performing with powerful discharges.

According to the notes, Tesla devoted the greatest proportion of his time (about 56%) to the construction of the transmitter, about 25% to developing receivers for small signals, about 16% to measurements of the capacity of the vertical antenna, and about 6% to other miscellaneous research. He developed a large high frequency fast-spark oscillator with three oscillatory circuits, i.e., a typical Tesla coil plus extra coil connected to the high potential terminal of the secondary (Figure 8.8). With this arrangement he obtained voltages of the order of 10 million volts, starting from a 40 kV low frequency voltage source. Tesla used "extra coil" in series with the secondary to obtain additional rise of voltage at the antenna terminal. He did not explain the theoretical basis of this arrangement but certainly he proved his expectation by experimental means. Assuming that the primary (L_pC_p), and secondary (L_sC_s) operate as the coupled resonant circuits, and that the extra coil with the antenna capacitance $(L_{ex}C_a)$ is represented by a lumped element system, as shown in Figure 8.8, we could calculate the output voltages from the two arrangements for driving a generator of zero internal resistance placed across the spark gap. This simplified approach allows us to plot the frequency response of the two and three resonant circuits for the values of the circuit parameters similar to those given in the notes. Because of

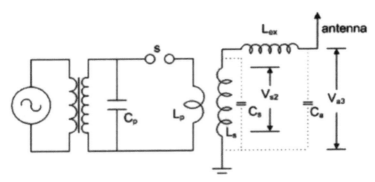

Figure 8.8. Lumped elements three-circuit Tesla coil.

strong coupling between the primary and secondary ($k = 0.57$), two peak resonant frequencies of the two-circuit system are far apart (Figure 8.9). The three-circuit system has a more pronounced peak very close to the resonant frequency f_c (equal to all three resonant circuits), which does not exist with the two-circuit system. It seems that at the time of writing the notes, Tesla did not know that his classical spark oscillator (with two resonant circuits) generates two damped sinusoidal signals, but he suspected that something is causing some problems in the primitive measurements that he could perform in using the resonance technique in his measurement of capacitances. In calculating inductances and mutual inductances of coupled coils, he used a 133 Hz alternating voltage as a source, and from the ratio of the voltage and current he found the modulus $|Z| = \sqrt{R^2 + \omega^2 L^2}$. He measured the resistance of the coils using a DC voltage source. In measuring the frequency, he used a self-resonating coil coupled with a one-turn coil connected to an incandescent lamp. The resonance was indicated by the maximum glow of the bulb and he assumed that the length of wire in the self-resonating coil to be quarter wavelength in free-space! The idea that a resonant secondary of his transformer has one-quarter wavelength length can be found in Tesla's US patent No. 593,138, dated Nov. 2, 1897, where he stated that:

> In constructing my improved transformers I employ a length of secondary which is approximately one-quarter of the wavelength of the electrical disturbance in the circuit including the secondary coil, ...

The idea of considering the Tesla coil as a helical transmission line resonator was elaborated in [35], and seemed to provide a more precise answer to the operation of Tesla coil rather than the simple tuned circuits model made of lumped elements.

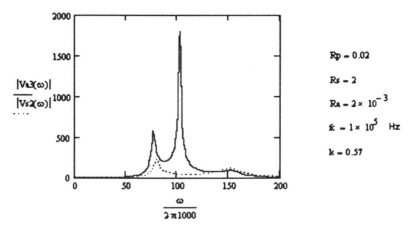

Figure 8.9. Relative voltages across output terminal of three coupled circuits (full line) and two coupled circuits (dashed line): primary circuit (L_pC_p), secondary circuit (L_sC_s) and circuit with extra coil ($L_{ex}C_a$).

Tesla tried out various modifications of the receivers with one or two coherers and special pre-excitation circuits. He made measurements of electromagnetic radiation generated from natural electrical discharges, and worked on the design of a special transmitting system using two carriers, on modulators for analog modulations of the carrier, shunt-feed antennas, etc. As almost all of these problems were new and hardly touched by the others, Tesla had an impression that he can continue with developing high power transmission, letting the completion of the easier problem of sending messages to others.

The last few days covered by the diary, indicates that Tesla devoted to photographing this laboratory inside and out. After publishing some 63 photographs thoroughly described in the diary, he caused much astonishment similar to those from his previous exciting lectures in the USA, England and France in 1891-1893. The famous German scientist Slaby wrote that the apparatus of other radio experimenters were mere toys in comparison to Tesla's Colorado Springs equipments.

Immediately after finishing his work at Colorado Springs, Tesla wrote a long article entitled "The problem of increasing human energy" [22] describing his work there. The article really did create a sensation, and was reprinted many times. At the beginning of the twentieth century Tesla believed that he is ahead of all others in developing wireless transmission not only for messages but also for energy by a large margin. While in Colorado Springs, he invented a multi-carrier transmitter with a special receiver tuned to all carriers [36]. In his own words:

> *this invention consists of generating two or more kinds or classes of disturbances or impulses of distinctive character with respect to their effect upon a receiving circuit and operating thereby a distant receiver which comprises two or more circuits, each of which is tuned to respond exclusively to the disturbances or impulses of one kind or class and so arranged that the operation of the receiver is dependent upon these conjoint or resultant action.*

Tesla's double circuit system is in a way a kind of spread spectrum system that are intended to protect message from intruders and at the same time decrease the disturbing effect of noise in the transmission. In Tesla's words this system "improves individualization and isolation of messages". Another interesting receiver is the one that rectify received signal and charge a capacitor periodically after discharging it through the receiving device [37]. Today, we call this type of receiver 'integrate and damp receiver' and it is used to improve signal to noise ratio in the detection of a signal. Tesla used it to magnify the input signal by integration. Around 1900 Tesla was already developing continuous radio systems at the time when others were developing the spark gap oriented telegraphy!

8.5   MARCONI AND BRAUN RESEARCH

Marconi was the first to send a message across the ocean in 1901, and thus, he was certainly responsible for 'developing' the radio, but he did not invent it. The apparatus and the system of four tuned circuits he used in sending messages was not his invention and he never claimed that. For the invention of radio, he and Professor Karl Ferdinand Braun from Strasbourg University, Germany, received the Nobel Prize in Physics in 1909 "in recognition of their contribution to the development of wireless telegraphy". In his Nobel Lecture Marconi described his background and "association with radiotelegraphy" [26]. His first tests were carried out with an ordinary Hertz oscillator and a Branly coherer as the detector. Important improvement he achieved after connecting one terminal of Hertzian oscillator "to earth and the other terminal to a wire or capacity area placed at a height above ground, and in also connecting at the receiving end one terminal of the coherer to earth and the other to an elevated conductor". From 1898 Marconi connected the coherer through a transformer (Tesla coil or transformer as called by others) and tuned it to the transmitter frequency. This arrangement is described in British patents No. 12,326, June 1, 1898 and No. 6,982, April 1, 1899. In 1900 he constructed and patented a complete system of transmitters and receivers,

> which consisted of the usual kind of elevated capacity area and earth connection, but these were inductively coupled to an oscillation circuit containing a condenser, an inductance, and a spark generator, the conditions which I found essential for efficiency being that the period of electrical oscillations of the elevated wire or conductor should be in tune or resonance with that of the condenser circuit, and that the two circuits of the receiver should be in electrical resonance with those of the transmitter.

He stated further

> By the adjustment of the inductance inserted in the elevated conductor and by variation of capacity of the condenser circuit, the two circuits were brought into resonance, a condition, which, as I have said, I found essential in order to obtain efficient radiation. Part of my work regarding the utilization of condenser circuits in association with the radiating antennae was carried out simultaneously to that of Prof. Braun, without, however, either of us knowing at the time anything of the contemporary work of the other

Other names which appear in the lecture are Prof. J. A. Fleming, Prof. J. A. Zenneck, Mr. R. N. Vyvyan, Mr. W. S. Entwistle, Prof. S. A. Arrhenius, Sir J. J. Thomson, Prof. G. Artom, Messrs. Bellini and Tosi, and in references: Prof. A. Slaby, Dr. O. Lodge, B. Blondel and G. Ferrie. There is no reference to Tesla's lecture in 1893 or to Marconi's US patent No. 763,772 of June 20, 1904 [19] (which was later invalidated by the United States Supreme Court), or mention of Tesla and Stone.

In the specification of the reissue of Marconi's first US patent (Figure 8.5) he stated:

> *I am aware of the publication of Professor Lodge of 1894 at London, England, entitled "The Work of Hertz," and the description therein of various instruments in connection with manifestations of Hertz oscillations. I am also aware of the papers by Professor Popov in the 'Proceedings of the Physical and Chemical Society of Russia' in 1895 or 1896; but in neither of these is there described a complete system or mechanism capable of artificially producing Hertz oscillations, and forming the same into and propagating them as definite signals, and reproducing, telegraphically such definite signals; nor has any system been described, to my knowledge, in which a Hertz oscillator at the transmitting station, and an imperfect-contact instrument at a receiving station, are both arranged with one terminal to earth and the other elevated or insulated; nor I am aware that prior to my invention any practical form of self-recovering imperfect-contact instrument has been described.*

Early coherers were sensitive but they had a serious drawback in that once they become conductive they have to be tapped to be ready for receiving the next pulse. It was Popov who made important improvements in the simple receiver with a coherer by using an arrangement for automatically tapping back the filings to a sensitive condition. In January 1896, A. S. Popov communicated a paper to a Journal [27] in which he described experiments with a coherer used in studying the phenomena of atmospheric electricity, and stated that his apparatus was set up in July, 1895, to December 1897, working well as a lighting recorder. Similar device for a tapping coherer was filed by Marconi on June 2, 1896 (British patent No.12,039). Marconi's results were presented to W. H. Preece, who on June 4, 1897 delivered a lecture at the Royal Institution saying,

> *Marconi has produced from known means a new electric eye, more delicate than any known electrical instrument, and a new system of telegraphy that will reach places hitherto inaccessible.* [28]

8.6   LONG ISLAND LABORATORY

Immediately after returning to New York in 1900 Tesla took energetic steps to get backing for the implementation of a system of "World Telegraphy" [9]. He erected a building and an antenna in Wardenclyffe, Long Island, and started fitting out a new laboratory (Figure 8.10). From his subsequent notes we learn that he intended to verify his ideas about the resonance of the Earth, and he referred to a patent of 1900 [29]. The experiments he wanted to perform were not in fact carried out until the sixties of the last century, when it was found that the Earth resonates at 8, 14 and 20 Hz [30]. Tesla predicted that the resonances would be 6, 18 and 30 Hz. His preoccupation with this great idea slowed down the construction of his overseas station, and when radio transmission across the Atlantic was finally achieved with a simpler apparatus, he had to admit that his plans included not only the transmission of signals over large distances, but also an attempt to transmit power without wires. Commenting on Tesla's undertaking, one of the world's leading experts in this field, J. Wait, has written:

> *From an historical standpoint, it is significant that the genius Nikola Tesla envisaged a world wide communication system using a huge spark gap transmitter located in Colorado Springs in 1899. A few years later he built a large facility on Long Island that he hoped would transmit signals to the Cornish coast in England. In addition, he proposed to use a modified version of the system to distribute power to all points of the globe. Unfortunately, his sponsor, J. Pierpont Morgan, terminated his support at about this time. A factor here was Marconi's successful demonstration in 1901 of transatlantic signal transmission using much simpler and far cheaper instrumentation. Nevertheless, many of Tesla's early experiments have an intriguing similarity with later developments in ELF communication [18].*

In the period 1900 to 1906 Tesla did not write continuous notes about his research. However, in the Nikola Tesla Museum in Belgrade, we found

Figure 8.10. Long Island Laboratory.

among his papers, notes from during that period describing calculation of the radiation from a dipole antenna situated above a sphere and an associated globe based on the Hertz's model. From these calculations he found that the operating frequencies should be several tenths of cycles per sec in order for the radiation of Hertz's waves to be minimal. Tesla thought that this is essential if high power is to be transmitted through the earth instead of being lost by radiation!

In a letter to Morgan [32] early in 1902 Tesla explained his research, in which he envisaged the following

> *...distinct steps to be made:*
>
> 1. *The transmission of minute amounts of energy and the production of feeble effects, barely perceptible by sensitive devices;*
> 2. *The transmission of notable amounts of energy dispensing with the necessity of sensitive devices and enabling the positive operation of any kind of apparatus requiring a small amount of power; and*
> 3. *The transmission of power in amounts of industrial significance. With the completion of my present undertaking the first step will be made.*

For the experiments associated with transmission of large power, he envisaged the construction of a plant at Niagara to generate about 100 million volts [33]. However, Tesla did not succeed in getting the necessary financial backing, and after three years of effort to finish his Long Island Station, he gave up his plans and turned to other fields of interest. He remained convinced till his death that the wireless transmission of energy would one day become reality. Today, when we have a proof of the Earth's resonant modes (Schumann's resonance), and it is known that certain waves can propagate with very little attenuation, setting up standing waves in the Earth-ionosphere system, we can judge how right Tesla was when he said that the mechanism of electromagnetic wave propagation in "his system" was not the same as in Hertz's system with collimated radiation. Naturally, Tesla could not have known that the phenomena he was talking about would only become pronounced at very low frequencies, because it seems he was never able to carry out the experiments which he had so brilliantly planned, as early as in 1893 [8]. It is gratifying that after so many years Tesla's name is rightfully reappearing in papers dealing with the propagation of radio waves and the resonance of the Earth. For example, Jackson [34] in his electromagnetic book stated that:

> *This remarkable genius clearly outlines the earth as a resonating circuit (he did not know of the ionosphere), estimates the lowest resonant frequency as 6 Hz (close to 6.6 Hz for a perfectly conducting sphere), and describes generation and detection of these waves. I thank V. L. Fitch for this fascinating piece of history.*

The last patent in connection with the radio transmitters was submitted by Tesla on Jan. 18, 1902 [38], renewed on May 4, 1907 and issued on Dec.1, 1914. This patent is an extension of patents from 1897 [10,11], with improvements that enable safe operation of apparatus for transmission of electrical energy with an antenna charged to a high potential.

In the Archives of Nikola Tesla Museum in Belgrade [40] there are several drawings of antennas specially designed to work under high voltages without corona discharges. Some of these were published in 1993 [39].

8.7   CONCLUSIONS

History of radio has been the subject of many researchers following the historic experiments of Hertz's in 1887. The work of Hertz proved the theory of Maxwell, who was the first to comment on the nature of electromagnetic wave propagation. Early historians of radio science witnessed radio development and followed the researches of Dolbear, Branly, Lodge, Tesla, Popov, Marconi, Slaby, Fessenden, De Forest, etc. At that time the theory of radio wave propagation was in infancy and that explains why some of them carefully presented Marconi's and Tesla's patents leaving readers to comment on priorities [Sewall, 1904], others thought that Tesla's and Marconi's systems should be presented as different systems [Erskine-Muray, 1913], some discussed pioneers of wireless [Hawks, 1927] and mention Tesla's contribution to radio developments with the following comment:

> Before passing on to consider the final phase of wireless communication, as represented by Marconi, mention must be made of one other pioneer, Tesla, who worked perhaps in a less spectacular manner than those who gained a certain amount of publicity. Tesla's early efforts are often overlooked, and it is only fair that his name should be mentioned so that he may share in the credit due to the early investigators. Tesla experimented early in the problem of transmitting energy without wires. In February 1893 he advanced a plan of wireless transmission and expressed his conviction in a lecture at the Franklin Institution that "it certainly is possible to produce some electrical disturbance sufficiently powerful to be perceptible by suitable instruments, at any point of the earth's surface". [31]

In 1943 the United States Supreme Court gave priority to Tesla's patents applied in 1897 as a four-circuit system consisting of "an open antenna circuit coupled through a transformer, to a closed charging circuit at the transmitter, and an open antenna circuit at the receiver similarly coupled to a closed detector circuit". This judgment was delivered after the death of both Marconi and Tesla. It seems that this judgment did not alter much the views of many historians who neglect to mention the important role played by Tesla in the early development of radio.

Some of Tesla's statements about low frequency propagation and Earth resonances were proved relatively recently [Wait, 1974 and Jackson, 1975]; publication of Nikola Tesla Colorado Springs Notes disclosed many original inventions [Tesla, 1978]. However, it seems that all known, or recently disclosed about Tesla's contributions to early radio development did not help to consider Tesla as the one of the radio pioneers [EBU, 1995].

8.8   ACKNOWLEDGMENTS

This paper came after a long time of collecting relevant material and discussions with many people. My special thanks are to Professor Antonije Djordjevic who took time from his busy schedule to read and help me in preparing the paper. I am thankful to Dr. James Wait, Dr. James Corum, late Professor Branko Popovic and Dr. Djuradj Budimir, for inspiring discussions on Tesla's work. I am indebted to the personnel of Nikola Tesla Museum in Belgrade, who helped me with the Archives materials.

REFERENCES

[1]   Nikola Tesla: *Experiments with alternate currents of very high frequency and their application to methods of artificial illumination*, lecture delivered before the A.I.E.E., at Columbia College, May 20, 1891; republished many times after publication in *Electrical Engineer*, New York, July 8, 1891, pp. 25-48.

[2]   Nikola Tesla, *Method of operating arc lamps*, U.S. patent, 447,920, Mar 10, 1891, Applied on Oct. 1, 1890.

[3]   Nikola Tesla, *Alternating electric current generator*, U.S. patent, 447,921, Mar. 10, 1891, Applied Nov. 15, 1890.

[4]   H. G. J. Aitken, *The Continuous Wave: Technology and American Radio, 1900-1932*, Princeton, NJ: Princeton University Press, 1985.

[5]   Nikola Tesla, *System of electric lighting*, U.S. patent, 454,622, June 23, 1891, Applied April 25, 1891.

[6]   W. H. Eccles, *Wireless*, London: Thornton Butterworth Limited, 1933.

[7]   Nikola Tesla, *Experiments with alternate currents of high potential and high frequency*, lecture delivered before the Institute of Electrical Engineers, London, February 3, 1892 and Royal Institute, London, February 4. First published in Journal of Institution of Electrical Engineers, London, Vol. 21, No. 97, 1982, pp. 51-163, and repeated many times in USA and Europe. The lecture was also delivered in Paris, France and published in La Lumiere Electrique, 1982; republished many times.

[8]   Nikola Tesla, *On light and other high frequency phenomena*, lecture delivered before the Franklin Institute, Philadelphia, February 24, 1893, and before the National Electric Light Association, St. Louis, March 12, 1893, republished many times after publication in the Journal of the Franklin Institute, July, Aug., Sept., Oct., Nov., Dec. 1893.

[9]   Nikola Tesla, *Colorado Springs Notes 1899-1900*, Nolit, Beograd, 1978. Introduction and commentaries by Aleksandar Marincic.

[10] Nikola Tesla, *System of transmission of electrical energy*, U.S. patent, 645,576, March 20, 1900, Applied Sept. 2, 1897.

[11] Nikola Tesla, *Apparatus for transmission of electrical energy*, U.S. patent, 649,621, May 15, 1900, Applied Sept. 2, 1897.

[12] Nikola Tesla, *Electrical transformer*, U.S. patent, 593,138, Nov. 2, 1897, Applied Mar. 20, 1897.

[13] L. P. Wheeler, *Tesla's contribution to high frequency*, *Electrical Engineering*, New York, August 1943.

[14] H. Hertz, Sitzb. Berl. Akad. Wiss., Feb. 2, 1888; Wiedem. Ann. 34 (1888), p. 551.

[15] *"Six Great Pioneers of Wireless"*, EBU Technical Review, No.263, Spring 1995, pp. 82-96.

[16] Nikola Tesla, *My inventions*, Reprinted many times from *Electrical Experimenter*, New York, Feb. to Oct., 1919, last published by Nikola Tesla Museum, Belgrade, 2003.

[17] J. Erskine-Murray, *A Handbook of Wireless Telegraphy*, Crosby Lockwood, London, 1913, chap. XVII.

[18] J. R. Wait, Historical background and introduction to special issue on extremely low frequency (ELF) propagation", *IEEE Trans. on Communications*, Vol. COM-22, No. 4, April 1974.

[19] *"Marconi Wireless Telegraph Company of America v. United States"*, Cases adjudged in the Supreme Court of the United States at October term, 1942.

[20] Nikola Tesla, *The stream of Lenard and Roentgen and novel apparatus for their production"*, Apr. 6, 1897 (Nikola Tesla Museum, Belgrade).

[21] Nikola Tesla, *High frequency oscillators for electro-therapeutic and other purposes*, Read at the eight annual meeting of The American Electro-Therapeutic Association, Buffalo, NY, Sept. 13 to 15, 1898. LPA, Nikola Tesla Museum, Belgrade, 1956, p. L-156.

[22] Nikola Tesla, *The problem of increasing human energy*, Century *Illustrated Monthly Magazine*, June 1900.

[23] NikolaTesla, *Method of and Apparatus for Controlling Mechanism of Moving Vessels*, US Patent No. 613,809, Nov. 8, 1898, application filed July 1, 1898.

[24] Aleksandar Marincic, A Century of Remote Control, *Microwave Review*, No. 5, Belgrade, 1998, pp. 1-6

[25] Letter of Nikola Tesla to Benjamin F. Miessner, Sept. 29, 1915, Nikola Tesla Museum Archieves, Belgrade.

[26] Guglielmo Marconi, *Wireless telegraphic communication*, Nobel Lecture, Dec.11, 1909, pp. 196-222, http://www.nobel.se/physics/laureates/1909/marconi-lecture.pdf

[27] A. S. Popov, *Journal of Russian Physical and Chemical Society*, Vol. 28, Jan. 1896.

[28] J. A. Fleming, *The Principle of Electric Wave Telegraphy and Telephony*, Third ed., 1916, Longmans Green & Co., London.

[29] Nikola Tesla, *The art of transmission of electrical energy through the natural mediums*, US Patent No. 787,412, Apr. 18, 1905, application filed May 16, 1900.

[30] J. Galeys, *Terrestrial Propagation of Long Electromagnetic Waves*, New York, Pergamon Press, 1972.

[31] Ellison Hawks, *Pioneers of Wireless*, London: Methuen & Co. Ltd, 1927.

[32] Letter of Nikola Tesla to J. P. Morgan, Jan. 9, 19, Archive of Nikola Tesla Museum, Belgrade.

[33] Nikola Tesla, The transmission of electrical energy without wires, *Electrical World and Engineer*, March 5, 1904.

[34] J. D. Jackson, *Classical electrodynamics*, John Wiley, 1975, New York.

[35] J. F. Corum and K. L. Corum, "The application of transmission line resonators to high voltage RF power processing: history, analysis and experiments", *19th Southern Symposium on System Theory*, March 15-17, 1987, pp. 45-49.

[36] Nikola Tesla, *System of signaling*, US patent No.725,605 Apr. 14, 1903, applied July 16, 1900.

[37] Nikola Tesla, *Apparatus for utilizing effects transmitted from a distance to a receiving device through natural media*, US patent No. 685,955, Nov. 5, 1901, applied Sept. 8, 1899.

[38] Nikola Tesla, *Apparatus for transmitting electrical energy*, US patent No. 1,119,732, Dec. 1, 1914, applied Jan. 18, 1902.

[39] Aleksandar Marincic, "Nikola Tesla contribution to the development of radio", *IEEE Microwave Theory and Techniques Society Newsletter*, 133, 1993, pp. 19-22.

[40] Nikola Tesla Museum, 11000 Belgrade, Krunska 51, Serbia and Montenegro, www.tesla-museum.org.

9

AN APPRECIATION OF J. C. BOSE'S PIONEERING WORK IN MILLIMETER AND MICROWAVES

TAPAN K. SARKAR, *Syracuse University, Syracuse, NY;*
DIPAK L. SENGUPTA, *University of Detroit Mercy, MI*

9.1   INTRODUCTION

The pioneering work in the area of millimeter and microwaves, performed by J. C. Bose, a physicist from India, during 1894–1900, is reviewed and appraised. Various measurement techniques and circuit components, developed by him are still being used. The development of the electromagnetic horn, the circular waveguide, the point-contact detector, and the galena (semiconductor) detector of electromagnetic waves are attributed to the original research of J. C. Bose. His galena detector is the forerunner of semiconductor diode detectors. One of his goals was to demonstrate that light was indeed electromagnetic in nature by measuring the relationship between the dielectric constant and the permittivity of the materials as predicted by Maxwell, in the high frequency regime.

Jagadish Chunder Bose (1858–1937) was a great experimental physicist from Calcutta, West Bengal, India. As a professor of physics at the Presidency College of Calcutta, India, he conducted many of his pioneering and fundamental research investigations, and thereby initiated the tradition of scientific research in India towards the end of the last century. His life story, outlook of life, and scientific views are discussed in [1-5].

During 1894–1900, Bose utilized what then were called the Hertzian waves [now called electromagnetic (millimeter) waves] to perform a variety of pioneering research in the area of microwaves and millimeter (mm) waves. His original research papers in this and other related areas may be found in [4,6]. Although Bose's work in the millimeter wave area has been recognized by the microwave community [7-10], the impact of his contributions on millimeter wave and microwave detection is not adequately publicized and appreciated by the larger electronic community in general. The present article therefore discusses and appraises the work of Bose, specifically in the area of millimeter wave detection that led to the development of the galena detector [11], which is the forerunner of semiconductor detectors. A brief biodata of Bose is presented at the end of this chapter.

9.2   HISTORICAL PERSPECTIVE

It is now recognized that the discipline of microwaves and millimeter (mm) waves began in the year 1888, when Heinrich Hertz (1857–1894) published the results of his famous experiments in a paper entitled, "On Electromagnetic Waves in Air and their Reflection" [9]. These epoch-making results provided experimental confirmation of the existence of electromagnetic waves in air, theoretically predicted by James Clerk Maxwell (1831–1879) in 1864. By generating, radiating, and receiving electromagnetic waves (of wavelength $\lambda \simeq 66$ cm), Hertz firmly established the validity of Maxwell's theory. Hertz's original experimental setup and findings are well documented in [12,13].

Hertz's work inspired a number of scientists in different countries to get involved in research with Hertzian waves during the years 1890–1900. The physicist Bose was one of them. He decided to use Hertzian waves of lengths smaller than those used by Hertz. In fact, he used Hertzian waves having a wavelength of $\lambda \simeq 2.5$ cm to 5 mm ($f = 12$ GHz to 60 GHz) to experimentally verify the optical properties of electromagnetic waves like reflection, refraction, polarization, etc. Bose rightly believed that it would be advantageous to use millimeter waves due to the fact that the physical sizes of various components required for the experimental setup would be smaller. To this end, Bose succeeded in generating mm waves, and systematically used them for a variety of quasi-optical measurements. He developed many mm wave circuit components, which, in one form or another, are still being used in modern times [7, 10]. In fact, Bose essentially perfected a millimeter wave transmission and reception system at 60 GHz. He improved upon and developed various devices required to generate, radiate, and receive (detect) mm wave energy, and demonstrated that the newly discovered Hertzian waves, i.e., electromagnetic waves at mm wave frequencies, behaved like light waves. In 1896, Bose give a lecture demonstration of his work on (mm wave) electromagnetic radiation at the Royal Institution, London. Figure 2.29 shows a picture of Bose along with his experimental setup. In the following sections, Bose's work on the improvement of mm wave detection that lead to galena detectors will be discussed, and for completeness, we also include a brief discussion of his mm wave experimental setup of which the galena detector was a part.

9.3   A 60 GHZ TRANSMISSION SYSTEM

During the time of Hertz, electromagnetic waves were generated by spark gaps which generally produced wide bands of frequencies. Unless the receiving arrangement was shielded carefully, signals of unwanted frequencies interfered with the measurement setup and in addition, reflections from the room walls also produced undesirable effects. Lodge [14, 15] made some improvements on the above generating system by using an external resonator to filter out some frequencies. He accomplished this by inserting a metal ball in-between the sparking elements. However, the problem with this arrangement was that the surface of the metal ball became rough after a few sparks, and it started radiating spurious waves thereafter.

Bose improved the above setup by covering the metal ball with platinum and interposing it between two hollow metal hemispheres. Electrical oscillations were produced by sparking between the hemispheres. He succeeded in casting a solid metal ball and two beads of platinum which enabled him to obtain oscillatory discharge by exciting it with a Ruhmkorff coil excited by a battery. However, he had to discard the use of a vibrating interrupter to avoid irregular discharge. This improved version of the radiator is shown in Figure 9.1. The two ends of the primary coil of the excitor were connected with a small storage cell through a small key. The coil, small storage cell, and the key were enclosed by a tin box. In front of the box there was an opening through which a metal radiating tube projected out. This entire system was housed in a box which Bose referred to as the *radiator*, **R**, which is further described in [16,17]. This device increased the energy of radiation. It is conjectured that the use of the solid metal ball as an external radiator provided him with radiation of electromagnetic energy located at an extremely narrow frequency band. Later on, this setup helped him to measure the wavelength of the radiated fields. This was a significant improvement, as now; Bose could transmit a narrowband of frequencies as opposed to an extremely wide band generated by the spark gap. He found that the wavelength of the radiated waves from his radiator was approximately twice the distance between the sparking surfaces. He succeeded in generating and radiating electromagnetic waves having wavelength, λ, ranging from 2.5 cm to 5 mm.

He shielded the transmitting setup, first, by using a copper layer to minimize stray electric fields, and then, by using a soft iron box on top of it to minimize the effects of stray magnetic fields. It should be noted that Bose used a rectangular metal tube (waveguide) to guide the waves generated and eventually to radiate them through the open end. He placed a lens in front of the opening to

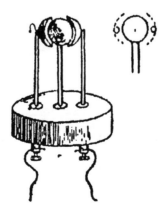

Figure 9.1. Sparkgap arrangements for generating wavelengths from 2.5 cm to 5 mm. [4].

focus the electromagnetic energy. To minimize multiple reflections inside the transmitter box so that alternating sparks could pass without roughing or oxidation, Bose used blotting paper soaked in an electrolyte to act as an absorber of these waves. To the best of our knowledge Bose was the first to utilize such an absorber of electromagnetic waves. In order to accomplish this he had to measure the dielectric constants of sulfur and other materials at 60 GHz. For this purpose Bose developed an electromagnetic quasi-optic spectrometer system as shown in Figure 9.2.

A number of items used in Figure 9.2 should be noted in the context of microwave and millimeter wave technology. The radiator, **R**, used the sparking method mentioned earlier to generate the desired waves; as shown in the figure, a metal tube (now called a waveguide) was used to guide these waves which were then radiated through its open end. Also, the item **F**, referred to as the *collecting funnel* by Bose, to receive the radiated waves, was a pyramidal horn which is still being used for such purposes. It is believed that Bose was the first to use such a device to radiate and receive electromagnetic waves.

9.4   DEVELOPMENT OF THE RECEIVER

Professor Branly (Figure 2.25), of Catholic University College of Paris, developed a *Radio conductor*. This was a glass tube filled with iron filings, as shown in Figure 9.3. On the two ends of the glass tube an electromotive force (EMF or voltage) was applied through a variable resistor so that an adjustable current would flow. Now, with millimeter waves incident on the iron filings, there appeared to be a diminution of the resistance in the circuit as indicated by a galvanometer associated with the circuit. But Branly found that the Hertzian waves, which could not produce appreciable induction in the filings, enormously

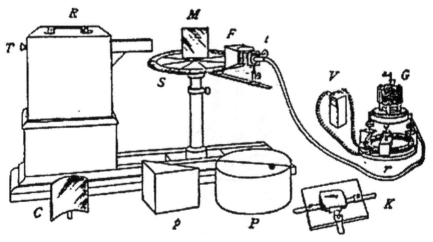

Figure 9.2. Electromagnetic Horn receiving antenna on microwave spectrometer used for quasi-optical demonstrations by Bose (1896). [Scanned from Longman's and Green Co].

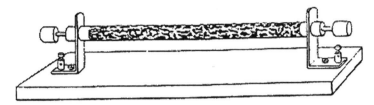

Figure 9.3. Conventional receiver of Hertz wave.

reduced their resistance sometimes even to a millionth. A problem with this Radio conductor was that after an interval, when the metal filings had acted as a receiver for some time, a tap was necessary to shake them back to their former state.

Lodge not only used this device, but offered an interpretation of its action as due to the fusing of minute points of contact of the filings by the inductive effect produced by the Hertzian waves, and for this reason, he renamed it a *coherer* – which for some reason remained with the English speaking community. Branly however maintained the original name, with his explanation that the Hertzian waves merely modified in some way the non-conducting film upon the surface of the filings. As discussed below, Bose's receiver was a great advance over that of Branly and Lodge.

According to Bose [16], the greatest drawback in conducting experimental investigations on the optical properties of electric radiation was the difficulty of constructing a satisfactory receiver for detecting the radiation. Here we quote Bose's own words [1]:

> *For this purpose I at first used the original form of coherer made of metal filings as devised by Professor Lodge. It is a very sensitive detector for electric radiation but unfortunately its indications are often extremely capricious. The conditions for a satisfactory receiver are the following:*
>
> i. *Its indications should always be reliable*
> ii. *Its sensitiveness should remain fairly uniform during the course of the experiment*
> iii. *The sensitivity should be capable of variation, to suit different experiments*
> iv. *The receiver should be of small size, but preferably linear, for accurate angular measurement.*
>
> *From a series of experiments carried out to find the causes of the erratic behavior of the receiver, I was led to suppose that the uncertainty in this response is probably due to the following:*
>
> (i) *Some of the particles of the coherer might be in too loose contact against each other, whereas others might be jammed together, preventing proper response.*

(ii) *The loss of sensitivity might also be due to the fatigue produced on the contact surfaces by the prolonged action of radiation.*

(iii) *Since the radiation was almost entirely absorbed by the outermost layer, the inner mass, which acted as a short circuit was not merely useless but might introduce complications.*

For these reasons, I modified the receiver into a spiral spring form (as shown in Figure 9.4). *Fine metallic wires (generally steel, occasionally others, or a combination of different metals) were wound in narrow spirals and laid in a single layer on a groove cut in ebonite, so that the spirals could roll on a smooth surface. The spirals are prevented from falling by a glass slide in the front. The ridges of the contiguous spirals made numerous and well-defined contacts, about one thousand in number. The useless conducting mass was abolished, and the resistance of the receiving circuit almost entirely was concentrated at the sensitive contact surface exposed to radiation. If any change of resistance, however slight, took place at the sensitive layers, the galvanometer in circuit would show strong indications...*

The sensitivity of the receiver to the radiation, I found, depends (1) on the pressure which the spirals are subjected and (2) on the EMF acting on the circuit...

The receiver thus constructed is perfectly reliable; the sensibility can be widely varied to suit different experiments, and this sensibility can be maintained fairly uniform. The sensitiveness, when necessary, can be exalted to almost any extent, and it is thus possible to carry out some of the most delicate experiments (specially on polarization) with certainty.

In working with these types of coherers, made of iron or steel, Bose encountered some difficulties in the warm and damp climate of his native state of Bengal. The surface of the metals soon got oxidized and this changed the sensitivity of the coherer. Bose then decided to use the coatings of materials which are less susceptible to oxidation. Since he found the sensitivity of the

Figure 9.4. Bose's Microwave Detector.

coherer depended on the metal coating and not on the substratum, he decided to make a systematic study of the action of different metals with regards to their detection properties [18]. These also included metalloids, nonmetals, amalgams and compounds. He found that in some of the materials the resistance decreased when irradiated but in others it increased.

For example, in potassium, the effect of radiation produced an increase of resistance, and the receiver recovered instantaneously on the cessation of radiation [18]. With sodium and lithium, the same situations occurred but to a lesser degree. With magnesium, zinc, and cadmium, he found a decrease of resistance. He experimented also with nickel, cobalt, manganese, chromium, bismuth, antimony, arsenic, aluminum, tin, lead, thallium, molybdenum, uranium, platinum, osmium, and rhodium, along with copper, gold, and silver.

Since the receiver made with potassium and the related alkali metals exhibited an increase of resistance by the action of radiation, he had difficulty explaining how a cohering action could increase the resistance. In addition, the receiver showed a remarkable power of self-recovery when the radiation source was withdrawn. He also increased the pressure between the potassium granules in his coherer till the receiver grew insensitive. All along, he observed an increase of resistance, even when one piece was partially flattened against the other.

The resistance of the receiving circuit was almost entirely concentrated at the sensitive contact-surface. When electric radiation was absorbed by the sensitive contacts, there was a sudden decrease of the resistance and the galvanometer was deflected. The sensitivity of the apparatus according to M. Poincaré *"is exquisite: it responds to all the radiations in the interval of an octave. One makes it sensitive to different kinds of radiations, by varying the electromotive force which engenders the current which traverses the receiver"*. Bose was successful in inventing other types of receivers which recovered automatically without tapping. He thus made himself the best equipped among the physicists of his time in the field of investigation. Thus, with the most perfect generation of the mm waves under his full control, he was able to produce a well defined beam of half-inch section. Furthermore, his receiver not only surpassed previous ones in sensitivity, but also, and more importantly, in its certainty and uniformity of action.

9.5   DEMONSTRATION OF PROPAGATION

Bose seemed to have used electromagnetic waves for signaling purposes. In 1895, in a public lecture in Calcutta, Bose demonstrated the ability of his electric rays to travel from the lecture hall and through the intervening room, passed to a third room 75 feet (~22 m) from his radiator [1-3]. In fact, the waves passed through solid walls on the way, as well as, through the body of the chairman (who happened to be the Lieutenant Governor of Bengal). The waves received by the receiver activated a circuit to make a contact which set a bell ringing, discharged a pistol, and thereby exploded a miniature mine. For an antenna, he used a circular metal plate at the top of a 6 m pole.

However, Bose was not interested in long-distance wireless transmission. Since the optical behavior of the waves could be best studied at short wavelengths, he concentrated on millimeter waves. Bose used the spectrometer setup shown in Figure 9.2 and some of its modified versions, extensively, to conduct a variety of quasi-optic measurements at mm wave frequencies. It can be seen in Figure 9.2 that he used hollow metal tubes as waveguides and expanded the open end to form a radiator, thereby producing the first horn antenna. The flared rectangular guide he called a *collecting funnel* was the forerunner of the pyramidal horn.

9.6   DEMONSTRATION OF THE PHENOMENON OF REFRACTION

Bose developed prisms made from sulfur, and he made lens antennas of sulfur. For this purpose, he had to measure the index of refraction of various materials, and he demonstrated the principle of total reflection. After the development of Maxwell's equations, there was some controversy as to the relation between the index of refraction of light and the dielectric constant of insulators. Bose eliminated these difficulties by measuring the index of refraction at mm wavelengths.

He determined the index of refraction at mm wavelengths by determining the critical angle at which total reflection took place. The apparatus he called the *electric refractometer* was used to measure the refractive index. This is shown in Figure 9.5. A rectangular aperture shielded the transmitter from the refracting cylinder and the sensitive receivers. He rotated the refracting cylinder on a turntable until he could measure the critical angle, and from the critical angle he determined the index of refraction. He showed that the values of the refractive index differed considerably from the values measured at visible light. He used his spectrometer arrangement to measure accurately the refractive indices of a variety of solid and liquid materials at mm wave frequencies.

With his spectrometer Bose utilized total reflection effects to measure the dielectric constant ε_r (refractive index n) of a number of materials [19]. For

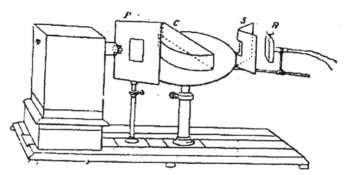

Figure 9.5. The electric refractometer developed by Bose [4]. **P** – the plate with a diaphragm; **C** – semi-cylinder of glass; **S** – the shield (only one shown in the diagram); **R** – the receiver.

glass he measured $n = 2.03$ and 1.53, at 10 GHz and at the frequency of sodium light, respectively. These results are surprisingly accurate, and further discussion of the results for other materials is given in [19]. With this work Bose not only verified experimentally the then controversial theoretical prediction made by Maxwell that $n^2 = \varepsilon_r$, but he also investigated the frequency dependence of n or ε_r of certain materials.

9.7   DEMONSTRATION OF THE PHENOMENON OF POLARIZATION

Bose demonstrated the effects of polarization by using three different types of polarizers:

a)   Polarizers made of wire gratings;
b)   Polarizers made of crystals, like Tourmaline or Nemalite; and
c)   Jute or vegetable-fiber polarizers.

His electric-polarization apparatus is shown in Figure 9.6.

Bose found a special crystal, Nemalite, which exhibited the polarization of electric waves in the very same manner as a beam of light is polarized by selective absorption in crystals like Tourmaline. He found that the cause of the polarization was due to different electrical conductivity in two different directions. The rotation of the plane of polarization was demonstrated by means of a contrivance twisted like a rope, and the rotation could be produced to the left or right. These and other results are discussed in [16,20]. The findings of his research were communicated to the Royal Society by his teacher Lord Rayleigh.

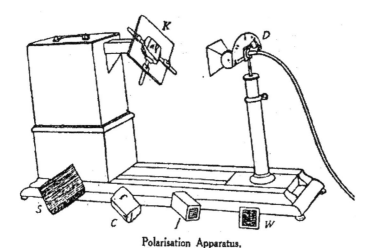

Polarisation Apparatus.

Figure 9.6. The polarization apparatus developed by Bose [4]. **K**- Crystal Holder; **S** – a piece of stratified rock; **C** – a crystal; **J** – jute polarizer; **W** – wire grating polarizer; **D** – vertical graduated disc by which rotation is measured.

Many of the leading scientific men wished to show their appreciation of the value of Bose's work in a practical way. Their natural spokesman Lord Kelvin, strongly realized the all but impossible conditions under which that work hitherto had been carried out, and he wrote to Lord George Hamilton, then Secretary of State for India "... to establish a laboratory for Bose in Calcutta. Following on this letter a memorial was sent... which was signed by Lord Lister, then President of the Royal Society, Sir William Ramsay, Sir Gabriel Stokes, Professor Silvanus Thompson, Sir William Rucker and others" [1, p.68]. This clearly shows that Bose's work underwent a thorough scrutiny by the best intellectual brains of that time and they not only appreciated his work, but tried to help him in every possible way they could.

9.8   DEMONSTRATION OF PHENOMENON SIMILAR TO THE PHOTO ELECTRIC EFFECT

Bose succeeded in detecting the effect of light in producing a variation of contact resistance in a galena receiver. One and the same receiver responded in the same way when alternately acted on by visible (light) and invisible electromagnetic radiation. He then proceeded to show the remarkable similarity of the curves of response produced under electromagnetic radiation and light. He tried to explain this phenomenon in terms of molecular strains produced in the materials due to electromagnetic radiation. He fabricated a strain cell to demonstrate that in providing angular torsion in a metal wire, the conductivity changed in a fashion similar to the case when the same structure was illuminated by electromagnetic radiation.

It is interesting to point out that after a Friday Evening Discourse at the Royal Institution, London, the publication *Electrical Engineer* expressed surprise that no secret was at any time made as to the construction of various apparatus so that it was open to the entire world to adopt it for practical and possibly money-making purposes. At that time Bose was criticized as being impractical and uninterested in making a profit from his inventions. However, Bose had his own ideas. He apparently was painfully disturbed by what seemed to him symptoms of deterioration, even in the scientific community, by the temptation of gain, and so he had made a resolution to seek no personal advantage from his inventions [1,3,21].

9.9   MEASUREMENT OF WAVELENGTH

The next problem faced by Bose was to verify that the radiated wave was actually of frequency 60 GHz, i.e., there was predominantly a single frequency wave radiated by the system and not the wide band of frequencies initially generated by the sparking system. To this end, Bose developed an accurate method to determine the wavelength of millimeter waves by using curved optical grating [22]. A sketch of the horizontal plane cross-section of the experimental system is shown in Figure 9.7. The arrangement consists of a cylindrical grating

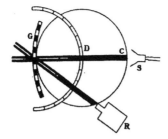

Figure 9.7. Sketch of experimental setup to measure wavelength [23].

placed vertically on a wooden table with its center at **C** – where a receiving horn and a spiral coherer (receiver) **S** are located. During the experiment the radiator **R** and the receiver **S** were always kept on the focal curve as shown. The graduated circle was used for the measurement of the angles of incidence, diffraction, etc. With the receiver placed at **C**, Bose used the following standard expression to measure wavelength:

$$(a+b)\sin\theta = m\lambda ,$$

where:

$(a+b)$ is the sum of the breadths of the alternate open and close spaces
      in the grating,
θ is the angle between the transmitter and the receiver.
m is an appropriate integer indicating the order of diffraction,
λ is the wavelength of the electromagnetic waves under consideration.

In the work cited in [22], he measured wavelengths of 1.84 cm and 2.36 cm. In fact, the accuracy of his wavelength measurement enabled him to measure the frequency dependence of dielectric constant (refractive index) of materials, described earlier. The results of this research were communicated to Lord Rayleigh, his teacher of physics during the graduate studies. It is interesting to note that at the initiative of Lord Rayleigh, the University of London awarded the degree of Doctor of Science to Bose on the basis of this work [1]. The quality of the research was so impressive that the University of London made an exception so that Bose was not required to defend his thesis in person by appearing at the University.

9.10   DEVELOPMENT OF THE GALENA DETECTOR

By the late 1890s, Braun had replaced a coherer with crystals to see if one-way rectification would also work in wireless circuits, instead of vacuum tubes. It did, and many researchers started experiments with them. By 1904, J.C. Bose had used a lead sulfide crystal for receiver detection. Shortly thereafter, additional crystal detectors were designed by Henry H.C. Dunwoody (carborundum) and by

G.W. Pickard (silicon). Bose found that the effect of radiation was to produce opposite effects on two classes of substances when he worked on the improvement of the coherer. He called one set positive, where the resistance decreased with incident radiation, and the other negative, when radiation produced an increase of resistance. He also found that the sensitivity was confined to the outer surface in contact and did not extend to the substratum. He used the term Electric Touch [23], in the restricted sense of sensitiveness to electric radiation. A summary of his observations is shown in Figure 9.8 [1, p.74] which identifies some of the test materials as having positive or negative electric touch.

It then appeared to Bose [23] that the observed effect was not due to a single cause, but was due to many causes. He then started a long and tedious process of successive elimination to find out the causes which were instrumental in producing the observed effects. He looked into many directions. We will focus only on the topic that is relevant to this special issue, namely on the change of sign of the response in a receiver due to a variation of radiation intensity.

He found that the resistivity of the receiver did not decrease/increase monotonically with the intensity of radiation. For example, in some experiments when the intensity of the transmitted beam was cut off due to total reflection, some interesting phenomenon happened. Bose observed that when the intensity of the beam became very feeble, the receiver indicated an increase instead of the normal diminution of resistance. He also found that Professor Lodge mentioned in one of his papers that an iron filing coherer exhibited an increase of resistance when acted on by feeble radiation. If the normal sign of response were reversed by a feeble intensity of stimulus, then negative substances may be expected to give a positive reaction with feeble radiation. He observed similar effects with arsenic and osmium receivers. Thus, the former belonged to a negative class and the latter to a positive class of materials, respectively.

This made him conjecture that electric radiation produces molecular change or allotropic modification in a substance. As one of the properties of a

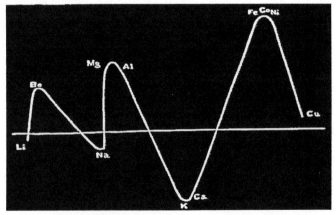

Figure 9.8. Periodicity of electric touch. Abscissa represents the atomic weight, ordinate the electric touch positive or negative [1,24].

substance is its electric conductivity, any allotropic change produced by radiation should be capable of being detected by a variation of the conductivity of the substance.

He carried out several experiments with different substances like mercuric iodide, two varieties of silver, arsenic, and so on, to demonstrate that the radiation produces molecular changes in matter which is reflected in the variation of conductivity. The nature of the change also depended on the intensity and duration of the incident electric radiation.

In addition, Bose discovered that the conductivity changes took place not only under very rapid Hertzian oscillations, but also under much slower electromotive variations [24]. The investigation thus resolved itself into the determination of the variation of conductivity of the particles as the mass was subjected to a continuously diminishing EMF from maximum back to zero. Bose also carried out experiments with both single and multiple contacts of the receiver setup. For the case of multiple contacts, a small quantity of metallic filings was put in a glass tube and the fragments compressed between two electrodes, the pressure being regulated by a micrometer screw. For a single contact device, the pressure of contact was adjusted by means of a fine micrometer screw.

In some cases, the contact ends were both rounded; in others, a pointed end was pressed against a flat-piece as shown in Figure 2.34, taken from the patent application [11] made by Bose in 1901. It is appropriate to mention here that the heading of the text of the patent reads thus:

> *United States Patent Office, Jagadis Chunder Bose, of Calcutta, India, Assignor of One-half to Sara Chapman Bull, of Cambridge, Massachusetts. – Detector of Electrical Disturbances. Specification forming part of Letters Patent No. 755840, dated March 29, 1904. Application filed September 20, 1901. Serial No. 77028 (No Model)*

Bose signed the five-page patent in the presence of two subscribing witnesses, R. E. Ellis and T. L. Whitehead [2].

Figure 2.34 shows a biased circuit consisting of a pair of point contacts (cat whiskers), in this case galena, in series with an EMF and a galvano-meter. In his patent application [11], Bose describes his galena detector thus:

> *By placing an ordinary glass lens… in the opening in the wall of the case-section… opposite the sensitive contacts… of the instrument and by throwing light upon the lens an immediate response is observed in the galvanometer, the needle of which is deflected in accordance with the spectral properties of the light thrown upon the sensitive contacts or artificial retina. With a glass lens the instrument will detect and record lights not only some way beyond the violet, but also in the regions far below the infra-red in invisible regions of electric radiation. We may thus style the apparatus a "tejometer" (Sanskrit tej=radiation) or universal radiometer…*

By removing the metallic and wooden casings and the lens the instrument may be used as a detector or so-called "coherer" for wireless or other telegraph...

What I claim, and desire to secure by Letters patent is - ... a coherer or detector of electrical disturbances, Hertzian waves, light waves, or other radiations comprising a pair of galena contacts...

As mentioned above, Bose referred to his galena point-contact detector alternatively, as *artificial retina*, sensitive receiver, universal radiometer, or *tejometer* [24]. A technical review of semiconductor research [25] clearly indicates Bose's priority in introducing a semiconductor as sensitive radio wave detector; in fact, it was recognized there that Bose's galena detector had the best sensitivity.

The response curves for a single point contact iron receiver obtained by Bose are shown in Figure 9.9 [26]. It is seen that the V-I characteristics are not straight but concave with respect to the axis representing the current [26,27]. Bose carried out similar experiments with magnesium and nickel. In addition he showed that the conductivity variation with the cyclic variation of EMF in iron receivers is similar to the hysteresis and were of the complete self-recovery type, as shown in Figure 9.9.

In summary, Bose developed point contact receivers that displayed nonlinear V-I characteristics and he provided an explanation of the nature of the

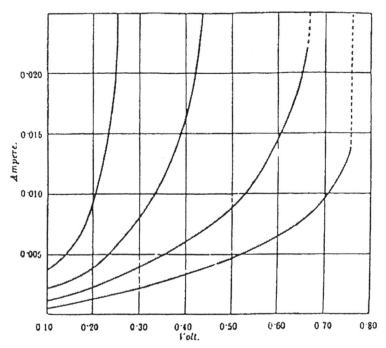

Figure 9.9. V-I characteristics of a point contact device [4].

change in the conductivity due to the impressed EMF. Some of them were point contact devices where one pointed end pressed against a flat end. The materials were of all types including metals, alkalis, amalgams, and various other combinations. As he pointed out in his patent, galena, tellurium, magnesium, lead, tin, potassium, silver, chromium, and zirconium are all sensitive agents and can be used as a point contact detector as show in the three figures of his patent application. He was also aware that the conductivity of the point contact changed with the magnitude of the applied EMF [26].

It is appropriate to mention here that in a paper presented to the Royal Society of London [26], Bose reported that the conductivity of certain polarizing substances changed significantly with the direction of transmission and absorption of electrical signals — the greatest conductivity occurred in the direction of absorption and the least in the direction of transmission. In that paper he also reported obtaining maximum conductivity changes of about 14 to 1 in the two significant directions. It is argued here that these differential conductivity results for materials eventually led Bose to the development of the galena detector.

Bose's goal was to develop a stable millimeter wave detector (similar to Figure 2.34) which does not change with time or climate. In fact, his original equipment, built over one hundred years ago, is still available at the Bose Research Institute, Calcutta, and is in good working condition, a photo of which is shown in Figure 9.10. Since Bose was interested in understanding the changes that took place in metals, alkalis, and semiconductors, he turned his research to plants and other biological tissues to see how they would respond to similar stimulus.

Bose succeeded in detecting the effect of light in producing variation of contact resistance in the galena receiver. One and the same receiver responded in the same way when alternately acted on by visible (light) and invisible electromagnetic radiation. He then proceeded to show the remarkable similarity of the curves of response produced under electromagnetic radiation and light. He tried to explain this phenomenon in terms of molecular strains produced in the materials due to electromagnetic radiation. He fabricated a strain cell to demonstrate that in providing angular torsion in a metal wire, the conductivity changed in a fashion similar to the case when the same structure was illuminated by electromagnetic radiation.

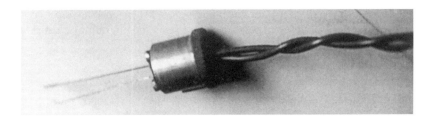

Figure 9.10. Detector Electrical disturbances built by Bose [11].

9.11 BIOLOGICAL EFFECTS OF MM WAVES

After 1900, Bose moved away from millimeter-wave research, and continued his research in the area of the electrical response of living and nonliving matter. His research on millimeter waves was carried out during 1894–1900. His research on the development of the millimeter-wave receiver naturally led him to a later topic of research, called "On Electrical Touch and the Molecular Changes Induced in Matter by Electric Waves" [23].

According to Bose,

> *I found that under continuous stimulation by the electromagnetic radiation, the sensitiveness of the metallic detectors disappeared. But after a sufficient period of rest it regained once more its normal sensitiveness. In taking records of successive responses, I was surprised to find that they were very similar to those exhibiting fatigue in the animal muscle. And just as animal tissue, after a period of rest, recovers its activity, so did the inorganic receiver recover after an interval of rest.*
>
> *Thinking that prolonged rest would make the receiver even more sensitive, I laid it aside for several days and was astonished to find that it had become inert. A strong electric shock now stirred it up into readiness for response. Two opposite treatments are thus indicated for fatigue from overwork and for inertness from long passivity.*
>
> *A muscle-curve registers the history of the fundamental molecular change produced by the excitation in a living tissue, exactly as the curve of molecular reaction registers an analogous change in an inorganic substance.*

As an example, he demonstrated that introduction of a low dose of poison acted as a stimulant, whereas a high dose killed it – which was the absence of any electrical response (which is the modern definition of death!).

In the Friday Evening Discourse, at the Royal Institution in 1900, he showed that on application of chloroform, plant response disappeared, just as it does for an animal; with timely "blowing off" of the narcotic vapor by fresh air, the plant too revived and recovered to respond anew. He also experimented with metals. He showed that a low dosage of oxalic acid on tin stimulated its response, while a large dosage "killed it," as shown in Figure 9.11. He concluded by saying that there was no line of demarcation between physics and physiology [1,26,27].

9.12 CONCLUSION

The brief description given above pertains mainly to the pioneering and significant contributions made by J. C. Bose, during 1898–1900, towards the advancement of the discipline of (microwave) millimeter wave physics and techniques. Further details about the investigations performed and the variety of experimental arrangements developed by him may be found in [6]. We conclude by highlighting the following significant accomplishments of Bose during the end of the last century:

(i)   Generated, radiated, and received electromagnetic waves having wavelengths ranging from 2.5 cm to 5 mm.

(ii) Developed an mm wave spectrometer to investigate the reflection, refraction, diffraction, and polarization of electromagnetic waves.

(iii) Experimentally measured the wavelength of mm electromagnetic waves.

(iv) Measured the refractive indices (and hence, the dielectric constants) of a variety of materials at mm wave frequencies.

(v)   Contributed to the development of the microwave lens.

(vi) Used blotting paper soaked in electrolyte as a lossy artificial dielectric to absorb millimeter waves.

(vii) Along with others, used metal tubes to guide and radiate electromagnetic waves. However, Bose made the significant step of flaring the walls of a rectangular waveguide, and thereby developed the pyramidal horn, which he called a *collecting funnel*.

(viii) Invented the *galena receiver*, which was the first use of a semiconductor as a detector of electromagnetic waves.

(ix) Developed the point-contact detector, which was the forerunner of the diode.

This led him to the development of the point contact galena detector for the reception of millimeter waves at 60 GHz. He was aware that such devices

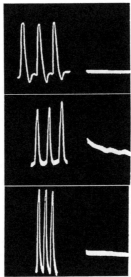

Figure 9.11. Action of poison in abolishing response of muscle (uppermost record), plant (middle record) and metal (lowest record) [4].

produced V-I characteristics that were not straight – namely, the conductivity changed nonlinearly with voltage.

9.13   EPILOGUE

Bose's work on wireless was primarily from 1895–1900. People say that the change in his research from wireless to biological aspects was due to the fact that *Lord Rayleigh and Sir W Crookes both told him that while the perfection of his methods was unquestioned, no one had yet been able in 1901, to repeat his experiments of 1895–96* [R. Tagore, *Chithipatra*, Vol. 6, p. 150]. Bose probably believed that something was amiss as he did not include in his Collected Physical papers [6], the paper on his thesis [22] on measurement of wavelength for which he obtained a doctorate in absentia, from University College, London. As we know now today, starting with Hertz and Marconi, all used the Tesla spark gap generator for their experiments. Therefore the wavelength they measured was related to the highly resonant transmitting and receiving systems they used and not related to that of the spark gap. So the measurement of wavelength at millimeter waves in Bose's thesis was connected to his transmitting and receiving apparatus and not that of the spark gap source, as he claimed. Hence, it was difficult for anybody to duplicate his work as not only the dimensions of his devices were very small (of the order of a few millimeters and of unknown tolerances) but also on the nature of the broadband source, namely the spark gap.

9.14   BIOGRAPHICAL SKETCH

As mentioned earlier, after 1900 Bose devoted his research interest into the field of animal and plant physiology where his outstanding contributions were much in advance of his time. He introduced new experimental methods and invented many delicate and sensitive instruments, such as his crestograph [6] for recording plant growth, magnifying a small movement as much as 10^6 (one million) times. He also devised apparatus for demonstrating the effects of air, food etc. on plants and demonstrated a parallelism between the responses of plants and animals. However, this is not the appropriate forum to discuss these topics. Instead, we give here a short description of Bose's educational and personal background which should be found helpful in appreciating him and his work.

   Jagadis Chunder Bose was born on November 30, 1858, at Mymensingh, now in Bangladesh. He joined St. Xavier's College, Calcutta, India, in 1875 and passed the First Arts Examination of Calcutta University in 1877. He passed the B.A. Examination of Calcutta University in 1880. During 1880–1881 he studied Medicine in London for one year. He had to give up medicine because of his health related problems. He entered Christ's College, Cambridge in 1881, and was interested in physics. In 1884 he graduated from Cambridge with a Natural Tripos, and in the same year he also passed the B.Sc. Examination of London University. He was appointed Officiating Professor of Physics at the Presidency College, Calcutta, where he was made a permanent Professor later. In addition to teaching he conducted his research there and communicated his research findings to the Royal Society of London through Lord Rayleigh the eminent physicist. It should be noted that Lord Rayleigh (Figure 2.21) taught Bose physics at Cambridge, and later acted as his mentor during his research in physics. As mentioned earlier in the text, it was primarily on Lord Rayleigh's recommendation that London University awarded the Doctor of Science degree to Bose. Until 1900, Bose conducted much of his research in the area of millimeter and microwave physics [10] where he is considered to be one of the pioneers. The semiconductor detector he developed grew out of his various investigations to establish the optical nature of electromagnetic waves. He retired from the Indian Educational Service in November, 1915, and was appointed an Emeritus Professor of the Presidency College with full salary for a period of five years. Knighthood was conferred on him in 1917. In 1917, Bose founded the Bose Research Institute in Calcutta, which is the first of such an institute in India. Bose is considered to be the initiator of scientific research tradition in India. He died on November 23, 1937, at the age of 79.

Bose was a prolific writer. He also traveled frequently to Europe and the United States on various missions, and he often gave technical lectures. On a personal level, Bose believed in the free exchange of scientific knowledge, and he strongly believed that knowledge grows by sharing it with fellow scientists. The idea of commercialization of science was so repugnant to him that in the founding charter of the Bose Research Institute he included a clause so that no member of his institute may be allowed to apply for a patent for an idea and/or device that he or she developed. He was also a humanist.

REFERENCES

[1]   P. Geddes, *The Life and Work of Sir Jagadish C. Bose*, New York: Longmans, Green & Co., 1920.

[2]   V. Mukherji, Jagadish Chunder Bose, Publications Division, Ministry of Information and Broadcasting, Government of India, New Delhi, India, 1983.

[3]   A. Home (ed.), *Acharya Jagadish Chunder Bose*, Birth Centenary Committee, 93/1 Upper Circular Road, Calcutta, 1958.

[4]   P. Bhattacharyya and M. Engineer (Ed), *Acharya J. C. Bose, - A Scientist and A Dreamer*, Volumes 1 & 2, Bose Institute, Calcutta, 1996.

[5]   P. Bhattacharyya, A.K. Dasgupta and M. Mitra (Ed), *Science and Society – Reflections, A Collection of Acharya J. C. Bose Memorial Lectures (1938-1996)*, Bose Institute, 93/1 Acharya Prafulla Chandra Road, Calcutta, 1996.

[6]   J. C. Bose, *Collected Physical Papers*, New York: Longmans, Green & Co., 1927.

[7]   J. F. Ramsay, "Microwave Antenna and Waveguide Techniques Before 1900," *Proceedings of the IRE*, 46, February 1958, pp. 405-415.

[8]   J. C. Wiltse, "History of Millimeter and Sub-millimeter Waves," *IEEE Transactions on Microwave Theory and Techniques*, MTT-32, 9, September 1984, pp.1118-1127.

[9]   A. W. Love, *Electromagnetic Horn Antenna*, New York: IEEE Press, 1976.

[10]   D. L. Sengupta and T. K. Sarkar, *Microwave and Millimeter Wave Research Before 1900 and the Centenary of the Horn Antenna*, presented at the 25th European Microwave Conference, Bologna, Italy, September 4-7, 1995, pp. 903-909 (Invited Paper).

[11]   J. C. Bose, *Detector for Electrical Disturbances*, U.S. Patent No. 755840, Patented March 29, 1904, (Application filed September 30, 1901, Serial No. 77,028).

[12]   H. Hertz, *On Electromagnetic Waves in air and Their Refraction*, London: MacMillan and Co., 1893, and New York: Dover, 1962.

[13]   J. H. Bryant, *Heinrich Hertz, IEEE Catalog No. 88th 0221-2*, The IEEE Service Center, 445 Hoes Lane, Piscataway, NJ 08854, 1988.

[14]   O. J. Lodge, The History of Coherer Principle, *Electrician*, November 2, 1897.

[15]   O. J. Lodge, Signaling Through Space Without Wires, *Electrician*, London, England, Printing and Publishing Co., Ltd., 1898.

[16]   J. C. Bose, *Electromagnetic Radiation and Polarization of the Electric Ray*, pp.77-101 reference [6].

[17]   J. C. Bose, "On Polarization of Electric Rays by Double-Refracting Crystals," Asiatic Society of Bengal – May 1985, pp. 1-7 of reference [4].

[18]   J. C. Bose, *On a Self Recovering Coherer and the Study of the Cohering Section of Different Metals*, Proceedings of the Royal Society, Volume A65 pp. 166-172, 1899.

[19]   J. C. Bose, *On the Determination of the Index of Refraction of Sulphur for the Electric Ray*, pp. 21-30 of ref. [6].

[20]   J. C. Bose, *The Rotation of Plane of Polarization of Electric Waves by a Twisted Structure*, pp. 102-110 of reference [6].

[21]   J. C. Bose, *On the Continuity of Effect of Light and Electric Radiation on Matter*, pp. 163-191 of reference [6].

[22]   J. C. Bose, *The Determination of the Wavelength of Electric Radiation by Diffraction Grating*, Proceedings of the Royal Society, Volume 60, pp. 167-178, 1896.

[23]   J. C. Bose, *On Electric Touch and the Molecular Changes Produced in Matter by Electric Waves*, Proceedings of the Royal Society, Volume A66, pp 452-474, 1899.

[24]   J. C. Bose, *The Response of Inorganic Matters to Mechanical and Electrical Stimulus*, pp. 259-276 reference [6].

[25]   G. L. Pearson and W. H. Brattain, *History of Semiconductor Research*, Proceedings of the Institute of Radio Engineers, Volume 43, No. 12, pp. 1794-1806, December 1955.

[26]   J. C. Bose, *On the Selective Conductivity Exhibited by Certain Polarizing Substance*, pp. 76-77 of reference [6]; also, Proceedings of the Royal Society, Volume LX, pp. 433-436, 1987.

[27]   J. C. Bose, "On the Change of Conductivity of Metallic Particles Under Cyclic Electromotive Variation", pp. 223-252 of reference [6].

10

SIR JOHN AMBROSE FLEMING – HIS INVOLVEMENT IN THE DEVELOPMENT OF WIRELESS

JOHN MITCHELL, HUGH GRIFFITHS, IAN BOYD

Department of Electronic and Electrical Engineering, University College London, UK

10.1 INTRODUCTION

Professor Sir John Ambrose Fleming was one of the great pioneers of electronics and radio telegraphy. Best known as the inventor of the thermionic diode valve, he also made significant contributions to teaching and research in photometry, radio, and electrical measurement technology. He published more than 100 papers and books; his name is indelibly linked to the left- and right- hand rule mnemonics used by students worldwide to remember the relative directions of field, current and force in electrical machines.

10.2 THE EARLY YEARS

John Ambrose Fleming (Figure 10.1) was born in Lancaster on 29 November, 1849. The eldest of seven children of James Fleming, then minister of the local Congregational Chapel, Fleming showed a number of childhood traits in common with many other scientific greats – a thirst for scientific knowledge and an intuition for practical invention – that would lead him from a bedroom experimenter to a renowned engineer and pioneer of the electronic age.

After relocating to London's Tufnell Park when his father took charge of the Kentish Town Congregational Chapel, Fleming's interest in the industrial age was fuelled by train journeys to his maternal grandfather's Portland cement factory in Greenhithe in Kent. John Bazley White, one of the pioneers of Portland cement, was just one of a number of public luminaries in his close family. Others included Mrs Ellen Ranyard, founder of the Ranyard Biblewomen's Mission, and the Rev. Edward White, author and preacher.

His early education left the young Fleming with the desire for more practical science and formed his view of education that learning by rote was not

Figure 10.1. Professor Sir John Ambrose Fleming (picture circa 1905).

an ideal form for enlightening young minds to the wonders of science. At about 11 years old he was given a box of magnetic toys, while a little later he began to practise photography, coating and fixing his own glass plates to be used in his home built camera made from a cigar box. This thrill of practical science was reinforced by regular visits to the "Polytechnic" in Regent Street, whose aim it was to interest the public in Science. Most importantly, the experience of a lecture by a family friend, Mr Dixon, introduced him to the wonders of electricity, and led to him devoting his pocket money to jam pots, zinc and copper plates, copper wire and sulphuric acid in order to construct voltaic batteries, Leyden jars and electromagnets. He also took great pride in giving 'public' demonstrations of his experiments to friends and family, including asking the audience to link hands so that a small shock could be sent around the line.

At the age of 14, Fleming was sent to University College School in London, a progressive establishment which unusually for the time prided itself on conducting education without corporal punishment. It was while here, as his parents' thoughts turned to his future, that Fleming, with his love of machines decided that a career in Engineering would be most suitable. However, the common route of being articled to an engineering firm was not an option due to the often large payment involved. Instead he prepared for the London Matriculation examination conducted by the University of London. With this completed at the age of 16, enrolment into University College London (UCL) to study for the degree of Bachelor of Science (BSc) followed. Here the influence of great men of science and mathematics such as Augustus de Morgan (Mathematician, best known for his definitions in Logic), Alexander Williamson (Chemist, who provided the Theory of Aetherification) and Carey Foster (Physicist, now best known for the Carey Foster Bridge, a more accurate version of the Wheatstone bridge) took effect. In the years up to his graduation in 1870 Fleming took employment, first copying plans in a shipbuilders in Dublin and later as a clerk on the London Stock Exchange. After being awarded the degree in the First Division, he took a position as science master at Rossall College, a boys' public school near Blackpool. However, it was not long before the yearning to enhance his scientific knowledge brought him back to London as an assistant to Dr Edward Frankland at the Science Schools in Exhibition Road SW1. It was here he was to meet Oliver Lodge who was to become a life long friend (Sir Oliver Lodge contributed the preface to Fleming's autobiography) despite a number of disagreements over the work of Marconi.

After two years Fleming returned to remunerative teaching work, this time as a science teacher at Cheltenham College, which was to encourage and nurture a distinctive characteristic of his teaching, the use of laboratory experiments in the teaching of science. Despite entering a poorly equipped laboratory he soon constructed chemical and physical apparatus for his use in class. At this time a second-hand copy of the three volumes of Faraday's Researches on Electricity was inspiring further free-time experimentation and in turn this introduced the work of James Clerk Maxwell. The impression Fleming left on the staff was evident from a testimonial commenting on the difficulties that the school faced in trying to find a worthy successor: "Men of the highest attainments in science knowledge are plentiful enough, but in combination with teaching power and tact in the management of pupils are scarce."

Once again the cycle of transitions from teacher to student would repeat, this time with Fleming winning a scholarship to St John's College Cambridge to study in the Cavendish laboratories under Maxwell, often being one of only two students who attended Maxwell's lectures.

In November 1879 Fleming was to be dealt a double blow. On 5 November his great mentor Maxwell died having been ill for most of the year, and 6 months after giving his last lecture. Five days later, Fleming's father died after a short illness. This was all only a few months after Fleming had been awarded the degree of Doctor of Science by the University of London on the subject of "Electricity Treated Experimentally". In order to fulfill his now increased family obligations, he took an appointment as demonstrator in Mechanics under Prof. James Stuart, which allowed him to stay at Cambridge.

Soon, however, Fleming would once more return to teaching, taking the first Professorship of Physics and Mathematics at the newly-inaugurated University College, Nottingham. This was a position that he would only hold for a few months. He was nevertheless able in this short time to give public lectures, including a lecture on Solar Eclipses using coloured lantern slides. In the previous few years a number of exciting developments had furthered the use of electricity and opened new opportunities for men such as Fleming. In 1876 Bell (who had also studied at University College London prior to immigrating to Canada) had patented the telephone; in 1878 David Hughes invented a carbon microphone, while a year later the electric lamp was introduced.

So it was that in the early part of 1882 Fleming took the appointment of "electrician" to the Edison Electrical Light Company, who were constructing large dynamos to drive electric lighting systems, and in particular the 1882 Electrical Exhibition at the Crystal Palace. During this time he advised several city corporations on new Ferranti alternating current systems and lighting installations, designed some of the first electric lighting systems for ships, and also visited the United States.

However, Fleming was still continuing his acclaimed lectures and was invited to give a set of lectures at University College London. In 1884 this was followed by an invitation that would dominate the rest of Fleming's working life.

10.3   RESEARCH OF THE UNIVERSITY PROFESSOR

In 1884 Fleming was appointed as Professor of Electrical Technology at University College London. Engineering had been taught at UCL since 1827 when W.J. Millington was appointed as "Professor of Engineering and Application of Mechanical Philosophy to the Arts", but at the end of the 19th Century the discipline was expanding and consequently dividing into new specialties. One of these areas was Electrical Engineering, which up to that point had been included in Dr Alexander Kennedy's Mechanical Engineering Laboratory. Although this, the first Chair in the subject in England, offered great new opportunities, the resources put at Fleming's disposal were modest to say the least. Later he recalled that all that could be provided were a blackboard and a piece of chalk!

However, within a year Fleming was able to obtain a grant of £150 which allowed him to purchase a small gas-engine and a storage battery of twenty five cells as well as small experimental equipment such as a photometer. He was also able to take two rooms within the University, one to be used as a teaching laboratory for student demonstrations and another as a working laboratory for his research. The use of practical laboratory experience as a means of teaching was particularly important to Fleming. Although an expert lecturer, his College lecture theatre would see little of him; for many years he gave just a single one hour lecture on Fridays from 11 to 12. To aid the laboratories he developed a system of twenty separate printed laboratory sheets, each of four pages of foolscap paper [1]. These would describe the principles of the experiment, the procedures and the quantities to be recorded. This system of instruction has since become common in most engineering schools.

Fleming had a supreme insight into the physical mechanisms underlying electrical technology, but more importantly was accomplished in being able to offer representations and explanations that would illuminate complex subjects to both students and the general public. He began writing on the subject of alternating currents and transformer theory, coining the term "power factor" to describe the ratio of true Watts to the product of the Volts × Amperes. However, perhaps the best known example of this skill would be the mnemonics that are still taught the world over today: Fleming's left and right hand rules. The rules were devised to assist students in remembering the relative direction of field, current and force in electrical machines, the left hand referring to motors and the

---

[1]   These are held in the archives of the Institution of Electrical Engineers, Savoy Place, London.

right hand (Figure 10.2) to generators. In a dynamo, the thumb represents the motion, the first finger the field and the second finger the current.

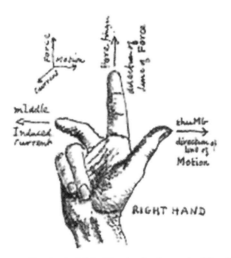

Figure 10.2. Fleming's Right Hand rule, drawn by Fleming.

In 1893, with a £800 donation he furnished his first full laboratory in the South Wing of the Quadrangle of UCL. This was further enhanced by £400 per annum from the London County Council on the proviso that Fleming gave public lectures during the winter terms.

Major growth in the Department's facilities was enabled in 1897 when the Pender Memorial Committee donated £5,000 of the £6,277 that had been collected to UCL to improve the electrical engineering laboratories. Sir John Pender was the founder of what is now the Cable and Wireless company when, in 1856, he was one of three contributors who risked £1,000 in the laying of a transatlantic cable. In honour of this contribution the college inaugurated the Pender Chair and founded the Pender Laboratory. This allowed more staff to be appointed to the department including W.C. Clinton and J.T. Morris.

So far, we have discussed Fleming's work which encompassed many areas, although not the area of wireless telegraphy. But it was his reputation as an experimental scientist and expertise in areas such as power engineering that were to lead him to his greatest achievements.

10.4   SCIENTIFIC ADVISOR TO THE MARCONI COMPANY

In the mind of many, the most famous pioneer of wireless is undoubtedly Guglielmo Marconi. However, in common with many great technological developments, a raft of other contributors made these achievements possible. In 1896, Marconi arrived in England with his "secret box" much to the fascination of the public and the disdain of the upper echelons of the British Scientific

Establishment. Fleming's first opportunity to see a practical demonstration of wireless telegraphy was in April 1898 while on holiday in Bournemouth. In his memoirs he recalls his astonishment when a telegraphic instrument printed the Morse code message "Compliments to Professor Fleming" which had been transmitted across 12 miles of sea. In April of the following year, Fleming wrote a long letter to *The Times* describing his inspections of Marconi's experiments and by 9 May, he had been unanimously elected as Scientific Advisor to the Marconi Company, with a fee of £300 per annum. Although having not worked greatly in the wireless area, but having studied under Maxwell, it was not surprising that Fleming had taken an interest in the theory of electromagnetic waves, and had even replicated the experiments performed by Hertz in 1887.

At first, Fleming's main role was to add scientific authority to Marconi's work by using his public lectures to give demonstrations of Marconi's wireless systems. In September 1899, the British Association held its annual meeting in Dover, at which Fleming was invited to give the principal evening lecture. He used this lecture to demonstrate the Marconi wireless system using a mast and aerial wire erected on the Dover Town Hall. During the lecture, wireless messages were sent to the Goodwin Sands Lightship and to Boulogne, where the French Association was also meeting. Soon Marconi's mind turned to trying to extend the reach of his system to across the Atlantic. Fleming describes how

> ...he had only up to that time used apparatus which might be called laboratory apparatus. But as I added some experience in the use of powerful high tension alternating current in electric lighting, the Marconi Company engaged me to design the power station and the plant that was necessary for long distance wireless transmission. A site was selected at Poldhu, a lonely spot on the coast of Cornwall.

At the time the greatest distance that Marconi had achieved was a little less than 200 miles, so to cross the Atlantic, a distance of over 2,000 miles, a significant increase in power was required to overcome the free-space loss, which increases with the square of the distance. Marconi calculated that a two inch spark would be required, which suggested to Fleming that 100,000 Volts would be necessary, considerably more than had ever been produced using a single transformer.

Marconi was particularly concerned with the secrecy of the project, and he would scoff at the idea that he was about to attempt transmission across the Atlantic, in case the assertion of many scientific men were true and radio waves could not be received over the horizon.

Fleming was tasked with designing the power plant capable of such a mammoth transmission (Figure 10.3). The plant included a 25-horsepower Mather and Platt alternator, a 32-horsepower engine, and two transformers which were capable of producing 20,000 Volts. Included in the design was also a form of large capacity condenser, devised by Fleming, made of zinc sheets separated by glass plates in which alternate sheets were connected together. These plates were placed in large wooden boxes filled with linseed oil. A significant step was

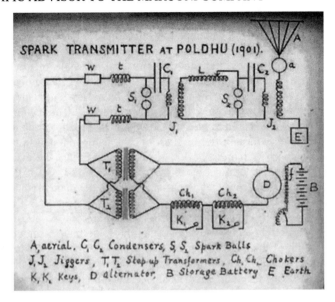

Figure 10.3. Circuitry used in Poldhu, 1901.

the use of a transformer in place of the inductor that had been used for most of the stations at that time.

The station also included a "jigger", patented by Marconi in 1900, to link the high tension discharge circuit and the antenna circuit. It soon became clear that the increase in power required created additional issues for the circuitry. In particular, controlling of the sparks, which would usually be done through a simple key, is considerably more complex at very high powers and with condensers that need to be charged for controlled periods. Fleming produced a series of schemes to solve the problem, and with each one offered improved practicality and reliability. The circuitry used to create the signals by today's standards looks relatively straightforward, but for the time, it required a significant advance in power engineering. The waves were created by first charging a large bank of condensers (capacitors) and then discharging them to form a spark between two metal balls (known as a spark gap transmitter). This spark was connected across the primary of a transformer which was used to tune the circuit by supplying matching to the antenna. To develop the power required for the Poldhu station, two stages of sparks were required. The first stage produced a low frequency with a relatively small spark, which was then stepped-up through a transformer to produce a much larger spark, before being connected to the antennas via the jigger transformer.

After more than a year of experimentation in the laboratory at UCL and at the test site in Poldhu, full tests started in July 1901. It was soon clear to all those near the transmitter site that something significant was occurring. Using the very high power created by Fleming's design, the station produced sparks that were likened to claps of thunder by the locals. Despite problems with aerials

(Figure 10.4, they were highly susceptible to wind damage), by early November 1901 Marconi was ready to attempt transmission. He sent a telegram to his assistants in Poldhu to begin transmitting the test sequence, "SSS" on 11 December. Due to bad weather, no reception was possible on this first day. However, at 12.30 pm and again at 1:10 and 2:20 on 12 December 1901 the message "SSS" was received at Signal Hill, St. John's, Newfoundland, Canada.

This propelled Marconi to even greater fame. Fleming, however, felt that the credit was not shared as well as he perhaps deserved. Marconi was quick to reassure Fleming with an offer of shares in the company, and in response he continued to support Marconi's work. However, even before the event Fleming had already been informed in a letter from Major Flood-Page of the Marconi Wireless Company that "if we get across the Atlantic, the main credit will be and must forever be Mr. Marconi's" (Figure 10.5).

Even now, 100 years after the event, there are still sceptics who doubt that Marconi truly did receive the signals that day. The main issue concerns the exact wavelength of the transmission, which even with modern calculations still forms a point of debate. This is not helped by both Marconi and Fleming stating differing values at different occasions. At the time, the measurement of capacitance was a difficult task, and the measure of inductance with any accuracy was almost impossible. And as Fleming commented himself, he was not able to

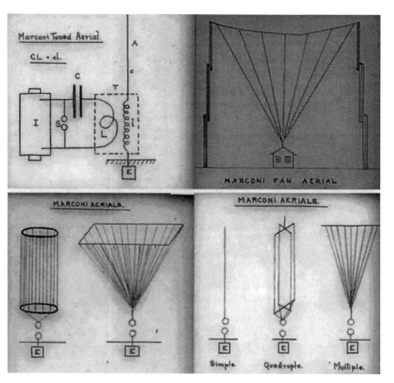

Figure 10.4. Examples of Marconi's designs for wireless aerials.

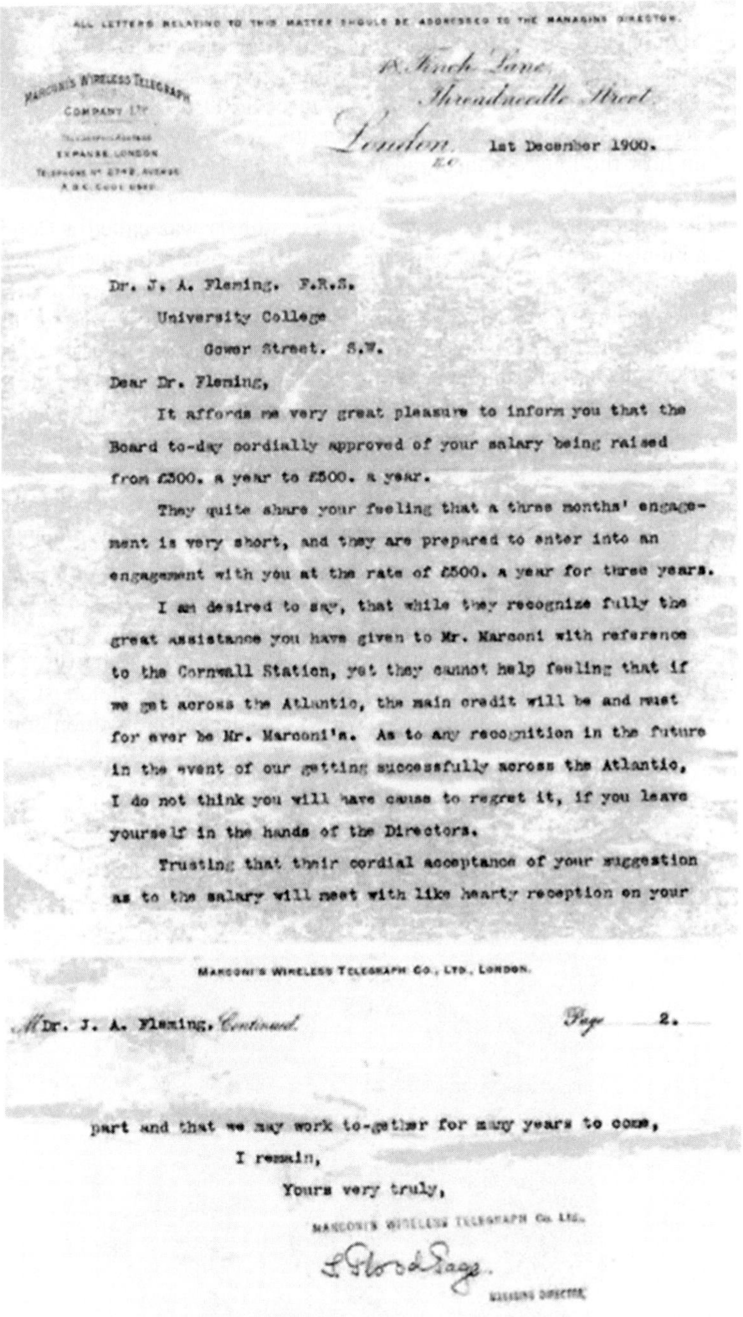

ALL LETTERS RELATING TO THIS MATTER SHOULD BE ADDRESSED TO THE MANAGING DIRECTOR.

MARCONI'S WIRELESS TELEGRAPH
COMPANY LT?

EXPANSE, LONDON
TELEPHONE N° 2742, AVENUE
A.B.C. CODE USED.

18 Finch Lane,
Threadneedle Street,
London. 1st December 1900.
E.C.

Dr. J. A. Fleming. F.R.S.
University College
Gower Street. S.W.

Dear Dr. Fleming,

It affords me very great pleasure to inform you that the
Board to-day cordially approved of your salary being raised
from £300. a year to £500. a year.

They quite share your feeling that a three months' engagement is very short, and they are prepared to enter into an
engagement with you at the rate of £500. a year for three years.

I am desired to say, that while they recognize fully the
great assistance you have given to Mr. Marconi with reference
to the Cornwall Station, yet they cannot help feeling that if
we get across the Atlantic, the main credit will be and must
for ever be Mr. Marconi's. As to any recognition in the future
in the event of our getting successfully across the Atlantic,
I do not think you will have cause to regret it, if you leave
yourself in the hands of the Directors.

Trusting that their cordial acceptance of your suggestion
as to the salary will meet with like hearty reception on your

MARCONI'S WIRELESS TELEGRAPH CO., LTD., LONDON.

Dr. J. A. Fleming. Continued. Page 2.

part and that we may work to-gether for many years to come,

I remain,

Yours very truly,

MARCONI'S WIRELESS TELEGRAPH Co. LTD.

L. Flood Page.

MANAGING DIRECTOR

Figure 10.5. Letter to Fleming from Major Flood-Page, UCL Collections MSadd122.47.2.

measure the wavelength of the transmission as he had not yet invented his wave-meter, or "cymometer", as he termed it (Figure 10.6). Marconi suggested in 1902 that the wavelength was around a fifth of a mile, giving a frequency of around 900 kHz, while later in 1908; he suggested it to be 1200 ft (800 kHz). What is certain nevertheless is that the original transmitter was very wideband and this may explain how the long distance transmission of a "daylight wave" may have come about.

The device used to receive the wireless signals was called a Coherer. Although a number of different designs were used, the basic principle involved a small tube of metal filings that would conduct (or cohere) when excited by a radio wave. A particular difficulty of the design of these devices was to find a way to 'break up' the filings between symbols. Marconi's early design incorporated a clockwork 'tapper' to tap the tube and break the connection between Morse symbols. This was one of the main improvements made by Marconi on the Coherer device that had originally been invented independently by Sir Oliver Lodge and Edouard Branly for detecting the presence of electric waves. However, his most notable addition to the Poldhu system was the use of long vertical wires attached to high masts which would effectively radiate and collect radio waves.

The coherer, although it was the cornerstone of the first successful receiver, did not provide high sensitivity and was inherently unreliable due to its delicate mechanical components. In addition it was very slow-acting, restricting the speed of messages to just a few words per minute. Fleming, amongst many others, sought to develop an improved device. The deeper motivation for his

Figure 10.6. Fleming's Direct Reading Cymometer.

work has been discussed by a number of authors. Some have argued that Fleming's increased deafness that had affected much of his life, led him to look for a device that could be easily viewed on a galvanometer. More recently, Hong suggests that following an incident where a public demonstration by Fleming was sabotaged (the Maskelyne Affair) [1], the Marconi Company was unsure of Fleming's future use to the company. His focus on finding a new device was therefore perhaps an attempt to renew his relationship with the Company.

Whatever the reason for his interest, the result of his work came to fruition in 1904, as one of the most significant single advances in the history of wireless.

10.5   THE THERMIONIC VALVE

The story of the invention of the valve goes back to the early 1880s, when several scientists became concerned with the discoloration of the inside surface of the glass envelope of incandescent lamps during their operation. Fleming had just become scientific advisor to the Edison Company, and his instinctive interest in physical phenomena led him to begin to study the underlying phenomena of this bulb darkening, and more specifically why the carbon filaments broke so easily. Electric lamps had come into being in 1879 when, independently, Joseph Swan in Britain, and Thomas Edison in the United States used a liquid mercury pump invented by Hermann Sprengel to create a high vacuum, enabling the first working bulbs to be produced.

In Fleming's experiments of 1882 and 1883 he noticed that inside most of the burned-out bulbs, a fine line was present in the plane of the filament that did not get coated. This well-defined strand of clear glass had obviously been masked by the shadow of the un-burnt filament from the carbon evaporating from the hot spot, which eventually led to the lamp failure.

Edison himself performed experiments in 1883 with metal plate inserts in an attempt to minimise this blackening effect originating from the filament. He found that while no current flowed in the plate circuit when connected to the negative terminal of the filament supply, a small current flowed if it was connected to the positive electrode. In 1884, Sir William Preece also studied this "Edison Effect", and concluded that it was associated with the evaporation of carbon molecules in straight lines. However, the effect was not fully explained, nor did anyone attempt to make use of the phenomenon.

In 1888, some three years after his appointment at UCL, Fleming found a small amount of time to further explore this line of study using metal plates and also cylinders. As Fleming had shown throughout his career he held an academic interest that led to a desire to understand the fundamental theory of the phenomenon that he was witnessing. However, he also held a strong sense of the practical, with a desire to create instruments that were of use to the scientific community. At this time, one of his primary interests was in the standardization of electrical measurements and he is credited with leading the movement to form the National Physical Laboratory in Britain which oversaw these standards. This led him to look for a photometric standard, a lamp of a set and repeatable brightness that did not deteriorate markedly with use.

By altering the position of these plates around the filament he found he could vary the intensity of particle emission and even stop it altogether. Furthermore, he also discovered he could actually rectify alternating currents under specific conditions. He was honoured with election as a Fellow of the Royal Society for these Edison Lamp studies.

The Edison Effect was explained using various terms by a number of those who investigated it although it could not be fully explained until 1897 when J. J. Thomson discovered the electron. In an evacuated bulb, with a filament and an electrically insulated plate, it was seen that if the plate was connected to the positive end of the filament, a current flowed, but only a very small current would flow if a connection was made to the negative end.

Fleming reported his experiments in a paper to the Royal Society in 1890, which states:

> It has been known for some time that if a platinum plate or wire is sealed through the glass bulb of an ordinary carbon filament incandescent lamp, this metallic plate being quite out of contact with the carbon conductor, a sensitive galvanometer connected between this insulated metal plate enclosed in the vacuum and the external positive electrode of the lamp indicated a current of some 4 to 5 milliamperes passing through it when the lamp is in action, but the same instrument when connected between the negative electrode of the lamp and the insulted plate indicates no sensible current – it is less than 0.0001 of a milliampere. This phenomenon was first observed by Mr Edison in 1884 and further examined by Mr W. H. Preece in 1885.

However, Fleming had made the following discovery:

> When the Lamp is actuated by an alternating current a continuous current is found flowing through the galvanometer connected between the insulated plate and either terminal of the lamp. The direction of the current through the galvanometer is such to show that negative electricity is flowing from the plate through the galvanometer to the lamp terminal.

The Edison Effect ceased to attract any further attention in the ensuing years, and by 1899, Fleming had become scientific advisor to the Marconi Company and as we have described, he devoted most of his time to designing and building the huge transmitter required for transatlantic wireless propagation. Despite the success of the powerful Fleming-designed transmitter, the detection sub-system was still limited to the use of the rather primitive and insensitive coherer device, and Fleming began to realise this could severely limit the application of wireless communication. Alternative chemical rectifiers and magnetic detectors were explored, but none were satisfactory.

The problem facing Fleming was "to discover how to change this feeble electrical oscillation [from the receiving aerial] into a feeble direct current which could work the ordinary cable-recording instruments".

Fleming then writes that in 1904 he thought, "Why not try the lamps?" He asked his assistant, Mr G. B. Dyke to put up the arrangements for creating a high frequency current. Taking one of his experimental lamps from a cupboard where it had been stored for nearly 20 years, and incorporating it into a specially tuned circuit with a galvanometer, he found that a steady state current easily registered the oscillations emitted by an identically tuned circuit positioned nearby. He had once again rectified high frequency currents, but this time for a necessary application.

Soon after this Fleming informed Marconi of his discovery, although not with the fanfare that might have been expected. In a letter to Marconi, he speaks of his other invention of the time, the "cymometer", which would now allow them to measure wavelengths of up 20,000 feet. Almost as an afterthought he adds at the end of the letter,

> *Also quite lately I have made another interesting discovery. I have found a method of rectifying electrical oscillations – that is making the flow of electricity all in the same direction. So that I can detect them with an ordinary minor galvanometer. I have been receiving signal on an aerial with nothing but a minor galvanometer and my device, but at present only on a laboratory scale. This opens up a wide field for work as I can now measure the effect of the transmitter. I have not mentioned this to anyone yet as it may become useful.*

Fleming made several modifications and improvements to the design and patented the invention in Great Britain on 16 November 1904 in patent No 24,850 (Figure 10.7). Subsequently it was also patented in the United States in patent No 803684 filed on 19 April 1905. Over the following years, he continued to modify and optimise the detector, to include tungsten filaments and a protecting shield, and it was incorporated into the Fleming-Marconi receiver. In fact the main contribution of the patent was the circuit in which the device was used. Fleming called the detector an oscillation valve, and later the Fleming valve (Figure 10.8). Although the term diode valve is more commonly used for this particular device, part of the family of what are called vacuum tubes in the United States, Fleming avoided the term diode, he preferred "valve" as it conveyed better the action of the device.

In 1905, Lee de Forest, who had been made aware of Fleming's device through his assistant C. B. Babcock, patented various detectors including a two-electrode valve. This started a bitter row that dragged through the courts for nearly 20 years. In early 1906, de Forest applied for a new patent for an "Oscillation Responsive Device" that shared many similarities to Fleming's diode valve. This caused Fleming to initiate the first salvos of this war, accusing

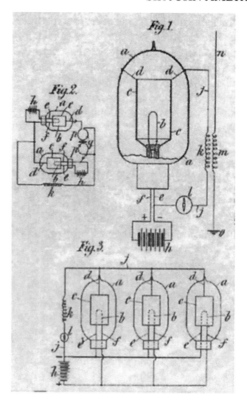

Figure 10.7. Valve Circuit from GB Patent No 24,850, 16 November 1904.

Figure 10.8. Fleming's first wireless valve (right) compared with Marconi's coherer (left); which it replaced. *Photos reproduced by courtesy of the Director of the Science Museum, South Kensington.*

de Forest in the press of plagiarism. However, de Forest was already working on another improvement. He believed that controlling the flow of current in the device would lead to higher sensitivity, and to this end he tried several modifications. This resulted in him inserting a third electrode to form a grid between the filament and the plate. This triode device he termed the "Audion". Although this advance on Fleming's original discovery was to produce one of the most vital commodities in electronic engineering, amplification, this property was not immediately recognised. Until 1913, the Audion was used as a radio detector in a similar fashion to Fleming's diode valve, and little was known about its exact principles of operation. It was not until a group of scientists at AT&T Labs overcame many of the problems of the earlier Audions, which included creating better vacuums and improved filaments, that the real potential of this device came to the fore.

By this time the Marconi Company, which held the rights to Fleming's Patents in Britain, and the Radio Corporation of America had both brought legal actions against the de Forest Company for infringements. The cases centered around the Marconi Company's claim that the addition of a grid was not an invention in its own right, while de Forest claimed that Fleming's patent was invalid as he had only repeated what was inherent in Edison's patent of 1883. Fleming admitted he had no claim to the Audion, writing that:

> *In one of my bulbs I had placed a carbon filament and a zigzag wire of platinum with the object of discovering whether this wire with it apertures was as effective in catching the electrons as a solid plate of the same overall area. But, sad to say, it did not occur to me to place the metal plate and the zigzag wire in the same bulb and use an electric charge of positive or negative on the wire to control the electron current to the plate.*

In 1916, a US District Court ruled that the addition of grid, though a contribution of value, was not independent of Fleming's prior fundamental invention. This was subsequently upheld in the Court of Appeals in 1917.

Despite being cited as the original inventor of the valve, and for laying the foundations that de Forest and many others would improve on, Fleming, in his memoirs still did not happily reminisce about the acknowledgement he received for his role in this invention. He says that:

> *...one firm has sold valves for many years made exactly in accordance with my patent specifications, but which they advertise and mark as "Marconi Valves,"... other firms advertise "Cossor Rectifier Valves", "Phillips Valves,", "Osram Valves".*

The final word in the saga was to come in 1943, when the US Supreme Court proclaimed that Fleming's patent was "rendered invalid by an improper disclaimer".

10.6   LATER LIFE

Fleming held his Chair at UCL until 1927, retiring at the age of 77. In that time he continued to actively research and teach, as well as give many public lectures. In particular he was proud of work with Sir James Dewar who was the Resident Professor at the Royal Institution in London. They worked together on measurement of electrical resistance at low temperatures, which was aided by the invention of the Dewar Flask. It is little known that it was due to Fleming's connections with the Edison-Swan factory that the first prototypes of these flasks were made.

Even after retirement Fleming kept extremely busy, becoming an eager advocate of the new technology of Television, giving lectures as part of his role as the President of the Television Society. He retired to Sidmouth on the south coast of England, to a house he shared with two of his sisters and his new wife when he married for the second time in 1933. In the basement he still maintained a small laboratory to continue his experimental work. He died in Sidmouth on 18 April 1945, aged 95.

He was rewarded with many distinctions for his invention, including the Gold Medal of the Royal Society of Arts, the Hughes Medal of the Royal Society, the Kelvin and Faraday Medals of the Institution of Electrical Engineers, the Gold Medal of the Institute of Radio Engineers, and the Franklin Medal of the Franklin Institute. He was knighted in 1929.

REFERENCES

[1]   Sungook Hong, *Wireless; From Marconi's Black-Box to the Audion*, Cambridge, MA: MIT Press, 2001.

[2]   Gavin Weightman, *Signor Marconi's Magic Box*, New York: Harper Collins 2003.

[3]   Obituary Notices of the Fellows of the Royal Society, Volume 5, November 1945.

[4]   Ambrose Fleming – His Life and Early Researches, *Proceedings of the Royal Television Society.*

[5]   Keith R. Thrower, *History of the British Radio Valve to 1940*, MMA International Ltd, 1994.

[6]   Percy Dunsheath, *A History of Electrical Engineering*, London: Faber and Faber Ltd., 1962.

[7]   J T. MacGregor-Morris, *The Inventor of the Valve: A Biography of Sir Ambrose Fleming*, The Television Society, 1954.

[8]   Sir J. Ambrose Fleming, *Memories of a Scientific Life*, Marshal, London: Morgan & Scott Ltd, 1934.

[9]   Sir J. Ambrose Fleming, *Fifty Years of Electricity*, London: Iliffe & Sons, 1921.

11

HISTORICAL GERMAN CONTRIBUTIONS TO PHYSICS AND APPLICATIONS OF ELECTROMAGNETIC OSCILLATIONS AND WAVES

MANFRED THUMM, *Forschungszentrum Karlsruhe, Association EURATOM-FZK, Institut für Hochleistungsimpuls- und Mikrowellentechnik Postfach 3640, D-76021 Karlsruhe, Germany and Universität Karlsruhe, Institut für Höchstfrequenztechnik und Elektronik, Kaiserstraße 12, D-76128 Karlsruhe, Germany*

11.1   INTRODUCTION

The chapter reviews a series of the most important historical contributions of German scientific researchers and industrial companies in Germany to the physics and applications of electromagnetic oscillations and waves during the past 140 years and intends to point out some relations to Russian scientists. The chronology highlights the following scientists: Philipp Reis (1834–1874): first telephone; Hermann von Helmholtz (1821–1894): unification of different approaches to electrodynamics; Heinrich Hertz (1857–1894): fundamental experiments on electromagnetic waves; Karl Ferdinand Braun (1850–1918): crystal diode, cathode ray tube, transceiver with coupled resonance circuits; Christian Hülsmeyer (1881–1957): rudimentary form of RADAR; Robert von Lieben (1878–1913): triode as amplifier in transmitter; Heinrich Barkhausen (1881–1956): Barkhausen-Kurz oscillations, first transit-time microwave tube; Manfred von Ardenne (1907–1997): first integrated vacuum tube circuits; Hans Erich Hollmann (1899–1961): multi-cavity magnetron, principle of reflex klystron; Oskar Heil (1908–1994): principle of the klystron, multi-stage depressed collector, patent of field effect transistor (FET); Walter Schottky (1886–1976): tetrode electron tube, theory of shot noise, Schottky effect, Schottky barrier; Herbert Kroemer (1928): III-V semiconductor heterostructures; Jürgen Schneider (1931): quantum electronic model of electron cyclotron resonance maser.

The purpose of this chapter is to present a chronology of historical German contributions to the physics of electromagnetic oscillations and waves

327

and their applications to communication, radio, television, RADAR, computer systems and heating.

Often an invention is attributed to one or two persons, the names of whom vary from country to country, depending on the country of origin of the authors. This chapter will illustrate that simultaneous development was going on all over the world and to point out some relations of German and Russian scientists.

From 1926–1929, Alexander A. Andronov (1901–1952) was a post-graduate student at the Faculty of Physics and Mathematics of the Moscow State University under the supervision of Leonid Isaakovich Mandelstam (1879–1944) and developed in his PhD thesis the most general approach to the theory of auto-oscillators. On the other hand, L.I. Mandelstam got his education from 1899–1914 as PhD student, Assistant Professor, and University Lecturer in the Institute of Physics of Karl Ferdinand Braun (1850–1918) at the University of Straßburg. K.F. Braun was the Nobel Prize Laureate of 1909 in Physics of Electric Oscillations and Radio Telegraphy together with Guglielmo Marconi (1874–1937). At the University of Straßburg, L.I. Mandelstam became an excellent experimentalist and gifted lecturer. He conducted original research on radio transmitters and receivers and performed fundamental works in optics: scattering in optically uniform and turbid media, theory of dispersion, scattering at liquid surfaces, theory of optical microscope imaging, radiation of sources near the boundary of two liquid media and optical measurements in analogy with radio experimental investigations. In 1912 L.I. Mandelstam became a member of the German Society of Natural and Physical Scientists. In July 1914, just before the beginning of World War I, he and N.D. Papalexy, who also had been an Assistant Professor of K.F. Braun, moved back to Russia where they founded their scientific schools.

These close relations between German and Russian radio telegraphy scientists have been the major motivation for the present article.

11.2   CHRONOLOGY OF HISTORICAL GERMAN CONTRIBUTIONS

11.2.1   Phillip Reis: First Telephone

Johann Phillip Reis (1834–1874), Figure 11.1, reported on October 20, 1861, at a seminar in the Senckenberg Museum at Frankfurt/Main, "On the propagation of tones over arbitrary distances via galvanic currents". He demonstrated his apparatus by transmission of the following sentence: *"The horse does not eat cucumber salad"* [1]. However, this work was not highly regarded in Germany and Reis did not apply for a patent. On February 14, 1876, Alexander Graham Bell (1847–1922) took out a patent for the telephone in the United States. Two hours later, on the same day, Elisha Gray (1835–1901) also applied in the United States for a patent on the telephone, but he was refused. Figure 11.2 shows a comparison of the Reis-Telephone and the Bell-Telephone. In 1877 the Siemens & Halske Company in Berlin started the series production of telephones, and in

Figure 11.1. (left) Phillip Reis, (right) Reis testing his telephone in 1861.

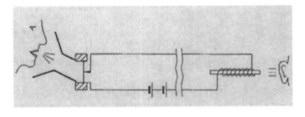

Figure 11.2. (above and upper right) Reis-Telephone (1861) with "Knitting Needle Receiver", (below and lower right) Bell-Telephone (1876) with permanent magnet rods and coils

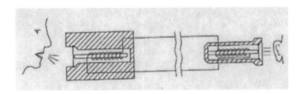

1882 they built and offered the first wall-mounted device {Werner von Siemens (1816–1892) & Johann Georg Halske (1814–1890)}.

11.2.2. Hermann von Helmholtz: Unification of Different Approaches to Electrodynamics

Hermann Ludwig Ferdinand von Helmholtz (1821–1894) was perhaps the last real "Universal Scientist" in the tradition of Gottfried Wilhelm Leibniz (1646–1716). From 1842 to 1849, he worked as a so-called Eskadron-Surgeon in the Charité at Berlin, in several guards' regiments at Potsdam, and as Lecturer for Anatomy in the Berlin Academy of Arts. His brilliant career as a University

Professor started in the spring of 1849 when he became Professor of Anatomy and Physiology at the Albertina in Königsberg (now Kaliningrad). From 1855 to 1858, he had the same function at the University of Bonn. His call to Bonn was initiated by Alexander von Humboldt. In the year 1858, Helmholtz was appointed to the Chair of Physiology at the University of Heidelberg where he was a colleague of the ingenious Robert Wilhelm Bunsen (Chemistry, 1811–1899) and Gustav Robert Kirchhoff (Physics, 1824–1887). During these productive years in Heidelberg he finished his intensive research work in the field of Physiology and published his famous "Handbook of Physiological Optics" (3 volumes), treating the 3-color theory, and his book on the "Theory of Sound Perception". After Heinrich Gustav Magnus (1802–1870) passed away, Helmholtz was appointed, in 1871, to the Chair of Physics and Mathematics at the University of Berlin (Figure 2.14) where he later also was engaged in Philosophy. In the year 1883, he was ennobled. His extraordinary contributions to electrodynamics are described in [2-4].

In 1847, Helmholtz suggested, in his work "On the Conservation of Forces", electrical oscillations 6 years before this process was theoretically calculated by William Thomson (Lord Kelvin, 1824–1907) (1853) and 10 years before it was experimentally verified by B.W. Feddersen (1857).

From 1870 to 1874, Helmholtz tried to unify different approaches to electrodynamics since its overall picture was difficult, incoherent, and not finished. The German Schools of Franz Ernst Neumann (1798–1895) and Wilhelm Eduard Weber (1804–1891) had developed a comprehensive explanation of both electrodynamics and electromagnetic induction in the classic Newtonian framework of a direct action at distance between charges and currents (i.e. moving charges). They tried to derive induced currents from an energy principle using an electrodynamic interaction potential of two linear conductors (currents). Helmholtz introduced a parameter, k, as additional term, with, $k = -1$ (Weber), $k = 0$ {James Clark Maxwell (1831–1879)}, and $k = 1$ (Neumann). The problem in deciding for the correct version was that integration along a closed current loop eliminates the k-dependence.

In 1878–1879, Helmholtz initiated a student competition of the Philosophical Faculty of the University at Berlin. The winner with distinction was his most outstanding student Heinrich Hertz who proved that electrical charges in time dependent currents do not exhibit inertia, which means that Weber's theory is wrong.

In 1879, Helmholtz initiated an international competition of the Prussian Academy of Sciences: "Do dielectric and galvanic currents have equivalent electrodynamic forces?" The winner again was Heinrich Hertz who confirmed Maxwell's theory with his brilliant epoch making experiments at the Polytechnical University of Karlsruhe (1886–1888, see section 11.2.3). Even today, the equations describing electromagnetic waves in homogeneous media (μ and ε are scalars) with no charge and current densities ($\rho = 0$, $\bar{j} = 0$) are called Helmholtz equations.

11.2.3   Heinrich Hertz: Discovery of Electromagnetic Waves

Heinrich Rudolph Hertz (1857–1894) (Figure 2.23) was a student of Kirchhoff and Helmholtz at the University of Berlin where he earned, on March 15, 1880, his Doctorate in Physics with his dissertation entitled "On induction in rotating spheres". After his Habilitation in May 1883 at the University of Kiel and intensive work on electrodynamics, he got the call to be a Full Professor of Physics at the Polytechnical University of Karlsruhe (1885–1889) where he conceived and performed his brilliant fundamental experiments confirming Maxwell's predictions and theory (see section 11.2.2) [4]. His classical experimental set-ups are shown in Figure 11.3 and Figure 11.4. Hertz generated, radiated (transmitted) and received (detected) electromagnetic waves at frequencies in the range of 50-500 MHz.

His initial experiment was on November 13, 1886, proving wireless transmission between two open circuits over 1.5 m and the decisive paper on the finite velocity of propagation of electromagnetic waves in air ($v_{phase} = c = 1/\sqrt{\mu_o \varepsilon_o}$) was published in 1888. His epoch making experiments conclusively proved the optical properties of electromagnetic waves such as: frequency, wavelength, amplitude (power), phase, polarization, reflection, refraction, diffraction and interference. He used reflectors at the transmitting and receiving positions to concentrate the waves into a beam.

Maxwell's ideas and equations were expanded, modified and made understandable after his death by the efforts of Hertz and the three "Maxwellians"; George Francis FitzGerald (1851–1901), Oliver Lodge (1851–1940) and Oliver Heaviside (1850–1925) [5]. It is important to note that Hertz and the Maxwellians were not aware of each other's work until Hertz published his 1888 work. The Maxwellians appreciated Hertz's brilliant experiments and their implications and gave them the widest possible publicity and labeled them from the beginning as a decisive new confirmation of Maxwell's theory. The four

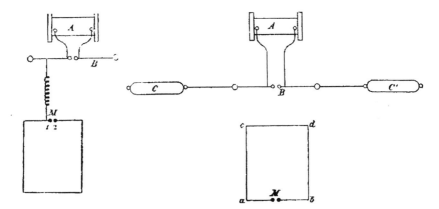

Figure 11.3. Experimental set-ups of Heinrich Hertz (**A**) Inductorium, (**B**) Spark gap in transmitter circuit, (**C**) Metal spheres at both ends of transmitting antenna, (**M**) Spark gap in receiver circuit ("Nebenfunken").

Figure 11.4. Original Hertzian oscillator at Institute of Physics of Polytechnical University of Karlsruhe (1886-1888).

equations in vector notation containing the four electromagnetic field vectors are now commonly known as Maxwell's equations. However, Einstein and Heaviside referred them as Maxwell-Hertz and Maxwell-Heaviside-Hertz equations, respectively. Since Hertz did not know anything about modulation of high frequency electromagnetic waves at low frequencies, he stated that waves with frequencies in the audio range (kHz) have too long wavelength and cannot be focused by reflectors so that they cannot be used for wireless telegraphy.

In 1887, Hertz also discovered the photo electric effect. He observed that the length of the spark between two electrodes increases when ultraviolet light falls on the negative electrode of a spark gap.

In the autumn of 1886, Hertz was offered chairs at the Universities of Gießen, Berlin, and Bonn. His choice was Bonn where he became, in April 1889, the successor of Rudolf Emanuel Clausius (1822–1888) and worked on "Principles of Mechanics" (1891). However, on January 1, 1894, he died at the age of only 37 years owing to a severe ear-, nose-, and throat infection connected with a bone disease. H. von Helmholtz stated in his touching obituary: *"He is a victim of the envy of the gods"*.

The Russian Alexander Popov (1859–1906), Figure 2.27, demonstrated, in 1895, his so-called *"Thunderstorm Recorder"* using aerial, coherer {invented in 1890 by Edouard Branly (1844–1940)}, and electromagnetic relay. He succeeded in the transmission of the words *"Heinrich Hertz"* over a distance of 250m. The antenna was mounted to a balloon. A few days later, also in 1895, Marconi transmitted and received a coded message over a distance of 1.75 miles, and one year later, he applied for the first patent in wireless covering the use of transmitter and coherer connected to a high aerial and earth. On October 7, 1897, Professor Adolf Slaby (1849–1913) of the Technical University of Berlin-Charlottenburg and his Assistant Georg Graf von Arco (1869–1940) succeeded in wireless telegraphy over a record distance of 21 km from Berlin Rangsdorf to Schöneberg employing a system similar to that of Marconi.

11.2.4. Karl Ferdinand Braun: Crystal Diode; Cathode Ray Tube, Wireless Telegraphy

The chronology of the professional activities of Karl Ferdinand Braun (1850–1918) shown in Figure 2.20 is as follows: 1870–1874: Assistant Professor at the Universities of Berlin and Würzburg (Habilitation supervised by H. v. Helmholtz); 1874–1877: Teacher at the Thomas Gymnasium Leipzig; 1877–1879, 1880–1882: Professor at the Universities of Marburg and Straßburg; 1883–1885: Full Professor at the University of Karlsruhe (Predecessor of H. Hertz); 1885–1895: Full Professor at the University of Tübingen and 1895–1918 Full Professor at the University of Straßburg. In 1874 K.F. Braun discovered conduction and rectification in metal sulfide crystals that occurred when the crystal was probed by a metal point (whisker). On November 14, 1876 he demonstrated this rectification effect of a metal-semiconductor contact at Leipzig to a broad audience but his work was not recognized at that time. Later, his discovery led to the development of crystal radio detectors (1899–1906) instead of fritters in the early days of wireless telegraphy and radio. On February 15, 1897, he invented the cathode ray tube (CRT) with magnetic deflection [6] which in Germany is called "Braun's Tube", Figure 11.5. On September 20, 1898 K.F. Braun discovered the transceiver with two coupled resonance circuits (Figure 11.6) (Patent DRP 111578 of October 14, 1898) which act as an impedance transformer allowing much more power compared to Marconi's transmitter. Braun used a loop aerial for transmission and reception of wireless signals [7]. He shared the Nobel Prize for Physics in 1909 with G. Marconi for his contributions to the physics of electric oscillations and radio telegraphy, but during his scientific life he could not verify his dream *"Funken ohne Funken"* which means "Wireless Telegraphy without Sparks".

11.2.5. Christian Hülsmeyer: Rudimentary Form of RADAR

Christian Hülsmeyer (1881–1957), a German inventor fascinated by Hertzian waves, was far ahead of his time. One of many contributors to the development

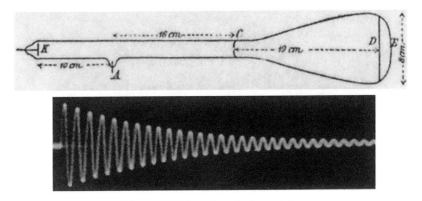

Figure 11.5. Braun's cathode ray tube.

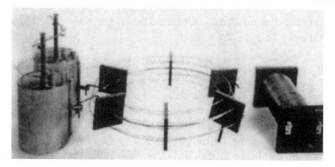

Figure11.6. Braun's transmitter with two Leyden jars and two coupled resonance circuits.

of electromagnetic waves for wireless communications, he got the idea for a different application: "Seeing ships through fog and darkness by transmitting waves and detecting the echoes". On April 30, 1904 he applied for a German Patent on a "Means for reporting distant metallic bodies to an observer by use of electric waves (DRP No. 165546 and later DRP No. 169154), a rudimentary form of RADAR. In Figure 11.7 the drawing *Fig. 1* of Hülsmeyer's patent is given, showing what application he mainly had in mind. Figure 11.8 shows a schematic cross section of the quasi-monostatic system with single frequency operation (1 m wavelength) and tunable pulse repetition frequency (PRF), which he called *"Telemobiloscope"*. Although he was able to demonstrate a range of 3000m, neither shipping nor naval leaders were interested in his invention. Even

Figure 11.7. Drawing from Hülsmeyer's patent of April 30, 1904, showing the application he mainly had in mind.

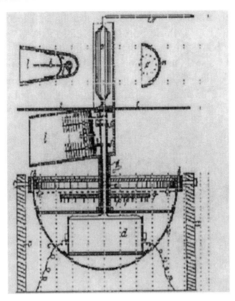

Figure 11.8. Schematic of Hülsmeyer's quasi-monostatic RADAR system. Quasi-monostatic system with single frequency operation, tunable PRF, 1m wavelength, **a:** Support structure, **b:** Cardian Join azimuth, **c:** Cardian joint elevation, **d:** Induction coil, **h:** spark gap, **t:** Large metallic disc to reduce coupling between transmit and receive. Transmitting Arial: **m:** Concave Mirror, **l:** Projection Case, Receiving Arial 'o': **n:** Cage Arrangement Reflector.

Telefunken rejected an offer to buy his patent. Around 1930 Hülsmeyer's idea was taken up again, or independently arrived at. At least eight countries developed RADAR systems, but for warning of aircrafts attacks rather than for ship navigation. Robert Alexander Watson-Watt (1892–1973) of Scotland patented such a system in 1935. The term RADAR, an acronym of radio detection and ranging, was not proposed until 1940.

11.2.6. Robert von Lieben: The Triode as an Amplifier in a Transmitter

The development and production of high vacuum electron tubes started in Germany at Telefunken, the company having been created in 1903 by the joint efforts of AEG and Siemens. Later in 1911 the so-called "von Lieben Konsortium" was founded by AEG, Siemens, Telefunken, and Felten & Guilleaume specifically to evaluate the von Lieben patents [8].

Robert von Lieben (1878–1913) was born in Vienna, Austria. In 1906 he obtained a patent for an inertia-less relay using a gas filled amplifier tube which can be denoted as a deflection grid controlled triode. In the same year, in the United States, Lee de Forest (1873–1961), the so-called *"Father of Radio"* invented the transmission grid controlled triode, a triode with a cold grid-like electrode between cathode and anode. This allowed control of the flow of electrons from the heated cathode. Calling it a Three-Electrode Audion (patent issued on February 18, 1908), de Forest referred to it as a *"device for amplifying feeble electric currents"* but, as von Lieben, until 1912 he used the triode only for detecting radio waves. On December 20, 1910 Robert von Lieben applied for a patent on the so-called *"Lieben Tube"* (see Figures 11.9, 11.10a, and 11.10b), DRP No. 249142, a transmission grid controlled triode with a Wehnelt cathode. In 1912, engineers were coming to realize that the triode had other uses besides detection of radio waves. Lee de Forest, Fritz Loewenstein, and Irving Langmuir in the United States as well as Robert von Lieben and Otto von Bronk in Germany realized that it could be used in a transmitter and could work as an oscillator. These functions were soon put to use. The three-electrode vacuum tube was included in designs for telephone repeaters in several countries.

Also in 1912, Edwin Howard Armstrong (1890–1954), a student at Columbia University in New York City, found that he could obtain much higher

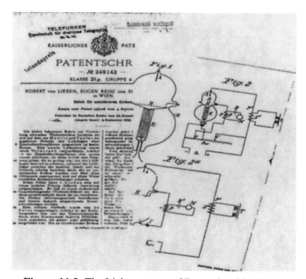

Figure 11.9. The Lieben patent of December 20, 1910.

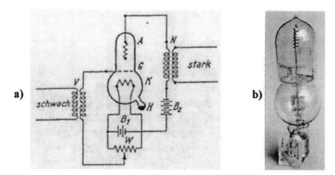

Figure 11.10. a) Circuitry for the Lieben tube, **b)** The big Lieben tube (height: 315 mm).

amplification from a triode by transferring a portion of the current from the anode back to the signal going to the grid (regenerative receiver). He also found that increasing this feedback beyond a certain level made the tube into an oscillator, a generator of continuous waves (CW). At about the same time, others, including Alexander Meißner (1883, Vienna –1958, Berlin) at Telefunken in Germany, Henry Round in England, and Lee de Forest, created similar circuits. Armstrong himself went on to make other fundamental contributions to radio science such as the superheterodyne circuit (1918) (see section 11.2.10) and frequency modulation (FM) techniques. The triode became the basic component for radio, RADAR, television and computer systems until transistors began replacing vacuum electron tubes in the early 1950s [8].

11.2.7. Heinrich Barkhausen: First Transit Time Microwave Tube

The retarding-field tube (or reflex triode) can be regarded as the first transit-time tube. It was invented by Heinrich Barkhausen (1881–1956), Figure 11.11a, in 1920. During some measurements on a triode with a positive grid and a negative anode (Figure 11.11b) Barkhausen and K. Kurz noticed irregularly fluctuating anode currents. Barkhausen interpreted them as self-excited oscillations generated by the tube [8,9]. Later they were known as *"electron dance oscillations"* on account of the oscillatory motion of the electrons around the wires of the grid. Due to the existence of a retarding field between the grid and the anode the name *"retarding-field tube"* was generally adopted. It is interesting to note that the three effects characteristic of transit-time devices are already present in the retarding field tube: they are velocity modulation, bunching (i.e. conversion of velocity into density modulation) and power transfer from the beam to the circuit. The principle can also be described by extraction of *"wrong phase"* electrons and negative absorption by a stationary ensemble of non-isochronous oscillators [10].

This was the first tube in which the unavoidable transit time effect was put to good use. In 1920 the shortest wavelength which could be reached using commercially available triodes was 43cm. Owing to their simple design; such

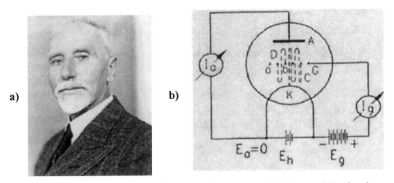

Figure 11.11. a) Heinrich Barkhausen, **b)** Retarding-field tube and its circuit.

reflex triode oscillators became very popular, especially among university institutes. They were mostly used as high frequency local oscillators and sources of oscillations for various measuring instruments (0.3-6.4 GHz at 5 W – few mW). In order to achieve higher frequencies and higher output powers early variants of the original Barkhausen tube, which had no separate resonant circuit, were investigated. The world's first decimeter transmitter and receiver was built and operated in 1931 by Hans Erich Hollmann [11] at the Heinrich Hertz Institute in Berlin. The unit worked by using a symmetric opposed, resonant Lecher circuit excited by a so-called *"hammer"* retarding field tube (see Figure 11.12). Later only reflex klystrons were used as local oscillators. The so-called *Vircator* (Virtual Cathode Oscillator) is a modern version of the retarding-field tube [10].

11.2.8. Manfred von Ardenne: First Integrated Vacuum Tube Circuits

Manfred v. Ardenne (1907–1997) was younger than 15 years when he first met Siegmund Loewe (1885–1962), the owner of the Loewe Radio Company in Berlin-Steglitz who later introduced the quartz crystal as a frequency standard in electronic circuits.

M. v. Ardenne and H. Heinert developed in Loewe's Company the so-called Loewe-3 fold tube 3NF (3 triodes) [12] in which the audion (receiver), the resistive amplifier (RC amplifier), the output amplifier and the coupling capacitors and anode resistors where integrated into a single vacuum tube (patent 1924, first 3NF 1925). Only a few months later they developed the Loewe 2-fold

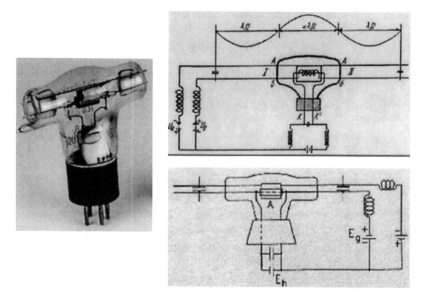

Figure 11.12. (left) Retarding-field tube RS296 and its circuit, (upper right) Hollmann 1930 and (lower right) Kühle 1932 at Telefunken, *"Hammer Tube"* [8,11].

tube 2 HF (2 triodes with common space-charge grid) for broadband amplifiers (in 1926). The circuit diagram of the *"five-in-two"* receiver and the photograph of a selection of Loewe tubes are shown in Figures 11.13 and 11.14, respectively. Note that the two tubes are coupled through the tuned-radio-frequency transformer composed of coils L_3 and L_4 and the condenser C_2. The first stage of RF amplification is resistance-coupled. Figure 11.15 shows the autodidactic M. v. Ardenne together with Lee de Forest in 1926, and with Siegmund Loewe and the first broadband amplifier in 1928.

11.2.9. Hans Erich Hollmann: Multi-Cavity Magnetron, Principle of Reflex Klystron

In 1921, Albert W. Hull at General Electric Co. investigated the motion of electrons in a cylindrical diode under the influence of a homogeneous axial magnetic field. He noticed the possibility to control the electron current to the

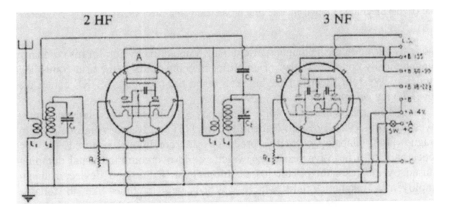

Figure 11.13. Circuit diagram of the *"five-in-two"*-receiver, note that the two tubes are coupled through the tuned-radio-frequency transformer composed of coils L_3 and L_4 and the condenser C_2. The first stage of RF amplification is resistance-coupled.

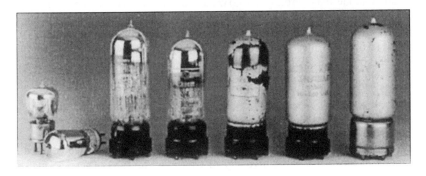

Figure 11.14. Selection of Loewe tubes.

Figure 11.15. Pictured above are M.v. Ardenne together with (left) Lee de Forest (1926), and (right) Siegmund Loewe and the first broad band amplifier (1928).

anode by variation of the magnetic field. Hull wanted to develop for his company a magnetically controlled relay or amplifier in competition to the grid controlled triodes of Western Electric Co., but also noted the possibility of RF called his novel device "*magnetron*".

The magnetron for high frequency oscillations was independently investigated in 1924 by Erich Habann in Jena [13] and Napsal August Zázek in Prague [14]. Habann correctly predicted the conditions required for the appearance of a negative resistance which would overcome the usual damping caused by the resonant circuit losses. In contrast to the Hull device, Habann employed a magnetic field which was constant in time like in today's magnetrons. Using his split-anode magnetron, Figure 11.16a, Habann was able to generate oscillations in the 100 MHz range. Zázek developed a magnetron with a solid cylindrical anode and generated frequencies up to 1 GHz. The breakthrough in generation of cm-waves by magnetrons came in 1929 when K. Okabe operated his slotted-anode magnetron (5.35 GHz) at Tohoku University in Sendai, Japan. Hans Erich Hollmann filed in Germany on November 27, 1935, a patent on the multi-cavity magnetron. US Patent 2,123,728 was granted on July 12, 1938, Figure 11.16b, well ahead of J. Randall's and H. Boot's work in February 1940. The operation principle of a reflex klystron was anticipated by Hollmann as early as 1929 who patented a "*double-grid retarding-field tube*" [11].

11.2.10. Oskar Ernst Heil: Field Effect Transistor, Principle of Klystron

Oskar Ernst Heil (1908–1994) was a peripatetic German scientist, who earned his Doctorate in Physics at the University of Göttingen in 1933. There he met and married Agnesa Arsenjeva (1901–1991), a promising young Russian physicist (Figure 11.17) who also got her PhD from the University of Göttingen

a)   b)

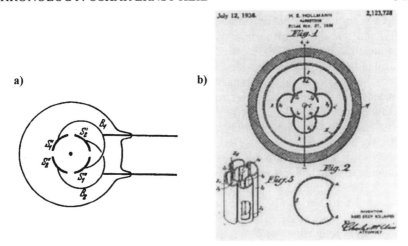

Figure 11.16. Pictured above are schematics of **a)** Habann's split-anode magnetron, and **b)** Hollmann's multi-cavity magnetron.

(1928).Together, the Heils travelled to England and worked with Lord Rutherford at the Cavendish Laboratory in Cambridge.

On March 2, 1934, O. Heil applied for a patent on "Improvements in, or relating to electrical amplifiers and other control arrangements and devices" (British Patent No. 439457) which can be seen as the theoretical invention of the capacitive current control in field effect transistors (FETs). During a trip to Italy A. Arsenjeva-Heil and O. Heil wrote, in Bormio, a publication entitled "On a new method for producing short, undamped electromagnetic waves of high intensity" (in German), which was published in the *Zeitschrift für Physik* 95, 1935, pp. 752-762. This publication gives the first description of the fundamental principles behind modern high power linear beam microwave electron tubes.

They described a transit-time tube in which the three characteristic features; velocity modulation, phase focusing, and energy transfer were designed to occur in three separate regions, an arrangement which is also characteristic of a klystron. Further they demonstrated, in my opinion for the first time, that it is

Figure 11.17. Oskar Heil and Agnesa Arsenjewa-Heil in Bormio, Italy (1935).

necessary, in order to achieve high RF power output, to use a linear electron beam, and that the beam must be positioned in such a way as to prevent the electrons from landing on RF electrodes – they must only be allowed to penetrate the fringe field of the RF electrodes, finally landing on a separate electrode, now called the collector. This arrangement made it possible to separate high frequency from beam guiding electrodes, thus permitting the use of high power electron beams (Figure 11.18, left).

In addition, the authors propose a two-beam electron gun and a multistage depressed collector (Figure 11.18, right) for efficiency enhancement by the use of a step-wise reduced voltage collector, a technique which is now commonly used in high power microwave tubes.

Almost immediately afterwards, the by now classical paper of R.H. Varian and S.E. Varian (1939) [15] became available in Germany and klystron research was begun at several industrial and government research laboratories.

It is unlikely that W.W. Hansen and the Varian brothers, Russell, a physicist, and Sigurd a former barnstormer and Pan-American pilot in the United States were aware of the Heils' work on velocity modulation. However, approximately two years after the Heils' paper was published in Germany, Russel *"had an idea in the middle of the night"* in which he visualized the movement and bunching of cars at different speeds on a highway. This amounted to the velocity modulation concept. Using reentrant versions of the "Hansen Rumbatron" cavity [16], the Varians constructed, at the Physics Department of Stanford University, several models of a two cavity oscillator, and the modern

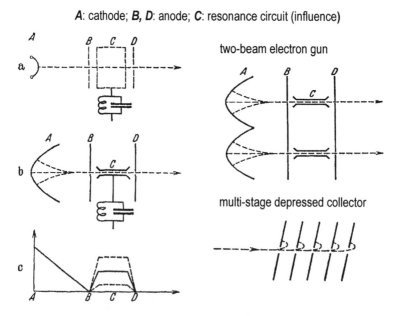

Figure 11.18. Original drawings from the Heils' publication, figures 1 (left), 9 (upper right), and 10 (lower right).

microwave tube was born. It was named *"klystron"* after an ancient Greek verb indicating waves washing on a shore.

O. Heil apparently joined A. Arsenjeva-Heil when she returned to the Leningrad Physico-Chemical Institute in the USSR. The research on velocity modulation was carried out there, although it did not result in a working device. Presumably because his wife was not allowed out of the Soviet Union again, Heil returned to the UK alone and continued his work on "coaxial-line oscillators," as the British named them, at Standard Telephone and Cables (STC). Just before World War II broke out, he slipped back into Germany without finishing his work at STC. He was apparently successful in completing development of his microwave oscillator at C. Lorenz AG in Berlin Tempelhof, Figure 11. 19. The Germans used his so-called *"Heil Generator"* tube in WW II [8, 17]. After 1947 O. Heil worked at different universities, research institutes, and companies in the United States.

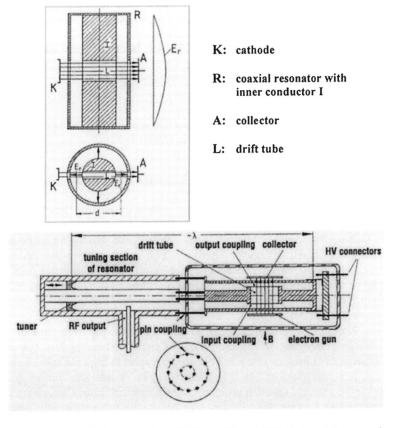

K: cathode

R: coaxial resonator with
       inner conductor I

A: collector

L: drift tube

Figure 11.19. The Heil Generator (coaxial-line oscillator) [17], (bottom) Cross-section of Heil Generator RD12La (C. Lorenz AG) $\lambda = 21.5 - 24$ cm, PRF = 15 W, $U_0 = 500$ V, $I_0 = 200$ mA, B = 0.15 T.

11.2.11. Walter Schottky: Tetrode, Theory of Shot-Noise, Schottky Barrier

Walter Schottky (1886–1976), Figure 11.20, was born in Zürich, Switzerland. During 1912–1915 he worked as an Assistant Professor with Max Wien in Jena, from 1916–1919 as a Research Scientist with Siemens & Halske Company in Berlin, 1920–1922 as Privatdozent (Habilitation in 1920) at the University of Würzburg with Wilhelm Wien (1864–1928) from 1923–1927 as Full Professor for Theoretical Physics at the University of Rostock and from 1927–1951 as Research Leader with Siemens & Halske in Berlin and Siemens-Schuckert-Company in Berlin and Erlangen. We can see in him a real pioneer of electronics. His most important scientific achievements are [18]:

1914:   Discovery of the Schottky Effect: Reduction of electron work function by an electric field

1915/16:   Development of space-charge-grid tube and screen-grid tube (Tetrode)

1918:   Invention of superheterodyne detection principle, independent of E.H. Armstrong (see section 6), Theory of shot noise and thermodynamics of electrons

1929:   Experimental verification of barrier layer in metal-semiconductor contact (Schottky Barrier)

1938/39:   Development of space-charge and edge-sheet theory of crystal rectifiers.

We should not forget that a great part of his work was accompanied by his "Mathematical Assistant", Eberhard Spenke (1905–1992).

11.2.12. Herbert Kroemer: III-V Semiconductor Heterostructures

The 2000 Nobel Prize in Physics honored three scientists for their work in information and communication technology. Herbert Kroemer, Figure 11.21 [19], and Zhores Ivanovich Alferov [20] were recognized for developing semiconductor hetero-structures used in high-speed microwave and opto-electronics and operating continuously at room temperature, ranging from satellite communications to mobile phones and lasers used in CD players. Jack St. Clair Kilby was honored for his part in the invention of the integrated semi-conductor circuit (Ge).

Kroemer, who holds a Ph.D. in theoretical physics from the University of Göttingen, Germany, was born in 1928 (Weimar, Germany). His dissertation on Germanium transistors discussed electron transport in high electrical fields.

Kroemer's career began in 1952, when he became the "house theorist" in the semi-conductor research group of the telecommunications laboratory of the German Postal Service. During this time, he began to wonder why emerging junction transistors did not compare in speed to the earlier point transistors. This led him to the question, *"How can an electric field be built into the base region of a junction transistor?"* Kroemer realized that one possible way to do this was

Figure 11.20.   **Figure 11.21.**
Walter Schottky (1961).   Herbert Kroemer.

by not using a single semiconductor, but a graded base region that started with one material and ended in another with a continuous transition between the two. Despite his brave idea, nothing materialized initially because at that time no substantial data had been gathered on semiconductors.

In 1957, while at RCA's David Sarnoff Research Center in Princeton, NJ, Kroemer originated the concept of the heterostructure bipolar transistor. He found his earlier ideas could be expanded when he learned that composition gradients acted as forces on electrons. Kroemer left RCA in 1957 and returned to Germany becoming head of a semiconductor group at Philips Research Laboratory in Hamburg where he pushed for work on GaAs.

In 1959 Kroemer went to work for Varian Associates in Palo Alto, CA. From that year to 1966 his work yielded the invention of the double heterostructure laser. After submitting in 1963 a paper containing his ideas to the journal *Applied Physics Letters* and having it rejected, Kroemer presented his Nobel Prize-winning work to the *Proceedings of the IEEE*, where it was published under the title "A Proposed Class of Heterojunction Injection Lasers". He also filed for a patent (issued in 1967, expired in 1985) and received an inventor's award of one-hundred US dollars!

> *I was told not to work on light-emitting semiconductors because my ideas were judged on the basis of then existing applications,"* Kroemer noted. *"When you look at the history of technology, you see that the principal applications do not evolve incrementally, but are created by the technology. Until you come up with such applications,"* he said, *"you cannot judge how promising technology is. It is foolish to ask immediately what a new technology is good for. ...*

In 1964 Kroemer was the first to publish an explanation of the Gunn Effect. After working for the University of Colorado, Boulder, Colo., Kroemer joined the University of California, Santa Barbara, Calif. (UCSB), in 1976. Initially, Kroemer thought UCSB would never catch up to the leaders in silicon technology, but on the advice of Electrical Engineering and Computer Science Department chair Ed Stear, he was asked to focus all his efforts on developing compound semiconductor technology. The result has earned him a Nobel Prize.

11.2.13. Jürgen Schneider: QE Model of Electron Cyclotron Maser

Jürgen Schneider (1931), Figure 11.22, who got his Ph.D. in Physics from the University of Freiburg, Germany, in 1957 moved in the same year to Duke University in Durham, North Carolina, , in order to perform research on basics in the field of microwave spectroscopy. There he published his work "Stimulated Emission of Radiation by Relativistic Electrons in a Magnetic Field", *Phys. Rev. Lett.* 2 (1959), 504, which, together with a paper of the Australian R.Q. Twiss (1958) [21], is the first publication on the quantum electronic (QE) model of the electron cyclotron resonance maser interaction. Neglecting the electron spin, he solved the relativistic Schrödinger equation for the electron motion perpendicular to a magnetic field Bo and obtained a discrete spectrum of the kinetic electron energy, the so-called Landau levels of a non-harmonic oscillator:

$$E_n = m_o c^2 \left[\sqrt{1 + 2\left(n + \frac{1}{2}\right)\frac{\hbar\omega_{co}}{m_o c^2}} - 1 \right], \qquad (11.1)$$

where: m_o is the electron rest mass, c the velocity of light, $\hbar$ is Planck's constant divided by 2π, and $\omega_{co} = eB_o/m_o$ the non-relativistic electron cyclotron resonance frequency. These Landau levels are non-equidistant with a decreasing energy gap for increasing quantum number n:

$$E_n - E_{n-1} > E_{n+1} - E_n \qquad (11.2)$$

with

$$\omega_{n+1} \approx \omega_{co}\left[1 - n\frac{\hbar\omega_{co}}{m_o c^2}\right]. \qquad (11.3)$$

An external transverse electric RF field with $\omega_{RF} \geq \omega_{n+1}$ can lead to stimulated emission (negative absorption, see Figure 11.23) of cyclotron radiation (bremsstrahlung) since the quantum electronic probability for electric dipole transitions scales with ω^3. Due to the very small different energy gaps between neighbouring Landau levels and their natural widths, multi-photon transitions can produce powerful microwaves at short wavelengths. However, at a typical electron energy of 80 keV and a frequency of 100 GHz we have $\hbar\omega_{co} = 0.41$ meV and $n \approx 2 \cdot 10^8$ so that Bohr's Correspondence Principle allows a classical description of high power cyclotron resonance masers such as gyrotrons [22, 23].

Gyrotron oscillators (gyromonotrons) are mainly used as high power millimeter wave sources for electron cyclotron resonance heating (ECRH), electron cyclotron current drive (ECCD), stability control and diagnostics of magnetically confined plasmas for generation of energy by controlled thermonuclear fusion [24]. The maximum pulse length of commercially available

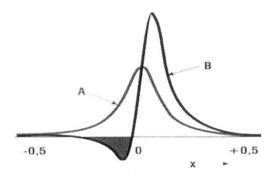

Figure 11.22.
Jürgen Schneider.

Figure 11.23. Cyclotron resonance absorption of electrons ($x = (\omega_{n+1}-\omega_{RF})\ \tau$), non-relativistic (A), relativistic (B).

1 MW gyrotrons employing synthetic diamond output windows is 5 s at 110 GHz (CPI and JAERI-TOSHIBA), 12 s at 140 GHz (FZK-CRPP-CEA-TED) and 9 s at 170 GHz (JAERI-TOSHIBA), with efficiencies slightly above 30%. Total efficiencies of 45-50 % have been obtained using single-stage depressed collectors (for energy recovery). The energy world record of 160 MJ (0.89 MW at 180 s pulse length and 140 GHz) at power levels higher than 0.8 MW has been achieved by the European FZK-CRPP-CEA-TED collaboration at FZK where the pulse length restriction to 180 s is due to the HV power supply at $I_{beam} \approx 40$ A. At lower beam current ($I_{beam} = 26$ A) it was even possible to obtain 506 MJ (0.54 MW for 937 s). The longest shot lasted for 1300 seconds at 0.26 MW output power. These very long pulses were limited by a pressure increase in the tube. A maximum output power of 1.2 MW in 4.1 s pulses was generated with the JAERI-TOSHIBA 110 GHz gyrotron. The Russian and the Japanese 170 GHz ITER gyrotrons achieved 0.54 MW at 80s pulse duration and 0.5 MW at 100 s, respectively. Diagnostic gyrotrons deliver $P_{out} = 40$ kW with $\tau = 40$ μs at frequencies up to 650 GHz ($\eta \geq 4\%$). Gyrotron oscillators have also been successfully used in materials processing. Such technological applications require gyrotrons with the following parameters: $f \geq 24$ GHz , $P_{out} = 10\text{-}50$ kW, CW, $\eta \geq 30\%$.

11.3   ACKNOWLEDGMENTS

The author is particularly grateful to Prof. Michael I. Petelin from the Institute of Applied Physics (Russian Academy of Science) in Nizhny Novgorod for encouraging me to prepare this chapter and for giving me a lot of valuable information about L. I. Mandelstam and A. A. Andronov. This work could not have been done without the help, stimulating suggestions, and useful discussions of Prof. Herbert Döring (1911–2001) from the RWTH Aachen, and Dipl-Ing Max Münich from the Max-Planck-Institute of Plasma Physics, Garching. The author also wishes to express his deep gratitude to Prof. Edith Borie for critical reading, and to Mrs. Christine Kastner and Mrs. Martina Klenk for careful typing this manuscript, and to Mrs. Ursula Feisst for her kind help in drawing of the figures.

REFERENCES

[1]   R. Bernzen, Das Telefon von Ph. Reis, Verlag Marburg (1999) (in German).
[2]   H. Kant, *Physik in unserer Zeit* 25 (1994) 284 (in German).
[3]   T. K. Sarkar, M. Salazar-Palma and D. Sengupta, A Chronology of Developments of Wireless Communications and Electronics, *Proc. Millenium Conf. on Antennas and Propagation (AP 2000)*, Davos (2000).
[4]   W. Wiesbeck, ed., *Proc. Heinrich-Hertz-Symposium "100 Jahre elektromagnetische Wellen"*, Karlsruhe, VDE-Verlag Berlin (1988).
[5]   D. L. Sengupta and T. K. Sarkar, "Maxwell, Hertz, The Maxwellians and the Early History of Electromagnetic Waves", *Proc. Millenium Conf. on Antennas and Propagation (AP 2000)*, Davos (2000).
[6]   K. F. Braun, *Annalen der Physik und Chemie N.F.* 60 (1897) 552 (in German).
[7]   F. Braun, *Drahtlose Telegraphie durch Wasser und Luft,* Veit & Co., Leipzig (1901) (in German).
[8]   H. Döring, *Int. J. Electronics* 70 (1991) 955.
[9]   H. Barkhausen and K. Kurz, *Physikalische Zeitung* 21 (1920) 1 (in German).
[10]   M. I. Petelin, *IEEE Trans. Plasma Science* 27 (1999) 294.
[11]   H. E. Hollmann, *Physik und Technik der ultrakurzen Wellen, Vol. 1 und 2*, Springer, Berlin 1936.
[12]   H. Börner, *Funkgeschichte* 66 (1989) 4 (in German).
[13]   E. Habann, *Zeitschrift f. Hochfrequenz* 24 (1924) 115 (in German).
[14]   A. Zázek, *Casopis pro Pest. Matem. a Fys.* 53 (1924) 378 (in Czech).
[15]   R. H. Varian and S.E. Varian, *J. Appl. Phys.* 10 (1939) 321.
[16]   W. W. Hansen, *J. Appl. Phys.* 9 (1938) 654.
[17]   H. Döring, *Funkgeschichte* 80 (1991) 5 (in German).
[18]   O. Madelung, *Physikalische Blätter* 42 (1986) 238 (in German).
[19]   H. Kroemer, *Die Naturwissenschaften* 40 (1953) 578, *Proc. I.R.E.* 45 (1957) 1535, *RCA Rev.* 18 (1957) 332, *Physica Scripta* T68 (1996) 10.
[20]   Zh. I. Alferov, *Fizika i Tekn. Poluprovadn.* 1 (1967) 436 and 1579, 3 (1969) 1328, *Physica Scripta* T68 (1996) 32.
[21]   R. Q. Twiss, *Ausralian. J. Phys.* 11 (1958) 567.
[22]   A. V. Gapanov, *Radiophys. Quantum Electron.* 2 (1959) 195.
[23]   A. V. Gapanov, M. I. Petelin and V. K. Yulpatov, *Radiophys. Quantum Electron.* 10 (1967) 794.
[24]   M. Thumm, *Fusion Engineering and Design* 66-68 (2003) 69.

12

THE DEVELOPMENT OF WIRELESS TELEGRAPHY AND TELEPHONY, AND PIONEERING ATTEMPTS TO ACHIEVE TRANSATLANTIC WIRELESS COMMUNICATIONS

JOHN S. BELROSE, *Communications Research Centre Canada*
Ottawa, Ontario

12.1   INTRODUCTION

The revolution of wireless technology began with earnest at the end of the nineteenth century, and recently, various organizations throughout the world have been promoting and sponsoring conferences and public demonstrations celebrating 100 Years of Radio, beginning with the Institution of Electrical Engineers (IEE) international conference of the same title held in London, UK, 5-7 September, 1995. References to several of the papers presented at that conference are made in this chapter. Since that date, many papers have been published by historians lauding various pioneers that have had a part in contributing to the technology of wireless telegraphy and telephony, including the postage stamps displayed in Figure 12.1.

Marconi has generally been considered to be the *"Father of Wireless"*. This description attributed to him by Popov (Popoff), a contemporary Russian scientist, who was one of the many people studying the work of Hertz in the latter part of nineteenth century. Marconi was not an inventor, and he discovered relatively few things in his early work. He used initially Righi's spark transmitter, a version of the Branly and Lodge coherer, and the vertical aerial of Dolbear (also used by Tesla). In follow on experiments he used resonant tuning circuits based on the work of Tesla, Braun and Lodge. But Marconi had the necessary ability to make things work, and because of his single minded purposed devotion to the idea of using Hertzian waves for telegraphic communications, his many demonstrations of wireless communications, and his formation in 1897 of the Marconi Wireless Telegraph Company, his name, for the general public, became associated with all aspects of the invention of wireless. The world did not understand what Maxwell, Hertz, Tesla, Lodge,

Branly, Popov, Fessenden, and many others had done, and their achievements had not been publicized in the non-scientific press.

In recent years many articles have been published, reminding us about forgotten pioneers, *cf.*: Unsung Genius [1]; Right Against the World [2]; Who was Fessenden [3]; Oliver Lodge-The Forgotten Man of Radio? [4]; Nicola Tesla: The Unsung Inventor of Modern-Day AC Electric Power Systems and Radio [5]; and other papers are appearing, not only in professional and non-technical journals, but also on internet web pages in increasing numbers. Many historians are reminding us that no one person is the father of radio.

The purpose of this chapter (based on several previously published papers by this author, John Belrose [6,7,8,9,10]) is to describe briefly the technologies used during the first decade of the twentieth century for wireless telegraphy and telephony communications, and to give credit where credit is due, with an emphasis on long distance (transatlantic) communications. In this latter endeavor two names stand out, *viz.* Marconi and Fessenden.

Guglielmo Marconi (1874–1937), who had worked initially under the guidance and advice of Righi (in 1895), had, with a keen eye for commercial opportunity, realized that there was a market for a wireless telegraphy system. He came from a wealthy well connected Italian family, and so he could afford to spend his time making his early wireless apparatus work. Marconi spent his entire life developing wireless communication into a practical reality ("practical" for him in the early history of wireless). In 1897, he formed the Marconi Wireless Telegraph Company and set about experimenting, developing, promoting, demonstrating, manufacturing and installing wireless systems, which importantly could be used for marine communications. His well-publicized first transatlantic experiment on December 12, 1901, caught the attention of the

Figure 12.1. Canadian Stamp 1987 (left) remembering Fessenden the Father of AM Radio (first voice over radio December 23, 1900, and later first radio broadcast (speech and music) on Christmas Eve 1906); and UK Stamp 1995 (right) remembering Marconi, on the occasion of the 100th anniversary of his first wireless experiments on his home estate in 1895.

world, and sparked the challenge to achieve reliable transatlantic wireless communications.

Marconi was not a systems designer, he was a systems developer, a hard headed entrepreneur, and he was an expert in concealing what he did so that others could not copy him. As a consequence, it is difficult to be technically certain of what Marconi did, and of when and what success was achieved in terms of the quality and reliability of his communication systems. For Marconi, the early years saw a continuous struggle to keep the performance of his wireless systems ahead of anything his competitors could achieve.

Reginald Aubrey Fessenden (1866–1932) was a brilliant inventor and a practical experimenter, whose original inventions contributed significantly to the development of radio as we know it today. The apparatus he designed is described in detail in his numerous patents. And, his published papers give a detailed reference to the state of the art at the time, recognizing what others had done before. His many patents covering inventions made and developed are filed under his name, or under the joint names of co-operators. Fessenden is a principal pioneer of radio, as we know it today, and the *"Father of AM Radio"*.

12.2   A BRIEF HISTORY OF THE BIRTH OF WIRELESS

Many scientists and engineers have contributed to the early development of electromagnetic theory, the invention of wireless signaling by radio, and the development of electromagnetic antennas needed to transmit and receive the signals. Some of the early radio inventors were Loomis, Henry, Edison, Thomson, Tesla, Dolbear, Stone, Fessenden, Kennelly, Alexanderson, de Forest, and Armstrong in the United States; Hertz, Braun, and Slaby in Germany; Faraday, Hughes, Maxwell, Heaviside, Poynting, Crookes, Fitzgerald, Lodge, Jackson, Marconi, and Fleming in the UK; Ampere and Branly in France; Popov in the USSR; Oersted, Lorenz, and Poulsen in Denmark; Lorentz in Holland; Righi in Italy; and Bose in India.

The very possibility of wireless communications is founded on the research of James Clerk Maxwell, since his equations form the basis of computational electromagnetics. Heinrich Hertz established their correctness, when in 1887 he discovered electromagnetic radiation (EMR) at UHF frequencies, as predicted by Maxwell. Since the pioneering work of Maxwell, beginning in the middle 1850s, and of his followers, a small group that became known as the Maxwellians, which included UK's Poynting and Heaviside (for a modern review see Sengupta and Sarkar [11, also Chapter 5]); his equations have been studied for over a century, and have proven to be one of the most successful theories in the history of radioscience. For example, when Albert Einstein found that Newtonian dynamics had to be modified to be compatible with his special theory of relativity, he found that Maxwell's equations were already relativistically correct. EM field effects are produced by the acceleration of charges, and so Maxwell had automatically built relativity into his equations.

But the history of wireless can be traced back much earlier. It should be pointed out that in the experimental verification of the results foretold by

Maxwell's theory use was made of the results of experiments in pure physics that William Thomson Kelvin had made forty years previously [12]. Kelvin had set himself the task of investigating the way in which a Leyden jar discharged, and found that under certain conditions, the discharge gave rise to alternating currents of very high frequency. Joseph Henry, an early experimenter with wireless telegraphy, was not only the first to produce such high frequency electrical oscillations for communication purposes, he was the first to detect them at a distance, albeit a very short distance, from an upper floor room to the basement, using what was later known as a magnetic detector [13,14]. Thomas Edison, Elihu Thomson, and Edwin Houston [15] made many experiments on these transmitted waves, and reports on their experiments can be found in Edison's papers in technical journals of that time.

Marconi was awarded what is sometimes recognized as the world's first patent for radio, his British patent 12,039, "Improvements in Transmitting Electrical Impulses and Signals and in Apparatus There-for" on July 2, 1897, but there were others before him. The first wireless telegraphy patent, however impractical, was issued in the United States on July 20, 1872, to Mahlon Loomis, fifteen years before Hertz. His patent no. 129,971 was for "Improvement in Telegraphing", and covered *"aerial telegraphy by employing an 'aerial' used to radiate or receive pulsations caused by producing a disturbance in the electrical equilibrium of the atmosphere"*.

In October 1866, in the presence of United States Senators from Kansas and Ohio, Loomis set up a demonstration experiment on two mountain peaks in the Blue Ridge Mountains of Virginia, 22 kilometers apart (Figure 2.17 right). He flew kite-supported wire aerials, 183 meters long, connected at their support ends to a galvanometer, which itself was connected to a plate buried in the earth. Each kite had a piece of wire gauze about 38 square centimeters attached to the underside.

With a prearranged time schedule, signals were sent from one peak to the other by making and breaking the aerial connection of one galvanometer and noting the response of the other galvanometer on the other peak. The operation at each station was reversed. Military engineers duplicated the Loomis' simple antenna-galvanometer-ground arrangement and its workability substantiated.

Between 1870 and 1888 von Bezold, Fitzgerald and Hertz had clarified to a considerable extent the nature of the phenomena being observed, and Hertz's work had shown that the experimenters were in fact dealing with EM waves.

Dolbear and Edison had been using vertical grounded antennas for telegraphing wirelessly, though the effects they obtained were thought to be (at that time) mainly electrostatic, and not propagating EM waves.

Crookes in the *Fortnightly Review* for February, 1892, proposed that resonant circuits should be used to select-out messages from different stations. Nicola Tesla, who had been doing a great deal of work in producing HF oscillations, proposed in 1892 a system for transmitting wirelessly using the vertical antenna of Dolbear [16] and tuned transformer circuits at the sending and receiving ends. Tesla in 1893 described his wireless system to the Franklin Institute, in Philadelphia, in March, and later in the same year in St. Louis, MO, before the National Electric Light Association. During the St. Louis presentation he demonstrated sending wireless waves through space, complete with a spark transmitter, grounded antenna, tuned circuits, a Morse key, and a receiver with a Geissler tube as an indicator [17,18,19].

Tesla's stroke of genius was to use two tuned circuits in the transmitter, and two in the receiver, inductively coupled, and so, move the energy storage capacitor (or 'discharge capacitor') to the primary side, and add a ground connection. Tesla was the first to inductively couple the secondary side, the antenna side where the capacitances must be small, to a primary tuned circuit where the energy storage capacitance could be huge by comparison. This made possible the generation of RF signals immensely more powerful than those generated by the Hertz type of apparatus which others were using at that time [5].

Historians over the years have attributed the invention of tuning to Marconi, his so-called master tuning patent. Certainly, Marconi filed and held (in early years) several patents on tuning. His original patent filed in 1896, described sending and receiving stations with no tuning at all. Marconi's second patent (British patent 586,193) granted on June 4, 1901 was for a two-circuit system, one circuit in the transmitter and one circuit in the receiver, again, a very inefficient system. This was in fact a Lodge circuit, claimed to be assigned to Marconi. Marconi's "famous" four-sevens patent (British Patent No. 763,772), a four-circuit system, two circuits in the transmitter and two in the receiver, was granted on April 26, 1900. This patent was reissued as an American patent on June 28, 1904. Litigation that began after World War I and extended to 1943 was over a Marconi Company claim that the American Government used wireless equipment around WW I that infringed on his 1904 American patent without paying royalties. Finally in 1943, the United States Supreme Court largely absolved the American Government by retroactively invalidating Marconi's 1904 American patent on the grounds that its tuning was not original, prior art set forth in a Braun British patent (filed on January 26, 1899); and a Lodge patent (British patent No. 609,154), filing date August 16, 1897; but primarily the Tesla American patent No. 645,575, filing date September 2, 1897. The United States Supreme Court decision in 1943 that struck down the Marconi patent ruling in Tesla's favor (for detail see in particular the preface to reference [20]), clearly established that the Tesla patent is the key to early long distance wireless communications, although Tesla himself never demonstrated long distance wireless communications.

Marconi was also in trouble with respect to his Lodge patent, and in the same year 1943 – of particular importance is the ruling of the United States Supreme Court that the only valid patent of the three held by the Marconi company in the area of resonance or tuning was the one that the company had acquired from Lodge in 1911 [4].

Although it took the courts several decades to figure this out, the facts were well understood by impartial technical men of the day that Tesla, not Marconi, was the inventor of *king-spark* (a term used by many as this form of signaling developed) or damped wave method of wireless transmission. Robert H. Marriott, the first president of the Institute of Radio Engineers, once said that

Marconi had *"...played the part of a demonstrator and sales engineer. A money getting company was formed, which in attempting to obtain a monopoly, set out to advertise to everyone that Marconi was the inventor and that they owned that patent on wireless which entitled them to a monopoly"*.[Radio Broadcast, December 1925 (Vol. 8, No. 2), pp. 159-162.].

Continuing, Lodge [21], and Popov [22], who had used a vertical grounded antenna, coherer and tapper-back, pointed out in 1894/95 that their apparatus might "be adapted to the transmission of signals to a distance".

In fact, some historians consider Oliver Lodge to be the first to transmit Morse code letters before a learned audience, from his induction coil and a spark gap transmitter over a distance of some 60 meters on August 14, 1894 (Austin [4]). Lodge's receiver consisted of a coherer, a device fabricated (some say designed) by Lodge, which was connected to either a Morse recorder which printed onto paper tape, or a Kelvin marine galvanometer, the deflected light spot made viewing by the audience easier. Several notable scientists in their own right, including Prof. J. A. Fleming, viewed this demonstration at a meeting of the British Association, in Oxford, England. Lodge made no attempt to protect the use of his apparatus by others by a patent at that time, he did so three years later, in 1897; or to publicize and promote the idea of wireless telegraphy. Although he described the experiment at the time as *"a very infantile form of radio telegraphy"*, a statement reflecting his modesty, but an undoubtedly significant one, because it established what he had actually done when the induction coil was actuated by a Morse key by his assistant E. E. Robinson.

There is also on the basis of historical research, indirect evidence (however not disclosed until 30-years after the event) that Aleksandr Popov, a contemporary of Lodge and Marconi, gave a demonstration of a wireless telegraph link to a meeting of the St. Petersburg Physical Society on March 12, 1896. It is said that he transmitted the words "Heinrich Hertz" and that the code characteristics were received, the chairman translated them into letters and chalked them on the blackboard. A description of the equipment he used had been published prior to the demonstration, but no verbatim record of the demonstration survives (IEEE Spectrum, August 1969, p. 69, see Süsskind [23]).

In July 1896, Marconi gave a demonstration to the English Post Office at Salisbury Plain, where he succeeded in increasing the range from its previous 800 meters obtained by other experimenters to a distance of about 3 km [24]. And, as a result of this demonstration, history has accredited him with the invention of an early form of wireless telegraphy. In the same year Captain Jackson, later Admiral, of the British Navy found that considerably greater distances could be obtained using the Dolbear-Edison-Tesla arrangement of vertical antennas and tuned sending and receiving transformers at both transmitting and receiving ends.

Such was the state of the art at the beginning of the 20th century. Hertz was not interested in the commercial exploitation of Maxwell's equations. Application of Hertz's work was left to Lodge, who also did little to exploit practical application, and to Fessenden, de Forest, Braun, Marconi, and many others.

12.3   EXPERIMENTS ON SPARKS AND THE GENERATION OF ELECTROMAGNETIC WAVES

12.3.1   The Basic Spark Transmitter Local Circuit

The simplest method of producing high-frequency oscillations is to give an electrical shock to an oscillator circuit consisting of an inductance and a capacitance in series, see Figure 12.2a. The capacitor C (the *discharge* capacitor) is charged to a high voltage by an induction coil (e.g. Ruhmkorff). Eventually the spark gap breaks down and the resulting spark acts like a closed switch to complete the L-C *oscillator circuit*. The potential energy stored in the discharge capacitor "kick-starts" an oscillatory response in this L-C circuit, at a frequency determined by the values of L and C. (Note: the conducting spark has negative resistance properties, a fact not known at that time, which may account for the extraordinary low damping associated with the L-C discharge circuit.)

12.3.2   The Plain Aerial Spark-Gap Transmitter System

Hertz, Popov, Marconi, and others, placed the spark-gap across the terminals of the aerial (Figure 12.2b). Hertz used an end-loaded dipole aerial. Popov and

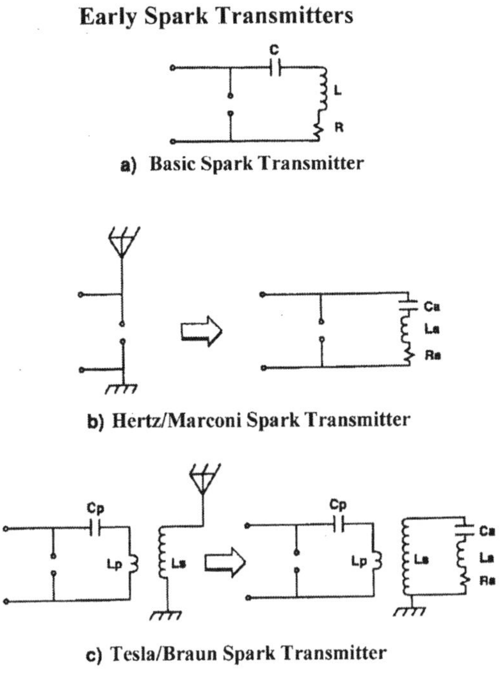

Early Spark Transmitters

a) Basic Spark Transmitter

b) Hertz/Marconi Spark Transmitter

c) Tesla/Braun Spark Transmitter

Figure 12.2. Sketches illustrating the circuitry for a spark transmitter; **a)** the basic circuit, **b)** plain aerial transmitter and, **c)** primary oscillatory circuit, are inductively coupled to the antenna circuit.

Marconi used a wire fed against ground. Figure 12.3 shows a circuit diagram for the transmitter-receiver used by Marconi in 1897.

The disadvantages of a plain aerial transmitter system are that the burst of damped oscillations was very brief, Figure 12.4a, since the oscillator circuit is the antenna itself (oscillation frequency determined by the antenna parameters La-Ca-Ra), and as soon as the spark ceased the oscillations ceased; the radiation efficiency was very low; and great difficulty was experienced due to the extremely high potentials that the aerial was subject to. In spite of these early discovered disadvantages, some contemporary mathematicians concluded on the basis of their theoretical studies that no aerial system could radiate unless the spark-gap was placed across the terminals of the antenna.

12.3.3   Spark-Gap and Local Oscillatory or "Tank-Circuit"

The invention of *king-spark* is historically attributed to Tesla, a Serbian-American physicist working in the United States. In 1893, Tesla was the first to publicly demonstrate the transmission and reception of wireless signaling using tuned transmitter and receiver systems. He was the first to patent the idea of using a transformer with the spark-gap and capacitor in the primary circuit inductively coupled to the aerial circuit (US patent 645,516, Sept. 2, 1897).

The Tesla type spark transmitter (previously depicted in Figure 12.2c) was a considerable improvement over the plain aerial transmitter used initially by Marconi. Here the oscillations generated by the conducting spark take place in a *primary circuit* that is inductively coupled to the secondary circuit, the *aerial circuit*.

When the induction coil was working properly the condenser was charged up, and when the potential across it was sufficiently high to break down

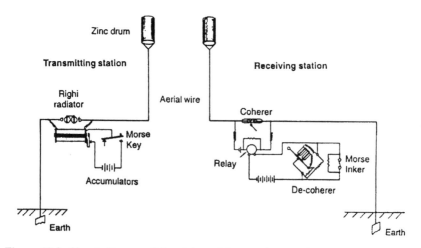

Figure 12.3. Circuit diagram of the plain aerial transmitter-receiver system used during Marconi's Lavernnock Point trials in May 1897 (after Lemme and Menicucci [27]).

the insulation of air in the gap, a spark then passed. Since this spark has a comparatively low resistance, the spark discharge was equivalent to closing of the oscillatory Lp-Cp circuit. The condenser now discharged through the conducting spark, and the discharge current took the form of a damped oscillation, at a frequency determined by the resonant frequency of the tuned circuit. The RF energy flowing in the primary circuit was inductively coupled to an antenna, which was tuned to the same frequency as that of the oscillator circuit. The induced oscillation in the antenna circuit was also a damped wave, but the duration of oscillation was significantly longer than the oscillation in the primary (Figure 12.4b), since when the oscillations in the primary ceased, the oscillation in the antenna circuit could continue, determined by the damping in the antenna circuit. In effect, comparing with a present day transmitter, the primary is the *tank circuit* and the secondary the *antenna circuit* – Fessenden used these words in his 1908 Lecture.

Both circuits should be tuned to the same frequency, and the coupling between the oscillator circuit and antenna circuit should not be too high, since a double peaked amplitude frequency response will result. Such an amplitude-frequency response was undesirable, and in effect resulted in the radiation of a *double wave*. In fact, early radio regulations were introduced encouraging *single wave* or *sharp* emissions, by limiting the amplitude of the second wave to

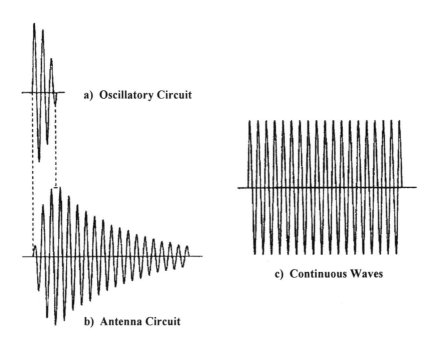

Figure 12.4. Idealized waveforms in the oscillation circuit **a)**; and the antenna circuit **b)**. The spark generated damped waves are much more noisy and broadband compared with **c)** undamped or continuous waves.

approximately one-tenth the amplitude of the stronger, desired wave. Tesla's *tank circuit* was coupled inductively to a secondary consisting of the antenna in series with a coupling coil in which the driving electromotive force was induced and which provided a continuous path from the antenna to ground. Except for the later insertion of a transmission line between the antenna and the coupling coil, the Tesla circuit arrangement provided the complete electrical equivalent of the present day transmitter tank circuit and base driven monopole antenna. In fact, the similarity is closer than one might think. The amount of energy stored in the fly-wheel oscillations in the plate tank circuit of a present day vacuum tube transmitter that is delivered to the output terminals of the transmitter is continuously provided for by the DC current pulses flowing in the transmitting valve circuit (excepting here there is a current pulse for each RF cycle).

Several pioneers were experimenting with local-oscillatory-aerial circuit arrangements. Professor Elihu Thomson discovered that by using a transformer, with the spark-gap and a capacitor in the primary circuit, and with the secondary suitably tuned, a great rise in potential could be obtained [Electrical World, Feb. 20 and 27, 1892]. This same method was later used by Tesla for his 1893 wireless demonstrations (and later for his Tesla Coil experiments). In 1898, Fessenden also was using this device, *"suitably modified for wireless telegraphic purposes, so as to give instead of a continuously cumulative rise of potential, an initial rise of potential followed by a gradual feeding in of energy from the local oscillatory circuit to supply the energy lost from radiation"*, Fessenden's words. He experimented using various methods of connection between the local oscillatory circuit and the aerial (US patents 706,735, and 706,736, December 1899).

Karl Ferdinand Braun, a German physicist, patented a similar circuit (British Patent filed January 26, 1899) in which the spark gap was in a separate primary circuit in series with an appropriate coil and condenser. He realized and demonstrated that *loose* coupling between the spark gap oscillator circuit and the antenna circuit produced significantly less damping of the pulses of oscillations. And further, the effect of low damping was highly beneficial in that more energy was radiated, and energy was distributed over a narrower band of frequencies [25].

Braun later realized that if loose coupling was beneficial between oscillator and antenna, it might also improve the performance of the receiver. In 1902, he carried out experiments that demonstrated that transferring energy from the receiving antenna to the detector through loose coupling resulted in sharper resonance effects as well as increased received signal strength.

When Marconi filed his famous *four sevens* tuning patent (filed April 26, 1900), Braun immediately felt that it was remarkably similar to his own patent filed earlier, and in personal discussion, Marconi admitted that *"he had borrowed"* from Braun's ideas. The Braun-Siemens Company (the name of the company which Braun associated) should perhaps have immediately sued Marconi. When a suit was filed later, the company found that its delay had severely weakened its legal position [25].

The contributions of Marconi to the development of wireless telegraphy largely obscured those made by Braun. The reason for this was mainly due to personality differences between the two men. Unlike Marconi, Braun avoided publicity and sought no personal recognition for his work. Braun saw his work solely in terms of advancement of science. History has for the most part forgotten Braun, since his patent was predated both by Tesla's patent, filed on September 2, 1897, and by Tesla's demonstrations, before public audiences, of the sending and receiving of wireless signals in 1893. The significance of Braun's work was however recognized by the joint awarding of the 1909 Nobel Prize in Physics to Braun and Marconi [26].

12.3.4 Power Sources for Spark-Gap Transmitters

Now we will consider the power source(s) for the spark transmitter. The induction coil, Figure 12.5a, had a low voltage primary winding and a high voltage secondary winding, the voltage used to charge the discharge capacitor. The low voltage primary winding was driven by a battery and an interrupter, which made-and-broke the connection of the primary winding to the battery at about one-hundred breaks per second.

Marconi's early telegraphy experiments were made using such a power source. The interrupter (e.g. a Wehnelt type interrupter) was a mechanical device, operating at a rate of say 100 breaks/second. Thus each time the key was pressed sparks would be generated at 100 spark/sec, and the listener at a remote receiver would *"hear"* a buzz (ignoring for the moment that a suitable detector was not present so that the operator could actually hear the sound of the transmitted signal). The audio sound heard was the interrupter rate frequency accompanied by the ragged and irregular noise of the spark generated signal.

Most early radio experimenters followed or improved upon the Marconi method of signaling, because in their view a spark was essential to wireless. But later experimenters employed an AC generator and a high voltage step-up transformer, rather than an induction coil and a battery. For this power source see Figure 12.5b.

It is essential to charge the capacitor to a very high voltage, since stored potential energy is $0.5\ CV^2$. The transformer steps up the voltage of the alternator to values sufficient to generate sparks. But such a simple arrangement is not very satisfactory, since at the most one spark/cycle or per half cycle, or a few irregularly spaced sparks, could be obtained, and so, the idea of a rotating gap was conceived (Figure 12.6a).

12.3.5 The Synchronous Rotary Spark-Gap Transmitter

Early practical spark-gap transmitters employed rotating gaps, more specifically *asynchronous rotary gaps.* That is the speed of rotation of the wheel is entirely independent of the AC frequency of the generator. A spark will occur whenever the gap between a fixed electrode and a point on the moving electrode

Power Sources for Spark Transmitters

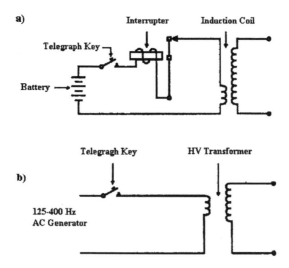

Figure 12.5. Power sources for spark transmitters.

approaching it is small enough. This being the case, it is obvious that there may be several sparks during one cycle of the AC generator. Not only is it possible (see Figure 12.6b) to miss a spark altogether, but the interval between sparks is not constant. If the charging current after one breakdown is such that the condenser voltage before the next discharge is greater than usual, the spark will take place over a longer gap, i.e. when the studs are a little bit further apart than usual. If the condenser voltage is less than its average value, the spark will take

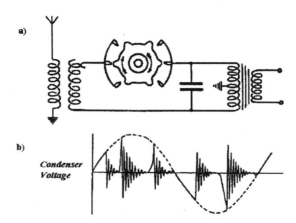

Figure 12.6. a) A synchronous rotary gap arranged to generate sparks on, **b)** both positive and negative excursions of the AC voltage.

place a little bit later than usual. The energy stored in the discharge capacitor and the proportion radiated in the separate wave-trains is therefore variable, and the note heard at the receiving station will be impure.

Fessenden's work in wireless was important not only for the results he secured, but also because of its originality. From the outset, he sought for methods to generate and receive continuous waves, not damped waves that started with a bang and then died away quickly. However, for his early experiments he had to make due with spark transmitters, the only means known at that time for generating appreciable power. So, he set his mind to make his type of transmitter more CW like. This led Fessenden to the development of a *synchronous rotary spark gap transmitter* (Figure 12.7).

An AC generator was used, which, as well as providing the energy for the spark transmitter, was directly coupled to a rotary spark gap so that sparks occurred at precise points on the input wave (Figure 12.8). The spark was between a fixed terminal on the stator and a terminal on the rotor, in effect the rotor was a spoked wheel, rotating in synchronism with the AC generator. With this arrangement two sparks per cycle per phase occur at a precise point on the AC voltage wave form (at the maximum voltage points for generating the greatest power). Figure 12.8 corresponds to a single phase generator. For a 3-phase AC generator, the resulting spark rate would be 6-times the frequency of the AC generator.

Thus a higher spark rate was achieved, high compared with the frequency of the AC generator. Another important advantage was realized; in effect his rotary spark gap apparatus was a kind of mechanically *quenched spark gap transmitter*. The oscillations in the primary circuit ceased after only a few cycles since when the rotating gap opened, the spark ceased. However (as previously described), the antenna circuit continued to oscillate with its own damping. This quenched spark gap was more efficient and the signal generated was less noisy and narrower in bandwidth compared with the unquenched gap.

Figure 12.7. Fessenden's synchronous spark transmitter.

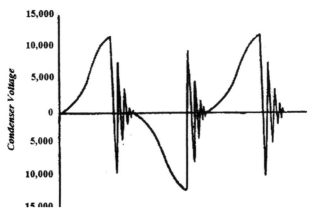

Figure 12.8. Synchronous spark transmitter 2-sparks/cycle precisely at voltage peaks of the AC waveform.

Any of the spark methods of excitation inherently involve consumption of energy in the spark in addition to the energy losses occurring in the antenna circuit. Many forms of quenched spark gap transmitters were devised, described as *Wein* transmitters, but the Fessenden synchronous rotary spark gap transmitter was certainly one of the best. With a synchronous spark gap phased to fire on both positive and negative peaks of a 3-phase waveform, precisely on the peak for maximum efficiency, a 125 Hz AC generator would produce a spark rate of 750 times a second. These rotating gaps produced clear almost musical signals, very distinctive, and easily distinguished from any signal at the time. It was not true CW but it came as close as possible to that; the musical tone was easily read through atmospheric noise and interference from other transmitters. (Using a laboratory version of a spark gap transmitter fabricated by the writer, one can hear genuine spark generated signals, for various spark rates at www.hammond museumofradio.org/belrose-fessenden.html.

Fessenden's Brant Rock station employed a synchronous rotary spark gap transmitter, the largest one built to date (previously shown in Figure 12.7). It was completed on December 28, 1905, a detailed discussion of this transmitter appears later in this chapter. The rotary gap measured 6 feet in diameter at the stator and 5 feet in diameter at the rotor. Its rotor had fifty electrodes (poles) and its stator had four. Coupled to this rotary gap was a 125 Hertz 3-phase 35 KVA alternator.

12.4 EARLY RECEIVING DEVICES

12.4.1 Hertz Resonator

The first and simplest detector of electromagnetic waves was the Hertz resonator. It consisted of an open ring, dimensioned so that it was resonant at the frequency generated by the transmitting apparatus, fitted with a small metal sphere attached at either end. A spark jump across the small gap for each spark generated by the transmitter. Hertz used a microscope to view the small weak spark. Such a detector would be hopelessly insensitive, and impractical for communication purposes, but using such an apparatus (beginning on October 25, 1886) Hertz carried out his experiments on induction and resonance between coupled circuits, and later on refraction, reflection, culminating in interference and standing waves.

12.4.2   Coherers

The coherer is a curious device, said to be the most sensitive method to detect electromagnetic waves at the time. Nearly all the early pioneers used this device until about 1902. Fessenden, who began his studies in 1892 in search of other methods to detect electromagnetic signals [28], considered the device useless, since no matter how sensitive such a device could be made, its use as a detector could never be employed for a practical telegraph system. Its performance was erratic, and the device could not distinguish between signals and noise. He considered that *"the invention of the coherer was a misfortune, tending to lead the development of the art astray into impractical and futile lines and thereby retarding the development of a practical detector"*.

Notwithstanding the coherer dominated the experimental studies of various pioneers for about a decade. In 1898 Lodge noticed the phenomenon of cohesion during experiments on lightning, in which two metallic balls would stick together after the passage of a spark [29], and become conducting only after such a spark. Initially he reserved the name for single point devices, but later when Branly, *cf.* reference [30], a French physicist, devised an entirely different more sensitive device, Lodge gave these new devices the name coherer. This name did not please Branly, who considered that the name coherence implied an incomplete investigation of the phenomenon (a lack of understanding which persists today). But the name coherer stuck.

The coherer consists of a glass tube (later evacuated) with end plugs filled with fine metal or carbon filings (Figure 12.9a). When an RF impulse was passed through the filings, the particles cohered and the resistance dropped. Branley noticed that the high resistance could be restored by a mechanical shock. The coherer could therefore be made responsive to a change of state, i.e. presence or absence of an RF signal, and the type of signal generated by a spark transmitter was the ideal type of signaling, since a spark generated signal is an impulsive signal, having a high amplitude right at the beginning. When a signal was present, a battery-powered circuit attracted the moving element of a relay and, consequently, the whole assembly could be used to receive Morse code signals from a remote transmitter. In order to separate the particles inside the tube (restore the high resistance), so that the apparatus could be operative again, some form of "de-coherer" such as a solenoid tapper back (used by Marconi) or a mechanical shaker (preferred by Lodge) was necessary. The DC current activating the tapper could also put a tick on a Morse Inker.

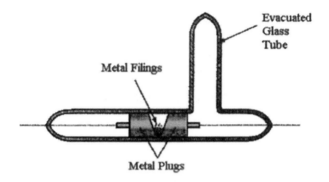

Figure 12.9a. Diagram of a coherer.

While Edouard Branly is generally considered to be the inventor of the device (his publications date 1891 [30]), it should be noted that the Anglo-American, D. E. Hughs also used a similar device (using powdered carbon) in his early demonstration of wireless transmission in 1879, although it had not been accepted for what is was and therefore remained unpublished until 1899 [31].

Lodge devised four versions of coherers, including the Branley type, and several curious devices, a single point coherer, a spiral wire coherer, and a mercury wheel coherer (see Wilson [32]). Certainly Lodge considered that he invented the device, but he was a latecomer [30]. One of the most advanced coherers was that developed by Marconi. He made the dimensions much smaller than most others (about 4 cm long and 0.5 cm wide), and he made the gap slightly wedge shaped (Figure 12.9b). Thus by rotating the tube about its long axis to lie in a wider gap, the user had some degree of sensitivity control. Marconi used filings consisting of 95% nickel and 5% silver, lying between polished silver plugs, the ends of which had been amalgamated with mercury. The coherer tube was evacuated of air to prevent oxidation of the particles, thereby stabilizing its long term performance [33].

The filings coherer was essentially a peak-sensitive device requiring 3-5 volts but only instantaneously to change its state. Once cohered, its resistance

Figure 12.9b. Marconi coherer. The metal plugs were generally silver, with platinum conducting wires [34].

dropped from about a mega ohm to a few hundred ohms; hence a 1.5 volt battery in the receiver could pass enough current through the coherer to operate a sensitive relay recording detection. The relay in turn operated a tapper to restore the coherer in time for the next (spark type) transmitter pulse. The coil resistance of the sensitive relay limited the current to less than 1 milliamp in order to reduce damage to the oxide layer on the filings. The physical mechanism of the coherer is still not completely understood [33, 34].

12.4.3   The 'Italian Navy Coherer'

The so-called 'Italian Navy coherer' is an entirely different device; it was in fact a metal oxide contact rectifier. In Figure 12.10, a copy of a sketch of the coherer, as patented (patent filed September 10, 1901) is reproduced. It consists of a glass tube containing a globule of mercury between two end plugs, usually of iron but sometimes carbon. One (fixed) end plug is treated so that the mercury wets it. The other end plug has a screw adjustment so that it just touches the globule.

In an information/hardware exchange with the Italian Navy, Marchese Luigi Solari, then (summer of 1901) a Lieutenant in the Navy, presented Marconi with a working model of a mercury coherer, and with his blessing Marconi (as was his usual) promptly patented it in his own name, see Phillips [35]. Some historians consider Jagadish Chandra Bose (India) to be inventor of this device.

Measured characteristics of the device reveal that its performance is very variable: a properly wetted end contact (the moveable screw adjustable contact) behaves as a resistance (as expected); if the contact barely touches the blob of mercury the device works (some of the time) like a metal oxide contact rectifier [34,35]; if the end contact does not touch the mercury, the device is merely an open circuit. The weak diode type rectifier characteristic can be obtained with no external bias applied. A somewhat stronger amplitude detector characteristic can sometimes be obtained by biasing the device by a few hundred

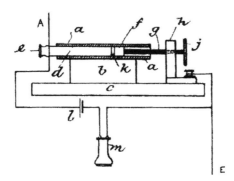

Figure 12.10. The Italian Navy Coherer as patented by Marconi in September 1901, British Patent No. 18,105 (after Phillips [35]); d = carbon plug, f = adjustable iron plug, k = drop of mercury, A = antenna, E = earth.

millivolts (to find the knee of the non-linear characteristic). The rectifier characteristic is very variable and is easily destroyed by large spark signals. When carefully (or by chance) adjusted, the rectifying action seems to be more pronounced on weak signals than on the larger ones.

Since the Italian Navy Coherer was one of the three coherers that Marconi said he used for his first transatlantic experiment (Vyvyan [36] said it was the detector he was using when he said he heard the signal), it is interesting to read what George Grisdale said in 1974, when he tested the device. George was then with the Marconi Research Laboratory [reference *"Tests on Marconi's Italian Navy Detector"*, Interim Technical Memorandum ITM 74/8, January 1974]. When the instrument (tested) was received from the IEE Archives there was no globule of mercury in the tube, only a silvery trace on the glass. A small amount of mercury was therefore added between the two electrode cylinders, one which could be adjusted longitudinally by a screw thread.

For some time the device showed no sign of rectification, it was either an open circuit or about 500 ohms, but after removing the mercury several times it suddenly showed signs of rectification. A conclusion reached by others that have tested such devices [34,35] is that the mercury should not be too clean.

With a pair of 2000 ohm earphones in series with the mercury detector, when by chance it was working, unmodulated input signal levels key on/off could be heard for input levels of 0.1 volt.

But the device was very temperamental. It could be changed from the non-linear to the high conducting state in several ways: a sharp tap could make the change in either direction; increasing the input voltage to 0.3 volt could cause a sudden change in conductivity state. On reducing the voltage the partial rectification could appear, either suddenly or gradually over several seconds. For reception of damped wave (spark) signals, the burst of oscillations needed to be long. An initial impulse spike adds nothing and (in fact) could disable the device [34]. Marconi shortly after his first transatlantic experiment gave up using the device. Its performance was too unpredictable.

12.4.4   The Magnetic Detector

From his early experiments with wireless, beginning in 1895 at the home of his parents, Villa Griffone, Italy; Marconi used various versions of the coherer. A coherer was one of the devices that he used to listen for signals from his sender at Poldhu, Cornwall, on Signal Hill, NL; and later on the *SS Philadelphia* (in February, 1902), sailing from Europe to America. But Marconi himself recognized (and stated) that the coherer devices he was using could not be further improved, that their performance was unreliable, and that coherers were certainly not suitable for long distance (transatlantic) communications.

In 1895, Ernest Rutherford, a scientist in England, was experimenting with magnetometers, a device for measuring the strength of the magnetic field. He was trying to improve the sensitivity of his magnetometer, to see if he could use an "earlier discovered phenomenon" as a means for detecting EM waves.

Rutherford's research was based on the work of Lord Raleigh, but principally on Joseph Henry, who in 1842 had succeeded in demagnetizing steel needles by Leyden jar discharges. Using a large Hertzian Oscillator, Rutherford attached an antenna to his magnetometer, and discovered he could detect signals at distances of about 1200 meters. The rods of his resonator were connected to a small bobbin of fine wire, in the center of which he placed a magnetized steel magnet. He measured the degree of magnetization before and after a spark had taken place at his distant transmitter, and he found that the received oscillator pulse had demagnetized it, and it had to be magnetized again before the experiment could be repeated. In 1900 he suggested to Dr. Ersken Murry that the oscillations might be able to record themselves on a steel band.

Marconi became aware of Rutherford's work, and he foresaw the possibility of utilizing such a device to detect his wireless signals, but it was not before January/February 1902 that he started to experiment with Rutherford magnetometers. Later in 1902, after carrying out meticulous experimental studies, Marconi filed a patent (Great Britain Patent No. 10,245) for two forms of magnetic detectors: in one his detection circuitry was subject to an oscillatory magnet field, a horseshoe magnet revolved by clockwork; in the other form the moving member was a band (a "rope") of continuously moving iron wires. Both of these devices have been called *Hysteresis Detectors* [37].

Marconi had no qualms about borrowing once more from earlier work. The device he patented is usually remembered as a Marconi invention, but earlier versions of it were referred to as a Rutherford-Marconi Magnetic detector [38]. In fact Rutherford also claimed to have invented this particular method [39].

Let us look at Marconi's device that was popularly called *the Maggi*, and speculate how it might have worked (Figure 12.11a). An endless loop of insulated iron rope (a) is kept moving by a clockwork drive (at speeds of 1 to 6 cm/second), past the like poles of two horseshoe magnets (d), so that any small portion of the wire passing beneath them is magnetized first in one direction,

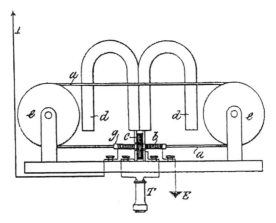

Figure 12.11a. Diagram from Marconi's Patent No. 10,245 of 1902 showing the magnetic detector.

then in the other. At the point where the magnetization changes direction, a single layer coil (**b**) is wound on a glass tube surrounding the wires, the ends of which are taken to aerial and earth. A secondary winding (**c**) made up of many turns of fine wire is wound on top of the primary winding, the ends of which are taken to the telephone receiver (**T**). Thus there is a constant supply of iron which is rapidly changing its magnetic state passing under the coils.

In the absence of received signals, no oscillatory currents flow in the primary circuit and the lines of force are, so to speak, pulled out of position by the moving band, which by its steady progression is being taken away from the poles before it had time to become fully magnetized. When iron is magnetized it does not become fully magnetized instantly, because there is an appreciable time lag behind the switching on of the magnetic force (the well known Hysteresis effect). If now an EM pulse of oscillatory signals passes suddenly through the primary windings (b) the iron becomes suddenly demagnetized, "suddenly" since the time lag apparently disappears, and the lines of force move back to the position they would occupy were the band stationary. This sudden movement of the lines of force creates a momentary current in the secondary coil (**c**), which causes a sound to be heard in the phones (see Blake, 1926 [37] and Phillips, 1980 [33]).

This device apparently works rather well, giving a response to the positive going mean amplitude of an oscillatory signal (in principle like a rectifier detector), and hence the listener (listening to a spark generated signal) could hear an audio signal corresponding to the spark rate. In fact, the device even responds to amplitude modulated (AM) signals, although not so well as a rectifier detector. Many of the early AM telephony transmissions were successfully received using Marconi's magnetic band detector [Phillips, private communications, 1994].

O'Dell [40] has published an alternative (more complicated) explanation of the phenomena. In O'Dell's view Marconi was making use of the effects of magnetic materials that depend on the motion of magnetic *domain walls*. The two magnets placed according to Marconi's instructions produce a quadrapole field that causes the iron rope to divide into two magnetic domains end to end. The existence for such domain walls, according to O'Dell [41], was not known until about 1907 when Pierre Weiss put forward the idea. In 1919 Barkhausen apparently found evidence for the existence of domain walls, but it was not until the 1930s that the first clear experimental work on moving magnetic domain walls was published by Sixtus and Tonks.

O'Dell tells us that Marconi himself put forward a very sensible account as to how his detector worked, which was written for the British Admiralty, but not released to the general public until 1950 [42]. O'Dell [40] built a working model of a magnetic detector, using 4-bar magnets, rather than 2-horseshoe magnets, with like poles together in the vicinity of the aerial/detection coils.

The Rutherford-Marconi magnetic detector had like poles together, but operators using the device found the *Barthhausen noise* rather annoying. Since the moving wire was in a constantly changing state of flux, a continuous hissing noise was heard in the headphones, which must have been very fatiguing to the ear during prolonged periods of listening for signals. Many writers [33] have found that this can be overcome if instead of putting like poles together one puts unlike poles together, the pole of one being some little distance up the limb of the other and the best position being found by experiment. With this arrangement however the sensitivity of the detector is reduced. It is not immediately apparent what this achieves; but for the interested reader Phillips shows what happens. An audio band pass or high pass filter would have been a better solution for this problem. In its commercial form (Figure 12.11b) Marconi's moving-band detector was found to be very satisfactory and trouble free and achieved considerable popularity. It was the standard detector on shipboard receivers (Marconi receivers) until vacuum tube detectors replaced it in about 1914.

12.4.5   Fessenden's Barretter — an Electrolytic Detector

Fessenden was convinced that the successful detector for wireless signals must be constantly receptive, instead of requiring resetting as was characteristic of the coherer. But this was more easily said than done. His experiments with wireless receivers (and spark-gap transmitters since he had to have a signal source) began when he was Professor of Electrical Engineering at Perdue University, West Lafayette, IN, in 1892–1893, and continued during the period when he was a Professor of Electrical Engineering in the Western University of Pennsylvania, Pittsburgh, PA. In 1896/97, in conjunction with two students Edward Bennett and William Bradshaw, he conducted numerous experiments in an attempt to develop an improved detector, the results of which were incorporated in their thesis [Western University of Pennsylvania, May 1897].

In the period 1897–1902 Fessenden tried dozens of methods, methods devised by others as well as by him. Indeed many of the curious devices tried have not been satisfactorily explained, and there is no way of assessing efficiency without a great deal of experimental work. One of the devices he used was a type of thermal detector, a very fine very short wire connected in series with a resistor and a battery (to provide a very small electric current) and an ear piece; it was

Figure 12.11b. Photograph of a commercial version of Marconi's magnetic detector device.

also connected between aerial and earth. The aerial current heated this hot-wire barretter, and the resulting change in battery current was audible in the ear piece.

Fessenden's hot-wire barretter, which consisted of a minute platinum wire a few hundred thousandths of an inch in diameter and a hundredth of an inch in length, was the device in use when he discovered by accident his liquid-barretter (an electrolytic detector). The filament for the "hot-wire" device was made of Wollaston wire, in which the silver coating was removed by a nitric-acid treatment. It was during such treatment that Fessenden observed that one of several such barretters, in this silver dissolving part of the process, was giving indications on a meter attached to the circuit of signals received from an automatic test sender sending D's. An examination revealed that this one had a broken filament, while the others were complete. A brief investigation disclosed the fact that this Wollaston wire dipping into the 20% nitric-acid solution was far more sensitive and reliable than any other known detector. Fessenden coined the word 'barretter' from his classical language background. The term is a derivation from the French word 'exchanger', inferring the change from AC to DC.

For proper operation the platinum-coated Wollaston wire needed to make point-contact, lightly touching the acid solution (reference US Patent No. 727,331, May 5, 1903, for the basic detector; and US Patent No. 793,684, December 1904, for a sealed detector for shipboard use). This detector was the standard of sensitivity for many years, until it was replaced by the germanium crystal detector and by the vacuum tube in about 1913. This detector, when used with a telephone receiver in a local shunt circuit (Figure 12.12), gave such accurate reproductions that radio operators could identify several wireless telegraphy stations in the pass-band of their receiver by the different characteristics of the spark transmissions, just as a friend's voice is recognized by its peculiarities of tonal quality. And, it subsequently made possible the reception of radio telephony (voice).

Fessenden at the time did not understand how the device worked. He rejected the opinion of some of his contemporaries that the action was electrolytic. He considered that the action was thermal. In fact, reading numerous and varying accounts of electrolytic detectors, which Phillips [33] refers to as fine-point-electrolytic detectors, seems only to have compounded the confusion prevailing at the time. But the device did work, it was an electrolytic detector, and it became the standard until the appearance of the galena crystal, germanium diode and the vacuum tube. For a modern analysis of the rectification properties of an electrode-electrolyte-interface, the reader is referred to the paper by Geddes, et.al [43]. The authors became aware of Fessenden's pioneering work only after acceptance of their paper by the Journal (see addendum to the referenced paper).

12.4.6   Heterodyne Detector for Wireless Telegraphy

Fessenden's rectifier type of detector was useless for reception of telegraphy by the on-off-keying of a CW carrier. A spark initiated signal was in effect a

Electronic Detector

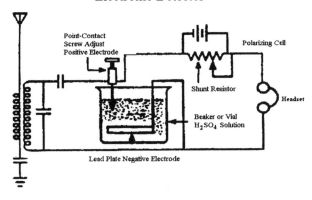

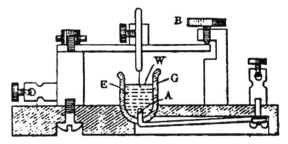

Figure 12.12. Depicted on **top** is an early version of a Fessenden receiver, which employed his Barretter, an electrolytic rectifier. On the **bottom**, the cross-section for this device shows that it contained a platinum-coated Wollaston wire making point-contact (W), lightly touching a 20% solution of nitric acid (after Elliott [3]).

modulated oscillatory damped wave, modulated by the spark rate. For intermittent continuous wave transmission all that a listener would hear would be clicks, as the Morse key was closed and opened.

Again, Fessenden's fertile mind worked around this problem. He devised the methodology of combining two frequencies to derive their sum and difference frequencies, and coined the word *heterodyne*, derived from the joining of two Greek words *hetro*, meaning difference, with *dyne*, meaning force.

His patent covering the heterodyne principle is described in US Pat. No.706,740, filed 28 September 1901 and granted 12 August, 1902. Further related heterodyne method patents were US No. 1,050,441, filed 27 July 1905 and granted 14 January 1913, and US Patent No. 1,050,728.

In this time period, however, heterodyning was way ahead of its time. It would take the addition of de Forest's triode vacuum tube, which was integrated with Fessenden's heterodyne principle in Edwin H. Armstrong's *superheterodyne* receiver of 1912, to make Morse code keyed CW telegraphy reception practical. Some historians, Belrose and Elliott [8], consider Fessenden's heterodyne principle to be his greatest contribution to radio. Armstrong's superheterodyne receiver is based on the heterodyne principle, and except for method improvement, his superheterodyne receiver remains the standard radio receiving method today.

Remarkably, Fessenden was able to conceive the heterodyne principle without the wonders of the vacuum tube. He had to work with what was available at the turn of the century — further development of the arc oscillator, primitive detectors, the HF alternator, etc.

His first patent considered using two transmitters, operating on slightly different frequencies, say differing in frequency by 1 kHz, so that the beat frequency heard when the signal was received would be a 1 kHz tone – such an arrangement operationally is quite impractical, but today a *beat tone* is often heard when two closely spaced independent CW carriers happen to be on the air at the same time. During 1902, Fessenden improved his heterodyne receiver by placing the *second-frequency-transmitter* right at the location of the receiver site. The second signal was now under the control of the receiver operator, making the system easy to manage. This lead to his advanced heterodyne method, US patent No. 1,050,441.

It is probable that Fessenden never did get this receiver to work very well (Figure 12.13). The Oscillating Arc was a noisy signal generator, and its broadband characteristic probably swamped possible reception of the wanted signal, unless it was quite strong. In the words of Sam Kintner [44], a member of the original Fessenden team:

> This (the heterodyne principle) was another bold stroke of Fessenden, in which he departed from the methods practiced by others. Like some other great inventions, it was made before he had suitable equipment with which to practice it. He required a source of local oscillations of adjustable frequency, and an HF alternator or oscillating arc was all he had.
>
> The term heterodyne was coined by (Prof.) Fessenden and has a technical meaning which may in general be

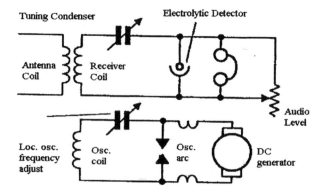

Figure 12.13. Sketch of Fessenden's advanced heterodyne receiver system.

explained as follows:

> *If currents of two different frequencies 'beat' in the same circuit they will produce a new frequency and are said to heterodyne (from the Greek words heteros – other or different, and dynamics – power). And the new frequency is called the heterodyne frequency. Further to simplify the matter if the frequencies being received are of such rapidity as to be beyond the (range of) human ear, such as 300 kHz, in order to hear the 'heterodyne' one would generate in the receiving apparatus a frequency of 301 kHz. The result of combining the two would be that the 'beat' note would equal to the difference between the two or 1 kHz, which could be easily transformed (in the earphones) into a tone suitable for the ear of the listener. Mysterious and magical no doubt but scientific fact nevertheless.*

> *Fessenden devised and named the heterodyne system and applied for a letter patent in 1905. His triumph was more theoretical than real for he was years ahead of the industry. His new CW (continuous wave) apparatus was the only system capable of using the heterodyne system. Not until 1912 when the triode tube of de Forest became practicable for the general public did Fessenden's heterodyne invention assume its importance in radio technology.*

John Hogan's 1913 Classic Paper, in which he described radio equipment used in the United States Navy's Arlington-Salem tests, explained the principle of heterodyning, and claimed that equipment employing this principle had greatly improved the sensitivity of wireless telegraphy receivers. This paper has recently been republished [45]. In the closing words of Hogan's paper: *"The maximum of credit is due to Prof. Fessenden, for his fundamental invention compared to which the improvements brought out by such as us as have continued the work are indeed small".*

12.5   CONTINUOUS WAVE TRANSMITTERS

Spectacular as was the Brant Rock transmitter employing Fessenden's synchronous spark gap transmitter, Fessenden, after achieving initial success in 1906, to be described, again turned his attention to the HF alternator, which he had been trying to develop in industry for some time. Fessenden had realized, as we have already noted previously, that the stop and go system, the spark transmitter, was incapable of transmitting satisfactorily voice and music. A means of sending and receiving continuous waves was required. The idea came to him during discussions with his uncle Cortez Fessenden, while visiting with him at his cottage on Chemong Lake, near Peterborough, Ontario, in 1897, and is described in his US patent No. 706,737, dated August 12, 1902. But it was not before the fall of 1906, when Fessenden's HF alternator was developed to a point where it could be used practically (frequencies up to 100 kHz were possible), that continuous wave transmission became feasible.

Marconi and others working in this new field of wireless ridiculed Fessenden's suggestion that a wireless signal could be produced by applying an HF alternating current to an antenna. All were unanimous in their view that a spark was essential to wireless; an error in reasoning that delayed the development of radio by a decade. Fessenden was right, but alone in his belief. *"The whip-lash theory however passed gradually from the minds of men and was replaced by the continuous wave one with all too little credit to the man who had been right"* [*New York Herald Tribune*, 46].

To document the reaction of his colleagues to this departure from conventional transmission methods, spark or damped wave transmissions, we can note that J. A. Fleming in his book "Electromagnetic Waves", published in 1906, said, in reference to Fessenden's Patent No. 706,737, that *"there was no HF alternator of the kind described by Fessenden, and it is doubtful if any appreciable radiation would result if such a machine were available and were used as Fessenden proposes"*. Fleming was totally wrong, since 1906, the year in which his book was published, was the year of Fessenden's great, long sought after achievement, using continuous waves generated by a HF alternator. One terminal of the alternator was connected to ground and the other terminal to the tuned antenna as described in his patent. Certainly the referenced statement did not appear in subsequent editions of Fleming's book.

Fleming seems to have thought that the sudden change in fields around the antenna produced by the spark discharge was essential for the generation of propagating EM waves. Perhaps he should have read more of Hertz's work. Hertz's EM theory of radiation from an oscillating dipole included diagrams of the radiation field out to many wavelengths from the source. Hertz used a spark source for his experiments to generate electrical oscillations at UHF frequencies (laboratory scale wavelengths), and he used resonant arrangements to encourage his signals to have a quasi-sinusoidal form.

Judge Mayer, in his opinion upholding Fessenden's patent on this invention, said, in effect that *it has been established that the prior art practiced, spark or damped wave transmission, from which Fessenden departed and introduced a new or continuous-wave transmission, for the practice of which he provided a suitable mechanism, which has since come into extensive use* [Kintner, 44].

12.5.1   Arc Transmitters

The spark across the gap in a spark transmitter is an arc, but a *spark transmitter* and an *arc transmitter* are very different. Up to this point, spark transmitters have been discussed, that is a spark discharge initiates the generation of an oscillatory signal in a tuned circuit, or in the plain aerial transmitter, the spark discharge shock excites the aerial system itself into oscillation; done without timing a sequence of damped waves result. But, if one could shock-excite a tuned circuit often enough, at a rate in phase with the damped oscillations, we could get close to generating a continuous wave. The multi-gap timed spark concept was pursued by Marconi, in the latter years of king spark, but timed spark transmitters generated a broad band noisy signal, and did not generate continuous waves.

An arc converter transmitter generated CW by shock exciting its tuned circuits continuously. Figure 12.14a shows a Poulsen Arc Oscillator with Magnetic Blast (US Patent No. 789,449, June 19, 1903). Figure 12.14b shows for purpose of discussion a practical arc transmitter. Here the DC power that maintains the arc between the two electrodes (**A**) and (**B**) also flows through field coils (better shown in Figure 12.14a) to produce a very strong magnetic field. The operator "punched" the negative terminal (**A**) contact (carbon) to strike (or start) the arc. A spring arrangement automatically pulled the contact (**A**) back to lengthen the arc. The magnetic field forced the arc (an ionized gas conductor) to loop outward, further increasing its length, the longer the arc, the greater its negative resistance. The arc's negative resistance offsets the damping effect of the transmitter's aerial circuit resistance(s). The resulting RF wave train is then continuous, that is not damped.

The key and chopper arrangements generate a frequency shift for

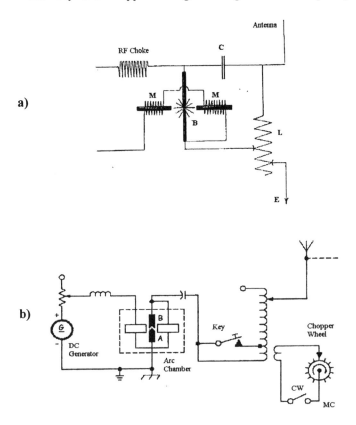

Figure 12.14. a) Poulsen Arc Oscillator with Magnetic Blast; **b)** a practical arc transmitter, for purposes of discussion (after Schrader[47]).

sending Morse code signals, since once started the arc must be maintained. The key shorts out a part of the inductor, increasing the frequency radiated by the antenna system. The *chopper wheel* can be used instead; it results in a frequency change that occurs at an audio rate, generating a modulated continuous wave (MCW) signal.

The method of producing high frequency oscillations using an arc to generate continuous current was however discovered much earlier, by Elihu Thomson (US Patent 500,630, July 18, 1892). Fessenden [28] in his 1909 Lecture to the AIEE said: *"the worker with high frequency oscillatory currents will soon discover that we are indebted to the genius of Professor Elihu Thomson for practically every device of any importance in this art"*. Yet with exception for the writings of Fessenden, and subsequently those writing about Fessenden, history has forgotten Elihu Thomson.

Between 1900 and 1902, Fessenden carried out numerous experiments using the arc as a source for wireless telegraphy and telephony. It was found, as noted earlier, that the arc could not be started and stopped as quickly as was necessary for telegraphic purposes, and the intensity of the oscillations and their frequency varied considerably. These were overcome by making some minor improvements, for example the difficulty in sending was overcome by permitting the arc to run continuously and using the key to change the electrical constants of the circuit (reference US patents 706,742, July 6, 1902; 700,747, September 28, 1901; 727,330, March 21, 1903; 730,753, April 9, 1903). The difficulty in keeping the intensity and frequency constant was overcome by substituting resistance for a portion of the inductance, and also using the arc under pressure (US patent 706,741).

In 1903, Poulsen (US Patent 789,449, June 19, 1903, see Figure 12.14a) modified the Elihu Thomson arc which consists in forming the arc in hydrogen instead of air or compressed gas as previously done, and while this was said to improve the performance of the arc transmitter, according to Fessenden [28], this modification was in fact less efficient and gave oscillations of varying amplitude and intensity, and accompanying strong harmonics. It was also prone to "exploding" during the setting up process of initiating the arc!

12.5.2   Fessenden-Alexanderson HF Alternator

It is reasonable to expect from the outset, *a priori*, that undamped oscillations of a high frequency could be generated by a machine, in the same way that commercial alternating currents of lower frequency are produced (50 Hz and 60 Hz mains frequencies). But to do this was an exceedingly difficult problem when frequencies of 100 kHz (or less) are to be obtained. Fessenden had this idea from the beginning, but it was not before 1906 that he succeeded in developing a suitable machine.

Initially Fessenden employed various forms of arc transmitters, and rotating spark gap transmitters with varying degrees of success. When he had perfected his HF alternator in 1906 Fessenden had achieved his goal, *viz.* a continuous wave transmitter, the frequency of which was not determined by aerial tuning but by the speed of the HF alternator — the aerial tuning only determined the power transfer from his transmitter to the aerial. Subsequently the HF alternator was replaced by vacuum tube transmitters, and nowadays by solid-state transmitters, but the basic requirement, namely the generation of continuous waves is the same as that today.

As early as 1890 Tesla built high frequency alternating current (AC) alternators. One, which had 384 poles, produced a 10 kHz output. He later produced frequencies as high as 20 kHz (reference Quinby [48]). There is no fundamental reason that such frequencies could not have been used for world wide wireless telegraphy communications. In fact in 1919 the first continuously wave reliable transatlantic radio service, with a transmitter installed in Brunswick, NJ, using a 200 kW HF alternator operating at a frequency of 21.8 kHz came into service [49]. However practical aerials used in the early days of wireless were not large enough to radiate efficiently at such a very low frequency, and so LF rather than VLF was used.

Fessenden contracted the GE Company to build a HF alternator operating at speeds of 50 kHz to 100 kHz. Alexanderson struggled for 2-years to develop such a machine, and in September 1906 GE delivered its best effort – which in Fessenden's view was a "useless machine".

The *'Alexanderson alternator'* did not meet Fessenden's specifications. This GE alternator was delivered with a letter stating that in the opinion of its engineers it *"could never be made to operate above 10,000 cycles"* [50]. It is not clear what had been achieved over the Tesla alternator.

So Fessenden took upon himself to rebuild the machine. He must have persevered, day and night, in the usual way he attacked a problem, since by November 1906, he had succeeded in developing a machine that could operate at frequencies in the 50–88 kHz band.

This first Fessenden HF alternator was a small machine (Figure 12.15) of the Mordey type, having a fixed armature in the form of a fixed disk, or ring, and a revolving field magnet with 360 teeth, or projections. At a speed of 139 revolutions per second, an alternating current of 50,000 Hz and a terminal EMF

Figure 12.15. Photograph of Fessenden's first HF alternator.

of 65 volts was generated. The maximum output of the alternator at the above speed was about 300 watts (Ruhmer [51]). Very little difficulty seems to have been obtained in running the machine at such a high a speed. A simple flat belt drive was used, powered by the steam engine at Brant Rock. A thin self-centering shaft entirely eliminated excessive vibration and pressure on the bearings. The belt and step-up gear box can be seen on the far right of the photograph; the alternator is on the left. The frequency of the alternator was determined by the speed of the steam engine, which had to be well regulated.

Fessenden later developed an HF alternator that had an output power of 50 kW. This machine was scaled up to 200 kW by the GE Company, and put on the market as the *Alexanderson alternator*, named after the man who supervised the job. Alexanderson certainly deserves his place in history, but history forgot that Fessenden developed the prototype.

Zenneck [52] has detailed the early efforts to develop this high frequency alternator, and he described its principle of operation. Figure 12.16a (copied from Zenneck's book) is a diagrammatic cross-section of one of these alternators. The excitation is obtained by means of a single large field coil (**S**), which is wound around the entire machine and supplied with direct current. The magnetic flux lines (**M**) of this coil pass through the iron cores (**E1** and **E2**) of the armature coils (**S1** and **S2**). The only moveable part (**J**) has teeth or projections (**Z**) of iron at its periphery. When one of these teeth is just between the small armature coils (S1 and S2), the magnet Flux (**M**) has a path almost entirely through iron, accepting only at the very small air-gaps between the teeth (**Z**) and the cores (**E1** and **E2**). In this position then the magnetic reluctance is a minimum, and the magnetic flux passing through the cores (**E1** and **E2**) is a maximum.

When the air gaps instead of a tooth lie between the armature coils, the reluctance is much larger, and so the amount of flux through the armature windings is very small. Hence as the moveable part (J) rotates, the magnetic flux passing through the armature coils varies periodically between a maximum and a minimum value, so that an oscillatory EMF, whose frequency equals the product of the speed in revolutions/second times the number of teeth, is induced in the armature windings. The rotor of the Fessenden-Alexanderson machine is shaped like the cross-section (J), in Figure 12.16a and has 300 teeth. The space between the teeth was filled with non-magnetic material (phosphor-bronze) so the surface of the rotor (J) was quite smooth, thereby preventing any loss due to air friction (windage).

The armature windings, in which the oscillatory EMF is induced does not, properly speaking, consist of coils, but of a single wire wound in a wave shape form (see Figure 12.16b). Any two consecutive U-formed wires may be considered as a pair of coils of one turn each, joined in series so as to oppose each other. In Figure 12.17 one half of the completed armature is shown.

The capacity of the machine shown with its DC motor in Figure 12.18 – increases as the air gap between the armature and the rotor is decreased. It generated 2.1 kW with the machine having a 0.37 mm air gap. The HF alternator

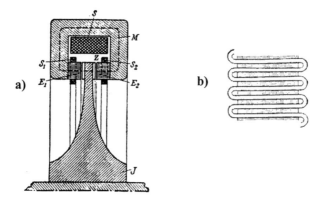

Figure 12.16. The Fessenden-Alexanderson HF alternator: **a)** diagrammatic cross-section of one of these alternators; and **b)** the armature winding in which the EMF was induced [after Zennick, 52].

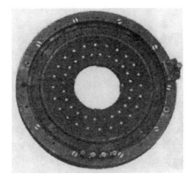

Figure 12.17. One half of the completed armature [after Zennick, 52].

Figure 12.18. Alexanderson-Fessenden HF alternator, with its DC motor drive. The alternator is gear driven. The belt operates an oil pump to keep the gear box and bearing well lubricated. It had an output power of 2.1 kW and a maximum frequency of about 100 kHz (credit North Carolina Division of Archives and History).

is directly driven by the DC motor. The belt system is for pumping oil to keep the gearing systems well lubricated.

12.6   ANTENNA SYSTEMS

The antenna (or aerial) systems used by Fessenden and Marconi for transatlantic communications were very different. Let us look at these differences and comment on how they came to be.

Initially simple wire aerials were used. Fessenden for his experiments on Cobb Island, MD, in 1900 used a simple T-type antenna (Figure 12.19). The mast height has not been published, but it was probably about 10 to 15 meters.

Marconi, from the very beginning, was struggling to achieve greater and greater transmission distances and in pursuit of that, he was putting up higher and higher aerial systems. Already, in 1897, Marconi and Kemp rigged up a 35 meter high mast (at Needles on the Isle of White, England) to support a simple wire antenna for communication trials (Figure 12.20). Signals were received on a tugboat on the sea and to the English coast over 22 km. The wire antennas Marconi used (very early) on steamships sailing the Atlantic were supported by masts 46 meters high (Figure 12.21).Both pioneers knew that for plain aerial transmitters an aerial having a large self-capacity was required. In Fessenden's important CW patent entitled "Wireless Telegraphy" (US Patent 706,787, filed May 29, 1901, approved August 12, 1901), he showed a mast surrounded by a cage of wires to increase the self-capacity of the aerial system. Unlike Marconi, Fessenden did not proceed to develop aerial systems having a high self-capacity. He was endeavoring to use continuous waves not a spark excited plain aerial transmitter. While his HF alternator was far from being developed, he may have foreseen that the generation of frequencies above 100 to 200 kHz would not be

Figure 12.19. Twin masts at Cobb Island, 1900.

Figure 12.20. The 115 foot (35m) mast at the Needles on the Isle of White, 1897 [24].

Figure 12.21. *SS Philadelphia*, showing her 150 foot (45m) masts (after Simons [24]).

possible using such a machine. In any case he clearly realized the need for a tall aerial mast. Fessenden never did use a plain aerial transmitter for his distance communication experiments.

Before 1900, Marconi used plain aerial transmitters (and receivers), and for a maximum range the self-capacitance of the aerial should be large. The self-capacity of the aerial is the discharge capacitor. A "requirement" for aerials having a large self-capacitance seems to have dominated Marconi's way of thinking judged by the aerial system he later put up. In his 1909 Nobel Lecture Marconi said "*my previous tests had convinced me that to extend the distance of communications, it was not merely sufficient to augment the power of the sender, but it was also necessary to increase the* **area** *and height of the transmitting antenna*". Marconi made no mention of the fact that increasing the size of the aerial decreased its self resonant frequency. He seems to have decided that for him 76 meters was a practical height, and the aerial systems he erected for long distance communication were all devised to have a large self capacity. After

1901 the plain aerial transmitter for Marconi fell into disuse, and after 1902 he discontinued using coherer detectors.

The cone aerial was perhaps the ideal device for generating the fast pulse at the beginning of each damped wave train. This was the signal that switched the coherer from its high impedance to a low impedance (signal detection), switching before the oscillatory current on the transmit aerial, determined by its self resonance characteristics, had time to begin. When detectors that required a longer oscillatory burst for detection became available, a high self-capacity broadband aerial was not needed – certainly not for CW or quasi-CW signaling. [1]

This author has found no record of Fessenden ever using a plain aerial transmitter for his distance communications experiments. He was from the beginning aware of the discovery by Elihu Thomson (in 1892) and Tesla (in 1893) that using a spark gap in a primary circuit that is tuned to resonance, and inductively coupled to the aerial circuit could generate far greater powers, since the capacitor in the local oscillator circuit could be enormous compared with the antenna system capacity.

Elihu Thomson did not patent his discovery, but the same method was used by Tesla in his experimental research (US Patent 645,516, September, 1897). Fessenden saw no need to go to extreme measures in trying to achieve aerial systems having a large self-capacity. He knew he needed a tall mast because he wanted an aerial system that could operate efficiently on frequencies below 100 kHz. He wanted to generate a continuous wave signal, and while his HF alternator was yet to be developed, he thought that 100 kHz might be a practical design goal. So, with this preamble, background, let us look at aerial systems used for transatlantic communications.

Marconi was a champion of multi-wire aerials; for numbers of wires, for speed in erecting them, and (in some cases) for lofty tumbling. His ambition at the turn of the 20th century, to demonstrate long distance wireless communications and to develop a profitable wireless telegraph service, led to his pragmatic proposal in 1900 to send wireless signal across the Atlantic. His conceived plan was to erect two super-stations one on each side of the Atlantic for 2-way wireless communications, in effect to bridge the continents together, in direct opposition to cable companies (the Anglo-American Telegraph Company). For the eastern terminal, he chose Poldhu Cove, in southern Cornwall. For the western terminal he chose the sand dunes on the northern end of Cape Cod, MA, at South Wellfleet.

The aerial systems were composed of 20 masts (Figure 12.22a), each 61 meters high, arranged in a circle 61 meters in diameter. The ring of masts was supported an inverted cone of 400 wires, each connected at the top and the bottom. Vyvyan, the Marconi engineer who worked on the 1901 experiment,

---

[1] Interested readers are referred to an interesting, but short, discussion by Fessenden in an article entitled "Wireless Telegraph", *Electrical World and Engineering*, Jun 29, 1901; and for a modern view on the subject refer to [34].

Figure 12.22a. Poldhu aerial systems: first inverted-cone aerial.

when shown the plan did not think the design to be sound [36]. Each mast was stayed to the next one and only to ground in radial directions. He was overruled, construction went ahead, and both aerial systems were completed in early 1901. However, before testing could begin, catastrophe struck. The Poldhu aerial collapsed in a storm on September 17th (Figure 12.22b), and the South Wellfleet aerial suffered the same fate on November 26, 1901.

At Poldhu, Marconi quickly put up a fan aerial of 54-wires spaced one meter apart at the top, and suspended from a triatic stay stretched between these two-masts at a height of 45.7 meters (Figure 12.22c). The aerial wires were

Figure 12.22b. Poldhu aerial systems: the effect of bad weather on the first aerial.

arranged in a fan shape, connected at the top to the conducting cable stay and connected together at the bottom. According to the view of the GEC Marconi Company historians, there were 54-wires and the photograph has been touched up. Twelve wires was a practical number to sketch in, and "blobs" were added at the top for emphasis, but there is a more interesting story concerning this touched up photograph below.

Figure 12.22a is considered to be genuine, but also touched up (a bit), e.g. only 19 masts are shown rather than 20 known masts. Notice the material spread out on the grass to dry. Now look at Figure 12.22b. There is no material spread out on the grass to dry. This photograph was taken later. The photograph in Figure 12.22c shows the same material spread out to dry as in Figure 12.22a, and, the two masts supporting the fan aerial are exactly sited as originally sited and rigged. That is the other 17-masts have been painted out. On closer inspection of Figure 12.22c, one can see further detail that what is said is true. For example one can see the shadow of one of the 17-masts that was painted out in front of the white building in the background on the right. The photograph in Figure 12.22c was published during the working lives of many of the engineers involved, and it probably portrays a general impression of what the temporary fan aerial would have looked like (Pegram et.al, 1997 [34]).

Marconi's initial aerial at Glace Bay, NS (Table Head site, 1902) also had many wires (400) in an inverted square cone (Figure 12.23, aerial system on the left). When Marconi enlarged this aerial in 1905, the original aerial was reinstalled at the center of a very large horizontal umbrella wire system of top loading masts (hundreds of wires 335m long). This aerial was erected at the Marconi Towers site, Glace Bay, NS. As mentioned earlier, Fessenden foresaw the need for a tall mast, and his mast for practical heights would be electrically short. In 1905, he erected two insulated-base umbrella top-loaded masts, 128

Figure 12.22c. Poldhu aerial systems: a temporary fan aerial (credit BAE Systems Marconi Research Centre, Chelmsford, Essex).

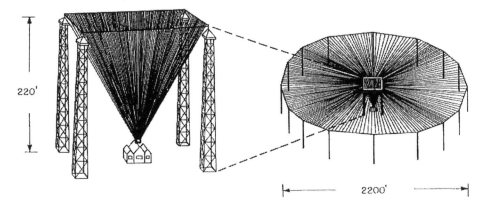

220'

2200'

Figure 12.23. Marconi's antenna systems used at Table Head, Glace Bay, NS (on the left); and (on the right) at Marconi Towers, Glace Bay, NS.

meters high, at Brant Rock, MA (Figure 12.24a), and at Machrihanish, Scotland (Figure 12.37b). The mast was a riveted steel tube with an internal diameter of 0.9 meters. It had inside ladder rungs for access to the top-loading structure that were designed to support the umbrella top-loading conductors. The top-loading consisted of four or more wires (cages) strung obliquely to ground from the top of the mast and insulated from ground at the bottom end. That is, the wire was broken by an insulator at some height above ground. Zenneck [52] in his book published in 1905 said that there were eight cage-like (diameter 1.2 m) umbrella wires 91 meters long. Fessenden's patent (Figure 12.24b), "Aerial for Wireless Signaling" US Patent 793,651, July 4, 1905; suggests that there were four. Photographs of the Brant Rock and Machrihanish stations (discussed in detail in section 12.7.1) certainly do not show cage-like umbrella wires (just wires very faintly), but the postcard in Figure 12.37a (date unknown) shows two (perhaps four) cage like conductors – clearly a touched up photograph.

Fessenden's 1905 patent describes an even more complicated antenna system in which the four sections of the mast could be, and may have been, insulated from each other so that inductors could be inserted to change the electrical length of the mast (perhaps connected as a continuous conductor). Fessenden certainly had ideas and presently used them to improve the performance of an electrically short mast, *viz.* inductively loading a mast with umbrella top-loading. The radiation efficiency that Fessenden achieved was high compared with the Marconi towers multi-wire aerial systems, since the radiation efficiency of electrically short antennas varies as height squared; and Fessenden's antenna structure was certainly a better design for an aerial system operated in a coastal area subject to icing conditions in winter months. Marconi must have had terrible problems with icing on his multi-wire antenna systems at Glace Bay, NS (Figure 12.25).

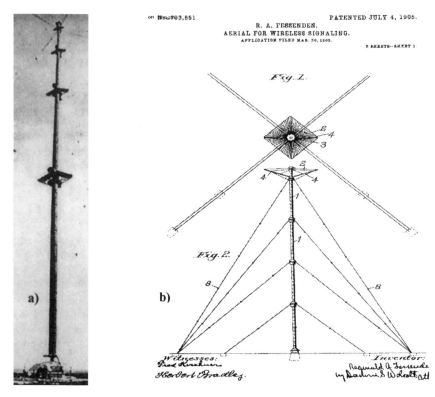

Figure 12.24. a) (left) Fessenden's Brant Rock, MA mast aerial used for his transatlantic experiments in 1906. **b)** (right) A sketch copied from Fessenden's 1905 patent.

Figure 12.25. The inauguration of the Glace Bay, NS station on December 21, 1902. Note the chilling backdrop against which Marconi and his staff are pictured (after Pam Reynolds [53]).

The photograph in Figure 12.26 shows an important difference in public relations between Marconi's work and Fessenden's. Marconi was heaped by international honors, titles, and personal recognition by exalted personages of the day. Conversely, Fessenden, in this time period, was ridiculed by journalists, businessmen, and even by scientists, for believing that voice could be transmitted without wires.

12.7   MARCONI'S FIRST TRANSATLANTIC EXPERIMENT

On December 12, 1901, signals from a high power spark transmitter located at Poldhu, Cornwall, were reported to have been heard by Marconi and his assistant George Kemp at a receiving site on Signal Hill near St. John's, Newfoundland. For this reception experiment Marconi used a kite supported wire aerial, an untuned receiver, a detector of uncertain performance, and a telephone receiver. The signals, if heard, would have traveled a distance of 3500 kilometers. Even at the time of the experiment, there were those who said, indeed there are some who still say, that he misled himself and the world into believing that electrostatic and atmospheric noise crackling was in fact the Morse code letter 'S' (dit-dit-dit). Precipitation static, due to electrostatic charge built up on his antenna system, a wire bobbing in as marine environment, would be expected; and such discharges tend to occur in groups of short pips.

12.7.1   The Poldhu Station

The aerial system used for this experiment, a temporary fan, previously described (Figure 12.22c), was driven by a curious two-stage spark transmitter (Figure 12.27a). There were many problems in getting it to work at the high power levels

Figure 12.26. A better to remember occasion for Marconi, when the Prince and Princess of Wales visited Poldhu (in 1903) with its newly designed aerial supports (after Pam Reynolds [53]).

Figure 12.27a. The two-stage spark transmitter system used for this experiment

desired (*cf.* the present day analysis by Thackeray [52]). Our principal concern here, in what follows, is the frequency generated by the Poldhu station. The frequency of the sender that Marconi used for his first transatlantic experiment has, by Marconi himself, been variously estimated.

Look at Figure 12.27b. The circuitry associated with the first spark gap (the one on the left of the figure) provides the power to operate the spark gap transmitter. This circuit has a strong influence on the achieved spark rate (and power), but the frequency of the sender is determined by the resonant response of the 0.037 microfarad (µfd) discharge capacitor together with the coupled impedance of the aerial circuit {the *aerial jigger* (Marconi's words)}.

The inductance values for the oscillation/aerial jigger transformer have long been debated since the original transformer is lost, there are no drawings,

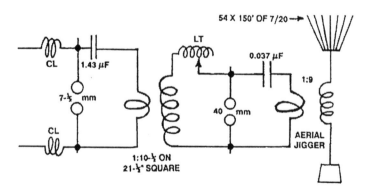

Figure 12.27b. Schematic (redrawn) showing component values, note the additional components CL (extra coils), and LT (a long adjustable tuning inductance of forty turns using 40 feet of wire), after Thackeray [52].

and reports about it differ. Fleming's notes record that the primary was two-turns paralleled, and the secondary had nine-turns; but Entwistle [55] said there were seven-turns. Possible limits for the size of the windings range from 45 to 60 cm on a square former. Thackeray [52] has postulated on possible values for this transformer, based on measurements by George Grisdale in 1985. Grisdale on a facsimile of the Science Museum's 20.3 cm open jigger, measured primary and secondary inductances of 1.05 and 18.8 microhenries (μH), and a turns ratio of 1:7. Scaling directly to a 50.8 cm square frame, multiplies all reactances by the ratio 50.8/20.3, which yields inductance values of 2.6 and 47 μH.

The writer has experimentally modeled (scale factor 75) a 54-wire fan monopole, wires connected to the triatic, and measured the self impedance frequency response (resistance and reactance) of this aerial on a large elevated ground plane. The experimental data, plotted as open and closed circles labeled Xa and Ra in Figure 12.28, exhibit resonances (the frequencies marked are full scale values) at 935 kHz and 3.8 MHz; anti-resonances at 2.4 MHz and 4.8 MHz; and approaching an anti-resonance between 7 and 8 MHz.

But the self impedance frequency response of the aerial itself is of interest only if the aerial is the oscillator (a plain aerial transmitter). Since the aerial is tuned (Figure 12.27b), to determine the operational impedance, we need to include the impedance of the secondary of the oscillator/antenna circuit transformer; it is this impedance that connects the aerial to ground, see discussion to follow.

The writer has numerically modeled, using a numerical electromagnetic code, NEC-4D, a 23-wire fan, included realistic sag for the triadic wire. In Figure

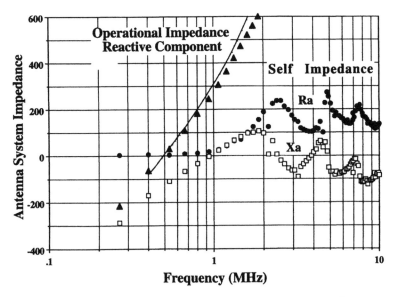

Figure 12.28. Impedance vs. frequency for Marconi's temporary fan monopole December 1901 (experiment and simulation)

12.28, the computed reactance versus frequency, operational value, i.e. including the 47 μH reactance of the secondary of the oscillator/aerial transformer, is plotted as a continuous line. The corresponding values inferred from the measured results are plotted as solid triangles. Clearly, there is a good agreement between simulation and experiment. The resonant frequency of the tuned aerial system is 500 kHz. And, for the tuned aerial system there can be no high frequency resonance, except for a spurious resonance at some high frequency corresponding to a self resonance of the coil (a high loss resonance at a frequency determined by the distributed capacity of the coil).

Note that the antenna circuit itself was not separately tuned (Figure 12.27b), but resonance was established. Fleming tuned the oscillator circuit by varying the value of the discharge condenser, a parallel series connection of 24 condensers, to maximize the RF aerial current. Vyvyan [36] has given the value of the discharge condenser for resonance as 0.037 μfd. Using the postulated value for the inductance of the primary of the jigger (2.6 μH), the resonant frequency of the oscillatory circuit is 511 kHz.

Since the resonant frequencies of the oscillatory circuit and the antenna circuit (according to the writer's modeling studies) are closely the same, the writer postulates that the radiation would be a broadband signal centered on a frequency of about 500 kHz.

Marconi himself had been evasive concerning the frequency of his Poldhu transmitter. Fleming in a lecture that he gave in 1903 said that the wavelength was 304.8m or more (984 kHz). Marconi remained silent on this wavelength, but in 1908 in a lecture to the Royal Institution he quotes the wavelength as 365.8m (820 kHz). And in a recorded lecture from the early thirties, he says the wavelength was approximately 1800m (166 kHz) and the power about 15 kW (see Bondyopadhyay [54]).

12.7.2   Reception on Signal Hill

For his transatlantic experiment, Marconi decided to set up receiving equipment in Newfoundland. In December 1901, he set sail for St. John's, with a small stock of kites and balloons to keep a single wire aloft in stormy weather. A site was chosen on Signal Hill, and apparatus was set up in an abandoned military hospital. A cable was sent to Poldhu, requesting that the Morse letter S be transmitted continuously from 3:00 PM to 7:00 PM Cornwall time. On December 12, 1901, under strong wind conditions, a kite was launched with a 155 m long wire (Figure 12.29). The wind carried it away. A second kite was launched with a 152.4 m wire attached. This kite bobbed and weaved in the sky, making it difficult for Marconi to adjust his new syntonic (tuned) receiver which employed the Italian Navy coherer. "Difficult", one can certainly agree with, but how he determined that "he could or could not tune it" is a mystery; because of this "difficulty", Marconi decided to use his older, untuned receiver. History has assumed that he substituted his metal filings coherer previously used with the untuned receiver for the newly acquired Italian Navy coherer, but Marconi never

Figure 12.29. Marconi and team launching kites on Signal Hill, Newfoundland.

said he did (Phillips [35]). Marconi referred only to his use of three types of coherers, as if there were something to hide: *"— one containing loose carbon filings, another designed by myself containing a mixture of carbon dust and cobalt filings, and thirdly the Italian Navy coherer containing a globule of mercury between two plugs".*

Clearly, there are uncertainties concerning the type and reliability of the detector used with the untuned receiver. Vyvyan gives rather a clear account in his book [36]:

> *It was impossible to use any form of syntonic (tuned) apparatus and Marconi was obliged to use the next best means at his disposal. He therefore used a highly sensitive self-restoring coherer of Italian Navy design, simply connected with a telephone and the aerial, and with this simple apparatus on Thursday 12 December, 1901, he and one of his two assistants (reportedly) heard the faint 'S' signals.*

Vyvyan is quite definite in his statement that it was the Italian Navy coherer that was used. Vyvyan's account casts a different light, too, on technical matters, as it is clear that if the letter 'S' was heard, it was due not to coherer action, but to the unwitting and unrecognized use of a simple diode rectifier. Thus we conclude that if Marconi did hear a signal, his mercury coherer, by chance [33] happened to be in the rectifying mode just at the time when signals were heard.

The unreliability of the device is made clear by present day experiments [Grisdale, previous cite, and references 34,35], and by Marconi himself. In his June 1902 lecture delivered to the Royal Institution he stated:

> *These no-tapped coherers [Marconi always referred to the device as a type of coherer] have not been found to be*

sufficiently reliable for regular commercial work. They have a way of cohering permanently when subject to the action of strong electric waves or atmospheric electrical disturbances and have an unpleasant tendency toward suspending action in the middle of a message.

Despite the crude equipment employed, Marconi and his assistant George Kemp convinced themselves that they could hear, on occasion, three clicks more or less buried in the static. And clicks they would be, because of the very low spark rate of his two-stage spark transmitter (estimated to be only a few sparks/sec). Marconi wrote in his laboratory notebook: *"Sigs at 12:30, 1:10 and 2:20 (local time)."* This notebook (Figure 12.30) is in the Marconi Company archives and is the only evidence today that the signal was received. It is ironic that the very low spark rate of his Poldhu transmitter at that time was compromised by Marconi himself, when in Newfoundland he put a telephone receiver to his ear to listen for three dots from Poldhu. Fleming must certainly have known about Tesla's widely published 4-circuit arrangement (2-circuits in the transmitter (a single stage spark) and 2-circuits in the receiver [20]). A higher spark rate could have been more easily achieved with Tesla's transmitter, and would have given the received signal, assuming that the mercury detector was working like a rectifier, a buzz-buzz sound, or a more musical sound, depending on the spark rate, rather than just a click, indistinguishable from an atmospheric or an electrostatic discharge.

12.7.3   Reception on a Ship

A little later, in February 1902, when Marconi was returning to North America on the SS *Philadelphia*, using a tuned shipboard aerial, Marconi received signals using his filings coherer from the same sender up to distances of 1120 km by day and 2500 km at night. These distances are remarkable considering the receiving apparatus he used.

Figure 12.30. A copy (of copy) of pages from Marconi's diary, December 12, 1901 (after Pam Reynolds [53])

12.7.3.1. The Enigma. Today we know that signals, depending on frequency used, can indeed travel across the Atlantic, and far beyond. But in 1901, anyone who believed that they could, and did, believed so as an act of faith based on the integrity of one man – Marconi.

If 500 kHz was the frequency of the sender, the tests took place at the worst time of day, because the entire path would have been daylight, and the daytime skywave would be very heavily attenuated, even though it was a winter day in a sunspot minimum period, and there were no magnetic storms at the time or before the experiment. From a knowledge only of propagation conditions, reception on Signal Hill is consistent with the observed limiting ranges of reception on the ship, only if the untuned land-based receiver was 47–55 dB more sensitive than the tuned receiver on the ship, see Table 12.1. This is unlikely. Table 12.1 establishes the sensitivity of the tuned receiver on the ship.

Historians have speculated that the transmitter might also have radiated a high-frequency signal as well, *cf.* Ratcliffe, 1974 [57], and Bradford, 2003 [58], since an HF signal would have been more suitable for transatlantic communications. The paper by Ratcliffe gives the early history and an analysis attempting to interpret what might have happened. This paper is very interesting to read, but his simulation of power radiated is incorrect. Ratcliffe's equivalent circuit for the aerial (a capacitor in series with a resistance) is not an equivalent circuit of a real aerial. Certainly an L-C-R circuit can represent the impedance characteristics of an aerial, but all three components are a function of frequency. And so, an equivalent lumped-circuit that exhibits a frequency response similar to that for Marconi's fan aerial is not easily determined (beyond the kin of the writer). The writer (in the time period mid 1990s) tried to develop an equivalent

Table 12.1.

Distance (kilometers)	Field Strength dB μV/m* (for 1 kW radiated)		
	Groundwave	Skywave	Total Field
Limiting Ship Ranges 1120 km Day 2500 km Night	24 −23	16 15.4 ± 4	23 ± 3.6 15.4 ± 4
Signal Hill, Newfoundland 3500 km Day	—	−32	−32

Computations are based on CCIR (now ITU Radiocommunications)
Report 265-7, Reports of CCIR 1990, Annex to Volume VI, Propagation in Ionized Media, Geneva, 1990, pp. 212-229.
Note: The ± symbols should not be considered to be an indication of accuracy, but results from the addition and subtraction of field strength vectors. The first and second hop skywaves at the distance of 2500 km at night are of comparable amplitude.

circuit for the transmitter-antenna system that Marconi used, including a realistic equivalent circuit of the antenna, to be analyzed by Desmond Thackeray, who had developed a computer program to evaluate the sine-wave-amplitude transfer response of circuits to a step-function voltage source, but that is a story for another day.

But what is clear, is that the tuned aerial system could not have exhibited a sharply resonant HF response, and so produce appreciable HF EM radiation.

12.7.3.2 So what might Marconi have heard?   The plane aerial receiver, previously shown in Figure 12.10, for the patented Italian Navy coherer receiver, shows the detector in parallel with the load (here the load is the impedance of the earphone). Grisdale [previous cite], in his testing of the device tried this arrangement, but it gave less output then the normal series arrangement. So this is another uncertainty about the Marconi experiment. Suppose that he was using the series arrangement. Some historians say, believing that Marconi was an experienced listener, that he could recognize signal from noise. But, Marconi had never before used a rectifier detector, and certainly had not listened with such a device, connected to a long aerial wire on a high hill in a Marine environment. He only acquired the Italian Navy coherer in the fall of 1901, and his Poldhu aerial came down in a storm on September 17, 1901. Unbeknownst to Marconi, and probably to most historians who have written about this experiment, is that he would have heard a natural atmospheric noise background in the frequency range that his ear could hear, and electrostatic discharge clicks as well, whether his temperamental detector was working or not. EM signals picked up by the aerial in this ULF (ultra low frequency) band would induce currents in the wire aerial, and so activate the earphone, which could be heard by the listener. In fact a louder sound would be heard if the device was not working, behaving like a low impedance or a high impedance, depending on whether the earphone was in series or shunt with the device, respectively

12.8   MARCONI'S STATIONS AT GLACE BAY

Fleming, who built the Poldhu transmitter used for Marconi's first transatlantic experiment in December of 1901, devised a curious two-stage spark transmitter. Undoubtedly, Marconi dictated this design approach. Fleming experienced such great difficulty in obtaining high power and an acceptable spark rate, that Marconi decided, for his Glace Bay and Clifden, Ireland stations, to return to a single stage spark transmitter; and, incredibly, he decided to use DC (high voltage direct current generators and a battery bank), rather than AC, to power the spark transmitter.[2]

Once initiated, a DC driven spark would of course continue, and a wicked current it would be, because of the low internal resistance of a battery source. To interrupt the spark and increase the spark rate, Fleming devised a rotary spark transmitter. Each time a pair of studs on the rotating disk passed between the terminals of the gap, sparks passed (Figure 12.31).

Marconi's Clifden and Glace Bay stations must have been something to behold. The power supply was DC, three 5000 volt DC generators in series (Figure 12.32), and a 12 kilovolt battery — six thousand 2 VDC 30 AH batteries — a room full of batteries that may well have been the largest battery in the world (Figure 12.33) [2]. The batteries were suspended by insulators from the ceiling. The electric power input of a few hundred kilowatts was provided by power houses burning coal at Marconi Towers, and peat, at Clifden.

The discharge capacitor itself was also almost beyond belief. It comprised a room full of steel plates, 1800 sheets of steel measuring 9 meters by 3.6 meters (Figure 12.34). This giant condenser was housed in a galvanized iron

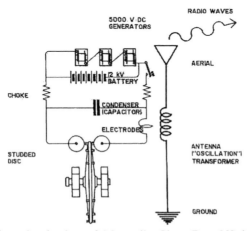

Figure 12.31. Schematic circuitry of Marconi's Glace Bay, NS high power spark transmitter

Figure 12.32. DC power source for the Glace Bay, NS station, four 5000 volt DC generators (one is a spare).

---

[2] The writer supposed that this decision was made because Fleming kept burning out alternators, even with protective resistors, during his struggle to achieve high power. No one, before or subsequently after, used a DC source for the high voltage spark.

Figure 12.33. View of the battery room, 6000 2-volt 30 ampere hour batteries (credit GEC Marconi Company, Chelmsford, Essex)

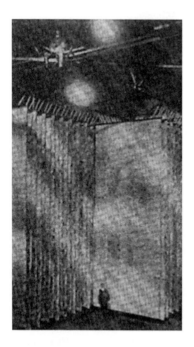

Figure 12.34.

(above) Top view of condenser room, 1800 sheets of steel measuring 9 meters by 3.6 meters

(left) bottom side view (after Pam Reynolds [53]).

clad building 107m long, 23m wide, with a height of 10m. The sheets were suspended 30cm apart, attached to porcelain rod insulators. When completed the condenser was tested up to 150 kilovolts.

Marconi's spark transmitter is shown in Figure 12.35. The large wheel in the middle (photograph on the top), driven by an electric motor behind it, is a part of the rotary-gap discharger. When a pair of studs on the rim of the wheel (see Figure 12.31) passed between the two electrodes, hidden from view, high voltage sparks jumped the gaps. Two large turns of the primary circuit (oscillator circuit) transformer can be seen in the background. The photograph on the bottom shows more clearly the oscillator/antenna circuit transformer (obviously the turns of the coil are fabricated differently). Note that the primary 'turn' was a large number of parallel wires wound on a doughnut shaped wooden form in a one turn spiral, so that all the wires were the same length. The secondary coil of the antenna transformer is to the right of the primary coil, and could be moved along its horizontal support to vary the mutual coupling.

How Marconi fine tuned such a colossal transmitter is a wonderment. The discharge capacitor that Fleming used at Poldhu in 1901 was a bank of paralleled capacitors mounted on shelves on the wall. He could select the number in parallel to maximize the aerial current. But, how one "tunes" a capacitor comprising sheets of steel filling a very large room, the writer does not know. The distributed capacity/inductance must have been large since the lead lengths were long.

Figure 12.35. Photographs of Table Head, Glace Bay, NS spark transmitter (credit Canadian Marconi Company).

12.8.1   Marconi's antenna systems

Marconi's antenna systems at Glace Bay were structures of many wires, as previously described and depicted in Figure 12.23. His Table Head antenna system was supported by 67m wooden towers; his Marconi Towers antenna used this same structure at the center of a very large capacity hat structure (670m in diameter).

The aerial system for the Clifden transmitter consisted of eight wooden masts each about 64m high. The masts, as for his earlier aerial systems, consisted of three wooden poles strapped and bolted together. The grounding system consisted of two sheets of heavy copper gauze, 183m long and 1.2m wide, buried in the ground in line with the aerial. About 61m of these strips were laid at the bottom of the lake. There were also steel wires in the ground connected to the steel frame of the condenser building. Marconi Towers in 1913 was using a frequency of about 37.5 kHz; Clifden transmitted on 54.5 kHz.

The receiving aerial systems originally employed two wires 640m long, supported at the tops of the transmitter aerial. They were later replaced by much longer wires at a remote location: the Louisbourg, Nova Scotia, receiver aerial wire was approximately 1000m long, supported by six 100m high towers; the Clifden receiver house was located in a disused quarry across the lake from the generator house, and comprised four wires each 120m long.

An AC generator and voltage step-up transformer should have been used, rather than DC generators and batteries (a room full of batteries). The spark current using batteries could be enormous, since batteries have a very low internal resistance, and so a great deal of power was probably consumed in the spark. The noise produced by the spark was so great that employees wore ear protection, and visitors to the station had to cover their ears. The noise could be heard to a distance of several kilometers around the stations, and the continuous flashing resembling lightning, could be seen, particularly at night. Undoubtedly the antenna system itself flashed under certain weather conditions.

Burning of the ball terminals and knobs, and heat dissipation clearly was a problem when Marconi ran his stations at full power (500 kW). The problem was solved initially by immersing the contacts in oil, and then by using rotating ball dischargers with streams of air from blowers to the spark, on the two sides of the studded wheel.

But Marconi seems to have liked using DC, and subsequently he devised a spark transmitter that (so he said) depended on a DC power source. The final version of his rotating disc discharger is shown in Figure 12.36. Here the middle disc is a part of the circuit, not merely an isolated element that initiates and breaks the spark (refer to the spark transmitter of Figure 12.31). Suitable brush or rubbing contacts made connection to the middle disc, and to the rotating ball dischargers. Initially the middle disc was smooth and, according to Marconi, the sparks between **A** and **C'** and between **A** and **C"** never occurred simultaneously, since otherwise the center electrode between the two capacitors **K** would not been alternatively positive and negative. The condenser **E** therefore, according to Marconi, discharged and charged alternately in reverse directions so

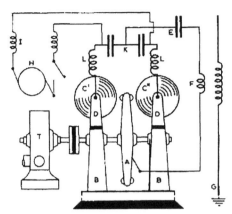

Figure 12.36. Final version of Marconi's rotating disc charger (1908), which was intended to simulate a CW like signal.

long as energy was supplied to the condensers **K** by the generator **H**. The signal generated was more or less continuous, or slightly damped oscillations.

But the sparking was irregular, and the spark duration must have been long, since the extinguishing of the spark depended solely on the rapid motion of the disc **A**, driving away the ionized air. To generate regular discharges, copper knobs or pegs were again used on the rotating disc to make and break the spark at regular intervals (timed spark), and so keep the received signal audible. But if the studs on the wheel were directly opposite each other as shown, sparks on both sides of the disc would occur at the same time. The rotating disc (in the view of the writer) must have had the studs on the wheel alternately staggered.

Marconi's timed spark transmitter was the end of the spark era, since others were using Poulsen arc transmitters or the Alexanderson-Fessenden alternator. Marconi stayed with spark to the end of king spark. Clearly, Marconi was struggling to generate a more continuous wave train of damped waves. And, certainly his Glace Bay, NS and Clifden, Ireland, transmitters can only be described as colossal "brute force" designs.

12.9 FESSENDEN'S BRANT ROCK STATION

Fessenden's technology and circuit arrangements were very different. His early work (1900–1903) was dominated by his interest in transmitting words without wires. But the news of Marconi's attempts to achieve transatlantic wireless telegraphy transmission had caught the attention of the world. And since the development of his HF alternator was taking longer than anticipated, Fessenden set his mind to make a more CW-like spark transmitter. This led, as we have said earlier, to his development of the *synchronous rotary-spark-gap* transmitter.

Fessenden's Brant Rock and Machrihanish stations employed a three-phase AC generator, driven by a steam engine. Figure 12.37a shows a general view of the Brant Rock Station, showing the smoke stacks for the steam engine that provided the power to operate. Note: This is a retouched postcard,

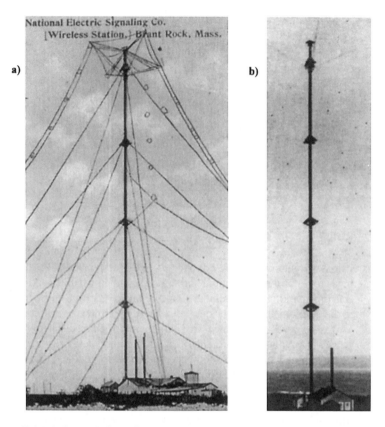

Figure 12.37 a) General view of Brant Rock, MA station, date not known but probably after 1906, showing the shore in the forefront. Notice the tall smoke stacks used by the steam engine to power the station. The Brant Rock tower might be standing today if it was not cut down during WW1, since the United States did not want this landmark visible to enemy ships at sea. For comparison **b)** shows the station at Machrihanish, Scotland

photograph showing tower with additional top loading, compare with Figure 12.24, the downward sloping active umbrella wires. The Machrihanish, Scotland station had the same mast, probably the antenna of Figure 12.24, perhaps with wire not cage top loading, and the buildings were smaller.

Fessenden's transmitter, the synchronous spark gap transmitter previously described, did not occupy several rooms as did the Marconi stations. It was installed in one room (Figure 12.38). The receiver and control equipment, using the same aerial for transmitting and receiving, was located nearby in an adjacent room Figure 12.39). When the Machrihanish station was installed and working, in January 1906, Fessenden is accredited with having established the first two-way wireless link across the Atlantic. The station operator was in simplex communication with the distant station (back and forth communications in real time).

Figure 12.38. Brant Rock transmitter room (see also Figure 12.7). The operator is Guy Hill (credit National Museum of American History, Smithsonian Institution, Washington, DC).

Figure 12.39. Pannill operating the Brant Rock Station (after Ernie DeCoste, 1992).

12.10   TRANSATLANTIC EXPERIMENTS IN THE FIRST DECADE OF THE TWENTIETH CENTURY

The struggle to achieve reliable transatlantic communications began with Marconi in 1902. The writer highlights below a few claimed communication achievements, and comments on what history tells us concerning the reliability and quality of some of these early communications experiments.

12.10.1 Marconi

- February 1902: East-to-west transmission, Poldhu, Cornwall to the *SS Philadelphia* returning to North America. Frequency 500 kHz.
- October 1902: First East-to-West transmission between Poldhu, Cornwall to Italian cruiser *Alberto* anchored in Sydney Harbour, NS. Frequency 272 kHz.
- December 5, 1902: First West-to-East transmission between Glace Bay, NS and Poldhu, Cornwall. Frequency 182 kHz.
- December 15, 1902: First Canada/UK transatlantic radio *message* sent from Glace Bay to Poldhu. It was a press message from a *London Times* correspondent to his home office.
- January 18, 1903: First American/British transatlantic radio *message* sent from South Wellfleet, MA to Poldhu, Cornwall. Message was from President Roosevelt to King Edward VII.
- October 17, 1907: The Transatlantic stations Clifden, Ireland and Glace Bay, NS, were opened for limited public service. Frequency about 45 kHz.
- February 3, 1908: The Transatlantic stations were opened for the general public.

12.10.2 Fessenden

- January 10, 1906: First *2-way transatlantic radio telegraphy* transmissions between Brant Rock, MA and Machrihanish, Scotland, and the beginning of the first reliable nighttime transatlantic communications. Frequency about 80 kHz.
- November 1906: First transatlantic *radio telephony* transmission (one way) between Brant Rock, MA and Machrihanish, Scotland. Frequency about 80 kHz.
- October 1, 1906 to December 5, 1906: First long period of *uninterrupted* transatlantic wireless telegraphy communications between Brant Rock, MA and Machrihanish, Scotland.
- December 24, 1906 (Christmas Eve): World's *first wireless broadcast* (voice and music) from Brant Rock, MA, received by ships at sea in the North and South Atlantic. Frequency about 80 kHz.

12.11   ON QUALITY/RELIABILITY OF MARCONI'S TRANSMISSIONS

During the week of February 24, 1902, when Marconi was returning to North America on the *SS Philadelphia*, history tells us that he

> *...silenced the skeptics by producing visible proof of his success. At night the simple letter "S" received from Poldhu,*

3377 km away, motivated the Morse inker for all to see, and the first readable transatlantic messages, as distinct from signals, was recorded at a range of 2496 km.

Marconi was using his tuned coherer receiver.

But the reports of the quality of the reception differ. According to Marconi himself [see *McLure's Magazine*, April, 1902, pp. 525-527] reception was excellent, and we can see dots-and-dashes (copy of tape recordings at 2496 km was reproduced), that is "words" not just pips corresponding to the letter "S". But Simons, 1998 [24], who has studied these heritage tapes has written: *"I am intrigued that the certified tapes of messages that we have, do not contain any recognizable plain language, or code, unlike the earlier records of experiments".*

In spite of what is said above, the view of the writer is that signals were heard, some of the time. Ratcliffe 1974 [57] gives the distances to which signals were heard to be 1120 km by day and 2500 km at night.

An important observation is that Marconi discovered the day night change in signal strength, but he did not understand why; even much later [reference his Nobel Lecture, December 11, 1909], it seems he still considered the effect to be a local phenomena occurring during daytime in the vicinity of his receiving transmitting aerials.

It is clear that Marconi succeeded in sending messages across the Atlantic prior to 1907, but the reliability and quality of his transmissions is a matter of conjecture. And, prior to 1907 non-Marconi stations monitoring the bands (operators at the Brant Rock station were very often listening) could not judge how well Marconi's one-way transmissions were being received. In the fall of 1907 Marconi's transmissions were 2-way in real time.

With reference to the transmission of the message from President Roosevelt to King Edward in January 1903, according to Fessenden [59] there are doubts about the reliability of the reception of the message. After receipt at Holland House the message was transmitted by cable (at 13:34 local time). Then the message was transmitted one-way (Marconi used the word "dispatched") by Marconi's station at South Wellfleet. Marconi then (without knowing whether his message had been received) sent a second cable (at 18:36 local time): *"President's message to King dispatched successfully from Cape Cod to Poldhu direct and also via Cape Breton you may publish the text immediately".*

Fessenden notes that the cablegram makes no mention whatever of any acknowledgement from Poldhu, and that Sig. Marconi cabled the London Office of the Marconi Company the necessary information that the message had been "dispatched", so that the text could be published as having been received. Marconi, as far as the writer knows, never controverted Fessenden's published correspondence.

And further, it should be noted that in his Nobel Lecture, Marconi makes specific reference to the fact that in the spring of 1903 transmission of press messages from America to Europe *were attempted* and that for a time the *London Times* published, during the later part of March and early April —. No mention is made of the message from Roosevelt to King Edward, in January 1903.

Concerning the October 1907 attempts to reliably bridge the Atlantic, Marconi immediately received glowing press releases, lauding his success.

> Commercial telegraphy had been inaugurated across the Atlantic Ocean — the operators were kept busy (so the media reports said) sending and receiving messages at a reported rate of 30 words/minute, compared with 22 words/minute for Atlantic telegraph cable.

But now Brant Rock operators could copy both the Clifden and Glace Bay stations, and so for the first time, the Brant Rock operators could listen in and judge how well Marconi's transatlantic stations were copying each other. Apparently Marconi's station operators were not hearing each other all that well. All messages had to be sent at least twice, usually more than twice, and sometimes after sending the same message six times, a "fill-in" for a few words missed was still required. And frequently reception was impossible due to atmospheric noise.

Fessenden's communication [59] clearly tells us that he took exception to the glowing success stories Marconi was receiving in the press, probably because of the negative media reports he had received a year earlier on the performance of his transatlantic stations (see below). He noted that the wave shape of Marconi's stations "would not be considered up to date, being much broader (in frequency) than advisable for cutting out interference and atmospheric disturbances". Fessenden could not use the full selectivity capability of his receiver at Brant Rock. After a careful study of these published reception reports it is clear that (in the autumn of 1907) Brant Rock operators were better able to copy messages than Marconi station operators could.

It is clear that Marconi was still struggling in 1908 to achieve reliable and private transatlantic communications. It is interesting to read a letter written on March 19, 1909 to Hon. Chauncey M. Depew, United States Senate, Washington, DC; signed by the five members of The Junior Wireless Club (now The Radio Club of America). The thrust of the letter was to comment on a proposed bill before the senate that would in effect restrict the use of the airwaves by radio amateurs, because of wrongly presumed malicious interference caused by radio amateurs. I quote from a part of that letter, which can be found in the *Seventy-Fifth Anniversary Diamond Jubilee Year Book of The Radio Club of America*, 1984:

> You probably have heard of the tests made last year between Glace Bay, NS and Clifden, Ireland, when the National Signaling Company (Fessenden's Brant Rock station) picked up the messages, which Marconi, on the test, was unable to deliver between his own stations, from both Glace Bay and Clifden, Ireland, in spite of the fact that the Marconi Company kept up a constant (purposely jamming) interference of dash-dash-dash, from their Cape Cod Station for 48 hours without interruption. But the National Signaling Company paid no attention to such interference and picked up all the messages,

which Marconi was unable to exchange between their own stations, and all these messages were handed over to Lord Northcutt at the Hotel St. Regis. ...

12.12   ON QUALITY/RELIABILITY OF FESSENDEN'S TRANSMISSIONS

Fessenden was not concerned with sending commercial traffic (but he certainly wished he could have been). His wireless transatlantic messaging was concerned with quality of reception, improving the operations of the Machrihanish station, sending personal messages, etc. Much has been written and said about the 1906 radiotelegraphy link between Brant Rock, MA and Machrihanish, Scotland. The 1906 year-end review by *Scientific American* carried the comment that NESCO's work on transatlantic communications was futile since commercial-use-capability was not reached.

In reply Fessenden had a letter published in the January 19, 1907 issue of *Scientific American*, which stated:

> *The work of NESCO on transatlantic telegraphy is so far from having been futile, that uninterrupted communications, with the exception of one day, was obtained between Scotland and Mass. from 1 October 1906 to 5 December 1906. Preparations were being made for placing these stations on a commercial basis when the tower at Machrihanish fell, owing to a defective joint in one of the guys made by an expert engaged from a Glasgow firm. ...*

Fessenden struggled for years to set up stations for wireless telegraphic messaging. He tried and tried to obtain permission to set up stations in Canada, Great Britain, Newfoundland, Sable Island, Jamaica and other West Indian Islands, Cuba, France, Spain, Australia and New Zealand, and Brazil. Excepting for Great Britain permission was refused, or no reply to his request was received. He certainly never got a transatlantic license from Canada. Only Marconi was licensed to erect towers in Canada and install radio equipment in Canada – a senseless government regulatory ruling that held back the competitive development of wireless in Canada for more than two decades.

12.13   MARINE WIRELESS COMMUNICATIONS

So as not to restrict his Company's future to one front only, Marconi decided to exploit the field of communications with ships at sea, a very fortunate decision, since there was clearly an urgent need for ships to be able to communicate with each other, and with shore stations for commercial and safety purposes. For further information see "Wireless Grows up at Sea", a chapter in a book published by the International Telecommunications Union, 1965 [60].

In order to monopolize the field Marconi decided, in 1900, to lease his apparatus rather than sell it outright, and he established schools to train operators to operate his equipment. This strategy did not work [61]. Competition developed in Germany (Telefunken Corporation), and in the United States {American de Forest and its successor United Wireless, and with The National Electric Signaling Company (Fessenden)} and Marconi was forced to sell rather than lease apparatus to the natives of various countries. He nevertheless retained numerous restrictions. At the height of this debacle English Stations worldwide refused to communicate with ships without Marconi equipment. This absurd situation had to change and coastal stations opened up to all senders in 1908 — but the belief that ships using Marconi equipment had priority of the airways hung on.

Reference [60] documents three dramatic happenings at sea, which showed the whole world the real value of wireless communications. On January 23, 1909 the ship *Republic* sailing in a deep fog, about 280 km east of the Ambrose light, off the coast of United States, struck the Italian steamer *Florida*. The wireless distress signal CQD from the *Florida*, summoned help, and the *Baltic*, guided by the *Republic*, saved all 1700 souls from the two ships.

Another landmark in the history of wireless communications at sea came during the visit of King George V and Queen Mary to India in 1911. It was wireless telegraphy that kept the King in touch with events at home, and even a Court Circular was dispatched on November 15, 1912 from Gibraltar.

But the most dramatic event of early wireless history at sea was the loss of the *Titanic*. She struck an iceberg on April 14, 1911, when on a Northern great-circle route to beat the record of an Atlantic crossing during her maiden voyage. The *Carpathia*, one of the rescue ships, heard the *Titanic* calling CQD and SOS at 11:20 PM. The ship's operator answered immediately, but it was not before daybreak that the *Carpathia* could reach the scene of the disaster. The last communications from the *Titanic* was at 01:25 AM on April 15: we are firing rockets, but there was no sign of a response after that brief terse message, The *Carpathia* was able to save 710 survivors, 1500 souls went down with the ship.

But there were some bitter lessons to be learned from the disaster. Several ships within radio reach did not know of the disaster, because they had no wireless. More distressing still was the story of the *California*, a small passenger vessel, also westbound, and in the evening hours of the 14th was only 31 km away. The Captain of the *California* had encountered the ice field, and her wireless operator sought to notify the *Titanic*, but the operator on the luxury ship was at the time exchanging messages with Cape Cod (Marconi's station, and the *Titanic* was using Marconi equipment). He told the operator of the *California* to "shut up", and keep out of the conversation.

The *California* was hove-to on the evening of 14th of April, because of the danger of the floating ice. But the Captain of the *Titanic* held his ship at full speed, 22 knots, because icebergs or no icebergs, his ship was unsinkable. The operator of the *California*, shortly after this brief unpleasant exchange with the *Titanic*, went to bed, after all he had been on duty for 16 hours, and the nearby presence of the *Titanic* with her powerful ship's wireless, made it impossible for him to use his wireless equipment. It was not before 4 AM, when he resumed duty that he learned of the awful disaster.

Of course this kind of thing was nothing new. As more and more ships had become equipped with wireless, more troubles began. Clearly there was an urgent need for International Regulations. But regulations were slow to come, not before the International Radio Convention of Washington, held in 1927.

12.14   WIRELESS TELEPHONY IS BORN

Fessenden realized from the very beginning of practical wireless communications (as we have said before) that to improve upon the Hertz-Henry damped wave spark generated transmission systems, with the Branley-Lodge-Edison bad contact (Fessenden's words) coherer detector system used by Marconi and others for the receiving wireless telegraphy signals, one needed a continuous wave signal. And, that for wireless telephony a CW signal was a necessity. Since there was no satisfactory means of generating a CW signal prior to 1903, his early work was concerned with trying to develop a more suitable receiver, which began in 1896, for Hertzian waves (CW or spark).

He knew that he needed a "continuously acting, proportional indicating receiver" (Fessenden's words). He tried dozens of methods, in the period 1896–1902, as we have previously noted.

In 1898 he was using a modified version (US patent 706,736 and 706,737, dated December 15, 1899) of Elihu Thomson's alternating current galvanometer (US patent 363,185, dated January 20, 1887). In the words of Fessenden, he describes how the ring of a short period Elihu Thomson oscillating current galvanometer rested on three supports, two pivots and a carbon block. A telephone receiver with a battery in series was used in the circuit with the carbon block. This primitive device must have produced small resistance changes associated with amplitude changes of the received RF signal, which were detected by the telephone receiver.

In November, 1899, while experimenting with this receiver, listening to a spark generated telegraphy signal, produced by a transmitter with a Wehnelt interrupter for operating the induction coil used for sending, he noted that when the sending key was held down for a long dash, that he could distinctly hear the peculiar wailing sound of the Wehnelt interrupter. This immediately suggested to him that by using a spark rate above audibility, and with some means to modulate, that is change the amplitude of the transmitted signal by speech, that wireless telephony could be accomplished. Recall that a method to generate CW was yet to be devised. So Fessenden decided to up the spark rate by a large factor, to better simulate a CW-like signal. Professor Kintner, who at that time was assisting Fessenden with his experiments, designed an interrupter to give 10,000 breaks/s. Mr. Brashear, a celebrated optician, constructed the apparatus, which was completed in January or February 1900.

But it was not before the fall of 1900 that this interrupter was used; the reason being that Fessenden was engaged in transferring his laboratory from Allegheny, PA, and in setting up new stations at Rock Point, MA, and on Cobb Island, MD.

It is clear that for his initial wireless telephony experiments in December 1900 that he was using a spark transmitter with the Kintner-Brashear interrupter, but the writer has found no mention of the type of receiver used. The detector must have been Fessenden's version of the oscillating current galvanometer, because, as noted above, he had little time to devise a better detector. Nor was the frequency for this first experiment mentioned, but since the transmission took place between two twin-mast aerial systems, on 15 m masts, 1600 m apart, the frequency could have been 5 MHz, probably much lower. The modulator for the spark transmitter was a carbon microphone inserted in the antenna lead, see Figure 12.40.

After many unsuccessful attempts, Fessenden was finally rewarded by success. Speaking very clearly and loudly into the microphone, he said: *"Hello test, one two, three, four. Is it snowing where you are Mr. Thiessen? If it is telegraph back and let me know".*

Barely had he finished speaking and put on the headphones, when he heard the crackle of the return telegraphic message. It was indeed snowing since Mr. Thiessen and Prof. Fessenden were only 1600 m apart. Intelligent speech by EM waves had been transmitted for the first time in the history of wireless. The received telephony transmission was described as *"words perfectly understandable"* except that the speech was accompanied by an extremely loud disagreeable noise due to the irregularity of the spark.

The writer using equipment similar to that used by Fessenden, except for the detector, has simulated the authenticity of that transmission [*www.hammondmuseumofradio.org/belrose-fessenden.html*, loc. cit.]

By the end of 1903 fairly satisfactory speech had been obtained by the more continuous arc method (more CW like compared with spark), but reception was still plagued by a disagreeable noise. The receiver in use at this time was

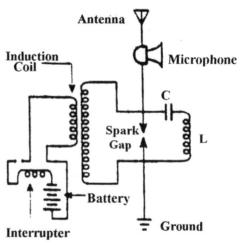

Figure 12.40. An early version of a transmitter for amplitude modulation shown with Fessenden's spark gap telephony transmitter, with a carbon microphone in series with the antenna lead

much improved, since it used Fessenden's electrolytic detector (Figure 12.12).

But Fessenden was still trying to develop an HF alternator giving a frequency high enough to be useful with practical antenna systems used at that time. Work on the HF alternator (Fessenden called this device a dynamo) was begun in 1900, but his instructions (in Fessenden's words) were not followed by the manufacturer, and when delivered in 1903 its highest operating frequency was 10 kHz.

A second alternator was delivered in 1905. A letter from the GE Company that built the machine stated that in the opinion of the Company, it was not possible to operate it above 10 kHz. So Fessenden scrapped this alternator, except for the pole pieces, and rebuilt the armature in accord with his design, in his Washington, DC, shop. By the autumn of 1906 he had developed a machine that gave him 75 kHz and a power output of half a kilowatt. Later machines gave frequencies as high as 200 kHz, and powers up to 250 kW. The problem had been solved. Fessenden could now transmit a pure CW wave.

His method of modulating his CW device, a HF alternator, was, as before, a carbon microphone inserted in the antenna lead. But with this apparatus he achieved important communication successes. In November 1906, on a night when transatlantic propagation was very good, Fessenden and his colleagues were conducting experimental wireless telephony transmissions between stations at Brant Rock and Plymouth, MA. Mr. Stein, the operator at Brant Rock, was telling the operator at Plymouth how to run the dynamo. His voice was heard by Mr. Armour at the Macrihanish, Scotland station with such clarity that there was no doubt about the speaker, and the station log confirmed the report.

Fessenden's greatest triumph was soon to come. On the 24th December, 1906, Fessenden and his assistants presented the world's first radio broadcast.

The transmission included a speech by Fessenden and selected music for Christmas. Fessenden played Handel's Largo on the violin. That first broadcast, from his transmitter at Brant Rock, MA, was heard by radio operators on board U.S. Navy and United Fruit Company ships equipped with Fessenden wireless receivers at various distances over the south and North Atlantic, as far away as the West Indies. The wireless broadcast was repeated on New Year's Eve.

This was the first radio broadcast. One can imagine the feelings of surprise to the lonely ship operators — accustomed to the cold colorless dot and dash of the Morse code — when music suddenly burst upon their ears, to be followed by understandable speech. Fessenden received many letters from operators on ships over the North and South Atlantic asking how it was done.

12.15   THE FIRST RADIO PROPAGATION EXPERIMENTS

There is little evidence that Marconi made any attempt to systematically investigate the characteristics of propagation at the frequencies he was using; except that he did make several interesting remarks, with no or meaningless interpretation as to the cause, in his Nobel Lecture December 11, 1909. He obviously must have had many interesting experimental observations since he was a careful experimenter. Fessenden on the other hand carried out many propagation experiments, beginning in 1899, and put forth hypothesis as to what he thought was causing the observed effects. The first record showing qualitatively the day-to-day variation of the intensity of transatlantic nighttime messages that were transmitted between Brant Rock, MA, and Machrihanish, Scotland, during the month of January 1906, is reproduced in Figure 12.41.

During that year Fessenden found that the absorption (attenuation) of signals at a given instant was a function of direction as well as distance, since on a given night the signals received by stations in one direction would be greatly weakened, while there would be less weakening of signals received by stations lying in another direction, and a few hours or minutes later the reverse would be the case. The measurements of signal strength on the path from Brant Rock to Machrihanish were found to have a definite correlation with variations of the geomagnetic field.

Experiments were made between Brant Rock and the West Indies, a distance of 2735 km, during the spring and summer of 1907. Frequencies in the band 50 kHz to 200 kHz were used. It was found that the absorption at 200 kHz was very much greater than at 50 kHz, and that messages could be successfully received over this path in daytime at the lower frequency. No messages were received in daytime with the higher frequency, though nighttime messages transmitted from Brant Rock at this frequency were officially reported as having been received at Alexandria, Egypt, a distance of 6436 km.

The fact that the experiments between Brant Rock and the West Indies were made during summer, and the receiving station was in the Tropics (high atmospheric noise levels), and the fact that the distance, 2735 km was practically the same as between Ireland and Nova Scotia was reported by Fessenden. After publication of the above results, Marconi in early October, 1907 abandoned his previously used higher frequencies and moved to even a lower frequency, 45

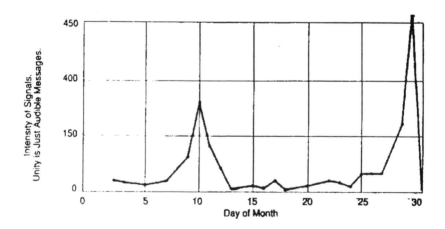

Figure 12.41. Curve showing variation of the intensity of transatlantic signals (path Machrihanish-Brant Rock, frequency about 80 kHz) for the month of January 1906.

kHz, and immediately succeeded in operating between Glace Bay, NS and Clifden, Ireland, a distance of more than 3000 km. The same messages were received at Brant Rock, MA, a distance of nearly 4825 km [61].

Certainly in the time period when these early propagation experiments were being conducted, little was known about the mechanism of propagation. For short to medium distances propagation was considered to be over the surface of the curved earth. In 1899, with the assistance of Prof. Kintner, a considerable number of propagation experiments were conducted, and published [61,62], in which a sliding wave theory, referred to by Elihu Thomson, was explained and illustrated. Fessenden later (1900) developed his mathematical model for such waves, guided over the surface of the ground, and showed that for transatlantic distances this (ground wave) signal would be negligible. The reception of signals across the Atlantic must therefore (in Fessenden's opinion) be due to a reflection from some conducting layer in the upper atmosphere. It should be noted that Kennelly and Heaviside were colleagues of Fessenden (Kennelly was a personal friend and colleague, and Fessenden corresponded regularly with Heaviside). And, certainly Fessenden was very familiar with their independent suggestions (in 1902) for the existence of such a conducting layer [63,64]. While Heaviside (in 1902) made the suggestion for a conducting layer "in the upper air", and that transatlantic propagation would in effect be due to a guidance by the sea on one side and the conducting layer on the other side, it seems that he thought (at that time) that transatlantic propagation was predominantly due to guidance by the conducting sea. Kennelly however (in 1902) was more specific. He gave a height for his conducting layer (about 80 km), and he suggested in some detail that long distance propagation was due to wave reflection in the upper atmosphere.

Evidence for the existence of such a reflecting layer was provided through the discovery by Fessenden in 1906 of what were called 'echo signals'. "On certain nights there appeared to be indications at the Brant Rock station that a double set of impulses from the Machrihanish station were received, one about a fifth of a second later than the other". Fessenden correctly interpreted that this delayed signal had traveled the other way around the great circle path [61]. Though this conclusion was severely commented upon at the time, we know now of the existence of around-the-world echo signals, and that such a conducting layer (the ionosphere) does exist.

12.16   FESSENDEN AND MARCONI, THE MEN

The writer has briefly outlined Fessenden's life history, and touched on his accomplishments in several earlier publications. For the interested reader it should be pointed out that the paper by William S. Zuill [65] is interesting to read. William Zuill's grandfather married Fessenden's wife's elder sister. This paper gives considerable personal detail about Fessenden.

Reginald Aubrey Fessenden was a most interesting radio pioneer, a man with a dynamic inspiring imagination. Chomping on his ever-present cigar (Figure 12.42), he would argue with anyone on any subject. With his razor sharp mind, his attempts to try to command all situations, his use of his classical

Figure 12.42. Fessenden about the time when he was working for The Submarine Signaling Company (after Erne DeCoste, private communications, 1992)

scholarship, and his lack of patience with slow minds, certainly did not agree with all who came face to face with him, or worked with him, but he could be charming. While he never graduated from university, his capacity for self-education was a remarkably successful substitute. Described by his contemporaries as "choleric, demanding, vain, pompous, egotistic, arrogant, bombastic, irascible, combative, domineering, etc.", when coupled with a notorious lack of patience, he could not help making waves constantly in every direction. When these characteristics emanated from a ginger-colored hair and bearded person, well over six feet tall, of large girth and wearing a flowing cape on his shoulders, topped with a seafarer's cap on his head, he must have commanded attention in any crowd [3].

Fessenden was clearly an outspoken skeptic of Marconi's claim to have received signals in Newfoundland from his sender in Poldhu, Cornwall, on December 12, 1901, and with reference to Marconi's wireless transmission in January 1903 of a message from President Roosevelt and King Edward.

His work involved with safety at sea (besides his contributions to the field of wireless, he invented SONAR, Sound Navigation and Ranging, used to measure the depth of oceans) won him the *Scientific American* Gold Medal in 1929. Other awards included the Medal of Honor of the Institute of Radio Engineers for his effort in that field, and the John Scott Medal of the City of Philadelphia for his invention of continuous wave reception.

He died on July 22, 1932 in his house by the sea in Bermuda. He was buried in St. Mark's Church cemetery, and over the vault was erected a memorial with fluted columns, on which is inscribed:

His mind illuminated the past
And the Future
And wrought greatly
For the Present

Beneath the scribed words, in the picture writing of the ancient Egyptians was: "*I am yesterday and I know tomorrow*". His son summarized his greatest achievements in one sentence: "*By his genius, distant lands converse and men sail unafraid upon the deep*".

The October 1932 issue of *Proceedings of the Institute of Radio Engineers* expressed the deep regret of The Board of Directors at the loss to radio engineering of this eminent pioneer (Reginald A. Fessenden) and constructive inventor.

Guglielmo Marconi was always a clear leader, but somewhat aloof (*cf.* Figure 12.43, which in the view of the writer is a typical pose). He was not cordial, in fact his manner was chilly and reserve. He always referred to his wireless systems in use not as Marconi systems, but to "our systems". He should be recognized for his organizing talents, having brought together a hundred contributing spectators, and detached discoveries, into a harmonious relation. The result of which was a system of wireless telegraphy, which was certainly susceptible to improvement, but it did obtain practical results.

He wore English dress, English in speech, and insisted on order of rank when his party sat down to a meal. He clearly enjoyed the large number of events, discussions, and awards given to honor him. It seems that every new event was followed by some public celebration, and newspapers of the period carried daily reports of his activities

In his Nobel Lecture, he said that he had never studied physics, or electrotechniques in a regular manner, but he was deeply interested in these subjects. He noted that he did attend one course of lectures under Prof. Rosa, at Livorno, and that he was fairly well acquainted (in that time period) with publications by Hertz, Branly, and Righi.

Figure 12.43. Marconi at Marconi Towers. James Holmes is receiving a wireless message [Public Archives of Nova Scotia].

He certainly had an unerring judgment in choosing men of brilliance to form the team with which he surrounded himself: men such as Dr. W. H. Eccles, Prof. J. A. Fleming, C. S. Franklin, A. Gray, G. Isted (who in 1991 wrote two reviews on Marconi [66]), Dr. Erskin Murray, H. J. Round, and R. N. Vyvyan. And while he did not inspire love, he did inspire a high degree of alliance and admiration in those who worked closely with him: G. Kemp and P. W. Pagent [24,51].

He devoted his entire life to wireless, and he himself put in a great personal effort in his work. International honors were heaped on Marconi throughout his life. News of Marconi's death, July 20, 1937, was carried to all parts of the world by wireless, and all wireless stations through-out the world shut down for two minutes of silence.

Marconi himself thought that his greatest contribution to the field of wireless was communications at sea. In 1932 he wrote: *"The greatest value of wireless, in my belief, is still demonstrated by its utility at sea"*.

12.17   CLOSING REMARKS

Fessenden made many contributions to the art and science of radio, including the first ever quantitative, scientific investigation of electromagnetic phenomena, wave propagation, and antenna design. His Brant Rock and Machrihanish umbrella top load vertical monopole antennas look like antennas used nowadays. His continuous waves, his invention of a new type of detector, which he called a liquid Barretter (an electrolytic detector), and his invention of the method as well as the coining of the word heterodyne, did not by any means constitute a satisfactory wireless telegraphy or telephony system, judged by today's standards. They were, however, the first real departure from Marconi's damped-wave-coherer system for telegraphy, which other experimenters were merely imitating or modifying. They were the first pioneering steps toward modern wireless communications and radio broadcasting.

Fessenden was at home in his laboratory, but out of his element when dealing with the business and political aspects of inventing. He never reaped, until late in his life, any financial reward for his many wireless inventions, and was compelled to spend much time and energy in litigation. His work with the United States Weather Bureau (1900–1902) came to an abrupt end in August 1902 over ownership of his patents. His partnership with two Pittsburgh millionaires, T. H. Given and Hay Walker began in September 1902 with the formation of the National Electric Signaling Company (NESCO), and collapsed in 1912 – the result of arguments about the direction the company should take, and again ownership of patents. Fessenden resented the financiers' efforts to meddle in his work, while they grew increasingly anxious for a return as their investment mounted.

In May 1912, Fessenden won a judgment of $400,000 from what remained of NECSO, but the company went into receivership before he could collect. Fessenden's patents were eventually purchased by Westinghouse in 1920 and then by RCA in 1921, prompting Fessenden to sue again. The legal suits that consumed much of his life finally came to an end on 31 March 1928, in an out-of-court settlement in which he received $500,000 from RCA, with $200,000 of this sum going to his lawyers.

Leaving Brant Rock (in 1912) did not impair Fessenden's creativity, but he made few major contributions to science and technology of radio after that date. During the period 1912–1921 he worked with The Submarine Signaling Company, where he developed the fathometer. During the 1920s, as radio exploded in popularity and new generations of inventors took on the task of improving it, the world's first broadcaster turned from the laboratory and devoted himself to research in ancient history. The products of these investigations were published privately under such titles as *"The Deluged civilization of the Caucasus"* and *"Finding a Key to the Sacred Writings of the Egyptians"* [65].

But, he continued writing for the popular press on his radio inventions, on the quality and reliability of his early communication systems compared with those of others, and as well on his own views concerning propagation [48,62].

It is a wonder that Fessenden was able to withstand, mentally and physically, the barrage of negative events that befell him, and yet continue to invent. The long-term grinding dissention however took its toll. But for the constant support of Helen, his wife, he might not have reached the year of 1932, when he passed away in Bermuda from a heart attack, on July 22. Helen in fact wrote a book on Reginald [67].

Marconi was not an inventor of wireless, but it was through his great personal efforts that wireless telegraphic communications became a "practical" reality.

There are those that say that Marconi's greatest triumph (the mother of all experiments) was when he succeeded in 1901 in passing signals across the Atlantic. There are those that say that he misled himself and the world into believing that electrostatic and atmospheric noise crackling was in fact the Morse code letter 'S'. Whether Marconi heard the three faint dots is really unimportant. His claim "sparked" a controversy among contemporary scientists and engineers about the experiment that continues today.

Certainly engineers and scientists of the present day are unanimous in admiring the bold and imaginative way in which Marconi attempted to take one spectacular step forward, to extend the range of wireless communications from one or two hundred kilometers to the 3500 kilometer distance across the Atlantic Ocean.

The world has acclaimed Marconi as the "father of wireless", although some say that Alexander Popov and Oliver Lodge were first in the field. History has accredited Marconi with the invention of an early form of radio telegraphy. Marconi did not "invent" wireless telegraphy (he never, himself, claimed to have done so). Starting with the work of others, he experimented until he had developed equipment arrangements which worked for him, and met commercial needs.

Since his company was in the competitive business of wireless communications, he therefore carefully (we can suppose) never explained exactly how his apparatus worked, or why he arranged things the way he did. Since he had little or no scientific training and no recognized academic qualifications, Marconi could not talk to academic detractors about his work, without risking derision, and so silence was his best policy.

We cannot say that any one person is the *Father of Radio;* "radio" has too broad a meaning. The development of wireless into what we know of as radio today was the result of many contributions over many years by many individuals. All we can do is try to place the key contributions which really are significant ones, and then give credit where credit is due. Certainly some of the early equipment used was primitive, and its purpose might not even be recognizable by radio communication engineers today — the writer has tried to shed some light on this as well.

12.18   ACKNOWLEDGMENTS

The author acknowledges discussion with many historians interested in the history of radio, particularly George Elliott, who for many years was Editor of an Amateur Radio Fessenden Society Bulletin; with Orn Arnason, then with The Brome County Historical Society; with Robert S. Harding, Curator with the Archives Center, National Museum of American History, Smithsonian Institution, City of Washington; with Fred Hammond (now deceased), who established an important personal radio museum; with Ernie DeCoste, Curator (now deceased), with The Canada National Science and Technology Museum, Ottawa, ON; and with Bruce Kelly, Curator (now deceased), of the Museum, The Antique Wireless Association, Bloomfield, NY; with Desmond Thackeray; with George Grisdale; with Alan Boswell, R. W. Simon and R. B. Molyneux-Berry, then with the Marconi Research Centre; with Jack Ratcliffe (now deceased), at the time he was writing his 1974 paper; and with Henry Bradford, who has been trying to find support for a Marconi Towers Historical Site.

REFERENCES

[1]   A. Waller, "Unsung Genius", *Equinox, the Magazine of Canadian Discovery,* No. 44, Mar/Apr., 1989, pp. 95-107.

[2]   J. F. McEvoy, "Right Against the World", *The Beaver,* Jun/Jul., 1990, pp. 43-47.

[3]   G. Elliott, "Who Was Fessenden?", *Proceedings of the Radio Club of America*, November 1992, pp. 25-37.

[4]   B. A. Austin, "Oliver Lodge – The Forgotten Man of Radio?", *Radioscientist*, Vol. 5, No. 1, March 1994, pp. 12-16.

[5]   W. E. Brand, M. Watts, and J.W. Wagner, "Nicola Tesla – The unsung inventor of modern-day AC power systems and radio", *Communications Quarterly*, Fall 1997, pp. 80-86.

[6]   J. S. Belrose, J. "Fessenden and Marconi: Their Differing Technologies and Transatlantic Experiments during the first Decade of This Century", IEE International Conference 100 Years of Radio, 5-7 September 1995,

Conf. Pub. 411, pp. 32-43. Note: 6 has been partially rewritten, on version of this paper which can be read on the WEB.

[7] J. S. Belrose, "Fessenden and the Early History of Radio Science", *The Radioscientist*, Vol. 5 No. 3, September 1994, pp. 94-110.

[8] J. S. Belrose, and G. Elliott, "Whose Heterodyne?", *Electronics World*, April 1997, pp. 293-297.

[9] J. S. Belrose, "A radioscientist's reaction to Marconi's first transatlantic experiment – revisited", Conference Digest, IEEE AP-S Symposium, Boston, MA, July 8-13, 2001, Volume 1, pp. 22-25.

[10] J. S. Belrose, "Reginald Aubrey Fessenden and the Birth of Wireless Telephony", IEEE Antennas and Propagation Magazine, Vol. 44, No. 2, April 2002, pp. 38-47.

[11] D. L. Sengupta and T. K. Sarkar, "Maxwell, Hertz, the Maxwellians, and the Early History of Electromagnetic Waves" *IEE Antennas and Propagation Magazine*, Vol. 45, No. 2, April 2003, pp. 13-18.

[12] J. A. Ratcliffe, *The Physical Principles of Wireless*, London: Methuen & Co., 1929.

[13] J. Henry, "On the Production of Currents and Sparks of Electricity from Magnetism", *Amer. Jour. Science*, Vol. 22 (appendix), 1832, pp. 403-408.

[14] "The Scientific Writings of Joseph Henry", 1832–1848, The Smithsonian Institution, Washington, D.C. Henry's important papers on electromagnetic induction are reprinted in *The Discovery of Induced Electric Currents*, edited by J.S. Ames, New York: The American Book Company, 1900.

[15] E. J. Houston, and E. Thomson, "The Alleged Etheric Force Test Experiments as to its Identity with Induced Electricity", *Jour. Franklin Inst.*, Vol. 101, 1876, pp. 270-274.

[16] A. Dolbear, US Patent No. 350,299, issued 5 October 1886.

[17] N. Tesla, "Experiments with Alternating Currents of High Potential and Frequency", *Jour. Institution Elec. Engs.* (London), Vol. 21, 1892, pp. 51-163.

[18] T. C. Martin, "Inventions, Researches and Writings of Nikola Tesla", *The Electrical Engineer*, New York, 1893; ch. 27, pp. 123 and 198-293. Experiments with Alternating Currents of High Potential and High Frequency by Tesla, February 1892. This book has been republished and is available from several sources.

[19] N. Tesla, *Experiments with Alternating Currents of High Potential and High Frequency*, New York: McGraw Hill Publishing Co., 1904.

[20] L. I. Anderson. (editor), "Nikola Tesla on his work with Alternating Currents and their Application to Wireless Telegraphy, Telephony, and the Transmission of Power - An Extended Interview", Sun Publishing, 1992, pp. 86, 123,-124, 160-162.

[21] O. Lodge, "The Work of Hertz", *Proc. Royal Institution* (London), Vol. 14, 1 June, 1894, p. 321.

[22] A. Popov, *Jour. Russian Phys.* Chem. Soc., 25 April, 1895. See also a recent review by R. Barrett, "Popov versus Marconi: the Century of Radio", GEC Review, Vol. 12 No. 2, 1997, pp. 107-116.

[23] C. Süsskind, "The early history of electronics – Pts. I-VI, *IEEE Spectrum*, August 1968, pp. 57-60; December 1968, pp. 69-74; April 1969, pp. 66-70; August 1969, pp. 78-83; April 1970, pp. 78-83; September 1970, pp.76-79.

[24] R. W. Simons, "Gugielmo Marconi and Early Systems of Wireless Communications", *GEC Review*, Vol. 11, No. 1, 1996, pp. 37-55.

[25] M. Lemme, and R. Menicucci, "From coherer to DSP", *EBU Technical Review*, Spring 1995, pp. 63-75.

[26] R. A. Fessenden, "Wireless Telephony", A published paper presented at the 25th Annual Convention of the American Institute of Electrical Engineers, Atlantic City, NJ, 29 June, 1908, pp. 553-620.

[27] O. J. Lodge, Signalling through Space without Wires", The Electrician, London, 1902.

[28] Susskind (April 1969), loc.cite [23], pp. 71-73

[29] J. J. Fathie, "History of Wireless Telegraphy", Blackwood, Edinburgh, 1899.

[30] J. P. Wilson, "Oliver Lodge and the Origins of Spark Transmissions", *IEE International Conference 100 Years of Radio*, 5-7 September 1995, Conf. Pub. No. 411, pp. 7-13.

[31] V. J. Phillips, "Early Radio Wave Detectors", Peter Preregrinus for the IEE, London, 1980.

[32] T. W. Pegram, R.B. Molyneux-Berry, and A.G.P. Boswell, "The Coherer Era – The Original Marconi System of Wireless Telegraphy", *GEC Review*, Vol. 12 No. 2, 1997, pp. 83-93.

[33] V. J. Phillips, "The 'Italian Navy coherer' affair: a turn-of-the-century scandal", *IEE Proceedings*-A, Vol. 140 No. 3, May 1883, pp. 175-185.

[34] R. N. Vyvyan, "Marconi and his Wireless", EP Publishing, 1974. First published in 1933 as "Wireless over 30 Years", Routledge and Keegan.

[35] G. G. Blake, "History of Radio Telegraphy and Telephony", Radio Press, London, 1993.

[36] W. Secor. "Radio Detector Development", *The Electrical Experimenter*, January 1917, pp. 652-656.

[37] E. Rutherforth, *Electrician*, 49, 1902, p. 562.

[38] T. H. O'Dell, "Marconi's magnetic domain that stretches into the ether", *Electronic World + Wireless World*, April 1993, pp. 666-669.38.

[39] T. H. O'Dell, Ferromagnetics, Macmillan, London, 1981, pp. 2-7.

[40] T. H. O'Dell, "Marconi's Magnetic Detector Physics", *Rivista Internazionale di Storia dela Scienza*, Vo. 25, 1983, pp. 525-584.

[41] L.A. Geddes, K.S. Foster, J. Reilly, W.D. Voorhees, J.D. Bourland, T. Ragheb abd N.E. Fearnot, "The Rectification Properties of an Electrode-Electrolytic Interface Operated at High Sinusoidal Current Density", *IEEE Trans. Biomed. Eng.*, Vol. BME-34, September 1987, pp. 669-672.

[42] S. M. Kintner, "Pitsburgh's Contribution to Radio", Joint meeting of the IRE and AIEE, at Pitsburgh, 7 April 1932 (abstracted in QST, July 1932, pp. 31-33 and 90).

[43] John L. Hogan, "The Heterodyne Receiving System, and Notes on the Recent Arlington-Salem Tests", *Proc. IRE*, Vol. 1, pt. 3, July 1913, pp. 75-102. Reprinted as a Proceedings Classic Paper Reprint, Proc. IEEE, Vol. 87, No. 11, November 1999, pp. 1979-1989.

[44] *New York Herald Tribune*, editorial at the time of Fessenden's death, on 22 July, 1932.

[45] B. Shrader, "Radio Gear of Yesteryear", QST, March 1994, pp. 41-43, 57.

[46] E. J. Quinby, "Nicola Tesla, Worlds Greatest Engineer", *A History of The Radio Club of America 1909-1984*, Vol. 50 No. 3, The Radio Club of America, Fall 1984, pp. 223-228.

[47] E. F. W. Alexanderson, "Transatlantic Radio Communications", Trans. AIEE, October 1, 1919, pp. 1077-1098.

[48] R. A. Fessenden, "The Inventions of Reginald A. Fessenden – Part 1, *Radio News*, January 1925 (Parts II to XII were published in the February through November issues of *Radio News*).

[49] E. Ruhmer, "Wireless Telephony", translated from German by J. Erskine-Murry, with an Appendix about R.A. Fessenden by the Translator, Crosby Lockwood and Son, Ludgate Hill, 1908. Author's Preface is dated 15 February 1907.

[50] J. Zenneck, "Wireless Telegraphy", Translated from the German by A.F. Seelig, McGraw Hill, 1915.

[51] Pam Reynolds, *Guglielmo Marconi*, booklet published by The Marconi Company Limited, 1984.

[52] D. Thackeray, "The First High-Power Transmitter at Poldhu", *The AWA Review*, &, 1992, pp. 29-45.

[53] W. S. Entwistle, *Year Book of Wireless Telegraphy & Telephony*, Wireless Press, 1922, pp. 1245-1258.

[54] P. B. Bondyopadhyay, "Investigations on Correct Wavelength of Transmissions of Marconi's December 1901 Transatlantic Wireless Signal", *IEEE Antennas and Propagation Society, International Symposium Digest*, 12, 1993, pp. 72-75.

[55] J. A. Ratcliffe, "Scientists' reactions to Marconi's transatlantic radio experiment", *Proc. IEE*, 121, 1974, pp. 1033-1038.

[56] H. Bradford, "Did Marconi Receive Transatlantic Radio Signals in 1901", *The Old Timer's Bulletin (OTB)*, Vo. 44, No. 1, February 2003; No. 2, 2 May 2003.

[57] R. A. Fessenden, "A Regular Wireless Telegraphic Service between America and Europe", *The Electrician* (London), 22 November 1907, pp. 200-203.

[58] "Wireless Grows up at Sea", chapter in book *From Semaphore to Satellite*, International Telecommunications Union, Geneva, 1965, pp129-142.

[59]   R. S. Harding, Registrar of the *George H. Clark Radio Collection* c180-1950, Second Edition, Archives Center National Museum of American History, Smithsonian Institution, Washington, DC, 1990.

[60]   R. A. Fessenden, "Wireless Telegraphy", *The Electrical Review* (London), Vol. 58, No. 1, 11 May 1906, pp. 744-746; and 18 May 1906, pp. 788-789.

[61]   R. A. Fessenden, "The Possibilities of Wireless Telegraphy". (A discussion). *Transactions of the American Institute of Electrical Engineers,* Vol. 16, mail edition 1899, pp. 635-642, regular edition pp. 607-614.

[62]   R. A. Fessenden, "How Ether Waves Really Move", *Popular Radio*, IV Number 5, November, 1923, pp. 337-347

[63]   J. Kennelly, "Research in telegraphy", *Elect. World & Eng.*, Vol. 6, 1902, p. 473.

[64]   O. Heaviside, "Telegraphy" in *Encyclopedia Britannica*, William Benton, 1902, p. 214.

[65]   W. S. Zuill, "The Forgotten Father of Radio", *American Heritage of Invention & Technology*, Vol. 17, No. 1, Summer 2001, pp. 40-47.

[66]   G. A. Isted, "Guglielmo Marconi and the History of Radio – Part 1, *GEC Review*, Vol. 7 No. 1, 1991, pp. 45-56; Part II, GEC Review, Vol. 7 No. 2, 1991, pp. 110-122.

[67]   Helen M. Fessenden, *Fessenden Builder of Tomorrows*, Coward-McCann, N.Y., 1940 (reprinted by Arno Press, N.Y., 1974).

[68]   International Conference on 100 Years of Radio, 5-7 September, 1995, IEE Conference Publication Number 411..

[69]   D. W. Kraeuter, *A New Bibliography of Reginald A. Fessenden*, The AWA (Antique Wireless Association) Review, Vol. 8, 1993, pp. 55-66.

[70]   "Letters from Guglielmo to his Father, 1896-1898", *GEC Review*, Vol. 12 No. 2, 1997, pp. 94-106.

[71]   Ormond Raby, *Radio's First Voice – The Story of Reginald Fessenden,* Macmillan of Canada, 1970.

[72]   E. N. Sivowitch, *A Technological Survey of Broadcasting's Pre-History – 1876-1920*, Antique Wireless Association Monograph (New Series) No. 2.

13

WIRELESS TELEGRAPHY IN SOUTH AFRICA AT THE TURN OF THE TWENTIETH CENTURY

DUNCAN C. BAKER, *University of Pretoria, Pretoria, South Africa*

13.1   INTRODUCTION

Looking at a map or globe of the world one is immediately struck by how far South Africa is from Europe and North America. The word "isolated" comes to mind. Indeed air travelers do not particularly relish the idea of flights lasting nine or more hours from Europe, and fourteen hours or more from North America. At the end of the nineteenth century this sense of isolation must have been particularly acute. Yet, despite this, some of the developments in wireless telegraphy at that time spilt over to South Africa. In some small way this country played a role in shaping the fledgling technology, which had drawn the interest of so many prominent scientists of the day.

This chapter provides some background to the local interest in wireless telegraphy at the end of the nineteenth century. It covers the independent experiments of one Edward Alfred Jennings in Port Elizabeth, early demonstrations in Cape Town, and the first application of wireless by an army in the field during the South African War of 1899–1902.

Much of the material is drawn from earlier publications [1-6], which contain most of the primary references. Where appropriate such primary references will be cited separately. The author has in addition taken the liberty of quoting large portions of text from some of the references verbatim because many of them will not be readily accessible to interested readers.

It is trusted that wider publication of the material than has hitherto been possible will give due credit to the pioneering efforts of Edward Jennings, who did his work in isolation at the southernmost tip of Africa.

13.2   THE CAPE COLONY

The following report appeared in the Eastern Province Herald of May 9, 1899. This newspaper is published in Port Elizabeth, a coastal city in the Eastern Province of South Africa, which boasts the oldest telephone exchange in the country.

421

WIRELESS TELEGRAPHY: LAST NIGHT'S LECTURE

At eight o'clock last evening, the hour for Mr Jennings to deliver his lecture on the highly engrossing subject of "Wireless Telegraphy", the Town Hall presented an animated appearance, the big Hall being crowded by an interested audience. The gallery was likewise fairly well packed. Mr M M Loubser occupied the chair and in a brief introductory speech said it must be very gratifying to Mr Jennings to see such a splendid audience present.

Whilst Marconi had been working in the wake of wireless telegraphy 7,000 miles away, Mr Jennings had been silently pursuing the study for several years of the same important scientific subject, and had arrived at results almost as satisfactory as those obtained by Professor Marconi. There was the highest credit due to Mr Jennings, and his pursuits had been carried on in his spare hours.

Mr A Marshall Hall, who read the paper on behalf of Mr Jennings while that gentleman demonstrated the operation of the new system of telegraphy, stated that Mr Jennings had omitted all reference to his own individuality and he accordingly mentioned a few facts as to the writer. Mr Jennings had nothing to guide him as to Marconi's discovery but newspaper reports, and the apparatus displayed that evening was the result of unremitting labour for several years and courageous facing of overwhelming difficulties of all kinds. For this reason he felt justified in asking the gathering to pay a deserved tribute of admiration to our young fellow townsman who had achieved such a great result.

The lecture opened with details as to previous efforts to that of Marconi towards transmitting intelligence to a distance without the usual connecting wires. In the fifties, a young Scotsman named Lindsay conceived the idea of sending messages under water by means of copper plates, immersed at each end of the circuit, but this did not meet with very good success. Edison and others experimented in a similar direction, and more recently much had been done by Mr Preece, Chief Electrician to the Postal Telegraphs of Great Britain, who determined that, by means of increasing the power of the magnetic waves, with a specially designed apparatus, it would be possible to transmit signals to much greater distances, and experiments were successful up to five miles. The apparatus was described, and the results demonstrated. Marconi's discovery in July, 1896, was then dealt with at length in a most interesting fashion, and then the results of Mr Jennings' researches were detailed.

The inventor and the lecturer had several instruments to illustrate the lecture, and it was clearly illustrated by various tests that the new invention bids fair to triumph in every possible way. One of the instruments was taken into the room at the rear of the platform and with the door closed a telegraphic message handed up by one of the audience was transmitted and received by the second instrument placed on the platform. Owing to the accommodation being slightly irregular, the results were not so good as others Mr Jennings had obtained. Mr Jennings said at a future lecture he hoped to show them even better results.

The Mayor proposed a hearty vote of thanks to the lecturer, which was seconded by Major Tamplin, MLA, in an admirable speech. Mr Jennings (Figure 13.1) briefly responded, and the meeting dispersed.

Edward Alfred Jennings was born in London in the UK during 1872. Today, sadly for South Africans, he is not remembered at all in the long history of Wireless Telegraphy and communications. However, under different circumstances his name could almost certainly have featured as prominently as those of the early pioneers who lived or worked in the United States, Germany, the UK, France, the USSR, Denmark, Holland, and Italy.

The fact that we know some of Jennings' background and story is largely due to the efforts of Eric Rosenthal, a well-known and much respected South African who made a lasting impression on several generations of South Africans in a number of radio programs for his incredible general knowledge. He also recorded many of the "footnotes" of South African history in a number of books. Rosenthal reported on an interview he had with Jennings in a book which was prepared to celebrate the 50th anniversary of broadcasting in South Africa in

Figure 13.1. E. A. Jennings (left) demonstrating his apparatus to Mr. Henschell, the Postmaster of Port Elizabeth, on board the *SS Gascon* at anchor in Algoa Bay during July 1899. (Photo reproduced by courtesy of the Estate of Eric Rosenthal).

1974 [7]. Almost all of what follows concerning Jennings is summarized from Rosenthal's book.

Rosenthal reports that Jennings started his working career in a company which supplied telegraphic equipment to railway companies. His skill as an instrument maker brought him to the attention of the firm Whitehead. There he worked in a department specializing in firing mechanisms. He later moved to the Government Instrument Factory at Holloway, near London. He was recruited to work in the Post Office of the Cape Colony and worked in Cape Town for two and a half years.

In 1896 he was transferred to Port Elizabeth, where he worked in the oldest telephone exchange in South Africa. This had been opened on 1882. The instruments of that time were very primitive and the microphones used carbon granules to convert speech into electric signals that could be sent over wires. The carbon in the microphones became compacted with time, thus reducing their efficiency even more. Jennings decided to try to improve these microphones. He filed silver off an old watch chain and devised a new microphone using a glass tube. Unfortunately, it did not work the way it was intended to.

However, it did produce an unexpected response. Whenever the electric bell at Jennings' house at The Hill was rung the experimental receiver, which he had set up, would emit a loud crackle. On closer examination he found that the silver filings adhered after these events and needed to be tapped gently in order to loosen them. In fact he had discovered the coherer, the somewhat primitive detector which was extensively used in the early days of wireless telegraphy. For more information on the coherer, the interested reader is referred to the chapter by Andrews in [8].

Jennings also observed that he also heard other loud crackles, unrelated to the ringing of the doorbell. He eventually made the connection between these and arcs drawn by electric trams, which had recently been introduced in Port Elizabeth, as they passed through a cross-over about 130 meters from his house. He called on the services of a number of well-known personalities with technical or scientific backgrounds to witness a demonstration. None could offer an explanation for what was observed.

In order to generate bigger sparks under his own control, he built a Ruhmkorff coil which took almost six months to finish.

> *The coil was a foot (30 cm) long and about half that in diameter. Over seven miles (11 km) of copper wire as fine as the bristles of a hairbrush were wound round 72 sections, each about one-sixth of an inch (4 mm) thick, and each in turn separated by a minute disc of vulcanite. All this had to be done on a treadle lathe. As for the all-important condenser, this was made up of 200 sheets of tinfoil, each eight inches (20 cm) by six inches (15 cm). The battery employed was a small four-cell accumulator, with a mercury interrupter. When the battery was switched on, a torrent of bluish-white sparks up to six inches (15 cm) in length was emitted with a deafening noise and a*

viciousness almost terrifying. When transmitting, the spark gap
was reduced to less than an inch (2.5 cm).

Jennings' first really successful experiment was to transmit a message from his house to Cooper's Kloof about 800 metres away. Interest in Jennings' activities rose when news of Marconi's experiments reached South Africa.

The Marquis of Graham, representing Lloyd's of London, visited South Africa in 1898. Lloyd's was interested in improving existing facilities for navigation at sea. It was suggested that a "telegraph without wires" might be established between the lighthouse facilities at Bird Island, some 60 km from Port Elizabeth in Algoa Bay. This was certainly a most ambitious idea at the time. Rosenthal recounts Jennings' recollections of this visit.

In July 1899, Jennings was able to achieve a record distance for transmission of a signal of about 13 km from the lighthouse in the Donkin Reserve to Cape Recife.

A formal request for funding to support Jennings' work was made in the Cape Parliament. The responsible Minister, John X. Merriman, was no more skilled in technology than many a present day politician, and essentially stifled the idea with the remark, "Life is troublesome enough with ordinary telegrams. With wireless telegraphy it will be unbearable." Imagine this comment being made today with reference to the ubiquitous cell phone!

Jennings nevertheless persevered with his experiments and eventually on May 8, 1899, gave the public demonstration referred to above in the extract from the Eastern Province Herald at the start of this section.

Further experiments included successful transmissions from Port Elizabeth to the mail steamer *Gascon* lying more than 4 km out in Algoa Bay in July of 1899.

But for the outbreak of the South African War of 1899–1902, Jennings' work may well have been recognized for the pioneering effort in the new field of Wireless Telegraphy that it most certainly was. Jennings later worked in Johannesburg, and eventually retired to Cape Town, where he died in 1951. On a personal note, Jennings' granddaughter, Mrs. Jean Boock, reports that Marconi would not let Jennings visit factories when he visited the UK, probably because of his reputation (Jean Boock, Private Communication).

Apart from Jennings' demonstrations in Port Elizabeth there was also another public demonstration in Cape Town on February 11, 1899. The event was attended by the Prime Minister of the Cape Colony, W.P. Schreiner. The audience included the Postmaster General, Sir Somerset French, several naval officers from nearby Simonstown, some journalists, and sightseers. The demonstration was conducted by a Scottish professor in physics Dr John C. Beattie, later first Vice-Chancellor and Principal of the University of Cape Town and, appropriately, Board Member of the South African Broadcasting Corporation.

For the demonstration Beattie had imported a large Ruhmkorff coil from his laboratory, along with a coherer from Oliver Lodge. Rosenthal reports that Oliver Lodge was Professor of Physics at the University College of Manchester, but he is revered as being from the University of Liverpool (Brian Austin, Private Communication). Beattie managed to send messages over a distance of more than 100 m. Schreiner's presence had stemmed from the possibility of linking lighthouses by wireless telegraphy. Certainly, safety at sea around the "Cape of Storms" would have been an important consideration at the time.

Rosenthal goes on to report on the capture of wireless telegraphy sets ordered by the South African Republic (Transvaal) prior to the outbreak of war. His book has an illustration of a captive hot air balloon for the support of antennas used during an attempt to evaluate the wireless telegraphy equipment. We will return to this aspect of the history later.

Rosenthal reports that the Cape Parliament was the first to enact legislation to regulate wireless telegraphy when in 1902 it amended the Electric Telegraph Act of 1902 [see also 9]. This forbade the operation of "any mast, standard or apparatus of any kind for the purposes of Aetheric Signalling, without wires, by means of electricity, magnetism, electro-magnetism or other like agencies, *except under license granted by the Governor.*" In 1903 the Parliament of the Colony of the Cape of Good Hope also became the first to introduce legislation concerning licensing of radio apparatus.

Rosenthal's book provides a valuable insight into the early days of plans and applications of radio for safety at sea around the South African Coast. In addition, Rosenthal also details some of the achievements of Hendrik van der Bijl, who did pioneering work in thermionic valves in Germany and the United States in the years leading up to the First World War [see for example 10,11].

13.3   THE SOUTH AFRICAN REPUBLIC

A considerable amount of material, which clearly illustrates the interest of the government of the South African Republic (*Zuid Afrikaansche Republiek - ZAR*) in Wireless Telegraphy, is housed in the National Archives and Records Service of South Africa, Pretoria, City of Tshwane, South Africa. Much of this interest was due to the visionary CK van Trotsenburg (Figure 13.2), the General Manager

Figure 13.2. C.K. van Trotsenburg, General Manager of Telegraphs in the South African Republic (Photo no. 2110 reproduced by courtesy of the National Archives and Records Service of South Africa, Pretoria, City of Tshwane, South Africa).

Figure 13.3. Members of the Telegraphy Department of the South African Republic in 1896. CK van Trotsenburg is seated in the centre of the middle row, facing slightly to his left. (Photo reproduced by courtesy of the National Archives and Records Service of South Africa, Pretoria, City of Tshwane, South Africa).

of Telegraphs in the ZAR at the time (Figure 13.3). Some background as summarized from [1,2,3,5] is in order.

In October of 1888, Paul Kruger, President of the ZAR, obtained the services of Paul Constant Paff, an employee of the Amsterdam Telegraph Department, on contract for a period of two years. The Field Telegraph Department, which was to form part of the State Artillery, was established in 1890. Paff was offered, and accepted, a commission as Head of the newly established Department (Figure 13.4). Paff's official duties included training the telegraphists who would man the Department (Figure 13.5), in the use of telegraph, heliograph, flags, and lamps to transmit messages in Morse code.

The diamond rush to Kimberley, near the border of the Cape Colony and the Boer Republics of the Orange Free State and the ZAR, had started in 1870. Gold had been discovered on the Witwatersrand in 1886. This "Ridge of

Figure 13.4. Captain Paul Constant Paff of the South African Republic's State Artillery (Photo no. 8585 reproduced by courtesy of the National Archives and Records Service of South Africa, Pretoria, City of Tshwane, South Africa).

Figure 13.5. Field telegraph unit of the State Artillery, South African Republic under command of Commandant Scheepers in 1895. (Photo no. 5143 reproduced by courtesy of the National Archives and Records Service of South Africa, Pretoria, City of Tshwane, South Africa).

White Waters" is the gold bearing reef which runs roughly East-West through the present Johannesburg. The discovery of gold resulted in an influx of foreigners ("*Uitlanders*" or Outlanders) bent on making a quick fortune. The British policy of the 1880s had been one of increasing isolation of the two Boer Republics. These Republics had good reason to feel threatened by British Imperial intentions. Pakenham [12] details the various actions on the part of Britain to support the increasing political demands of the *Uitlanders*, who were believed to outnumber the Boers in the ZAR and who had been essentially disenfranchised by Kruger.

Thus it was that Cecil Rhodes and Alfred Beit, two multi-millionaires who made their fortunes on the diamond fields of Kimberley, conspired to add the ZAR to the British Empire. Together they had founded a new British colony, Rhodesia (now Zimbabwe) north of the ZAR. Gold had enabled the ZAR to become the most powerful nation in Southern Africa, and also greatly supplemented Rhodes and Beit's fortunes. The Jameson Raid, an attempt to conquer the ZAR with mainly Rhodesian mounted police drawn from Rhodes' Chartered Company in Rhodesia and volunteers, was the net result of these efforts. The raid was led by Dr. Leander Starr Jameson and launched with some 600 men from Pitsani in Bechuanaland (now Botswana), and entered the ZAR near Ottoshoop on December 30, 1895. The expected support from revolutionaries in Johannesburg failed to materialize.

Although the raid was an abject failure and had resulted in sixteen of the raiders being killed, compared with only one Boer, this raid was in essence the real declaration of war in the coming conflict. Nothing could have signalled British Imperial intentions more clearly than the Raid itself, and the intelligence captured. With war being regarded as inevitable the government of the ZAR set about enlarging its army and seeing to its defenses.

Five forts, four around Pretoria, and one at Johannesburg were built at enormous expense. At the time these forts were considered to be impregnable, although they were to fall without a shot being fired. The history of the construction of the Pretoria forts is recounted in [13]. The ZAR imported the best Europe had to offer to arm the forts, including 40 cm Creusot guns and Maxims. The forts had their own electrical dynamos to provide lighting and power searchlights. Lightning conductors were installed to prevent accidental ignition of ordnance. Initially Fort Wonderboom was linked directly with the Artillery Camp and the Commandant General's office in Pretoria. The cost of laying 7 km of the underground telephone cable was 9000 Pounds Sterling, a very large amount in those days. Originally the intention had been that all the forts would be linked to Pretoria, and thus to each other, by underground cable.

In his official report dated March 2, 1898 to the ZAR Cabinet, van Trotsenburg deals with the difficulties experienced, and the risk of interception for such telephone cables, and continues [14]:

> *On account of the aforementioned and in view of high costs, I would not recommend the laying of an underground connection between the Artillery camp and Daspoortrand, but would suggest the erection of an overhead line, to be worked with an ordinary telegraph or telephone instrument or perhaps with both.*
>
> *For distances of about 6 miles telegraphic communications can be exchanged without wire. At present, experiments are being conducted in Europe on a large scale by Military Powers, and it appears to me that lately such improvements have been made to those instruments used therefore, that the system would probably answer well for the forts.*
>
> *I would suggest that I communicate with manufacturers and in case of satisfactory information being received, to order one set of instruments for trial.*
>
> *The costs connected therewith are comparatively low.*

Only a few days earlier, van Trotsenburg, no doubt with the blessings of his superiors, had written to Siemens Brothers in London to explore the possible use of Wireless Telegraphy as a means of providing communications between the forts, and with Pretoria [15] (Figure 13.6):

> *Gentlemen,*
>
> *A certain place "A" in a valley is surrounded by hills. I wish to correspond telegraphically without wires between this place "A" and those hills as marked in the margin 1, 2, 3, 4 ... Are there any difficulties, if so, which?, if not, can you supply us with the necessary instruments complete.*

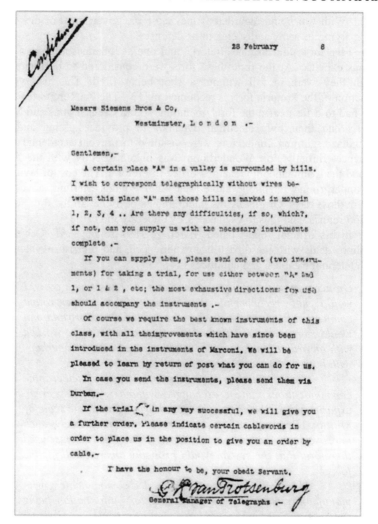

Figure 13.6. Copy of letter by from CK van Trotsenburg to Siemens Brothers in London dated 28 February 1898, enquiring about the availability of wireless telegraphy equipment (Copy reproduced by courtesy of the National Archives and Records Service of South Africa, Pretoria, City of Tshwane, South Africa).

> *If you can supply them, please send one set (two instruments) for taking a trial, for use either between "A" and 1, or 1&2, etc; the most exhaustive directions for use should accompany the instruments.*
>
> *Of course we require the best known instruments of this class, with all the improvements which have since been*

*introduced in the instruments of Marconi. We will be pleased
to learn by return of post what you can do for us.*

*In case you send the instruments, please send them via
Durban.*

*If the trial [is] in any way successful, we will give you
a further order. Please indicate certain cablewords in order to
place us in a position to give you an order by cable.*

I have the honour to be, your obedient servant,

C.K. van Trotsenburg.
General Manager of Telegraphs.

It is of interest to note that van Trotsenburg must have been well informed on wireless telegraphy. The file at the National Archives in Pretoria containing the Wireless Telegraphy material also includes information gleaned from a number of Journals current at the time. These included *Elektrotechnische Zeitscrift* (1897), *Electrical Engineer* (1897) and *The Electrical Review* (1898). An article in German on Wireless Telegraphy [16] appears to have been well used. There is also an article on Marconi's demonstration aboard the Royal Yacht *Osborne* [17].

Siemens responds to the enquiry in a letter dated 26 March 1898 [18]:

*We beg to acknowledge the receipt of your letter No. 1444 of
the 28th ultimo which has had our attention.*

*The exploitation of the Marconi system of Wireless
Telegraphy in this country is in the hands of a Company which
owns the Marconi patents. We have interviewed the Managing
Director of this Company on the subject of your letter and so
far as we can understand there should be no very great
difficulty in carrying out the installation you desire to effect but
there are some points which require consideration.*

*You do not mention the distance between the five
points shown in your diagram, but we have assumed these
would be reasonable distances. To establish communication it
is necessary to erect or arrange an insulated wire at each
Station in a vertical or nearly vertical position. The length of
this wire varies with the distance and may be affected by the
interposition of obstacles between the stations; it may be only a
few feet, or it may need to be as much as 100 yards. If the sides
of the hills surrounding the position A in your diagram are
abrupt the necessary wires could be run down them supposing
the distances or other conditions render long wires necessary.*

*Adjustments are possible by which it could be
arranged that the correspondence between any two stations
should not be overheard by another station, and these are
chiefly made in the disposition and connections of the vertical
wires.*

The Transmitter proper consists of an apparatus known as the Oscillator, worked by an induction coil and controlled by an ordinary Morse key. The size of the induction coil necessary varies with the distance etc. between the stations. The Receiver consists of an apparatus known as the Coherer, which is in circuit with the coils of an ordinary telegraph relay (Siemens polarized relay). On the circuit controlled by the relay contacts an ordinary Morse sounder or inker can be placed. We have seen a number of messages recorded on ordinary Morse slip which were transmitted by Marconi apparatus over a distance of 13 miles between a town on the English coast and the Isle of Wight. These signals were perfectly clear and distinct. A speed of 20 words per minute is quite practicable.

As regards the supply of the apparatus we find the company decline to sell outright. We discussed this point at considerable length but found they are at present unwilling to agree to any such arrangement. Their proposal is that a certain sum should be paid down on delivery of the apparatus and a further annual sum (to depend on the use made of the apparatus, distances over which worked, etc.) to be paid for a term of years. They also ask that they be permitted to send one of their assistants to install the apparatus and instruct users in the adjustment and care of it. The assistant's expenses to be, of course, paid for by the users of the system. The Company were anxious to know for whom the apparatus were required, but as you had marked your letter 'confidential' we did not feel at liberty to satisfy them on this point.

We shall be pleased to do anything more in our power to further your wishes on learning whether you are disposed to agree to the Company's terms, but we assume any agreement would have to be executed on your part by an official representative of your Government. Nothing has yet been said as regards the amount of the annual payment, since it entirely depends on conditions as to which we are at present uninformed, but the amount to be paid down on delivery of the apparatus would be reasonable.

We may add that we have supplied a good deal of apparatus to the Company - they are using our relays exclusively - and although the Coherers and some other essential parts are of their own make, the apparatus generally appears to be of substantial and reliable construction.

We are, Sir, Your obedient servants.

Siemens Brothers and Co. Limited
(Signed) Alex Siemens, Director.

It is of interest to note that in the fourth paragraph the Marconi Wireless and Telegraph Company anticipates that radiation patterns of antennas could be used to provide preferential radiation in a particular direction, although the basis for this assumption is not known. Dowsett, one of the Marconi engineers subsequently sent to South Africa to accompany Marconi systems provided to the British Military, reports that the idea of directional antennas using a two-element array was apparently first suggested by SG Brown in 1899 [19]. Also of interest is the fact that the Company is interested in a leasing arrangement for the equipment. In effect the client is expected to pay for the right to use the equipment, rather than to purchase it outright.

Van Trotsenburg's reply is dated 23 April 1898 [20], and provides the requested information. It is clear from the letter that van Trotsenburg is also concerned about interception of any transmissions.

Siemens Bros. in London had further discussions with Marconi's Company and in their letter dated 11 June 1898 replies as follows [21]:

We beg to acknowledge receipt of your letter No 2737 of the 23rd April and have been giving the subject careful attention. We have again seen the Managing Director of the Marconi Company but regret to say we can make no progress with them.

From the information you have communicated respecting the topography of the district in which you wish to use the new apparatus it would appear that the work should quite easily be done, but we have not shown the figures to the Marconi people for reasons which will appear. As we explained in our previous letter we are advised that adjustments can be made by which overhearing can be rendered practically impossible. This may be done either by special disposition of the connections and vertical wires, or by using a separate set of apparatus for each circuit and tuning the receivers to respond to differing rates of disturbance.

On the question of the cost of the apparatus we beg to say the sum to be paid down on delivery of the special parts of the apparatus to be supplied by the Marconi Company- i.e. the Oscillator and Coherer - would be about 35 Pounds Sterling. but it is now stated that the payment of this sum would only give the right to use the apparatus and that the apparatus must remain the property of the Company. (in addition to this special apparatus there would be the induction coil, costing about 25 Pounds Sterling. and the local sounder or Inkwriter with the necessary batteries).

The Company decline to name the sum to be paid as annual royalty unless they are made fully acquainted with the purpose for which the instruments are to be used. We have objected that they require to know too much and have pointed out that with no other class of telegraph instrument is the use of it by the customer subject to any restrictions, but the fact is (and this is practically admitted by the Company) they see no way to make a sufficient profit by the sale of apparatus and are thus endeavouring to obtain their return from a tax on its employment.

As regards the sending of an assistant to install the apparatus and instruct your staff in its use, the Company would undertake to charge only out-of-pocket expenses, which would amount to about 5 Pounds Sterling per week during the man's absence from London, plus the travelling expenses. It occurred to us it might suit your views better to send one of your own men to be instructed here and the Company would agree to that plan but would still claim to retain the right to send one of their men to investigate if any difficulties arose with the system.

It seems to us that the insistence of the Company upon the right to obtain control over the apparatus and your use of it is inconsistent with the free use you would naturally expect to obtain for your outlay, and if the conditions were fulfilled we do not think we need reply to the question towards the end of your letter.

Since writing our previous letter we have been making enquiries as to another system of 'space' telegraphy which has been designed by Prof. Oliver Lodge, F.R.S. We have been in communication with the makers of Prof. Lodge's apparatus, themselves of high repute in electrical circles, and although they do not feel able to recommend the apparatus at present for anything but purely experimental demonstrations for instructional purposes, yet they hold out hopes of soon being in a position to offer a commercial form of the apparatus.

We believe we can obtain apparatus of this type without any restrictions when it is ready.

We will keep the subject before us and advise you promptly of any useful result we attain.

The reference to the work by Lodge is of interest as the feeling was that his apparatus was only suitable for "experimental demonstrations for instructional purposes." This illustrates the commercial advantage that Marconi's Company held at that stage, and would maintain for some time to come.

Siemens and Halske in Berlin had referred van Trotsenburg's correspondence to their South African Agents in Johannesburg. In a letter dated June 21, 1898, they made the ZAR an offer to sell a number of units, and gave a short description of the proposed equipment. A translation of their letter of June 21, 1898 to van Trotsenburg reads [22]:

In the last mail we have received from our Office in Berlin a copy of your letter dated 23 April this year, with a request to give you the required information concerning spark telegraphy.

Therefore we take the liberty to offer you the electrical installation for five stations, at a total price of four hundred and eighty seven Pounds Sterling, delivered at Pretoria, excluding the cost of erection.

The installation consists of:

a. *Transmitter* (Authors' note – "Gever", literally "giver", translated as transmitter

> 5 *Spark inductors on case with condensors for 30 cm spark length.*
> 5 *Deprez current breakers with double contact.*
> 5 *Radiators on pedestal with switch clamp.*
> 5 *Keys (lever switches) on slate pedestal.*
> 5 *Accumulator batteries (12 volts, 8 amperes, 12 ampere hours)*
> 5 *Regulating resistances with five sections for regulating the primary current.*

b. *Receiver.*

> 5 *Receivers, consisting of*
>
>> 1 *polished mahogany pedestal, and mounted thereon 1 Coherer*
>> 1 *polarized relay*
>> 3 *Hellesen elements (Authors' note - presumably resistors)*
>> 5 *Morse sets (color print) with case on polished mahogany box pedestal, including horizontal paper roll*
>> 5 *Current switches of special construction.*

According to the experience gained by our parent company in spark telegraphy, each station must be installed on a high tower. Furthermore, it would eventually be necessary to take special precautions to guard against the incidence of weather and atmospherically induced electrical phenomena.

With regard to this matter it will be necessary to conduct an investigation to obtain information on local conditions etc.

In order to avoid any misunderstanding, we take the

liberty of attaching a technical description in German, especially with regards to the components of the transmitter and receiver.

We look forward to being of further assistance with obtaining information, or in other ways.

When one contrasts the total cost of 487 Pounds Sterling with the 9000 Pounds Sterling for the 7 km of underground cable, the advantages of Wireless Telegraphy even then must have been apparent.

Somewhere along the line van Trotsenburg must also have approached a French company in Paris, *Society Industrielle des Telephone*, for information on its system. Regretfully all that remains of this correspondence is their reply dated June 16, 1898, giving a quotation and a description of their equipment [23].

There now follows a gap in the correspondence on file at the National Archives of almost a year. No doubt there must surely have been more correspondence given the level of interest in the first place, and the gathering sense of foreboding that war was imminent. No record of what happened to this material exists.

Van Trotsenburg visited Europe in June of 1899. He held discussions with companies in London, Berlin and Paris. Of particular interest is a letter addressed to him by Marconi's Wireless Telegraph and Signal Company dated July 1, 1899 [20] (Figure 13.7):

...In connection with your visit to this office yesterday and the matter then discussed, I have pleasure in confirming what I then stated viz:-

...that we would be willing to supply instruments for one or more installations of wireless telegraphy to your Government. The cost of instruments complete would be as enclosed sheet, amounting in all to £95.10.0 (95 Pounds Sterling and 10 Shillings), Royalty for the use of the instruments would be as might be arranged, this depending on the distances dividing the stations for the installations. As I understood from you yesterday that five stations (which you mentioned as being the probable number you would require at once) would all be included within a radius of five miles, and that you would wish all these stations to intercommunicate if possible. The royalty for this number would be 500 Pounds Sterling a year, or 100 Pounds Sterling for each set of instruments. We would send out the instruments in charge of one or more of our trained assistants who would put these up and start them to work satisfactorily, the only extra we would charge for this would be salary and out of pocket expenses incurred in so doing.

Our assistant or assistants in charge would give the necessary instructions to your people how to work the

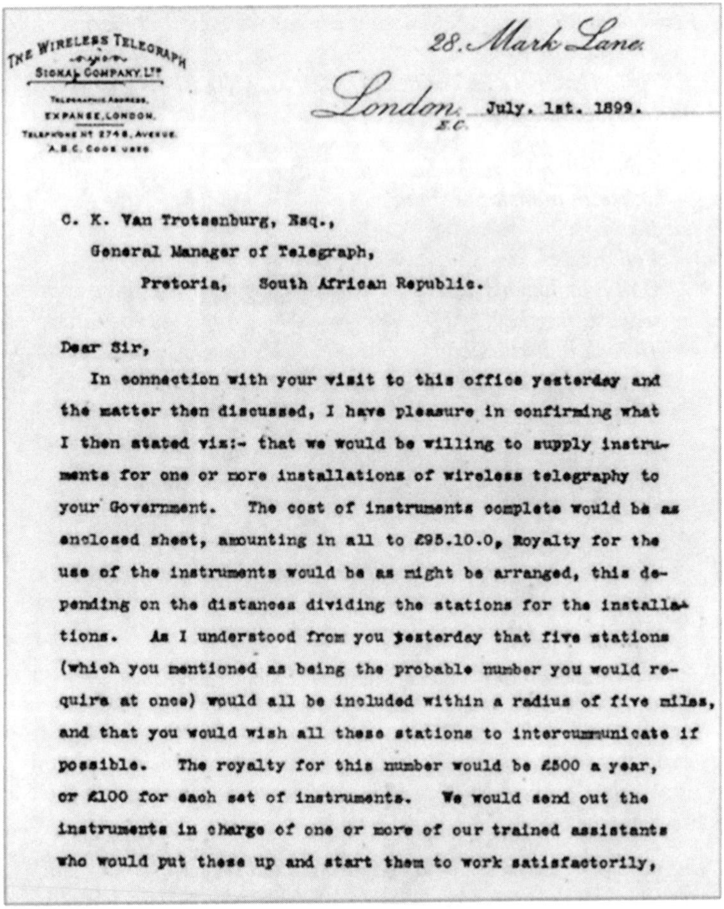

THE WIRELESS TELEGRAPH
SIGNAL COMPANY L.TD

Telegraphic Address,
EXPANSE, LONDON.
Telephone Nᵒ 2748, Avenue.
A.B.C. Code used.

28. *Mark Lane.*

London, July. 1st. 1899.
E.C.

C. K. Van Trotsenburg, Esq.,

   General Manager of Telegraph,

      Pretoria, South African Republic.

Dear Sir,

   In connection with your visit to this office yesterday and the matter then discussed, I have pleasure in confirming what I then stated viz:- that we would be willing to supply instruments for one or more installations of wireless telegraphy to your Government. The cost of instruments complete would be as enclosed sheet, amounting in all to £95.10.0, Royalty for the use of the instruments would be as might be arranged, this depending on the distances dividing the stations for the installations. As I understood from you yesterday that five stations (which you mentioned as being the probable number you would require at once) would all be included within a radius of five miles, and that you would wish all these stations to intercommunicate if possible. The royalty for this number would be £500 a year, or £100 for each set of instruments. We would send out the instruments in charge of one or more of our trained assistants who would put these up and start them to work satisfactorily,

Figure 13.7. First page of letter dated 1 July 1899 from The Wireless Telegraph Signal Company to C. K. van Trotsenburg, offering to sell a number of wireless telegraph units to the South African Republic (Copy reproduced by courtesy of the National Archives and Records Service of South Africa, Pretoria, City of Tshwane, South Africa).

instruments when set up, and the instruments being in our charge we would take full responsibility for their successful working.

   Trusting that we may soon be favoured with your orders for these instruments.

I am, Sir,
Your obedient Servant,

MANAGING DIRECTOR

ENCLOSURE

*ESTIMATE OF COST FOR ERECTING AND FITTING ONE
WIRELESS STATION*

		£	s	d*
Complete receiver in Screening Box	)			
and with adjustable stand.	)			
Inkwriter,	)			
Bell,	)			
Coil attachments and	)			
transmitting key.	)	35	9	3
10 inch Induction coil,		38	12	6
50 "M" size Obach Cells		11	0	0
2 Accumulator		5	9	0
Insulators		3	0	0
Pole and Rigging (not included)				
Aerial wire &c		2	0	0
		95	10	9

Electrician's time and travelling expenses   *(not included)*
Erection   "
Royalty.   "

(* Authors' note - Pounds Sterling, Shillings, and Pence)

On file in the National Archives in Pretoria is a memorandum, which
was presumably written by van Trotsenburg, to summarize his opinions of the
various installations viewed by him [25]. His reasons for selecting the Siemens
equipment are quite clear.

> *Information is scarce and unsatisfactory. More is still
> required. Inventions are being patented. Others endeavour to
> invent something else or are trying to improve on certain
> inventions and to patent the improvements.*
>
> *I saw the installation in Paris, in my opinion it works
> too slow. There is room for improvement, but how to improve
> has not yet been determined. No guarantee could be given.*
>
> *I saw the instruments of Marconi at the Wireless Coy.
> London. Specified tasks could not yet be worked. It is still a
> novelty. Full particulars about same are not yet known.*
>
> *Sometimes it works well, sometimes not when a strong
> wind prevails, there is a considerable difference of which no
> satisfactory explanation can be given.*
>
> *The purchase price for the instruments is cheap
> enough, but it was stipulated:*
>
> *That we should erect same at our own cost.*
>
> *That we should pay the sum of £100. for each
> instrument per annum, thus 600 Pounds Sterling per annum for*

6 instruments and that we should bind ourselves by contract for a certain number of years.

These instruments are now also being made in Germany, however these can also not be fully guaranteed, owing to circumstances which prevent the good working of same, as i.e. atmospheric electricity.

It is however possible, that the working here may be better than expected.

I would have preferred to experiment with 2 instruments, but should we require these instruments and they prove to be of use to us, we would want them to be fixed up at once.

There would not be sufficient time to indent for a repeat order. As these instruments would be required for the Pretoria forts, they are willing to give us a reasonable guarantee, especially as they take into consideration the present state of affairs. Whereas it was my intention to take two instruments for trial, I have now, under the abovenamed guarantee at once placed an order for 6 with Messrs Siemens & Halske of Berlin, even at a slightly increased price in order to ensure the shortest possible time within which to deliver.

These instruments cost 110 Pounds Sterling each, nothing further (no royalty.)

Three instruments will be delivered immediately under the guarantee given, should they work well, the three remaining instruments will be taken from their Johannesburg Agency.

The instruments must be delivered at Pretoria, payment will be made on delivery, consequently the instruments are not to be shipped as if for the S.A.R. but as for the Agency of Messrs 'Siemens Ltd' Johannesburg in order that the said Agency may, should occasion arise, make the eventual necessary declarations to ensure the free transit of same.

Should however, in case of war with England these instruments be confiscated or destroyed by the enemy, then the Agency will receive from us a reasonable refund for same.

The order has thus been made out as follows:
Six instruments complete with necessary spare parts and six telescope poles.

At the time of discussion, the result of which was the receipt of a Cabinet Letter dated April 20th. 98 No 255/1898 G.R. it was understood that I should order the instruments if I were satisfied with the information received.

> *On August 24 I again applied to the Cabinet and was informed that my idea was good, in order therefore to save time I am ordering the instruments without again referring the matter to the Government or to the Executive Council.*
>
> *If the instruments are to be of any use, they must arrive soon and a special point is, that the order be kept secret, as discretion could not be observed, if I were again to openly discuss the matter.*

Thus finally, in a procedure reminiscent of many current procurement processes, we arrive at an eventful moment in the long history of radio communications - possibly the first order for commercial radio equipment in the world! The accompanying figure is believed to be of van Trotsenburg's file copy (Figure 13.8). The order itself reads [26]:

> *With reference to your telegraphic communication of 20th. inst, stating:-*
>
> *We can deliver three stations within fourteen days, the rest within one month. The price at Berlin, one hundred and ten pounds each and a pole of forty metres will be required to work out a distance of fifteen K.M. We will then guarantee good working up to this distance, supposing there is good management and excepting atmospheric interruptions:-*
>
> *...and adverting to our personal conversation of yesterday, I now have the honour to inform you, that we accept your offer to supply 3 "spark-telegraph instruments" complete at £110 each at Berlin, payment to be made after the instruments have been erected in Pretoria and found satisfactory and in accordance with your guarantee.*
>
> *If these instruments prove satisfactory and answer our purpose, we are prepared to place an order for further 3 complete instruments at the same price and conditions as aforementioned, the six instruments to be forwarded from Berlin as stated in your wire:-*
>
> *I would further request your attention to the necessary poles for these instruments according to our conversation and especially concerning the following:*
>
> *(1) Material to be of light weight.*
> *(2) A simple way of erecting and breaking same down,*
>
> *perhaps your firm already has a simple method, if not, to enable us by a simple way of construction, to lower an erected pole.*
>
> *We should require the poles to be delivered with the instruments. Enclosed please receive the 'Electrical Engineer' London edition, No 14, 1898, page420.*

> In case we do not require to use the complete length of the pole and as in such case I would not like to use the pole higher than necessary, I trust that the pole will be constructed in such a way to enable us to do away with certain parts thereof if necessary.
>
> I would further request you to duplicate all such parts of the instruments subjected to hard wear and also those liable to breakage.

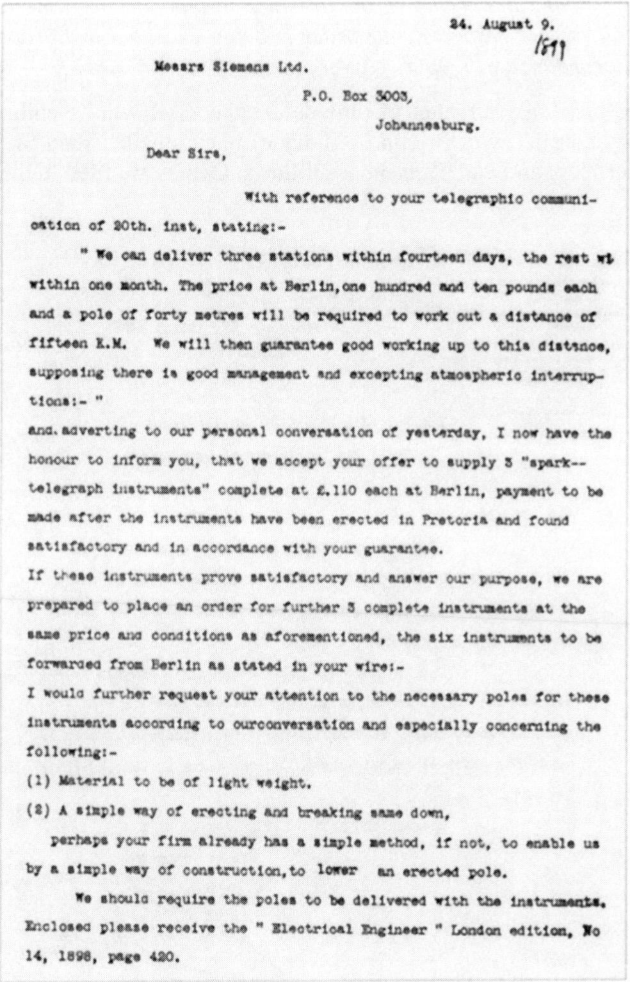

Figure 13.8. First page of letter dated 24 August 1899 from C. K. van Trotsenburg to the Siemens branch in Johannesburg ordering wireless telegraphy equipment for the South African Republic. (Copy reproduced by courtesy of the National Archives and Records Service of South Africa, Pretoria, City of Tshwane, South Africa).

Siemens Ltd. in Johannesburg, in their letter dated August 28, 1899, acknowledged receipt of the order [27] (Figure 13.9). After translation, it reads:

We have the honor to acknowledge receipt of your letter 1444/98 of 24th inst. and thank you for your order contained therein, which we have cabled to Berlin for immediate execution.

With regard to the poles we hope to be able to give you further information shortly.

We are trying to obtain suitable bamboo poles here. In all further respects your order is being executed in Europe in accordance with your request.

It is now only a matter of time before the ZAR will be embroiled in a bitter military conflict with Britain. A flurry of telegrams between Siemens and Halske in Germany and van Trotsenburg follows. Copies are filed in the National

Figure 13.9. Acknowledgement dated 28 August 1899 of the order placed by C. K. van Trotsenburg for wireless telegraphy equipment to be supplied to the South African Republic. (Copy reproduced by courtesy of the National Archives and Records Service of South Africa, Pretoria, City of Tshwane, South Africa).

Archives in Pretoria along with other material quoted and/or cited in this chapter. These telegrams deal with the availability of light bamboo poles in Durban to support the antenna wires to a height of 40 meters, and vertical steel masts which could act as radiators, and which would be ordered from Europe.

With tensions between Britain and the ZAR now escalating rapidly, the story of Wireless Telegraphy is overtaken by other events. On 28 September 1899 the ZAR mobilized. On October 2nd the Free State also mobilized its forces. At 5 pm on October 9, 1899, an ultimatum was delivered to the British. The fact that spring rains had fallen early, and that the spring grass needed to maintain a mounted army in the field was sprouting no doubt influenced Kruger's decision. In the ultimatum Britain is accused of breaking the London Convention of 1884 and interfering with the internal affairs of the ZAR by espousing the cause of the *Uitlanders*, and by massing its troops. Amongst others the demand is made that the troops which arrived after June 1st should be withdrawn from South Africa. Britain was given 48 hours to comply [12]. The ultimatum was ignored, and on October 11th, the ZAR and the Orange Free State declared war on Britain. Commandos were immediately dispatched to invade the British province of Natal. Britain would eventually have to draw on the vast resources of its goliath like Empire to crush the upstart David. But that is another story, ably retold by many historians in the commemorative books and journals which were published or republished a hundred years later.

The equipment ordered by van Trotsenburg arrived too late for the Boers to use. When the equipment finally arrived in Cape Town it was confiscated. The consignment was aboard five ships and had been traced through customs records. One Captain JNC Kennedy, an officer in the British Corps of Royal Engineers, records the events [28]. Kennedy comments on the Siemens equipment, and disparages the receiver "for not being encased with metal," perhaps one of the earliest references to a requirement for electromagnetic compatibility. Nonetheless, some of the equipment was regarded as superior to that of Marconi, which was intended for British use, and was duly "cannibalized" to provide some necessary bits and pieces. The British experience will be recounted in the next section.

But for the early declaration of war, the ZAR could well have become the first country in the world to have a "modern" wireless telegraph system to provide communications for its military.

The firm of Siemens Ltd. in Johannesburg was apparently subsequently compensated for the loss of the equipment due to confiscation along with other claims [13]:

> *Pay on account, Siemens of Johannesburg, value of Wireless Telegraphy, about £560 and field telegraphy apparatus £167.10.0. Enquire into balance on wireless telegraphy and remaining claims, and pay freight and other reasonable charges.*

War at the beginning of the 19th century must have been a peculiar mixture of viciousness, and gentlemanly conduct!

13.4   THE BRITISH EXPERIENCE

13.4.1   The Army

Marconi's Wireless Telegraphy patent had been issued in 1896. Although many prominent researchers in North America and in Europe had worked in this field, none can truly claim to have "invented" it [29]. One could argue that Marconi is remembered because he had the vision to capitalize on an exploitable commercial concept. In 1896 he had demonstrated his system to the British Military on Salisbury Plain. In 1898 he had the foresight to arrange for his system to be installed on the Royal Yacht during the Cowes Regatta. The Prince of Wales was thus able to send daily messages to Her Majesty, Queen Victoria, who was staying at Osborne House, on the Isle of Wight [30]. In the summer of 1899 three of the British Naval vessels involved in the summer manoeuvres had been equipped with Wireless Telegraphy. A maximum range of some 137 km (85 miles) was achieved. Summaries of the details of the exercise, the equipment and results achieved can be found in [1, and 19].

The outbreak of war presented Marconi with an ideal opportunity to promote his equipment. He contacted the British War Office and suggested that wireless telegraphy could be of considerable benefit to the British forces now on their way to South Africa. The intention was that Wireless Telegraphy could be used to control the flow of men and material into South African ports. Accordingly, a team of six Marconi Engineers under the guidance of one Mr. Bullocke, the Senior Marconi Engineer, and 5 wireless sets were dispatched to South Africa on a six-month contract, starting on November 1, 1899. Personnel from the Royal Engineers under Captain J. N. C. Kennedy were to assist the Marconi engineers [1]. Three of these portable wireless stations were assigned to Lord Methuen's column relieving Ladysmith, and two similar stations to the column under General Buller, who was tasked with the relief of Ladysmith [19]. As yet it appears that there is no evidence to suggest that the equipment was ever deployed at Ladysmith (Brian Austin, Private Communication). Heliographs and a searchlight from *HMS Terrible* mounted on a train were the only reliable means of long distance communications (Figure 13.10).

Upon their arrival in Cape Town on November 24, 1899, Marconi's Engineers were "invited" to volunteer for active service in the field (Figure 13.11). The equipment had originally been designed for permanent installation on board ship, and would have to be modified for mobile use [3,4]. The large battery supplies required were mounted in the bottom of a wagon. The sensitive relay and coherer, which comprised the receiver, were suspended from metal hoops in the middle of the wagon. The operator could work the Morse key while standing on the ground at the back of the wagon. On December 4, 1899, Captain Kennedy arranged for a demonstration of the system for military staff and foreign attachés at the Castle in Cape Town. These demonstrations were reported as being successful.

While the necessary modifications were being made to enable deployment of the Marconi equipment in the field, Kennedy inspected the Boer

Figure 13.10. Searchlight from *HMS Terrible* mounted on railway wagon for communications with the beleaguered forces at Ladysmith, Natal (Photo no. 26443b reproduced by courtesy of the National Archives and Records Service of South Africa, Pretoria, City of Tshwane, South Africa).

sets which had been ordered from Siemens and Halske, and which had been confiscated by customs on arrival in Cape Town. Kennedy cannibalized some of the Siemens and Halske equipment, but left the accompanying metal antenna masts because of the apparent complexity for deployment. Instead he opted to use bamboo poles in 30 ft (9 m) lengths to support the antennas, which on ship would have been suspended from the mastheads [31] (Figure 13.12).

On December 11, 1899, the engineers left for De Aar, some 220 km south of Kimberley. De Aar served as a railhead and dispersal point for troops and equipment. It was found that the rigid British wagons were unsuitable for the

Figure 13.11. Marconi company engineers in South Africa at the beginning of the South African War. (Photo no. 1006.262 from the Marconi and GEC Archives and Collection by courtesy of Marconi Corporation plc).

Figure 13.12. Early photograph of wireless telegraphy antenna believed to be at the Modder River encampment. This appears to have been constructed of bamboo poles, which caused problems because of splitting of the poles. (Photo reproduced by courtesy of the Royal Naval Museum, Admiralty Library Collection, MSS 277 Papers of Charles Beadnell).

wireless telegraphy equipment, and the equipment was transferred to Australian pattern sprung wagons.

Problems arose as soon as attempts were made to deploy the equipment. The untreated bamboo poles developed cracks, which rendered them unsuitable for use to create antenna masts. Since the antenna was a crucial part of the system, use was made of 6 ft linen kites (Figure 13.13) loaned from the Balloon Section of the Royal Engineers, and flown with conductive wires. Captain Kennedy reported on the difficulties experienced in flying two of these kites simultaneously at both ends of a wireless telegraph link [31]. Despite all the problems, wireless contact had been established over a distance of some 80 km, albeit by means of an intermediate relay station. Attempts to operate the equipment continued into early February 1900. These attempts were consistently thwarted by the extreme weather conditions encountered. These ranged from severe dust storms, to lightning storms, or winds which were either too light to allow the flying of kites, or strong enough to tear away balloons (Figures 13.14 and 13.15). Added to these was the poor conductivity of the ground around De Aar compared to that of Salisbury Plain where the original demonstrations were held. The field trials were regarded as unsuccessful.

In a paper delivered by Marconi on February 2, 1900, at the Royal Institute, Marconi defended his engineers and equipment and blamed the failure on the military authorities in South Africa. Not surprisingly, the Director of Army Telegraphs took umbrage at Marconi's remarks, and on February 12, 1900, he gave orders that the three sets in the De Aar/Kimberley area as well as the sets sent to support General Buller in his attempts to raise the siege of Ladysmith in Natal, be withdrawn from service.

Figure 13.13. George Kemp pictured outside a store of the National Museum of Science and Industry, holding a kite that was used during the transatlantic experiments, as part of the celebrations of the 25th anniversary of bridging the Atlantic by wireless. The kites used in South Africa may have been of a similar design. (Photo no. 1020.002 from the Marconi and GEC Archives and Collection by courtesy of Marconi Corporation plc).

Figure 13.14 (left) A very early photograph illustrating the use of wireless telegraphy in wartime. These aerials hanging from balloons served radio transmitters and receivers operated by the Royal Engineers in South Africa.

Figure 13.15. (right) Captive observation balloon of the type that may have been used to hold aerial wires aloft during wireless telegraph experiments at Modder River (Photo no. 26466 reproduced by courtesy of the National Archives and Records Service of South Africa, Pretoria, City of Tshwane, South Africa).

13.4.2 The Navy

For Marconi, this rejection by the Army was, in hindsight, a "fortunate" disaster. The naval manoeuvres the previous year had adequately demonstrated the viability of Marconi's equipment at sea. The British Royal Navy requested the transfer of the equipment rejected by the Army with a view to using the system in the naval blockade in Delagoa Bay. This is the bay in which Maputo (previously Lourenco Marques) is situated, and the closest harbour facilities for importation of war material for the Boer forces. The five Marconi sets were installed on the cruisers *HMS Dwarf, Forte, Magicienne, Racoo,* and *Thetis* by March 1900.

With the permanent installation made possible by shipboard use, the equipment soon came into its own. The ships were able to cover much larger search areas, out of sight of each other, and still maintain communications - regardless of whether it was day or night, and in just about all conceivable weather conditions. Moreover, with the *Magicienne* lying at anchor in Delagoa Bay and serving as a relay station, it was possible to communicate with the Naval Commander in Chief in Simonstown by means of a combination of wireless telegraphy and landline. Simonstown is more than 1800 km from Maputo along the coast.

For effective use of the equipment, the masts of the ships were extended to accommodate the long wire antenna used. Given the ideal conditions for radio propagation by ground wave offered by the highly conductive sea water, it was possible to communicate regularly over long distances. On April 13, 1900, for example, a range of some 85 km was reported to have been achieved.

By November of 1900 the war had entered a new phase, with the Boer forces resorting to Commando (guerrilla) tactics for another 18 months. There was in effect no continued need for the naval blockade, and by November the wireless telegraphy equipment was withdrawn from service and put into storage.

However, because of the undoubted success of Marconi's equipment in the Navy during the summer manoeuvres of the previous year, and during the blockade of Delagoa Bay, the British Navy equipped 42 ships and 8 shore stations with similar equipment by the end of 1900.

13.4.3 The Essential Difference

Austin [1] provides a valuable discussion concerning the reasons for the failure of Marconi's equipment when used by the British Army. Given all the facts one must conclude that the antennas, the geology and the meteorological conditions encountered around De Aar during summer all contributed to the failure.

The antenna was a crucial part of the equipment, and its length essentially determined the wavelength or frequency at which the antenna worked, in essence, the longer the wavelength, the better the propagation over different types of terrain. The reader is reminded of the problems encountered with trying to use the 9 m (30 ft) bamboo poles. It was thus unlikely that any two sets would have operated at close to the same frequency, thus contributing to the inefficiencies of the system.

The ground conductivity around De Aar and Kimberley was also significantly less than that which one would regard as typical at Salisbury Plain for the probable frequencies of operation. Attempts were made to improve the ground connection by using sheets of metal. Thus not only were the antennas less efficient in South Africa, but the propagation characteristics were also worse - all due to the lower ground conductivity. Given the comparatively high conductivity of sea water, and the good conditions prevailing on board ship as regards antenna rigging, the success achieved by the Royal Navy is hardly surprising.

13.5   AFTER THE SOUTH AFRICAN WAR

The Siemens equipment was sold at auction after the war by the Quartermaster General's Department. The purchaser was a young Post Office Engineer, F.G. Parsons. As the result of an article by Rosenthal, which was published in the (Johannesburg) Star newspaper, the equipment was acquired by the War Museum of the Boer Republics in Bloemfontein. In a letter dated December 19, 1931, and directed to Mr Marthinus van Schoor, who at the time was closely involved with the War Museum in Bloemfontein, Parsons states:

> *...As regards the authenticity of the statements made in the Radio Announcer I can only vouch for the fact that the apparatus was purchased by me from the British Military Authorities at a sale held at Fort Knokke, Cape Town during 1902.*
>
> *Mr. Rosenthal went to considerable trouble to find out the early history of the apparatus, and the information he gathered coincides with what I learnt myself from time to time in conversation with military and other personalities.*
>
> *With reference to the set being the earliest of its kind intended for use in warfare, it is safe to conclude that this is the case as Marconi's first patent was taken out only two years prior to the outbreak of the Boer War, and the design of the set in question is similar to that used by Marconi at that time.*
>
> *I think perhaps I could find the receipt for the purchase of the apparatus by me, if this would assist, but it will mean unpacking the apparatus.*
>
> *I might mention that I have had another offer for the set.*

The reference to the *Radio Announcer* was in fact to another article by Rosenthal [32] in which he describes the history and functioning of the early Siemens equipment referred to.

From the correspondence in the archives at the War Museum of the Boer Republics it is evident that the Ruhmkorff Coil, which formed part of this set as used by Parson's, was sold to a Carl Jeppe, who, it is understood, was working with X-rays at the Johannesburg General Hospital. Efforts by van Schoor to trace the missing Ruhmkorff Coil were unsuccessful. However, some time later, as reported in [33] he happened across a part of the apparatus in the Railway Museum at Johannesburg. The same article refers to the donation of a replica of the original Siemens receiver to the South African representatives of Siemens.

The equipment purchased by Parsons is now housed in the War Museum of the Boer Republics in Bloemfontein (Figure 13.16) and is described in [35], along with the provenance as established to date. The South African Corps of Signals Museum, South Africa has the only other original remaining Siemens receiver in South Africa, if not the world.

In various Annual Reports of the Postmaster-General of the Cape of Good Hope published in the years 1899-1903, it is clear that the Cape Government kept a close eye on the developments in the field of Wireless Telegraphy. The two factors of interest were its use as a means of regular communications, and communications to and from lighthouses around the treacherous Cape coast.

Klein-Arendt [36] describes the history of the application of wireless telegraphy in the German Colonies from 1904 to 1918. After describing the

Figure 13.16. Wireless telegraphy equipment (**upper left**) Refurbished Marconi Wireless Telegraph Co. Ltd. London Ruhmkorff coil (No. 31705) used for transmitting wireless telegraph signals during the South African War of 1899-1902. (**upper right**) Refurbished receiver consisting of a Siemens and Halske AG telegraphic relay (No. 23959) with coherer used to receive wireless telegraphy signals. This is believed to be one of only two original receivers used during the South African War of 1899-1902 still in existence. (**lower left**) Morse code inker (Farbenschreiber) (No. 50478) used to print Morse code signals on a paper strip for reading by wireless telegraphy operators. (**lower right**) Battery used as power source for the Ruhmkorff coil transmitter. (All photos by courtesy of the Museum of the Boer Republics, Bloemfontein, South Africa).

telegraph and optical communications in German Southwest Africa (now Namibia) he goes on to describe the establishment of a Wireless Telegraph Detachment for Southwest Africa (now Namibia) in April of 1904. Of particular interest to the technical reader is the description of some of the systems at that time.

13.6   IEEE MILESTONE IN ELECTRICAL ENGINEERING

The IEEE recognized the importance of the first use of wireless telegraphy equipment during military operations during the South African War of 1899–1902 as described above, and declared an Electrical Engineering Milestone to commemorate this event in September 1999. The citation reads:

FIRST OPERATIONAL USE OF WIRELESS TELEGRAPHY

The first use of wireless telegraphy in the field occurred during the Anglo-Boer war (1899–1902). The British Army experimented with Marconi's system and the British Navy successfully used it for communication among naval vessels in Delagoa Bay, prompting further development of Marconi's wireless telegraph system for practical uses

Two commemorative plaques were unveiled during September 1999. The first was at the Castle in Cape Town, near the site of the wireless telegraphy demonstration held by Dr John Carruthers Beattie on February 11, 1899, and itself the site of the demonstrations by the British Military during early December of 1899. The second plaque was unveiled at the TELKOM Exploratorium at the Victoria and Alfred Waterfront in Cape Town. The Waterfront is a popular tourist attraction. This second plaque also commemorates the historical links between TELKOM, South Africa's telecommunications utility, and the Port Elizabeth telephone exchange, where Jennings did his all but forgotten pioneering work.

13.7   ACKNOWLEDGMENTS

The author gratefully acknowledges the assistance of the National Archives and Records Service of South Africa, Pretoria, City of Tshwane, and of the Director and personnel of the War Museum of the Boer Republics in Bloemfontein South Africa. He is also deeply indebted to his colleagues Brian Austin at the University of Liverpool, UK, and Lynn Fordred, Curator of the South African Corps of Signals Museum for valued discussions and cooperation on earlier papers. The author also gratefully acknowledges the generous support of Marconi Communications South Africa to enable him to make a presentation in the special History Session of the AP2000 Conference in Davos, Switzerland during April 2000 on the topic of this chapter. Finally, a word of thanks to Max Schmitt in the United States, who sent the author a copy of the book "*Kamina ruft*

Nauen" by Reinhardt Klein-Arendt, which, amongst others, gives some of the early history of wireless telegraphy in what was German South West Africa, now Namibia.

REFERENCES

[1]   B. A. Austin, "Wireless in the Boer War," in *International Conference on 100 Years of Radio, IEE Conference Publication No. 411*, September 5-7, 1995, Savoy Place, London.

[2]   D. C. Baker, and B. A. Austin, "Wireless Telegraphy Circa 1898-1999: The Untold South African Story," *IEEE Antennas and Propagation Magazine*, vol. 37, no. 6, pp. 48-58, December 1995.

[3]   L. L. Fordred, "Wireless in the Boer War 1899-1902," in *Proceedings of IEEE AFRICON 1996, Stellenbosch, South Africa, 24-27 September 1996*, Vol. II, pp. 1133-1137.

[4]   L. L. Fordred, "Wireless in the Second Anglo Boer War 1899-1902," *Transactions of the S.A. Institute of Electrical Engineers*, vol. 88, no. 3, pp. 61-71, September 1997.

[5]   D. C. Baker, "Wireless Telegraphy during the Anglo-Boer War of 1899-1902," *Military History Journal*, vol. 11, no. 2, pp. 36-46, December 1998, Published by the South African National Museum of Military History.

[6]   D. C. Baker, B. A. Austin, and L. L. Fordred, "Wireless Telegraphy: Knocking on the door of the 20th century - the South African connection, in *Millennium Conference on Antennas and Propagation, AP2000, Davos, Switzerland, 9-14 April 2000*, p. 89 (abstract) - full paper on CDROM ESA SP-444.

[7]   E. Rosenthal, *You Have Been Listening ... A history of the early days of radio transmission in S.A*, Published by the South African Broadcasting Corporation, Purnell and Sons, 1974, SBN 360 00267 6.

[8]   "Oliver Lodge and the Invention of Radio," edited by Peter Rowlands and J. Patrick Wilson, PD Publications, 1994 (ISBN 1 873694 02 4).

[9]   Report of the Postmaster-General, Cape of Good Hope, for the year 1903, Cape Times, Ltd., Government Printers, G. 29-1904, 1904, p. 62.

[10]   D. J. Vermeulen, "The remarkable Dr. Hendrik van der Bijl *(Invited Paper)*," Proc. IEEE, Vol. 86, pp. 2445-2454, 1998.

[11]   H. J. van der Bijl, "Theory and operating characteristics of the themionic amplifier *(Classic Paper)*," *Proc. IEEE*, Vol. 86, pp. 2455-2467, 1998.

[12]   T. Pakenham, *The Boer War*, Jonathan Ball Publishers in Association with Weidenfeld and Nicolson, 1979, ISBN 0 297 77395 X.

[13]   J. Ploeger, and H. J. Botha, *The Fortification of Pretoria*, Publication No. 1, Military Historical and Archival Services, Government Printer, Pretoria, 1968.

[14]   Report by C.K. van Trotsenburg to L.W.J. Leyds, State Secretary, South African (Transvaal) Republic on Telegraph Communications between military camps and fortifications around Pretoria, March 2, 1898. File Reference TLD No. 1, National Archives, Pretoria, South Africa.

[15]   Letter from C.K. van Trotsenburg to Messrs. Siemens Bros. and Co. in Westminster, London, UK, stating wireless telegraphy communications problem, February 28, 1898, File Reference TLD No. 1, National Archives, Pretoria, South Africa.

[16]   *Telegraphie ohne Draht*, Zeitschrift fuer Electrotechnik, Jahrgang XV, Heft XXII, 15 November, 1897, pp. 651-654.

[17]   'Marconi Telegraphy,' The Electrical Review, Vol. 43, No.1082, August 19, 1898, pp. 264-265

[18]   Reply from Siemens Bros. and Co,. of Westminster London to C. K. van Trotsenburg's enquiry [15 above]; 26 March 1898. File Reference TLD No. 1, National Archives, Pretoria, Republic of South Africa.

[19]   H. M. Dowsett, *The History of Wireless Telegraphy*, The Gresham Publishing Co. Ltd., London, 1923.

[20]   Letter from C. K. van Trotsenburg to Siemens and Halske AG, Berlin, requesting whether they can supply wireless telegraphy equipment; 23 April, 1898. File Reference TLD No. 1, National Archives, Pretoria, Republic of South Africa.

[21]   Reply from Siemens Brothers, London to van C. K. Trotsenburg's enquiries dated 23 April 1898 [20 above]; 11 June 1898. File Reference TLD No. 1, National Archives, Pretoria, Republic of South Africa.

[22]   Reply from Siemens and Halske, Johannesburg to van C. K. Trotsenburg's enquiries dated 23 April 1898 [20 above]; 11 June 1898. File Reference TLD No. 1, National Archives, Pretoria, Republic of South Africa.

[23]   Letter from Societe Industrielle des Telephones, Paris to C. K. van Trotsenburg; 16 June 1898. File Reference TLD No. 1, National Archives, Pretoria, Republic of South Africa.

[24]   Letter from Wireless Telegraphy and Signal Company Ltd. London to C. K. van Trotsenburg, confirming discussions held on 30 June, 1899, and willingness to supply wireless telegraphy equipment to the SAR; 1 July, 1899. File Reference TLD No. 1, National Archives, Pretoria, Republic of South Africa.

[25]   Memorandum by unknown author, presumably C. K. van Trotsenburg, recording impressions gained of wireless telegraphy equipment while in Europe, and deciding on purchase of equipment; 24 August, 1899. File Reference TLD No. 1, National Archives, Pretoria, Republic of South Africa.

[26]   Order placed by C. K. van Trotsenburg on Messrs Siemens Ltd., Johannesburg for 6 wireless telegraphy sets, document no. 1444/98; August 1899. File Reference TLD No., National Archives, Pretoria, Republic of South Africa.

[27] Acknowledgment by Siemens Ltd., Johannesburg of C. K van Trotsenburg's order [19]; 24 August 1899. File Reference TLD No. 1, National Archives, Pretoria, Republic of South Africa.

[28] J. N. C. Kennedy, "Wireless Telegraphy - Marconi's System", Extracts from the Proceedings of the Royal Engineers' Committee, pp. 155-159, 1901.

[29] J. S. Belrose, "Who invented radio," The Radio Science Bulletin, No. 272, pp. 4-5, March 1995.

[30] 'Marconi Telegraphy,' *The Electrical Review*, Vol. 43, No.1082, August 19, 1898, pp. 264-265.

[31] Capt. J. N. C. Kennedy's Report to the Royal Engineers Committee 84/M/9586. 11-2-01. IGF. Royal Engineers Museum Archives, Gillingham, Kent.

[32] E. Rosenthal, "Our Boer War Radio Set - A relic of 1899," *African Radio Announcer*, pp. 300-301, October 26, 1931.

[33] "Salvaged from Past History - World's First War Radio Replica for its Makers," article in *The Friend*, 27 September 1965, p. 2 (Newspaper published in Bloemfontein, South Africa).

[34] D. C. Baker, "Wireless Telegraphy Artifacts 100 years old," *IEEE Antennas and Propagation Magazine*, Vol. 41, No. 4, pp. 60-63, August 1999.

[35] R. Klein-Arendt, *Kamina ruft Nauen! - Die Funkstellen in den deutschen Kolonien 1904-1918*, Willem Herbst Verlag, 1996, ISBN 3-923 925-58-1.

14

THE ANTENNA DEVELOPMENT IN JAPAN: PAST AND PRESENT

GENTEI SATO, *Antenna Giken Co., Ltd.;*
MOTOYUKI SATO, *Tohoku University, Center for Northeast Asian Studies, Sendai 980-8576, Japan*

14.1   INTRODUCTION

Just after the theoretical and experimental work on electromagnetics by Maxwell and Hertz, Japanese scientists recognized the importance of this work and began research on radio waves. Electromagnetic technology has played a very important role in the recent history of Japan.

This paper summarizes the historical development of this work including Meiji 36 type wireless communication equipment, Yagi-Uda antenna, split-anode magnetron and weather radar on Mt. Fuji along with some interesting descriptions of the use of this technology.

14.2   MAXWELL, HERTZ AND THEIR FOLLOWERS IN JAPAN

It was the ancient Greeks who, around 500 BC, first observed the phenomena of electromagnetism. But it was in the 19th century (after 1801) that serious research on electromagnetism began, chiefly among Western scientists.

At that time, Japan was under the rule of the Tokugawa shoguns based in Edo. In the year 1864, Iemochi Tokugawa was shogun, and Komei reigned as emperor. At this time, over two and a half centuries since the founding of the Tokugawa shogunate in 1603, an armed conflict broke out in the imperial seat of Kyoto between rival supporters of the emperor and the shogun. The same year also witnessed a devastating attack on the strategic port of Shimonoseki by a combined squadron of British, French, American and Dutch warships. The country was thus under dire threat from both within and without.

It was in this very year that the British scientist J. C. Maxwell (1831–1879) proposed his revolutionary electromagnetic theory including the concept of a displacement current. That was back in the most glorious days of the British Empire, when the Union Jack flew over the seven seas and England ruled the globe. Three years later, in 1867, the Tokugawa shogunate was overthrown in the

Meiji Restoration, and the emperor took personal control of affairs of state. In 1873 Maxwell published his immortal masterpiece "A Treatise on Electricity and Magnetism." The preface opens with the words: "The fact that certain bodies, after being rubbed, appear to attract other bodies, was known to the ancients." In Japan it was the sixth year of the Meiji era.

Maxwell went on to describe mankind's first encounter with electromagnetism: amber necklaces worn by women in ancient Greece would produce an electric charge by rubbing against clothing, thus attracting dust and small particles.

His electromagnetic theory was based on the concept of action through a medium, which was his own original idea. Yet the scholarly world was reluctant to accept his theory. It was even rejected by G. G. Stokes, whose lectures he had attended as a student at Cambridge, and W. Thomson of the University of Glasgow, renowned for his research on heat and his theoretical work on the transatlantic cable, which was successfully laid in 1850.

In the later years of his life, Maxwell served as the first director of the Cavendish Laboratory established at Cambridge University. One of the researchers at the laboratory, J. H. Poynting, introduced the concept of the Poynting vector based on Maxwell's theory, but that was in 1883, four years after the great scientist's death. Maxwell died of stomach cancer, as did his mother, at the age of forty-eight.

Today people who study electricity revere Maxwell as one of the towering giants of their field, but during his lifetime he was treated as just a scientist of ordinary ability. There is an oriental proverb, "A man's worth is settled only when he is laid in his coffin." The more innovative a man's ideas, the more difficult it becomes to assess the true extent of his achievement.

Maxwell's theory was taken up by a mere handful of German scientists. H. von Helmholtz of the University of Berlin was the first to recognize the validity of Maxwell's approach. He prompted his former student H.R. Hertz (1857–1894) to conduct a series of experiments on the existence of electromagnetic waves.

In 1886 Hertz proved the existence of electromagnetic waves with experiments at the University of Karlsruhe, thus confirming beyond any doubt the validity of Maxwell's theory, which had not until then been accepted. Those experiments were of immense significance in the history of science, indeed. But, seven years had already elapsed since Maxwell's death. It was twenty years since Maxwell had first postulated the existence of electromagnetic waves.

One can easily imagine the excitement of the thirty-year-old Hertz when he succeeded in validating Maxwell's daring but hitherto unproven hypothesis.

Hertz's diary provides a clear account of the progress of the experiment as follows (English translation was added by the authors):

Diary for 1886
> October 25
>> *Funkenmikrometer erhalten und Versuche damit angefangen.*

> *Obtained a spark-gap micrometer and started experiments with it.*

December 2

> *Gelungen, Rezonanzerscheinung zwischen zwei elektrischen Schwingungen herzustellen.*
> *Succeeded in producing resonance phenomena between two electric oscillations.*

December 3

> *Rezonanzerscheinung deutlicher, Schwingungsknoten, eigentümliche Wirkung auf Funken.*
> *Resonance phenomena becomes clear, wave nodes, peculiar effect on sparks.*

Diary for 1887

January 28

> *Augefangen die Versuche geordnet niederzuschreiben.*
> *Began to write down my experiments in proper order.*

March 19

> *Die Schwingungsarbeit fertig gemacht. Die Aufstellung der Apparate photographiert.*
> *Completed the oscillation paper. Photographed the experimental setup.*

Thus Hertz commenced the experiments on October 25, 1886, after having organized all the necessary equipment, and completed them barely five months later on March 19 of the next year. The results were published in May of the next year, 1887 in his paper,

> *Ueber sehr schnelle elektrische Schwingungen,*
> *On very fast electric oscillations.*
> *Wiedemanns Annalen, Bd. 31, S. 421, 1887*

The apparatus he used for the experiments is now on display in the German Museum of Natural Science and Technology in Munich.

Today's leading authority on antennas, Professor R.W.P. King of Harvard University, describes Hertz's experiment as follows:

> *Hertz's transmitter consisted of a straight copper wire end-loaded with large spheres or cylinders of metal and driven by a spark discharge across a gap at its center; its resonant frequency was 53.5 MHz. As a receiver Hertz used a small rectangular loop of wire with a micrometer spark gap in the middle of one side. A faint discharge across the gap as observed in a darkened room constituted "reception". Hertz built a bridge from mathematics to engineering, from the differential equations and boundary conditions of Maxwell to the wireless transmission of Rutherford and Marconi.*

Hertz's experiments on electromagnetic waves took place in 1886, the 19th year of the Meiji era in Japan. That same January Hidetsugu Yagi, inventor of the Yagi-Uda antenna, was born in Osaka.

In 1889, two year's after the publication of Hertz's paper, Hantaro Nagaoka (1865–1950), professor at Tokyo Imperial University, introduced this German scientist's work in the following series of articles in Japan,

Experiments of Herr Hertz,
Journal of the Scientific Society, Series 7, Vol. 61,62,63,66,
1889

Prof. Nagaoka later became President of Osaka Imperial University, and is known as the developer of the Saturnian model of the atom.

14.3   MARCONI AND THE FIRST JAPANESE WIRELESS COMMUNICATION

In 1895, nine years after Hertz's experiments, the Italian G. Marconi (1874–1937) conducted successful experiments on wireless telegraphy by using electromagnetic waves. He was barely twenty-one at the time. His father was a businessman in the northern Italian city of Bologna. For his experiments Marconi attached an induction coil to the same type of spark discharge gap as Hertz had used and caused it to oscillate. He attached one terminal to a lead wire and a metal cylinder atop a tall tree, while the other he buried underground attached to a metal plate. This was what we now call an antenna. He constructed a similar arrangement for receiving, connected with a form of the coherer invented by the Englishman O. J. Lodge, to detect radio waves.

In December of that year Marconi succeeded in transmitting radio waves from the rear window of his third-floor room in his father's villa in the suburbs of Bologna to a small hill that could be seen unobstructed a thousand-odd meters away, where they were received. The transmission of signals between two remote points without the use of electrical wires was truly revolutionary. The three-story building in which this historic experiment took place is still magnificently preserved today. Revered as the father of radio, Marconi was awarded the Nobel Price for Physics in 1909.

And so, since the ancient Greeks observed how the amber of a lady's necklace produced static electricity that attracted dust, and noticed how scraps of iron were drawn to magnetic ore, it has taken us almost three thousand years to harness these electrical and magnetic forces in the form of electromagnetic waves. The journey from the discovery of electricity and magnetism as they exist in nature to the practical use of electromagnetic waves has been long and arduous. Yet the ultimate success of this quest, which has brought immeasurable happiness to people around the world, is a tribute to mankind's glorious wisdom. Many brilliant scientists and engineers have contributed along the way.

The year of Marconi's invention of wireless telegraphy, 1895, was the 28th year of the Meiji era in Japan, when the war with Quinn that had broken out the previous year came to a close. The year marked the birth of Japanese antenna scientist Dr. Keikichiro Tani. Tani, a graduate of Tokyo Imperial University, served as a technical officer in the navy, where his specialty was antennas, transmitters and radars. He rose to the rank of technical rear admiral (Figure 14.1). He died at the age of one hundred. The following classical Chinese poem succinctly sums up his life:

Born with the wireless,
I have spent my years immersed in electrons.

Figure 14.1. Technical Rear Admiral Dr. K. Tani giving a talk at Sophia University in the 1970s.

In 1931 and 1934, Tani compiled two very detailed mathematical tables of Si and Ci functions. These are vital and indispensable for numerical calculations of parameters related to antennas. They were designated as military secrets in war time. These tables were more precise and covered a wider range than any previously published tables.

In the last few years of the nineteenth century, relations between Japan and Russia worsened rapidly and tensions were on the rise. The Japanese navy therefore planned on purchasing Marconi's radiotelegraph equipment in preparation for the outbreak of hostilities. The quoted price was so exorbitant, however, that the navy decided to develop its own technology. To that end it invited Shunkichi Kimura, who was a professor of physics at the Second High School in Sendai, to become a naval officer. At the time, Professor Kimura (Figure 14.2) was conducting research on wireless technology out of his own pocket, and he would demonstrate his experiments to students in the classroom. Incidentally, his father, Kaishu Kimura, was an envoy aboard the Kanrin Maru, whose captain was Kaishu Katsu, a famous politician. In 1860 the Kanrin Maru was the first Japanese cruiser to cross the Pacific Ocean to America. A monument commemorating this mission stands in Lincoln Park overlooking San Francisco Bay. The Second High School was incorporated into Tohoku University due to the postwar educational reforms. At the same time Tokyo Imperial University, Tohoku Imperial University, Osaka Imperial University and others were renamed simply Tokyo University, Tohoku University, Osaka University etc.

In his new post at the Navy Ministry, Professor Kimura became the central figure in the development of the Meiji 36 type radiotelegraph (Figure 14.3). "36" is the year of Meiji era, a Japanese calendar. At the end of 1903 around the clock, work commenced on equipping vessels with the new gear. War with Russia broke out at the beginning of the following February, 1904. To phrase it in modern terms, the project was initiated in response to an emergency demand, and was completed within the deadline, and successfully kept the costs down. [1]

14.4   SEA BATTLE OF THE TSUSHIMA STRAITS AND THE JAPANESE RADIOTELEGRAPH

The Baltic Fleet set sail from Russia intending to crush the Japanese navy after a voyage through the Atlantic and Indian Oceans. It finally arrived some six months later in the waters of the Far East. But after departing Camranh Bay in French Indochina it seemed to disappear off the face of the earth. How did it plan to link up with the Russian Pacific Squadron based in Vladivostok? Would the fleet sail out into the Pacific: and then through the Tsugaru Straits or the more distant Soya Straits? Or would it proceed via the Tsushima Straits and cross the Sea of Japan? The waiting Japanese navy was in complete confusion as to which direction the armada would come from - but that of course was all part of the Baltic Fleet's strategy (Figure 14.4). The converted cruiser *Shinano Maru*, which was patrolling the East China Sea near the mouth of the Tsushima Straits, sighted the Baltic Fleet advancing through deep mist early on the morning of May 27,

Figure 14.2. Professor Shunkichi Kimura, inventor of the Meiji 36 type radiotelegraph.

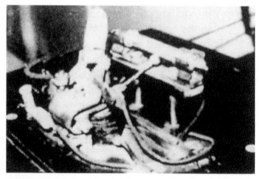

Figure 14.3. The Meiji 36 type radiotelegraph on the battleship Mikasa, Present day radiotelegraph, Replica at present.

---

[1] Each Japanese emperor has his own calendar; Meiji 1867–1912, Taisho 1912–1926, Showa 1926–1989, Heisei 1989–present.

Figure 14.4. Three supposed courses of Baltic fleet to Vladivostok.

1905. The vessel's telegraph officer promptly tapped out a signal "Enemy fleet observed at position 203, enemy moves into the east channel." Upon receiving the information, the Japanese fleet had ample time to make all the necessary dispositions to intercept the approaching force. The Japanese fleet could attack them from their most advantageous battle position, and they sank or captured most of the enemy, with only a few ships escaping to the Russian port including Vladivostok. As a result the Japanese navy utterly annihilated the enemy armada in the ensuing Sea Battle of the Tsushima Straits, one of the most overwhelming victories in the history of naval warfare. The wireless telegraph message that the enemy fleet had been sighted was a decisive factor in the Japanese victory.

Like the *HMS Victory*, the British flagship at the Battle of Trafalgar and the *USS Constitution* in the United States, the battleship *Mikasa* is still preserved in Japan today. The day, May 27th has been named Naval Memorial Day in Japan.

Meiji Japan, clearly recognized the importance of what we nowadays call electronics, and the development of the Meiji 36 type radiotelegraph was one result of this recognition. Japanese scientists, engineers and military officers studied the theories of the great European physicists Maxwell and Hertz. And they felt profound gratitude and respect for the inventor of wireless telegraphy, Marconi. Immediately after the Russo-Japanese War the navy proposed that a formal gesture of appreciation be made to Marconi. This idea was finally implemented in 1933, when he was invited to Japan as a guest of the state and awarded the Order of the Rising Sun, First Class, which is the highest honor medal that can be presented by the Japanese emperor (Figure 14.5).

Professor Kimura subsequently retired from the navy and founded Japan Radio Company. Due to reorganization of the school system in Japan, several colleges including the Second High School were later merged into Tohoku University. Most of the campus area of the former colleges is still used by Tohoku University. Katahira is the main campus of Tohoku University, where

Figure 14.5. Welcoming party for Mr. and Mrs. Marconi at Koyokan in Tokyo in 1933.

we still can find original buildings used in the 1920s. A commemorative inscription mark of the Second High School is also located in this campus.

14.5   YAGI-UDA ANTENNA

The Meiji period was followed by the Taisho era, beginning in 1912. The Second High School had been established in the city of Sendai in the Tohoku (Northeast Japan) region back in 1887. Tohoku Imperial University, the only national university in the region, opened its Faculty of Engineering in 1919. Among the faculty members of the new division was Hidetsugu Yagi (1886–1976) as shown in Figure 14.6 and 2.42. Professor Yagi poured his energies into highly original research. He chose as his field of specialization "weak-current engineering" – what is now called communications technology or electronics. Back in those days electrical engineering was dominated by so-called "strong-current" technologies such as power generation and its transmission, electric motors, generators, electric trains, lighting and so forth. As Professor Yagi once blisteringly remarked about the academic climate of his day, "Giving a lecture on

Figure 14.6. Professor Hidetysgu Yagi holding Yagi-Uda antenna.

electrical trains in the morning and doing research on electrical lights in the afternoon can not develop the engineering technologies in Japan".

Yuji Nishimura graduated from Tohoku University in March 1924, and was a classmate of Shintaro Uda (Figure 14.7). Nishimura had performed his graduate studies under the direction of Professor Yagi. The aim of his research was to measure the natural wavelength of a single turn coil using a very high frequency wave. The results were published after his graduation in 1925 as:

Y.Nishimura,
Measurements of natural wavelength of single-turn coil,
J.I.E.E.(Journal of the Institute of Electrical Engineering of
Japan) Sept. 1925

At the time, Nishimura was already an assistant at the department of Torpedoes in the Naval Arsenal in Kure, Hiroshima prefecture. The paper related to the measurement in which a VHF oscillator with a wavelength of about 2 meters was employed and a single turn of coil in the shape of a triangle, quadrangle and circle were placed in position a little ahead of a receiver, using a crystal detector connected at the middle of a receiving dipole antenna, that was placed 10 meters away from the oscillator. The natural wavelengths were measured by examining the rapidly varying values in the detected current of the receiving dipole and their dependence on the sizes of the coils. This experiment resulted in an interesting phenomenon. The presence of the single turn of a coil a little shorter than its natural frequency caused the current in the receiver to increase markedly — to exhibit a wave direction action. In contrast, the presence of a single turn coil a little longer than its natural wavelength caused the current in the receiver to markedly decrease, exhibiting a wave reflective action.

Then Yagi performed research that complemented and expanded this experiment, and published the results as;

H.Yagi,
On measurement of natural wavelength of single-turn coil,
J.I.E.E. Sept. 1925

In his experiment, he worked with both single turn coils and those nearly straight conductors which were thought to be the ultimate limit of a loop

Figure 14.7. Professor Shintaro Uda in his office.

type coil. The natural wavelength of this simple linear conductor was measured, and it was found that when the length of the conductor was shorter than the so-called natural wavelength, the current in the receiver increased, the director action was exhibited. On the other hand, when the length of conductor was longer than the natural wavelength, the current in the receiver decreased, the reflective action was exhibited.. His assistant, Takeo Sugimoto was the one who actually conducted this experiment.

The purpose of this experiment was to measure the natural frequency of the single turn coil and a linear conductor. But these phenomenon became the basis of the invention of using a director and reflector in the sharpest beam antenna. Then at the end of the Taisho era, Professor Yagi and lecturer Uda invented an antenna with outstanding directional capabilities. They published the results of their research (Figure 14.8) in English as:

H. Yagi, S. Uda,
Projector of the sharpest beam of electric waves,
Proc. Imperial Academy of Japan, Vol. II, No. 2, Feb. 1926

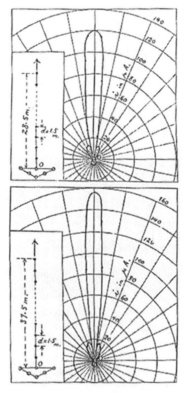

Figure 14.8. Characteristics of Yagi-Uda antennas. (H.Yagi and S. Uda, Proc. Imperial Academy, Feb. 1926) (S. Uda, JIEE, June 1927).

This paper was introduced at the meeting by Academy member Prof. Hantaro Nagaoka of Tokyo Imperial University. Only members of the Academy could present or introduce work at the meeting, and since neither Yagi nor Uda were members, Nagaoka acted as the introducer. Nagaoka has already appeared in these pages as the scientist who first introduced Hertz's experimental works on Journal of the Scientific Society in 1989. The paper began as follows:

Suppose that a vertical antenna is sending out electro magnetic wave in all directions around it. If a straight metallic rod of finite length be vertically erected within the field of its propagation, then the behavior of this metal rod will be as follows:

When the length of this rod is equal to or slightly longer than a half wave length, the current induced in it will be in phase with or lagging behind the E.M.F. caused by the electric wave, and the rod will act as a "Wave reflector."

If, on the other hand, the length be made somewhat less than a half wave length, the current induced in it will be leading before the E.M.F. and the rod will act as a "Wave director."

A single wave reflector placed behind a radiating antenna is sufficient to cause directive radiation of radio wave. It is especially efficient when placed a quarter wave length behind the radiating antenna. Again a wave director placed in front of and more than a quarter wave length distant from the radiating antenna is also effective in producing a directive radio wave.

When several director rods are arranged along a line with intervals equal to or more than a quarter wave length, the wave energy will be projected chiefly along this line, and the series of these wave directors forms what the authors will call "Wave duct" or a "Wave canal".

This antenna, which consists of a main projector with a director in front and a reflector behind, later came to be known as the Yagi antenna or Yagi-Uda antenna.

A month after the publication of the English paper co-authored by Yagi, Uda published the following paper in Japanese:

S. Uda,
On the Wireless Beam of Short Electric Waves (I),
J.I.E.E., March 1926

This was the first in a series of eleven papers detailing with the results of his research. The last paper came out in July 1929.

Two years after the publication of the first paper Yagi toured the United States giving lectures entitled and published as,

H. Yagi,
Beam Transmissions of Ultra Short Waves,
Proc. IRE, June 1928

One of the original Yagi-Uda antennas is preserved in Tohoku University. Figure 14.9 shows the antenna built in 1929 and used for wireless communication tests between Sendai and Ohtakamori, a 20 km distance. This apparatus was then exhibited at the world exposition in Brussels in 1930.

The Yagi-Uda antenna stirred only moderate interest in Japan, but its brilliant features caught the eye of scientists in the United States and Britain, where research and development work promptly got underway on how the technology could be applied to blind landings for aircraft and a new weapon - radar.

Figure 14.9. One of the original apparatus of Yagi-Uda antenna made in 1929.

14.6   KINJIRO OKABE AND HIS SPLIT-ANODE MAGNETRON

In July 1921, A.W. Hull of the General Electric Company in the US published a paper on a new discovery, that is, if a magnetic field is applied externally to a diode vacuum tube with cylindrical electrodes, the anode current will suddenly reduce at a particular magnetic field strength. Hull named this device the Magnetron. He subsequently went on to demonstrate the Magnetron's tremendous effectiveness as a switch for rapidly turning on and off a large current. He also showed how the Magnetron could be used as a detector and amplifier for radio waves.

At the end of the Taisho era, the Magnetron was selected as one of the subjects for student laboratory work in the Department of Electrical Engineering at Tohoku Imperial University. The results of these student experiments attracted the attention of associate professor Kinjiro Okabe (Figure 14.10), who went on to discover that the Magnetron produces slight oscillations of shorter wavelength under some conditions. He published his findings as,

K. Okabe,
A New Method for Producing Undamped Extra-short
Electromagnetic Waves,
Proc. Imperial Academy, Vol. III, No. 4, Apr. 1927.

Figure 14.10. Professor Kinjiro Okabe.

Okabe also described how powerful electric oscillations of shorter wavelength could be produced by splitting the anode into two or four separate units. This was the Multi Split-anode Magnetron (Figure 14.11). These findings appeared in,

> *K. Okabe,*
> *Production of Extra-short Radio Waves Using a Split-anode*
> *Magnetron (Part 3),*
> *J.I.E.E., March 1928*

In the latter half of the 1930s, several years after Okabe's discovery, international tensions rose rapidly in Europe, and then Britain, France and Germany began devoting considerable resources to developing a weapon that could harness the potential of the radio wave. Britain was the first to succeed in bringing this weapon into practical use. It was what is now called "radar", which is an acronym from "radio detection and ranging", though it was originally referred to as "radio locator".

Improving radar's capabilities required boosting the power output of the magnetron (Figure 14.12). The cavity magnetron, which is capable of producing large amounts of power, was invented by the American researcher A. L. Samuel, who patented the device in the United States in 1936. Meanwhile, in Britain, Professor J. T. Randall and Dr. H. A. H. Boot succeeded in developing their own high-power magnetron in 1939. In October 1940, under a technical cooperation agreement between the American and British governments, an experiment was

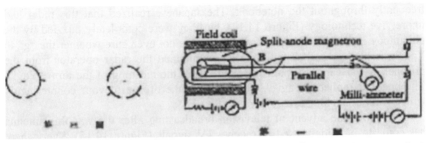

Figure 14.11. Split-anode magnetron, 2 and 4-segment type, 2-segment type magnetron with the oscillation circuit (K.Okabe, JIEE, march 1928).

Figure 14.12. Split-anode magnetron. (H.Yagi, Proc. IRE, June 1928), **(left)** 2-segment type, **(right)** 4-segment type.

conducted at Bell Laboratories in the US using an eight-split-anode high-power magnetron supplied by the British. The US went on to establish the Radiation Laboratory at the Massachusetts Institute of Technology, which carried out an ambitious program of military research and succeeded in developing high performance radars for practical use.

Meanwhile, in Japan, the Naval Technical Research Institute and a number of electrical companies were engaged in developing and manufacturing radar, but the technology was of questionable reliability and utility. During the Pacific War in WWII, the relative merits of the two combatants' radar systems was a decisive factor in determining the outcome of the battle for the skies and seas. The radar was one of the most important technologies in the battle in the Pacific Ocean during the war.

14.7   RADAR IN WORLD WAR II

More than a decade later, in December 1941, the Pacific War broke out. The following February the Japanese army occupied Singapore. The British radar equipment promptly fell into the hands of the Japanese forces, along with the notebook of a radar operator by the name of Newmann (Figure 14.13) who had the rank of noncommissioned officer. The words "Yagi array" appeared frequently throughout the notebook. The Japanese realized that this radar had impressive technology (Figure 14.14), but they were completely puzzled by the ubiquitous expression "Yagi array". They were not even sure whether the "g" in Yagi was hard or soft. They therefore summoned the radar operator from the prisoner-of-war camp to interrogate him about the meaning of the term "Yagi". The operator replied, "Yagi?", "That's the name of one of your countrymen!" with a wink of his blue eyes.

Since the advent of television broadcasting after the war this antenna has come to be widely used to receive TV signals (Figure 14.15). Everywhere across the globe rooftops now bristle with antennas as if a host of dragonflies had

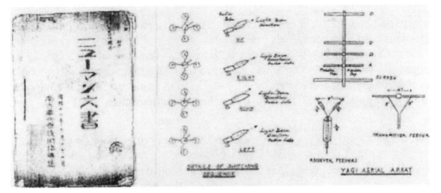

Figure 14.13. Newmann's note.

Figure 14.14. Yagi-Uda antenna employed in British radar for "Searchlight" (A Japanese translation of the book *Radar* written by Piere David, translated by Yukio Nakamura).

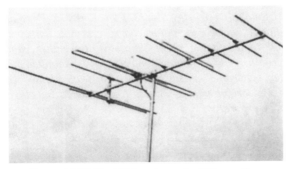

Figure 14.15. Yagi-Uda antenna for TV reception.

descended upon them. Thanks to the fact that it is so highly directional, this antenna is used in various fields in addition to TV.

On August 6, 1945, the atomic bomb was dropped on Hiroshima. On August 9, the second bomb was dropped on Nagasaki. Japan surrendered unconditionally one week later, on August 15. Both atom bombs were equipped with Yagi-Uda antennas to determine their height for detonation.

14.8   ELECTRICAL ENGINEERING MILESTONES IN JAPAN

Since the end of war, Yagi-Uda antennas have been widely used for television receiving and other wireless communications. For this great contribution, the IEEE (Institute of Electrical and Electronics Engineers) awarded its "Electrical Engineering Milestone" with title "Directive Short-wave Antenna" on June 1995. Its memorial is built at Katahira campus of Tohoku University that describes:

> *ELECTRICAL ENGINNERING MILESTONE*
> *DIRECTIVE SHORT-WAVE ANTENNA*
>
> *In these laboratories, beginning in 1924, Professor Hidetsugu Yagi and his assistant[2], Shintaro Uda, designed and constructed a sensitive and highly-directional antenna using closely-coupled parasitic elements. The antenna, which is effective in the higher-frequency ranges, has been important for radar, television, and amateur radio.*
>
> *June 1995, Institute of Electrical and Electronics Engineers*

This award was initiated in 1983 and give honor to an excellent achievement in the field of electrical engineering in the world. This was the 21st one, and the first in Asia.

The second Milestone for Japan was awarded to "Mount Fuji Radar System" (Figure 14.16). This weather radar could observe typhoons coming from the south to the main island of Japan, and contributed to giving notice and warning. Its inscription is:

> *Completed in 1964 as the highest weather radar in the world in the pre-satellite era, the Mount Fuji Radar System almost immediately warned of major storm over 800 km away. In addition to advancing the technology of weather radar, it pioneered aspects of remote-control and low-maintenance for complex electronic systems. The radar was planned by the Japan Meteorological Agency and constructed by Mitsubishi Electric Corporation.*

---

[2] At the time, Uda was not an assistant but a lecturer.

Figure 14.16. Mount Fuji radar system.

The specifications of the radar are as follows:

Frequency: 2880 MHz
Output power: 1500 kW
Pulse width: 3.5 μs

Antenna: Circular shape parabola type, 5 m diameter
Antenna revolution: 5 r.p.m. and 3 r.p.m.
Antenna radome: Metal flame type, 9m diameter

14.9   CONCLUSION

In the Russo-Japanese war electronics clinched the victory for Japan thanks to her superior technology, in the Pacific War it doomed her to defeat. Electronics thus literally decided the fate of the nation.

Electronics, as it is now called, assumed great prominence in the years after the war with Russia upon which Japan had staked her existence. Among the inventions based on the newly discovered principles of electronics that swayed the destiny of nations were the radiotelegraph and, in the Pacific War, the radar, the atomic bomb, and the radar-based proximity fuse.

Our forefathers in the Meiji era excelled in the field of electronics. But then, in the first half of the Showa period, Japan fell behind Europe and America and was reduced from the ranks of the five great powers on the world stage to the status of a defeated Third-World nation.

But Japan rose from the ashes, and today Japanese electronics is so sophisticated as to be a source of economic friction with the United States itself, hitherto the world's leader in state-of-the-art technology. The electronics industries in South Korea, Taiwan, Singapore and China are similarly poised for major advances.

At the outset of the Battle of the Tsushima Straits Admiral Heihachiro Togo signaled from the masthead of his flagship Mikasa, "The fate of the Empire hangs on this single engagement." Today it would be no exaggeration to say that the fate of the nation hangs on electronics.

REFERENCES

[1]   S. Uda, "On the wireless beam of short wave (7th report), " *Trans. Institute of the Electrical Engineers of Japan*, June 1927

[2]   G. Sato, "A recent story about the Yagi antenna," *IEEE Antennas and Propagation Magazine*, Vol.33, No.3, 7-18, June 1991.

[3]   K. Okabe, "Generation of short wave by Multi-anode Magnetrons (3rd report)," *Trans. Institute of Electrical Engineers of Japan*, March 1928.

15

HISTORICAL BACKGROUND AND DEVELOPMENT OF SOVIET QUASIOPTICS AT NEAR-MILLIMETER AND SUB-MILLIMETER WAVELENGTHS

A. A. KOSTENKO, A. I. NOSICH, *A. Usikov Institute of Radio-Physics and Electronics NASU Ul. Proskury 12, Kharkov 61085, Ukraine*;
P. F. GOLDSMITH, *Cornell University, Department of Astronomy, Ithaca NY 14853, USA*

15.1 INTRODUCTION

This article reviews the history and state-of-the-art of quasioptical systems based on various transmission-line technologies. We trace the development of quasioptics back to the very early years of experimental electromagnetics, in which this was pioneering research into "Hertz waves". We discuss numerous applications of quasioptical systems in the millimeter (mm) and sub-millimeter wavelength ranges. The main focus is on the work of scientists and engineers of the former USSR whose contribution to quasioptics is relatively little known by the world electromagnetics community.

15.2 QUASIOPTICS IN THE BROAD AND NARROW SENSE

After more than a century of its history, quasioptics (QO) can be considered to be a specific branch of microwave science and engineering. However, what is QO? Broadly speaking, this term is used to characterize methods and tools devised for handling, both in theory and in practice, electromagnetic waves propagating in the form of directive beams, whose width w is greater than the wavelength λ, but which is smaller than the cross-section size, D, of the limiting apertures and guiding structures: $\lambda < w < D$. Normally $D < 100\lambda$, and devices as small as $D = 3\lambda$ can be analyzed with some success using QO. Therefore QO phenomena and

473

devices cannot be characterized with geometrical optics (GO) that requires $D > 1000\lambda$, and both diffraction and ray-like optical phenomena must be taken into account. It is also clear that, as Maxwell's equations (although not material equations) are scalable in terms of the ratio D/λ, the range of parameters satisfying the above definition sweeps across all the ranges of the electromagnetic spectrum, from radio waves to visible light (Figure 15.1) and beyond. Therefore QO effects, principles and devices can be encountered in any of these ranges, from skyscraper-high deep-space communication reflectors to micron-size lasers with oxide windows. A good example of a universal QO device is the dielectric lens that was first borrowed from optics by O. Lodge for his 1889 experiments at $\lambda = 101$ cm [1], then used in microwave and millimeter-wave systems in the 1950-80s, and is today experiencing a third youth in terahertz receivers. Moreover, as the above relation among the device size, beam size, and wavelength is common in today's optoelectronics, it is clear that QO principles potentially may have a great impact on this field of science as well. Nevertheless, in the narrow sense, the term QO relates to the devices and systems

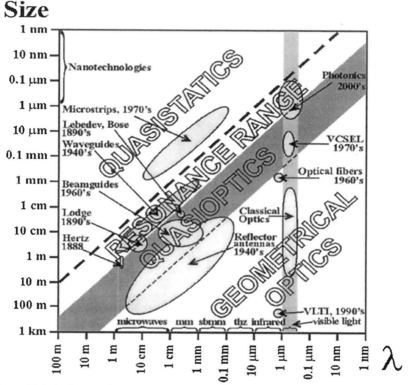

Figure 15.1. A diagram showing the place of quasioptical (QO) techniques with respect to geometrical optic (GO) and quasistatic techniques, in the plane of the two parameters - device size and wavelength. The frequency ranges are also indicated, as well as major related technologies and the dates of their emergence. The resonance range corresponds to device sizes between a fraction of a wavelength and several wavelengths.

working with millimeter (mm) and sub-millimeter (sub-mm) waves. F. Karplus apparently coined the term *quasioptics* in 1931 [2], and then it was forgotten for exactly 30 years before being used again in [3]. A parallel term *microwave optics* can be traced, however, in several remarkable books and review articles of the 1950–1960s [4-10] and others.

If compared with the classical optics of light, mm and sub-mm wave QO has certain characteristic features: here, electromagnetic waves display their coherence and definite polarization state, and they also display much greater divergence and diffraction, while direct measurements of their amplitude and phase are relatively easy.

It is difficult to find a publication in which the various historical aspects of QO are presented in a complete manner, tracing the development of this field and including an account of specific features of particular scientific problems and applications. A significant early Western publication dealing with QO is the collection of papers presented at the *International Symposium on Quasioptics* held in New York in 1964 [11]. It was L. Felsen, one of the organizers of the symposium, who should be credited with firmly establishing this term. Since then, several papers [8, 12-17] containing detailed reviews of QO principles and major applications have appeared. A book focusing on selected applications was published in 1990 [18]. In 1998, a comprehensive monograph [19] appeared, with a bibliography containing more than 700 titles. In this book, the theory of Gaussian wave beams is presented in a systematic way, together with the results of development of corresponding QO components. Here, specific solutions to many practical problems were considered, based on this important but not unique way of transmitting electromagnetic power and designing various functional systems. However, almost all of the referenced material was of Western origin.

Beginning in the early 1960s, active research and development into QO was undertaken in the USSR. There was a good background for this development: one of the most important mm-wave pioneers was Piotr N. Lebedev (1866–1912), who worked at Moscow University from 1892 to 1911 (Figure 15.2). Later, magnetrons were developed in many civil and military laboratories in the 1920–1940s. After World War II, the government of the USSR considered microwave radar to be the third most important defense

Figure 15.2. Piotr N. Lebedev in the 1900s.

technology, after nuclear weapons and intercontinental missiles. As we shall see, research into QO was done mainly in the laboratories of the USSR Academy of Sciences (now, Russian Academy of Sciences – RAS, and the National Academy of Sciences of Ukraine – NASU) located in three cities: Moscow, Nizny Novgorod, and Kharkov. It should be noted that the USSR microwave researchers always had good access to the Western scientific literature. However, after the late 1930s their papers were almost never published in international journals. Even if not classified, papers by Soviet scientists having a practical orientation had little chance to reach Western readers except through translations of Soviet journals having limited accessibility. Participation in conferences outside the USSR was virtually impossible.

The present article is thus an attempt to review the little-known QO technologies of the USSR based on the various transmission lines used, along with their numerous applications. Here, we have used several sources of information including useful reviews of the history of microwaves [20-23]. The 1960s were the "golden age" of QO, during which excellent reviews [24, 25] were published. Special credit should be given to the book [26], which contained comprehensive information on QO transmission lines of various types, and on the system design principles that corresponded to the components available in the USSR in the late 1960s. Additional information about the later developments based on hollow dielectric beam waveguides and metal-dielectric waveguides can be found in [27, 28]. We have also used interviews with the staff of R&D laboratories and reviewed formerly classified technical reports.

In order to make the proper positioning of the accomplishments of the Soviet scientists easier, we shall review them against the background of their Western counterparts. The basic idea of this review is to follow the development of QO transmission lines. Here, the following important topics will be touched only marginally: open resonators, filters based on various frequency-selective screens, diplexers and multiplexers, stabilization of solid state sources, power combining, power measuring devices of the absorption type, and cryogenic receivers. We shall mainly compare the characteristics of different types of transmission lines, and the opportunities for development of standard components, rather then consider specific devices and instruments. Nevertheless, we shall try to show major trends in research and development, and to emphasize the basic books, papers, and reviews in this field.

We shall begin with a brief survey of research into "Hertz waves" in the 1890s. This term was introduced by a handful of scientists who followed H. Hertz, carrying out early experiments with short electromagnetic waves. In fact, at that early stage they had already established all the fundamental QO principles, which were so widely and efficiently used in mm-wave technology 70 years later.

15.3   PIONEERING RESEARCH INTO THE "HERTZ OPTICS" (1888–1900) AND LEBEDEV'S CONTRIBUTION

The shaping of QO as a scientific field is closely tied to the early history of wireless communication. It was triggered by Hertz's famous experiments, which he presented on December 13, 1888 in the lecture "On the rays of the electrical force" at the meeting of the Berlin Academy of Sciences. In this presentation, Hertz convincingly proved that the nature of electromagnetic and light waves is identical [29].

When performing his experiments, Hertz tried to make the dimensions of the devices as small as possible: however, he still followed classical optical principles. Hertz used electromagnetic radiation with a wavelength of $\lambda = 66$ cm. "I succeeded," – he had written, – "to obtain the well-observable rays of the electrical force and to perform with their aid all the elementary experiments which are produced with light and heat rays". To concentrate electromagnetic power in a directive beam, Hertz had employed a parabolic cylindrical reflector made from a zinc sheet with an aperture of 2 m by 1.2 m and a focal distance of 12.5 cm (Figure 15.3). He placed a dipole at the reflector focal line, with a spark gap for connection with an induction facility of the Kaiser-Schmidt type (Figure 15.4a). The design of the receiving antenna was analogous, and a resonator in the form of two metal rods was placed at the reflector focal line. The internal ends of the rod were joined by wires passing through the reflector, and a micrometer screw was employed to regulate the spark gap (Figure 15.4b). The electromagnetic radiation was detected through the secondary spark discharge.

It is extremely impressive that although his "beam" and reflector were only 2 wavelengths in size, Hertz was able to confirm the laws of propagation, reflection and refraction formerly attributed only to "optics". He also studied polarization phenomena with reference to electromagnetic waves. Thus, in fact, he used for the first time all the QO principles that were to be employed in the

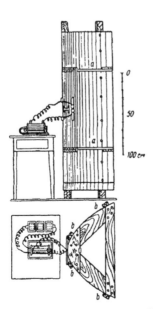

Figure 15.3. The parabolic reflector of Hertz's antenna used in experiments of 1888 with waves of $\lambda = 66$ cm (reproduced from [29]).

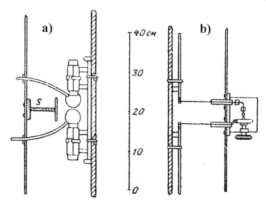

Figure 15.4. a) Transmitting and **b)** receiving dipoles of Hertz's antenna (reproduced from [29]).

future for microwave engineering, with the exception of, probably, a lens. His screens were made from tin foil, gold sheets, and wooden shields. To investigate the refraction of the beam passing from one medium to another, Hertz made a prism two wavelengths in size from asphalt having a mass of 1200 kg. Its cross section was that of an isosceles triangle having a base of 1.2 m, a height of 1.5 m, and an angle of refraction of 30°. In the polarization experiments he used a grating made from copper wires (diameter = 1cm, period p = 3 cm, thus p = 0.05λ) stretched across an octagonal wooden frame 2 m by 2 m (i.e., 3λ by 3λ) in size.

Hertz's experiments had fundamental significance and stimulated research into "optical" properties of electromagnetic waves and their practical applications. Hertz's followers, when reproducing and extending his experiments, tried to use shorter wavelength radiation, so as to improve the performance of the various components.

In 1894, A. Righi in Italy modified the Hertz dipole by introducing three spark gaps instead of a single one. This enabled him to obtain radiation with wavelengths λ = 7.5 and 20 cm [30]. One of his oscillators with a parabolic reflector is shown in Figure 15.5.

Waves of considerably shorter wavelength, λ = 6 mm, were experimentally studied in 1895 by P. Lebedev in Russia [31]. As he explained, "there appeared a need to make his (Hertz's) experiments on a smaller scale, more handy for scientific research". The turn to such short wavelengths was necessary to form and focus the "rays of electrical force" in the experiments on the interaction of electromagnetic waves with materials. Though in general, Lebedev's research program corresponded to Hertz's experiments, the dimensions of components developed by him were 100 times smaller and their technical realization at the time being was unique and was admired by his contemporaries. The primary radiator was a development of the idea proposed by Righi and consisted of two platinum cylinders 1.3 mm in length and 0.5 mm in diameter, placed at the focus of a circular-cylindrical reflector having an aperture

Figure 15.5. The design of one of Righi's oscillators with a parabolic reflector used in 1894 (reproduced by permission of the Museum of Physics, University of Bologna).

2 cm by 1.2 cm in size. The reflector was immersed in a tank filled with kerosene; the electromagnetic beam emerged from it through a mica window. The receiving antenna was made similarly: two straight-wire resonators 3 mm long were placed at the focus of the secondary reflector, where the indicator was not a secondary spark as it was for Hertz, but an iron-constantan thermocouple and a galvanometer which monitored the temperature rise and thus the incident power. In most of the experiments the distance between antennas was 10 cm.

The set of experimental components (Figure 15.6) developed by Lebedev included a wire polarizer (a grating of 20 thin wires tightened over a rectangular frame with dimensions of 2 cm by 2 cm), metallic reflectors of 2 cm by 2 cm, an ebonite prism (1.8 cm height, 1.2 cm base, angle of refraction 45°, 2 gram weight), and a quarter-wavelength phase-shifting plate made from birefringent crystals of rhombic sulphur. Thus the components made by Lebedev had dimensions of 2λ to 3λ, i.e. very similar to those of Hertz for longer waves. Besides reproducing Hertz's results, Lebedev's experiments enabled him, for the first time, to observe birefringence in anisotropic media, leading him to the conclusion of "the identity between the phenomena of the electrical oscillations and light in this more complicated case". Moreover, when taking a smaller wire radiator of length 0.8 mm and diameter 0.3 mm, Lebedev observed oscillations at a wavelength $\lambda = 3$ mm [32]. At the time these were the shortest electromagnetic waves obtained using an oscillating spark discharge. Lebedev's experiments anticipated the future development of QO methods for forming narrow directive beams and their transformation in various mm-wave systems. He wrote in 1895: "The short waves are promising in numerous applications because here, by using devices of moderate size and perfect in an optical sense, one can easily neglect diffraction phenomena, and very small quantities of the materials to be studied are quite sufficient for accurate measurements. Therefore, relatively simple experimental requirements common in optical research can be realized with the Hertz waves as well." [31]. Thus, we may state that in addition to his more widely known reputation of absolute pioneer in the experimental verification of the pressure of light on material obstacles, Lebedev was also an insightful pioneer of millimeter waves. In a letter to Lebedev dated 10.10.1899, A. Dubois

Figure 15.6. The components designed by P. Lebedev for his Hertz-type experiments at the wavelength $\lambda = 6$ mm in 1895 (reproduced from [31]).

wrote that, as a result of his work, "Russia becomes the world's small wave champion" [33]. This was not completely true as J. C. Bose in Calcutta, India, was already systematically experimenting with waves as short as $\lambda = 5$ mm [34, 35].

One of the most challenging areas was developing new radiating and receiving devices. In 1894, O. G. Lodge was the first to propose a waveguide-type feed by placing a wire radiator inside a section of a circular copper pipe closed at one end and open at the other (Figure 15.7). In his experiments he used electromagnetic waves of $\lambda = 7.5$ and 20 cm [36]. In 1897, analogous feeds based on open pipes of circular and square cross-sections were investigated by J. C.

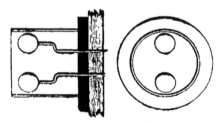

Figure 15.7. Lodge's waveguide radiator having the form of a section of a copper tube with circular cross-section used in 1894 (reproduced from [36]).

Bose in the range $\lambda = 0.5$ to 2.5 cm [34]. In the shorter wavelength portion of this range his "waveguide" was oversized (its transverse dimension was about 2.5 cm), and suppression of the higher-order modes was achieved by using an absorbing lining on the inner surface of the tube. The absorber was blotting paper dipped in an electrolyte. In 1900, in his research with waves of $\lambda = 20$ cm, J. A. Fleming proposed a rectangular-box feed [37], which can be considered to be a prototype of the rectangular open-end waveguide feed. He also employed a cylindrical lens made from paraffin for beam focusing (Figure 15.8a). In systems based on waveguide-section feeds, the investigators had used the filtering properties of the waveguide to suppress the low-frequency components of the broadband noise-like spectrum of the radiating source.

The possibility of concentrating electromagnetic power by using lenses attracted many early researchers as an alternative to reflector systems, which are not always convenient in laboratory situations. At first, in 1889, Lodge and G. Howard [1] used cylindrical lenses made from pitch. However they failed to achieve sufficient beam focusing because the lens size was smaller than the wavelength ($\lambda = 101$ cm). When in 1894 Lodge used a glass lens with a diameter of 23 cm for $\lambda = 7.5$ cm ($D = 3\lambda$), he immediately marked a noticeable focusing effect. Even greater focusing was found in Righi's experiments [38] with lenses (Figure 15.9) fabricated from paraffin and sulphur, having a diameter of 32 cm used at $\lambda = 3$ cm ($D = 10\lambda$). In 1897 Bose carried out very interesting experiments in the wavelength range between $\lambda = 5$ mm and $\lambda = 2.5$ cm [34]. He developed a number of QO devices including a shielded lens antenna (Figure 15.10), in which a feed and a cylindrical sulfur lens having 25 mm diameters

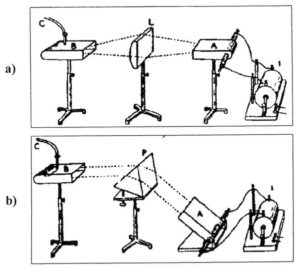

Figure 15.8. Fleming's experimental facilities in 1900 for the wavelength of $\lambda = 20$ cm: **a)** Radiator designed as a rectangular waveguide with a cylindrical focusing lens, **b)** Measurement facility for studying the refracting properties of a prism (reproduced from [37]).

Figure 15.9. Lenses made from sulfur used in Righi's 1897 experiments (reproduced by permission of the Museum of Physics, University of Bologna).

were combined in a single unit within a tube. To avoid undesirable multiple reflections, he used an absorbing lining at the tube inner surface, similar to that in his waveguide radiators. In addition, Bose was the first to employ a pyramidal waveguide horn as a receiving antenna [39].

As early as 1890, E. Branly proposed a detector based on the variation of conductivity of metallic powder under electromagnetic-wave illumination. Subsequently, most researchers used various versions of a similar device improved by Lodge in 1894 and named by him a "coherer". Bose also made a great contribution to the development of detecting devices. Trying to raise their reliability and stability, he modernized the coherer by using a spiral steel spring instead of metallic powder [34, 39]. Such a device was in fact a multicontact detector exploiting the semiconductor properties of the natural oxide coating of the spring. The detector design enabled one to control the sensitivity, allowing Bose to perform quite precise measurements with high reliability. His follow-on research on the conductivity of a number of materials under electromagnetic-wave illumination brought him to the development of a point-contact semiconductor detector based on lead sulfide. Bose's invention was registered in 1901 and later was recognized as the world first patent for a semiconductor device (dated March 29, 1904 [40]) [41]. This was, in fact, a QO device designed for experiments in the mm wavelength range. The point-contact detector (cat whisker) was located in a spherical case, and electromagnetic radiation was incident through a glass lens (Figure 2.33).

Together with the development of new radiating and detecting devices, researchers in the 1890s worked with other QO components. In particular, Lodge

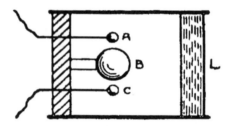

Figure 15.10. The design of Bose's shielded lens antenna of 1893 (reproduced from [34]).

[36] and Fleming [37] studied the refracting properties of paraffin prisms (refer back to Figure 15.8b). A possibility to control, over a wide range, the value of the ratio of the transmitted power to the reflected power by varying the distance between the faces of two dielectric prisms (Figure 15.11) was first noted by Bose, and this property led him to the invention of an original attenuator [34]. An important step was the development of polarimetric and interferometric systems based on available components. By using an interferometer developed in 1897, G. Hall carried out a measurement of wavelength ($\lambda = 9.12$ cm) with an error less than 1% [43]. Bose had designed a spectrometer (Figure 2.29 and Figure 9.2), which used a set of QO devices [39] in the wavelength range of $\lambda=5$ mm to 2.5 cm. additionally, he had developed a number of polarimetric systems [43] in which he used both wire diffraction gratings and metal-plate periodic structures as polarizers (Figure 15.12). The period of the latter was chosen to provide a cutoff regime for the principal mode in a waveguide-type system.

Successful research into the "Hertz optics" at the turn of XIX and XX centuries had led to the emergence of new methods of investigation of materials. Bose applied his set of measurement facilities to study the polarization properties of many natural materials and artificial substances, including various crystals and vegetable fibers [43]. He also performed the first experiments on the microwave modeling of the molecules of some optically active substances [44]. Righi proposed a quarter wave plate [38] based on the polarization-selective properties of vegetable fibers. In 1894, A. Garbasso and E. Ashkenass made a polarization-selective reflector formed by an array of dipoles, and also a prism formed by a system of dispersive dipoles ($\lambda = 7.4$ cm) [45] that can be considered to be the first device made from an artificial dielectric. In 1894, M. Birkeland fabricated "synthetic" materials, in particular, "ferro-paraffin" which consisted of iron filings (or powder) mixed together with quartz powder in paraffin [46]. The properties of water and ice were examined at various frequencies (Fleming [37], Cole [47]), as well as those of alcohol, castor and olive oils, petrol, and other liquids (Branley [48]).

Being of great fundamental value, research on the "Hertz waves" had attracted the attention of many bright electrical engineers to investigate wireless

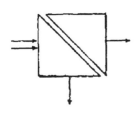

Figure 15.11. A design of Bose's attenuator based on two prisms with controlled clearance between faces of 1897. (reproduced from [34]).

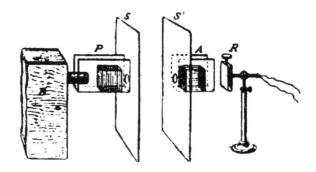

Figure 15.12. Bose's metal-plate polarizer used in 1895 (reproduced from [43]).

theory and technology, including G. Marconi, A. Popov, N. Tesla and others. In his early research Marconi was influenced by the work of his one-time teacher Righi, and used certain QO ideas of the latter. Thus, in his first trials on radiotelegraphy in 1894 Marconi employed a wavelength of 25 cm with a reflector antenna, to obtain a directive beam of radiation. This antenna [49] had the shape of parabolic cylinder (Figure 15.13) and enabled him to achieve communication over a distance of 6.5 km. A similar parabolic reflector, with a spark-gap oscillator, dipole feed, and coherer detector was used by the German engineer C. Hülsmeyer in his patent on "Device for detection of distant metal objects by using electric waves" (1904) [50]. This device was able to detect a boat by measuring the reflected signal: it was, in fact, radar. At that time, however, this invention did not attract the attention of the military, probably because of the small operating distance resulting from the small power of spark-gap oscillator and the low sensitivity of the coherer employed as the detector.

Research results obtained in 1888 to 1900 by using QO technologies had achieved the main goal of Hertz and his followers: they had proved that light and electromagnetic fields of other wavelengths had the same nature. Nevertheless, despite such a bright and promising start, in the following years, interest in both microwaves and QO methods diminished. Part of the reason was the available technology: the spark discharge oscillator approach was quickly exhausted in the effort to further reduce the wavelength of operation and to move to higher-frequency domains of research.

Lebedev was one of those who clearly realized that the development of monochromatic sources in the mm and sub-mm ranges would present the central and most complicated problem. It is worth citing his words written in 1901 [32]:

> *To obtain the waves between λ=3 mm and λ=0.1 mm, we must find a new source. To measure the lengths of these shorter waves by using interference and to observe them by a thermocouple will not pose any difficulty, however to obtain them by the already known ways is hardly possible; the wire*

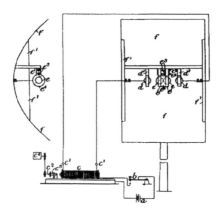

Figure 15.13. Marconi's antenna with a parabolic reflector (reproduced from [49]).

feed and resonator should be given the sizes such that if compared with them, the most perfect masterpieces of the watch-maker or jeweler would seem only clumsy metallic masses; the energy which could be accumulated on such a charged metallic wire would be negligibly small; moreover, it is absolutely unknown if it is possible to generate the waves by using a spark discharge of these negligible charges. Today we have no chance to foresee how we will succeed in solving this trouble; in any case, here we shall meet significant difficulties, and the technique of obtaining even shorter waves will be a very important step forward in the field of experimental physics.

However, there was a more fundamental reason – "Hertz optics" did not initiate any immediate demand from industry, military, governments or the public. For example, *Telefunken* experts rejected Hülsmeyer's radar as useless, causing the author to switch to mechanical engineering. In contrast, the idea of making telegraphy wireless was easily understandable and clearly had fantastic potential. Although Marconi met rejection in Italy, both from the military and from the king's court, his successful demonstrations of long distance communication in England quickly convinced important UK government and industry customers. As early as 1901, Marconi had succeeded in establishing a radio link between Europe and America across the Atlantic, over a distance of 3500 km. In 1904 both sides of the Anglo-Boer War were already equipped with wireless telegraphy. This refocused the interest of researchers, for more than 30 years, from microwaves to short, medium and long radio waves having wavelength a factor of 100 to 100,000 times greater.

15.4   EARLY SUCCESS: FREE-SPACE GAUSSIAN-BEAM QUASIOPTICAL TECHNOLOGIES

The 1920s saw only isolated research efforts undertaken with short wavelength electromagnetic radiation. In 1923, E. Nichols and J. Tear used special version of the spark gap source and produced a 0.22 mm wavelength signal [51]. The next year, A. Glagolewa-Arkadiewa in Moscow also developed a special type of spark source, which included an induction coil and aluminum fillings immersed in mineral oil. The waves were radiated by wires located near the focus of a parabolic reflector and collected by a thermal detector at the focus of a second reflector. The range of wavelengths produced by this source covered the band from 50 mm to an amazingly short 0.082 mm [52].

Large-scale research and development into microwaves was renewed only in the late 1930s, when monochromatic oscillation sources, as well as the sensitive receivers and amplifiers had become available. Another major factor was the development of hollow metallic waveguides and corresponding components. As waveguide technology advanced to shorter wavelengths, the methods of fabricating components were also improved, new materials were created, and measuring devices and techniques were developed. For nearly 20 years, waveguide technology was the dominant method of transmitting electromagnetic energy.

However, by the end of the 1950s, the decimeter, centimeter and in part the millimeter wavelength ranges were already well-developed. Scientifically and technologically the ground was therefore prepared for development of components and systems in the near mm and sub-mm ranges. Thanks to the shorter wavelength, the QO principles: (1) keep the beams well collimated so that the edge illumination of limiting apertures, focusing elements, and scatterers is relatively low, and (2) follow geometrical optics and physical optics rules for directing and focusing the beams, quickly proved to be very fruitful in these wavelength ranges and opened wide opportunities for the design of necessary components and circuits. During these years a breakthrough was achieved in the working out of the QO techniques for transmission and processing of electromagnetic waves propagating in the form of Gaussian beams.

As mentioned above, the idea of transmission of electromagnetic power from a transmitting antenna to a receiving one by a directive wave beam was tested in the initial studies of "Hertz waves". Therefore it was quite natural that these simplest free-space QO systems had been developed, studied and used in various applications before the appearance of other types of QO transmission lines.

15.4.1 Reflector and Lens Antennas

Despite low level of interest in microwaves during the 1920s and 1930s, scientists occasionally applied QO approaches when making antennas for communications with shorter waves. All of them were in fact one or another form of parabolic-dish antenna, whose principle of operation is clearly an optical one. The following are some highlights (from [53]):

- *1916.* Marconi and Franklin built a parabolic cylinder antenna for λ=15 m,
- *1922.* Marconi demonstrated a communication link, in New York, using two parabolic reflectors, λ =1 m,
- *1932.* Marconi experimented with an over-the-horizon communication system in the Mediterranean using a 50-cm diameter parabolic dish,
- *1934.* A. Clavier established the first wireless telephone link between France and the UK, λ =17 cm.

With the war steadily approaching, the new area of radar antenna design quickly developed:

- *1937.* G. Southworth reported on the focusing of centimeter waves [54],
- *1938.* A. Slutskin in the USSR designed the first 3-coordinate pulsed radar working at λ =60 cm with a 3-m (D = 5λ) dish reflector (Figure 15.14) [55].

Figure 15.14. Dipole-fed 3-m diameter reflector antenna of the first Soviet 3-coordinate pulsed radar operating at a wavelength of $\lambda = 60$cm used in 1938.

It must be noted, however, that early reflector antennas were designed with the simplest GO principles, and did not allow for wave optics effects such as edge diffraction, spillover sidelobes, and backward radiation. The feeds were also simple: half-wavelength wire dipoles placed in the GO focus that failed to provide the reduction of the reflector edge illumination that is standard today. As mentioned in [55], a computation of the radiation from Slutskin's radar antenna using an accurate numerical method shows that its 5λ reflector was able to provide the far-field power pattern having an 18° by 24° main beam size and sidelobe levels of -7 and -10 dB in the E and H planes, respectively.

The further history of reflector antennas is well known and shows that significant improvement of their performance was achieved only when both theory and practice started using essentially QO ideas and methods. In the USSR, many laboratories were involved in R&D on microwave and mm-wave reflector antennas. One of the major non-defense application areas was satellite-TV and long distance communication. A network of ground stations known as the *Orbita* system was created by the end of the 1960s. They were equipped with 12-m parabolic reflectors working in C-band [56], and provided TV coverage for the most distant parts of the USSR.

Another very specific R&D area related to QO was the development of large reflector antennas for radio astronomy observations of cosmic sources of microwave and mm-wave radiation. This large-scale and interdisciplinary problem has been addressed in numerous publications. Therefore we here restrict ourselves to work from the USSR that was closely related to or based on the principles of quasioptics (QO). In 1956, the scientists of the P. Lebedev Physics Institute, Moscow (now LPI RAS) led by A. Salomonovich and P. Kalachev started designing the first radio telescope in the USSR, the RT-22, having a 22-m diameter parabolic dish antenna. It was built in Pushchino near Moscow in 1959 [57] and for a while was the largest instrument of this type in the world (Figure 15.15). The first study carried out was of radio emission from the Sun at $\lambda = 8$ mm, and also from Venus at $\lambda = 8$ mm and 4 mm. Subsequently, the experiments

Figure 15.15. The radio telescope RT-22 used in 1959 with a 22-m diameter parabolic dish antenna in Pushchino near Moscow (reproduced from [57]).

using the RT-22 were greatly extended, especially after the telescope was equipped with more sensitive receivers [58, 59]. QO components of various types were used in the design of mm wavelength radiometers. The experience obtained with the first RT-22 in LPI RAS was later of great value when the Academy of Sciences decided to build a similar radio telescope on the coast of the Black Sea. In 1966, the second radio telescope RT-22 of the Crimean Astrophysical Observatory was built in Katsiveli, the Crimea, Ukraine [57], having a modernized antenna system. In particular, unique receivers using QO oversize waveguide components enabled radio astronomy investigations over a very broad range of frequencies including the sub-mm range. In the 1970s two even larger parabolic antennas were built — the 64-m TNA-1500 near Moscow and the 70-m RT-70 in Yevpatoriya, the Crimea [60], Ukraine. Here, one of the tools to achieve a record 0.74 efficiency was the use of a shaped subreflector [61]. In the 1988, a fixed 54-m radio telescope of the USSR Academy of Sciences, ROT-32/54, was built even further to the South, in Armenia. Similar to the famous Arecibo telescope, its primary reflector was a fixed section of a sphere. In the present case, the surface was composed of 3800 adjustable panels, and the beam could be pointed over a large range of angles by rotating the correcting secondary reflector, feed, and receiver, about the center of the sphere. The operating range of this instrument extended to $\lambda = 1$ mm, and QO components were widely used in the signal processing circuits [62].

Towards the end of the Soviet period, an absolutely unique 8-mm wavelength QO phased-array antenna was built near Moscow. Called *Ruza*, it was intended for space tracking and target recognition applications. It was assembled from as many as 144 separate antennas each a shaped as dual reflector system. Here, the feeding of radar from two 0.5 MW gyroklystrons was provided by the circular oversized waveguide (OSW) using the H_{11} mode. In the receiving circuit, free-space Gaussian-beam forming was applied [63]. Much of the above-mentioned work on QO antennas was led by or based on the theory developed by B. Kinber, who taught at the Moscow Physics and Technology Institute and worked at various research establishments, most notable of which is the R&D Institute "Radiofizika" of the USSR Ministry of Radio Industry (now a joint-stock company of the same name).

Lens antennas were suggested almost as early as reflectors. However, QO applications of classical homogeneous lens designs remained restricted to aperture phase correction in mm-wave horn feeds. At the same time, inhomogeneous lenses that were only theoretical curiosities at traditional optical wavelengths were realized in the microwave range and still attract much attention. These include the Luneburg lens, the Maxwell "fish-eye" lens, and others. The spherical or cylindrical Luneburg lens has a dielectric constant varying smoothly from 2 at its center to 1 at its outer boundary, which is the focal surface in the geometrical optics (GO) approximation. While lens dimensions are larger than microwave wavelengths, spheres with a dielectric constant varying smoothly on the scale of the wavelength are impossible to fabricate. Therefore various sorts of Luneburg lenses employing discrete dielectric layers have been devised. The oldest one consists of a finite number of spherical or cylindrical layers each with constant permittivity. Such lenses were investigated experimentally, e.g., in [64, 65]. In the USSR, dielectric lenses were developed by several organizations starting in the 1970s, as indicated in [66]; their applications were, however, restricted to defense. Today the Luneburg lens is an attractive candidate antenna for multibeam wideband mm wavelength indoor and outdoor communication systems [67], and for airborne surveillance radar applications.

It is necessary to note that reflector and lens antennas are normally fed with horn feeds. Here, an important engineering rule is that the feed should provide an illumination of the antenna edge at the level of −10 dB with respect to the central point. Then, the antenna performance in terms of directive gain is optimal. This empirical rule is approximately valid for any QO system, although for systems carrying out extensive manipulations of Gaussian beams a more conservative rule of −20 dB to −35 dB is common [19].

15.4.2   Circuits for Antenna Feeding and Gyrotron Coupling

The QO Gaussian beam approach was widely used in the development of the feeds for microwave relay stations and ground-based antennas for radio astronomy, as well as satellite and deep-space communication [68-70]. Power transfer by a narrow beam from a stationary receiving-transmitting unit to the main antenna eliminated the long coax or waveguide transmission line (and associated losses), and provided good performance over a wide frequency range. Here, the basic unit was a free-space transmission line formed by two reflector antennas whose beams were focused on each other (Figure 15.16a). QO feed systems often employed "periscopic" arrangements formed by various combinations of parabolic and flat reflectors (Figure 15.16b). Similar combined

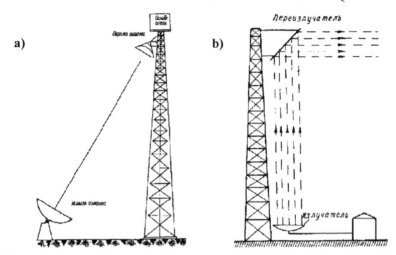

Figure 15.16. a) A microwave wireless power transmission line made of two focused reflector antennas, and **b)** a line of the periscopic type (reproduced from [68, 69]). Circa 1965.

reflectors were also used in laboratory antenna measuring ranges of the QO type [26].

In 1966, A. Gaponov of the R&D Institute of Radio Physics (RDIRP) in the Nizhny Novgorod State University, Russia proposed a new high-power source of microwaves and mm waves – the gyrotron [71]. Due to the specific geometry of this vacuum tube and the "whispering-gallery" field pattern of the operating mode, it was easier to deal with its output as a beam radiating into free space. Therefore a natural idea was to capture this beam and guide it with reflectors. Such a system was designed at RDIRP by V. Averbakh, S. Vlasov et al. [72,73]. This coupling system is able to provide efficient output by coupling up to 90% of power into the guided Gaussian beam. This design, often referred to as a Vlasov coupler, proved to be so successful that it is widely used in mm-wave systems based on gyrotron sources. Here, the most impressive application is definitely the heating of the plasma in controlled fusion machines with 110-170 GHz mm waves from 1-2 MW gyrotrons [74,75].

The principles considered above, i.e. beam focusing and −10 dB edge illumination, have been commonly used in the elements and devices of all two-aperture systems. Many of them have turned out to be applicable to other QO transmission lines. Nevertheless, it should be noted that the total length of the free-space Gaussian beam between the two apertures could not exceed the range of the Fresnel zone. This limited the area of applications of open-type QO systems and led to the emergence of periodic reflector beam waveguides (see section 15.6.2).

15.4.3 Components for Beam Manipulation

In the 1950s and 1960s, as a result of research into propagation between receiving and transmitting antennas with Gaussian beams, a number of specific QO devices were developed which were able to perform a variety of functions. A pair of prisms having a controlled small gap between their facets was an early idea of Bose [34]. Now it was used in the design of directional couplers, absorbing attenuators, SWR meters, polarization converters, etc. [8,10,76]. Another element, which found very wide application in QO technology, was a beam splitter based on a partially transparent dielectric plate tilted relative to the beam propagation direction. In particular, such an arrangement was used as a principal part of the design of a passive antenna duplexer [76].

An important stage in these developments came with the invention of various artificial dielectrics and periodic structures, as their specific properties enabled one to build a number of new QO devices and systems. The proposed artificial media can be subdivided to the structures consisting of scatterers and those of the waveguide type [8, 10]. In particular, a waveguide structure formed by a set of equidistant metallic plates was widely used in the design of polarizers, polarization-selective reflectors, polarization converters, etc. On the basis of a similar structure, a duplexer working with the circular polarization was made in the early 1960's by R. Fellers [76]. A laboratory device of this kind ($\lambda = 8.5$ mm) is shown in Figure 15.17.

Periodic structures of various types used as efficient polarization-sensitive elements have played an extremely important role in the development of QO techniques. Polarization discrimination is a property of dense grids (i.e., having small period, $p < \lambda/5$) made of highly-conducting wires or strips. Therefore, such a grid is rather a quasi-static device than a QO one. Experiments show that when a well-collimated beam is incident on finite-size periodic grid that has overall dimensions much larger than the beam size, its reflection and transmission characteristics can be predicted by using the theory of plane-wave

Figure 15.17. Open type antenna duplexer for operation in a circularly polarized mode at $\lambda = 8.5$ mm used in 1962 (reproduced from [76]).

scattering by an infinite grid. This greatly simplifies the design of grid polarizers. In the mid-1960s, on the initiative of the team of LPI RAS, led by academician A. Prokhorov (1916–2002; Nobel Prize winner 1964) (Figure 15.18) whose goal was developing mm-wave spectroscopy, the Central Design Bureau of Unique Instrument-Making of the USSR Academy of Sciences developed a technology for making tungsten-wire grids tightened on a metal-ring frame. Later such gratings were produced at the Moscow Electric-Bulb Plant having wide range of parameters: wire diameter $2b$ = 8 to 20 μm, periods p = 20 to 400 μm, and frame size $2a$ = 40 to 100 mm [77]. This enabled researchers of LPI RAS first to measure the properties of these grids [78] and then to design QO sub-mm wave Gaussian beam type measuring equipment [77], including a polarizer (Figure 15.19), Fabry-Perot interferometer, attenuator, polarization plane shifter, calibrated wavemeter, and others. The same grids were used in the first experimental polarizing devices in the sub-mm range based on hollow dielectric beam waveguide in the Institute of Radio-Physics and Electronics (now IRE NASU) in Kharkov (see section 15.6.1).

15.4.4 Measuring Systems for Spectroscopy and Plasma Diagnostics

Transmission of electromagnetic power between receiving and transmitting antennas by means of a directive beam was in use as early as the 1950s. It was utilized in laboratory measuring systems such as spectrometers and interferometers [79-83], their operation being implemented in an open type QO unit. At that time, together with designing devices for microwave bands [82, 83], similar systems were designed in the mm wave range [79-81]. A photo of the Michelson interferometer used for measuring the parameters of dielectrics in the 100 to 300 GHz frequency range [3] developed by F. Sobel in 1961 is shown in Figure 15.20.

Subsequently, similar devices found wide application as laboratory measuring systems. In particular, in 1967 a spectrometer based on QO circuitry was designed by N. Irisova *et al.* in LPI RAS for solid state research in the

Figure 15.18. Alexander M. Prokhorov, 1974.

Figure 15.19. Millimeter-wave wire grid polarizer of 1968 (reproduced from [77]).

Figure 15.20. Open type Michelson interferometer used in 1961 for the study of dielectric characteristics in the frequency range f = 100–300 GHz (reproduced from [3]).

wavelength range of λ = 0.5 to 2.5 mm at liquid helium temperatures. Here, a paraxial plane-polarized wave beam was first collimated by a lens, guided by the set of reflectors, and eventually injected into a helium cryostat through a Teflon window [84]. The second lens, the material to be studied, and the InSb-based receiver were placed inside the Dewar flask. This system enabled analysis of absorption spectra of solid state samples cooled to helium temperatures with magnetic fields from 0 to 5 kO in the aforementioned wavelength range. In the USSR, this area of research was nicknamed as "BWT-spectroscopy" due to the use of backward-wave tube oscillators (BWT) as sources. These mm and sub-mm wave BWTs were developed in the R&D Institute "Istok" in Fryazino near Moscow (now "Istok" State Co) by M. Golant and his team [85]. The main characteristics of these sources were quite large output power (1-10 mW), continuous wideband tuning (50-100%), and one-to-one correspondence between the supply voltage and output frequency and power. Their working band reached 1200 GHz, and to cover the whole sub-mm wave range one needed 5 or 6 separate BWTs. In the 1970–1980s, the work on mm-wave spectroscopy employing open QO systems was greatly advanced by A. Volkov *et al.* in the Institute of General Physics of the USSR Academy of Sciences in Moscow (now IGP RAS) [86,87]. These researchers made wide use of additional focusing lenses when building complex sub-mm-wave interferometric circuits. This research was very much alike and in line with work carried out by their Western counterparts [88,89].

     Besides spectroscopy, the second major application area of sub-mm waves until the 1990s was hot-plasma diagnostics. However, the development of corresponding technologies in this area followed different paths in the West and the USSR. Western plasma interferometers were based exclusively on open Gaussian-beam circuits, while *Tokamaks* in the USSR always used closed ones: first scaled standard waveguides then oversized metal-dielectric structures (see sections 15.6.1-15.6.3). This was because the *Tokamaks* built in the USA, Japan and Europe were constructed later than those in the USSR, and were designed from the start to accommodate large plasma cameras, and also because free-space components were well developed by that time. An example of successful application of the free-space (open) Gaussian-beam transmission line with periscopic circuitry was the 4-channel interferometer designed in the mid-1970s for the French experimental nuclear fusion machine *Tokamak-TFR*, with a HCN laser (λ=0.377 mm) as the source [90].

15.4.5   Long-distance Microwave Power Transmission

High-directivity narrow-beam antenna systems had been always attractive for microwave wireless power transmission (WPT). N. Tesla first suggested and tested this idea in 1899 [91]. In the USSR, a pioneer of this research was S. Tetelbaum (1910–1958) (Figure 15.21) at the Kiev Polytechnic Institute (now National Technical University "KPI"). In 1945 he considered the power efficiency of such a transmission line [92, 93]. He then became a head of the Laboratory of HF Currents at the Institute of Electrical Engineering of NASU (IEE NASU), whose staff worked during the period of 1948–1958 on several R&D projects related to WPT. The tasks of these projects encompassed the following: development of collimated beams and their focusing at given distances; determination of the efficiencies that could be obtained, taking account of atmospheric losses; development of microwave beams carrying CW power at the 10^3 to 10^4 kW level. To achieve these ambitious goals, Tetelbaum proposed and built in IEE NASU so-called "polyoscillators", in which the sources, integrated with in-phase radiators, were placed on a common surface [94]. Later he worked on the design of a city electric bus supplied with power from ground-based klystrons. However, the general level of vacuum electronics in the 1950s was not adequate for creating high-power systems. At the same time, the idea of "polyoscillators" can be considered as a prototype of the power combining arrays and phased antenna arrays which are now common in microwave technology. The projects performed in IEE NASU and KPI served as seeds for the development of sophisticated microwave sources in Kiev in the 1960–1990s. Here, it is important to note that the near-field gain of the aperture antennas implemented in microwave WPT was estimated with the aid of Physical Optics (PO) [95]. This analysis had shown, for example, that elliptical reflectors should be more efficient in this application than parabolic ones. It is interesting to find that in parallel to WPT with essentially Gaussian beams, a completely different idea had been actively pursued in the USSR until the early 1960s. This was long-distance microwave power transmission using a network of oversized circular waveguides [23] (see section 4.1). Many other researchers worked on microwave WPT [96, 97]. In 1968 P. Glazer in the USA realized that a potentially cheap and clean source of electric power could be an orbiting solar power satellite transmitting power to the Earth with a directive microwave beam [98]. In the USSR, theoretical feasibility investigations of this idea were also carried out in the 1980s [99]. Another fascinating proposal was connected with microwave-beam fed spacecraft. The originator of this idea was the Russian spacecraft and rocket pioneer K. Tsiolkovsky (1857–1935, Figure 15.22), who suggested it in 1924 [100]. His 80-year old paper is worth citing:

Figure 15.21.
Semion I. Tetelbaum, 1956.

Figure 15.22.
Konstantin E. Tsiolkovsky in the 1930's.

Finally, there is the most attractive way to acquire a velocity. It consists in the transmission of power to the shell from outside, from the Earth. The shell itself can avoid carrying a material power source, i.e. a weighted one, such as explosives or fuel. It will be transmitted as a parallel beam of electromagnetic rays of small wavelength. If its size does exceed several tens of centimeters, then such electromagnetic "light" can be directed as a parallel beam with the aid of a large concave parabolic mirror towards the flying airplane, and produce there the work needed for throwing away the particles of air or stored "dead" material, and for obtaining the space velocity already in the atmosphere.

Forty years later a microwave WPT system was proposed to bring a spacecraft into orbit around the Moon [101]. The ground-based antenna was envisioned to be an array of 1000 parabolic elements ($\lambda = 3$ cm, diameter 10 m), and the on-board antenna, when unfurled, was to be a single parabolic one with a diameter of 100 meters (Figure 15.23). It was estimated that the latter could collect more than 30% of the radiated power at a distance up to 1000 kilometers. However, realization of these technically complicated global projects can be hardly expected in the near future. The solutions of more specific small-scale WPT problems seem to be readily accessible and are supported by a number of experiments. In 1964, a small helicopter powered by an electric motor was demonstrated. It maintained an altitude of 17 m during a flight lasting 10 hours, fed only by a microwave beam [102]. In 1975, a successful experiment was reported on microwave wireless power transmission of 30 kW power over a distance of one mile, on a ground test range [23]. In the 1990s microwave WPT technology was actively pursued at Kyoto University, Japan, where a phased array of printed radiating elements was used for forming the microwave beam. A detailed review of the research in this area can be found in [103].

These could be considered as the first successful demonstrations of microwave WPT if not for an amazing story that has been recently published in the Ukrainian newspaper *Zerkalo Nedeli*, no. 20 (445), 2003 [104]. In 1945,

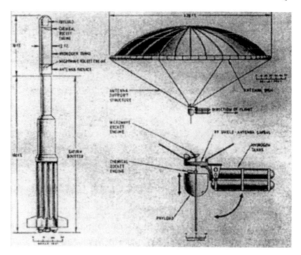

Figure 15.23. The design of parabolic antenna for the microwave engine designed to bring a spacecraft into orbit around the Moon (reproduced from [101]).

Soviet children at a summer resort *Artek* in the Crimea presented visiting American ambassador A. Harriman with a magnificent 1-m size wooden emblem of the American national bird. The feathers of the eagle had various colors due to the different sorts of wood employed. The emblem was placed on the wall in the ambassador's office in Moscow and during eight years survived four senior officials until it was found that a tiny microphone integrated with a small transmitter had been skillfully planted inside. It is hard today to imagine a realization of such project with components available in the mid-1940s, however the article insisted that the device was powered by microwaves beamed from a distant location, believed to be in a nearby building. Today this emblem is on display in the CIA "museum of spying" in Langley Virginia.

15.5   ALTERNATIVE: METALLIC OVERSIZED WAVEGUIDES (SINCE 1953) – QUASIOPTICS IN DISGUISE

In view of above-mentioned difficulties with free-space Gaussian-beam transmission lines, the interest of researchers was attracted in the early 1950s to metallic waveguides having cross-sections considerably larger than of the standard ones - so-called oversized waveguide (OSW). This means of transmission may have significantly lower attenuation than standard waveguides, but parasitic excitation and propagation of higher-order modes is possible and must somehow be handled. OSWs turned out to be quite promising and were used as basic building blocks for various systems and devices. In principle, one can study any closed waveguide by expanding the fields in terms of a modal series, and therefore we may say that OSWs are not essentially QO transmission lines. However, the duality between the modal and ray field representations was revealed by L. Felsen [105] and B. Katsenelenbaum [24] as early as in 1964. They demonstrated that summing up the modal series in multi-mode OSWs is asymptotically equivalent to using ray optics. For the analysis of long-distance propagation in regular OSW, ray optics is the superior approach for analysis and engineering, while diffraction from irregularities is more easily accounted for using modal expansions.

15.5.1 Circular Waveguide with the H_{01} Mode

The axially symmetric magnetic H_{01} mode in a circular metallic waveguide has an amazing property: as the cross-section is increased, its attenuation tends to zero considerably more rapidly than that of other modes. This is explained by the fact that the H_{01} mode does not excite longitudinal currents in the metallic walls, unlike all the other modes of hollow waveguides. In the 1950–1960s, this was the reason for increased interest in circular cross-section OSW with the H_{01} mode, as a promising multi-channel long-distance communication line. Large-scale research had been carried out on the physical properties of these waveguides, many functional devices had been developed, and experimental waveguide communication lines had been built. A review paper [106] summarized the results of a complex research project at Bell Telephone Laboratories. In the USSR this paper was published in a book [107], which collected a number of the key Western publications that comprehensively covered all aspects of the problem, including research at millimeter wavelengths. It concentrated on the examination of the H_{01} mode in circular waveguide, and on the experimental communication transmission line (length 125 m, diameter 120 mm). This line operated at a frequency of 9 GHz, i.e., with $S/\lambda^2 = 10.2$, where S is the cross sectional area of the waveguide. In the USSR, similar work started in the Institute of Radio-Engineering and Electronics of the USSR Academy of Sciences (now IRE RAS, Moscow) in 1953 (the year the institute was established), in the group led by Y. Kaznacheyev. The designers developed a waveguide transmission line in the wavelength range $\lambda = 5$ to 8 mm. Although this was a challenging problem, they had considerable success, developing all the necessary components for a circular H_{01} mode waveguide system: mode converters, matching transformers, angled bends, and directional couplers. They also developed measuring circuits, and studied the characteristics of components and the attenuation in the actual OSW. These results were published as a book in 1959 [108].

A number of problems concerning specific features of implementation of such waveguides with angled and smooth bends were highlighted in the later book of R. Vaganov, R. Matveev and V. Meriakri of IRE RAS [109]. Here a copper pipe waveguide of 60 m length was designed and constructed (Figure 15.24). Adjustment of the sections was carried out with the aid of precise optical instruments. In this OSW with inner diameter 60 mm ($S/\lambda^2 = 44.2$), the measured attenuation at $\lambda = 8$ mm was $2.5 \cdot 10^{-3}$ dB/m, compared with $1.85 \cdot 10^{-3}$ dB/m predicted by theory. The results of this project were released to the R&D Institute of the USSR Ministry of Communications in preparation for industrial

Figure 15.24. Experimental communication transmission line based on circular OSW using the H_{01} mode (f = 36 GHz) designed in 1957 at IRE RAS (reproduced from [108]).

production. The main intent was to provide a large (for the time) mm-wave bandwidth, typically from 50 to 100 GHz. However, in the early 1970s, research on circular OSW using the H_{01} mode as a communication line was terminated. The reason was the appearance of low-loss optical fibers, which solved the problem of long-distance communication on a new technological level. It should be noted that the operational principle of many devices used with a circular H_{01} mode waveguide system was the conversion of the latter into the H_{01} mode of rectangular waveguide, implementation of the required component(s), and then conversion back to the H_{01} mode of the circular OSW. In addition, the exciters of the H_{01} mode are large-size, complicated devices having appreciable loss. For these reasons circular H_{01} mode OSW failed to find wide application. The same fate awaited the proposal for long-distance microwave power transmission using the H_{01} mode in a network of large-diameter circular waveguides (1–2m in diameter) placed under ground [23].

15.5.2   Rectangular Waveguide with the H_{10} and H_{01} Modes

For rectangular cross-section H_{10} mode OSW, the common transmission line is a standard waveguide with more than tenfold increased cross-section relative to its single-mode counterpart [110]. In particular, using 3-cm range waveguide WG-16 (transverse dimensions 22.86 by 10.16 mm) at a wavelength of 2 mm (S/λ^2=57.5), one obtains an attenuation of 0.23 dB/m, or 20 times lower than with single-mode waveguide [111]. For 7.2 by 3.4 mm waveguide, the losses are as low as 1.4 dB/m in the wavelength range λ = 0.8 to 0.9 mm (S/λ^2=33.5) [112].

In the West, the properties of rectangular OSWs, including excitation difficulties, were examined in detail by the beginning of the 1960s [113]. Analysis of the linear smooth taper from a standard single-mode waveguide to the OSW [111,113] shows that, for efficient excitation of the principal mode in OSW, the taper length should provide, as a rule, a phase error, $\Delta\phi$, not greater than $\pi/8$ at the periphery of the taper aperture with respect to the phase at the aperture center. In the paper [114], the authors describe components in waveguide cross-section dimensions 7.2 mm by 3.6 mm for operation at a wavelength of $\lambda=1$ mm. They used a linear taper having a length of 152 mm that provided an error $\Delta\phi=\pi/12$. Using non-linear tapers and correcting the phase error with dielectric lenses and reflectors allows one to reduce the length of the taper by a significant factor [111,114]. When designing components for OSW, it is necessary to account for the possibility of propagation of higher-order modes. Thus, the application of waveguides of inhomogeneous cross section does not seem possible. In the paper [114], the use of OSW components in the wavelength range of $\lambda = 0.5$ to 8 mm built on the basis of a standard 3-cm band waveguide is described. These included pyramidal tapers, detectors, a cross-shaped divider based on a semi-transparent plate, a phase shifter, standard absorbing attenuators, and a Mach-Zehnder interferometer. Based on a waveguide of cross sectional dimensions 7.2 mm by 3.6 mm, a directional coupler (Figure 15.25), a tunable attenuator, and a phase shifter with a variable phase shift using a double-prism divider [114,115] were developed by J. Taub. Four section filters using stacks of fused-quartz plates were developed, having losses below 3 dB in the 130-145 GHz passband and attenuation greater than 20 dB in the 291-308 GHz stopband. Analogous components for wavelengths between 0.5 mm and 8 mm were presented in [116], including cross-section transformers, bends, phase shifters, attenuators, and also a $\lambda = 0.65$ mm Mach-Zehnder interferometer for plasma studies. The paper [111] considered a right-angle bend in rectangular OSW employing a plane reflector that was used in channel switch. Various combinations of faceted reflectors enabled the researchers to design phase shifters and balanced mixers [111,116].

In the USSR, rectangular OSWs were used as the basic transmission line in various millimeter and sub-millimeter systems developed in the RDIRP, and later in the Institute of Applied Physics of the USSR Academy of Sciences

Figure 15.25. Directive coupler on the basis of rectangular OSW with the H_{10} mode having a cross-section of 7.2 by 3.6 mm for $f = 345$ GHz (reproduced from [114]).

(now IAP RAS) in Nizhny Novgorod. In their research on rectangular OSW, the scientists from Nizhny Novgorod used not only the H_{10} but also the H_{01} mode. Both modes led to identical design principles for functional elements, but employing the H_{01} mode enabled them to achieve slightly lower loss in the waveguide. In 1968 L. Lubyako built an interferometer based on rectangular OSW with cross-section 23 mm by 10mm for the wavelength range 1 mm to 4mm [117]. Figure 15.26 depicts its principal component, a cross junction with a reflector formed by a metallic cube, acting as a 50-50 power divider for each of the beams incident from two input arms. In subsequent work in Nizhny Novgorod, QO interferometric circuits were widely used in various systems. Lubyako developed an original method of studying dielectrics and ferrites with a Michelson interferometer based on the mentioned OSW [118]. L. Fedoseev suggested using a Mach-Zehnder interferometer in superheterodyne radiometers at $\lambda = 1.1$ mm $-$ 1.6 mm (OSW cross-section 23 mm by 10 mm) and at $\lambda = 0.8$ mm $-$ 1.1 mm (OSW cross-section 11 mm by 5.5 mm) [119]. Such a device provided operation with an intermediate frequency in the microwave range, in addition to including a coupler for local oscillator injection, which also suppressed parasitic signals and attenuated heterodyne noise [120]. In Figure 15.27, one can see the components (a) and superheterodyne radiometers (b) based on the rectangular OSW that were used in numerous systems developed in IAP RAS by Y. Lebsky, e.g., in the radio-astronomical investigations of the Moon, Venus, Jupiter, and galactic sources [121,122], and in the study of the atmospheric ozone layer [123].

15.5.3   Circular Waveguide with the H_{11} Mode

Circular OSW with the H_{11} mode has characteristics comparable to the rectangular OSW. It had also been used since the 1970s, in the realization of antenna feed systems in the mm-wave range [124]. Investigations have shown

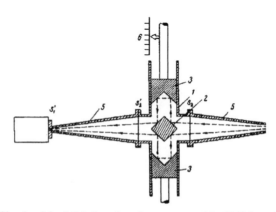

Figure 15.26. Circuit of the Michelson interferometer used in 1968 based on rectangular OSW with cross-section of 23 by 10 mm working with the H_{10} mode at $\lambda = 1$-4 mm designed at IAP RAS (reproduced from [117]).

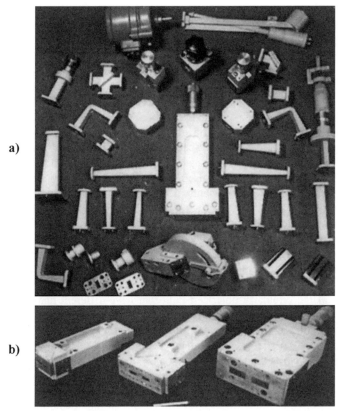

Figure 15.27. The components of 1970 **a)** and superheterodyne radiometers **b)** based on the rectangular OSW with the H_{10} mode developed at IAP RAS (photo provided by L. I. Fedoseyev).

that the losses in such a waveguide can be quite low. In particular, in the 3-mm wavelength range with waveguide diameter of 24 mm ($S/\lambda^2=50.3$), they do not exceed 0.1 dB/m [125]. For the excitation of these systems, one can apply both linear and nonlinear smooth tapers. Here, it is necessary to note that for non-axisymmetric modes in a circular waveguide, there is polarization degeneracy. As a result, a discontinuity (e.g., a section with small ellipticity), can produce transformation of the H_{11} mode into higher-order modes as well as excitation of the H_{11} mode with the orthogonal polarization. In the work [109], methods of minimizing H_{11} mode conversion in the circular OSW were considered. It was shown that together with reducing the deformation of the waveguide walls to a minimum, it is important to shorten the inhomogeneous section of the waveguide where the power coupling between the modes occurs. Therefore, application of similar systems in the case of extended lengths of waveguide section did not seem possible. At the same time, when the full length of the circuit containing all the elements meets the requirements formulated in [109], circular H_{11} mode OSW could be used in the design of various millimeter-wave systems.

In the 1970s and 1980s, these ideas were realized in the IRE NASU in Kharkov when building a number of antenna feed systems in the 2 mm wavelength range for pulsed and continuous-wave radar systems [126, 127]. Implementation of such QO systems became possible after V. Churilov, A. Goroshko and G. Khlopov proposed and studied, in 1970, an open-ended hollow dielectric beam-waveguide (HDB, see section 15.7.1) as an antenna feed [128, 129]. This development enabled them to avoid using standard waveguide components at the antenna front end. It also eliminated "locked-mode" resonances, and provided the desired amplitude and phase pattern across the antenna aperture [127]. A similar scheme was implemented when building antenna feed circuits on the basis of the circular waveguide with the H_{11} mode. Here, G. Khlopov and A. Kostenko used circular OSW of 20-mm diameter for a number of general-purpose components (power dividers, polarization converters, etc.) and some specialized devices such as reflector antenna feeds, scanners, rotating joints, and antenna switches. In particular, a ferrite antenna duplexer was designed and studied in the 2-mm wavelength range [130]. A specific feature was the application of impedance matching using multilayer ferrite structures [131, 132]. The matching of such a ferrite element made of two or more disks with controlled spacing, enabled one to eliminate reflections for two orthogonal components of the circularly polarized field that have different phase velocities in a longitudinally magnetized gyrotropic medium. This allowed the authors to achieve a remarkable performance of the receiving-transmitting system as a whole. A photo of this circuit is presented in Figure 15.28a, while general view of the system integrated with oscillators and placed behind a dual reflector antenna is shown in Figure 15.28b. Most of the components were designed as single compact units, which eliminated the waveguide flanges and shortened the total length of circuit, both extremely advantageous in the development of short wavelength systems. In the work [130], implementation of an optoelectronic antenna duplexer was considered. Here channel switching is under the control of a semiconductor laser, whose light pulse illuminates a semiconductor plate placed diagonally in the cross-section of the OSW cross-shaped junction.

It is important to keep in mind that all metallic OSWs have one intrinsic demerit - the lack of self-filtering of the higher-order modes – with the consequence that special design measures to suppress these modes are generally required.

15.6   COMPROMISE NO. 1: DISCRETE BEAM WAVEGUIDES AND EAST-WEST COMPETITION (SINCE 1961)

The limited distance of free-space Gaussian beam propagation between a pair of apertures can be overcome by arranging periodic or iterative correction of the phase distribution in the beam cross-section. Such an approach was considerably stimulated by the idea of using the Fabry-Perot interferometer as a laser cavity, suggested independently in 1958 by R. Dicke [133] and A. Prokhorov [134], and also by A. Shawlow and C. Townes [135]. Soon, a theory of resonators with spherical reflectors was developed, for which the natural oscillation modes were

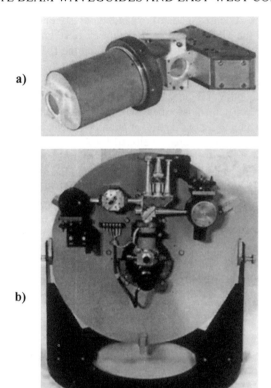

Figure 15.28. Single-antenna receiving-transmitting systems in the 2-mm wavelength range based on 20-mm diameter circular OSW with the H_{11} mode developed at IRE NASU in 1978: **a)** the single, compact receiving-transmitting unit with ferrite antenna duplexer; **b)** general view of the system on the rear of a dual-reflector antenna.

presented in the form of Gaussian wave beams [136-138]. An analogous modeling approach, in the Fourier-transform domain, was developed by G. Goubau and F. Schwering [139], who also proposed a beam transmission line formed by a system of equidistant lens-type phase correctors, and introduced the term "beam waveguide". The whole decade of the 1960s witnessed active competition between the Western and the USSR teams who were engaged in mm-wave reflector and lens type transmission line research. Here, researchers in the USSR enjoyed full access to the relevant Western publications while the opposite was not always possible. Broadly speaking, although theoretical results were relatively rapidly published in the open literature, measurements and practical designs frequently remained classified for a number of years.

15.6.1   Lens and Iris Beam Waveguides

The first experimental prototype of the lens beam waveguide [140] was designed by Goubau in 1961 for a wavelength of 1.25 cm and clearly demonstrated its potential. In this system, 20 lenses made of expanded polystyrene were spaced by 1m (Figure 15.29). In the USSR, several laboratories were engaged in similar research starting from 1962, one of the main ones being the IRE RAS, Moscow, with a team led by B. Katsenelenbaum (Figure 15.30). Another important team worked in Nizhny Novgorod led by V. Talanov (Figure 15.31). A lens beam waveguide in the frequency range 75 GHz, as well as a number of QO components: (waveguide-beam transformers with pyramidal and bimodal horns, mode converters, absorbing and polarization attenuators, dielectric beam divider, and reflection-coefficient gauge) were developed by A. Akhiyezer at the R&D Institute of Measures and Measuring Devices (now "Metrologiya" State Co.) in Kharkov, Ukraine. Several years later Akhiyezer collected his results in the book [141]. In the next several years this area of research was actively pursued both in the West and in the USSR. The lens beam waveguide was implemented when building QO measuring circuits for plasma diagnostics, materials science measurements, QO antenna-feeding systems for remote sensing and radar, and other important applications. An overview of the accomplishments of various Western teams can be found in the monograph [19]. In particular, Figure 15.32 shows a dual-wavelength receiver in plasma diagnostic system, and Figure 15.33 shows the transmitting and receiving circuits of a dual-frequency QO material measurement system.

Figure 15.29. Goubau's experimental lens type beam waveguide of 1961 (reproduced from [140]).

| **Figure 15.30.** | **Figure 15.31.** |
| Boris Z. Katsenelenbaum in the 1960s. | Vladimir I. Talanov in the early 2000s. |

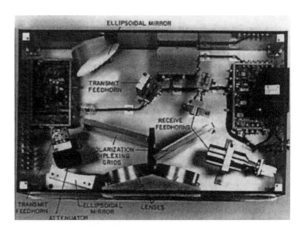

Figure 15.32. (top) Dual-wavelength quasioptical receiving circuitry used in plasma diagnostics system (reproduced from [19]).

Figure 15.33. (bottom) The incident and reflected power channels of quasioptical material measurement system used in 1988 (reproduced from [19]).

The shortcomings of the lens beam waveguide are the power absorption in the dielectric and the reflection from the lens surface. Though the single-lens reflection losses are relatively small (e. g., 0.02 dB was reported in [140]), when building extended multilens circuits, this factor may spoil the overall characteristics. One of the possibilities to eliminate reflections is use of the lens guide proposed by V. Shevchenko at IRE RAS in 1963 [142]. Here, the lenses were elliptic paraboloids inclined relative to the beam axis by Brewster's angle. This provided the phase correction equivalent to that of a conventional lens [143]. An experimental model of such a beam waveguide was built at IRE RAS in Moscow and tested in the sub-mm range ($\lambda = 0.9$ mm – 0.7 mm) [144]. The waveguide consisted of 22 polyethylene lenses of maximum thickness 3.2 mm, separated from each other by 10 cm distance. This beam waveguide provided losses no greater than 1.7 dB/m. Beam waveguides employing non-reflecting lenses were used later when designing measurement setups and developing spectroscopic techniques for the study of various materials including low-loss dielectrics, ferrites, semiconductors, and liquids, [27,145]. This work was also performed at IRE RAS and was led by Meriakri.

A diaphragm or iris beam waveguide is closely related to the lens one - its focusing effect depends on truncation of the beam, although it does not entail any explicit phase correction. This transmission line was also the subject of competition between Western and Soviet researchers. In [146], a beam waveguide having rectangular diaphragms was considered by Goubau for use in the wavelength range $\lambda = 4$ mm to 8 mm. In [144], results actually related to the work started in IRE RAS in 1964 were published, describing the studies of a beam waveguide, using metallic iris apertures to provide a smooth control of the aperture size (λ = 0.7 mm – 0.9 mm). In this system, the distance between the diaphragms was 12 cm, and the loss was 2.2 dB/m. The diaphragm beam waveguide has greater loss than does the lens waveguide. It becomes competitive only in the wavelength range shorter than 0.5 mm due to the increased loss in the lens material [27]. The loss due to absorption in the lens can be reduced by using novel, low-loss materials. For example, Teflon F-4 not exposed to thermoprocessing has a loss tangent tgδ = $0.25 \cdot 10^{-3}$ at λ = 0.6 mm while for standard Teflon the value is $0.7 \cdot 10^{-3}$. Dielectric reflection losses can be reduced by using artificial matching layers with grooves, but these are complicated to make, somewhat narrowband, and imperfect in their performance. Dielectric losses can be completely eliminated only by using a reflector analog of the lens beam waveguide.

15.6.2   Reflector Beam Waveguide

The reflector beam waveguide was patented in 1962 and published the following year by Katsenelenbaum of the IRE RAS [147]. Here, the reflector playing the role of a phase corrector was shaped as a section of the surface of an ellipsoid of rotation, although simpler shapes are acceptable (for instance, spherical reflectors [25]). The amplitude distribution in the transverse cross-section of the beam obtained in such a manner evidently has a more complicated structure than in the case of the axially symmetric lens-type beam waveguide. This is an important point to be kept in mind by the system designer. Practically at the same time, J. Degenford designed and measured a reflector beam waveguide in the 4-mm wavelength range using elliptic reflectors; the loss per single iteration being 0.015dB [148]. Another team in the USSR worked on reflector beam waveguides in the RDIRP in Nizny Novgorod. Paper [72] contained experimental data on such a device made of 8 reflectors shaped as rotationally symmetric ellipsoids. The total loss for this system was found to be 3.2 dB, including the loss due to the feed horns. The shortcomings of this transmission line were its quite large dimensions and very high sensitivity to the reflector adjustment. Therefore, the periscopic beam waveguide, in which reflectors are combined in pairs as fixed units, is preferable (Figure 15.34) [149]. If the spacing between the reflectors in

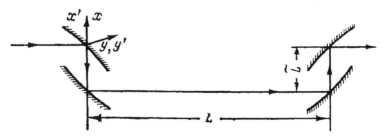

Figure 15.34. Periscopic reflector beam waveguide of 1965 (reproduced from [149]).

each pair is significantly smaller than their focal distance, than the phase correction of such a unit equals the sum of the two reflector corrections. Rotation of the device made of a pair of fixed reflectors yields a displacement of the beam parallel to its axis but does not produce a change in its direction. Hence the system as a whole is less sensitive to the misalignment of the phase correctors. An experimental test of such a line was performed in the optical range ($\lambda = 0.63$ μm) [150] and showed good potential. Although the marked technical shortcomings are still an obstacle to the wider application of reflector beam waveguides in the mm and sub-mm range, there are a few impressive implementations. One of them is used in the feed system of the dual-reflector scanning antenna of a deep-space satellite communication ground station (Figure 15.35) [70,151,152]. Here, a beam waveguide provides multi-frequency operation. Another important application is the transmission of mm waves in the 110-170 GHz range from high-power gyrotrons for nuclear fusion machines – tokamaks and stellarators [75], to provide electron cyclotron resonance heating of the plasma.

15.7   COMPROMISE NO. 2: CONTINUOUS BEAM WAVEGUIDES AS A WIDELY-USED USSR TECHNOLOGY (SINCE 1963)

In the design of certain measurement facilities, such as interferometers for hot plasma diagnostics in fusion machines or defense radar systems, the basic transmission line must have specific electromagnetic characteristics that are not provided by periodic beam waveguides. These include invariance of the field amplitude and phase along the guide, amplitude and phase symmetry in two orthogonal planes, linear polarization of the field, self-filtering of the higher modes, high degree of shielding, wide range of the working frequency, and low loss for the principal mode [28]. In addition, such transmission lines must satisfy a number of environmental requirements including mechanical rigidity and weather resistance. This relates to the transmission line, which is simultaneously a screen, a support structure for the functional elements, and a protection against external influences. All of these requirements are met by the "hollow dielectric channel" waveguide technology. This technology is based on circular or rectangular OSW, with inner walls completely or in part covered with a lossy dielectric or layered-dielectric lining (Figure 15.36). In spite of the difference in lining structures, these systems exhibit some common properties produced by the presence of the dielectric boundary of the channel, that justifies uniting them into a common class. In brief, the principal mode here has a field distribution which is nearly that of a Gaussian beam in cross-section, while the higher-order modes are filtered due to the increased absorption which they suffer.

15.7.1   Hollow Dielectric Beam Waveguide

This specific QO transmission line was conceived in 1963 by Y. Kuleshov (Figure 15.37) at IRE NASU in Kharkov. Earlier, his team had accumulated rich experience in developing standard waveguide measuring devices in the whole

Figure 15.35. Beam waveguide feed system of the dual-reflector antenna of a ground-based satellite communication station (reproduced from [151], Circa 1961).

Figure 15.37. Yevgeniy M. Kuleshov, in 1960.

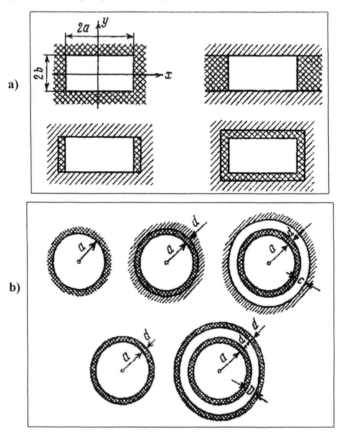

Figure 15.36. Waveguides of "the hollow dielectric channel" type with **a)** rectangular and **b)** circular cross sections (reproduced from [172, 175], Circa 1972).

mm-wave range – a series of projects had been performed with physically small single mode waveguides in the wavelength range between 1.5 and 8 mm. The requirements mentioned in the introduction to this section appeared quite natural for Kuleshov due to his involvement in the *Tokamak* plasma research underway at the Institute of Atomic Energy (now Federal Scientific Center "IAE") in Moscow. *Tokamak*s were extremely high-power devices, which worked in a pulsed mode and experienced very intense vibrations. Therefore a workable design for a mm-wave interferometer for precise measurements of the electron density of the hot plasma in these fusion machines had to provide rigid coupling between the sections of transmission lines and the measuring units. At first, Kuleshov had no idea of using hollow dielectric "beamguide" (or beam waveguide but called HDB) in *Tokamak*s; in fact the plasma density in the first fusion machines of this type was not very high (see section 6.3) and hence plasma diagnostics could be done with mm waves. *Tokamak*s requiring sub-mm wave interferometers came only after 1971, and then HDB was quickly selected as a basic technology.

The patent for HDB ("authors certificate" in USSR terminology) was entitled, "Dielectric beam waveguide of the sub-mm wavelength range". It was submitted, registered as classified, and declassified in 1969, 1971, and 1972, respectively, when the circuits based on HDB had already been in use for several years. The subject of the invention was formulated as follows [153]:

> *# 1. Dielectric waveguide of the sub-mm wave range, having the form of dielectric tube, whose distinction is that, in order to provide the necessary rigidity of the structure and avoid radiation, it is made of dielectric having the permittivity of 2.5 to 4 and loss tangent of 0.05 to 0.1 (e.g., phenoplastic), and fixed inside a flanged metal tube.*

> *# 2. In order to improve the filtering of spurious modes, the inner surface of the dielectric tube (see beamguide from # 1) is provided with longitudinal triangular ribs having depth smaller than a half-wavelength.*

It should be noted that a study of very great importance for HDB was published in 1964 by E. Marcatili and R. Schmeltzer [154]. They considered in greater detail the structure formally solved by J. Stratton in [155], i.e., a circular cylindrical channel in an unbounded dielectric medium, and found out that it could support quasi-single-mode propagation. Independently and half-a-year earlier, guided by engineering intuition and simple calculations, Kuleshov had selected a similar channel when designing a new basic transmission line for a set of broadband mm and sub-mm wave measurement circuits.

Although at first HDB had no clear area of application, this research was readily funded by the USSR Ministry of Radio Industry as an innovation potentially applicable to various systems. In 1964–1966 and 1968–1971, Kuleshov and his team developed and studied both HDB and a complete measurement setup covering the near-mm and sub-mm wave wavelength ranges. The first project was aimed at exploring the feasibility of developing a "kit" of HDB-based measuring devices in the wavelength range 0.7 to 1.7 mm. The HDB was fabricated as a Phenol plastic tube, the inner diameter being 20 mm or 40 mm, the thickness being 5 mm, which was placed inside a metallic tube with coupling flanges. Two variants of the beam waveguide had been developed: one with smooth inner surface and another with longitudinal ribs of triangular cross-section (Figure 15.38). The use of a relatively thick layer of Phenol, which had considerable loss, allowed elimination of the effect of the metallic tube, whereas the use of a ribbed surface lowered the effective permittivity ε_{eff} down to the value of approximately 1.5 that helped suppress the higher-order modes. The low-loss principal mode of HDB is the almost linearly polarized HE_{11} mode, whose phase front is almost flat in the waveguide cross-section, and whose amplitude pattern is very close to a Gaussian beam [156,157].

Experimental investigation of HDB was carried out in the wavelength range of 0.8 to 1.6 mm, and it was found that at sub-mm wavelengths ($\lambda < 1$ mm) the loss did not exceed 1 dB/m. Within the first project, a large number of innovative QO components based on HDB were developed, including waveguide-to-beam waveguide transitions, rotating joints, beam splitters with a semi-transparent dielectric plate, corner reflectors, transmission and reflection-type wavemeters based on Fabry-Perot open resonators, absorbing attenuators, phase shifters, thermistor mounts, absorbing loads, movable loads, movable reflectors. Despite the fact that most of the QO principles used here were known since the time of the "Hertz wave" pioneers, adapting them to the specific structure of HDB required considerable creativity and non-trivial technical solutions. Here, Kuleshov and Yanovsky were most frequently the authors of new ideas and configurations.

In view of an absence of immediate customers, the same USSR ministry funded Kuleshov's next project on HDB components only in 1968–1971. Its idea was to dwell on and refine the polarization principles in the measuring circuits for the wavelength range between 0.5 and 0.8 mm [158]. Working on polarization-selective QO devices, researchers showed that HDB of 20-mm diameter could be used over this whole wavelength range. Among new components were the following: double-lens transformers, Fabry-Perot wavemeter with metal-strips (1-μm thick silver or aluminum), grating reflectors made by vacuum deposition on the quartz substrate, phase shifters with dihedral

Figure 15.38. HDB sections of inner diameters 40 mm and 20 mm, with smooth and ribbed inner surfaces were developed in 1968 (photo provided by V. Kiseliov).

reflectors in the 90° angled bends of the beam waveguide. However, the main focus was on devices using gratings of 10-μm diameter tungsten wire, with periodicity between 20 and 60 μm, wound on 40-mm diameter frames.

By the middle of the 1970s, a complete set of HDB-based QO components had been designed, fabricated, and refined. It included waveguide to beam waveguide transitions, matched and movable terminations, reflectors, polarizers, transitions with controlled and fixed bending angles, telescopic and rotary junctions, wave meters, polarization attenuators, phase and frequency shifters, dielectric-plate and wire-grating beam splitters, reflectance gauges, matching transformers, amplitude modulators, electromechanical switches for the beam propagation direction, balanced mixers, duplexers, polarization transformers, and power meters (Figure 15.39) [28,159]. Together with general-purpose devices, specialized ferrite and semiconductor devices had been developed [160]. This *LEGO*-style kit of over 20 components and devices was later widely used in application research and new technologies. The kits were produced serially until the mid-1980s and purchased by more than fifty Soviet organizations that worked with mm and sub-mm waves. In all, over 6500 components and devices were produced. For example, a team led by Y.

Figure 15.39. Quasi-optical LEGO-like kit for a sub-mm wave engineer. HDB-based components used for building wideband (λ = 0.5-1.7 mm) measuring circuits developed in IRE NASU: 1- polarization attenuator, 2- polarization phase shifter, 3- tunable attenuator-power divider, 4- polarization plane rotator, 5- wave meter, 6- polarization converter, 7- tunable phase shifter, 8- matching tuner, 9- beam splitter, 10- cassette of polarization discriminator, 11- linear polarizer, 12- right-angled corner, 13- movable two-facet reflector, 14- rotating joint, 15- terminating load, 16- movable reflector, 17,18- two waveguide-to-beamguide transitions, 19- telescopic section, 20- HDB straight section (photo provided by V. Kiseliov, 1971).

Gershenzon of the Radiophysics Laboratory at the Moscow State Pedagogic University, jointly with Kuleshov's team of IRE NASU, developed measuring circuits for sub-mm wave spectroscopy of semiconductors and superconductors [161,162].

Successful development of HDB-based components and instruments stimulated their use in antenna feed systems short-mm wavelengths. As a result, several teams at IRE NASU built multi-functional transmit-receive systems for prototypes of 2-mm wave radars. These prototypes were used in large-scale experimental studies of the performance of such radars in field conditions with various types of terrain and vegetation. In 1971, Kuleshov, Yanovsky and their colleagues of the IRE NASU proposed a duplexing device [163] that coupled both the transmitter and the receiver to a single common antenna for circularly polarized signals. This principle was well known in standard waveguide technology and in QO antennas [76,164,165]. The designers developed a novel version of the device, in which the polarization transformer was a combination of a wire grid and an adjustable flat reflector placed into a 90° bend of HDB. This duplexer was used as a principal unit for the HDB and MDW-based antenna feed systems designed in the 1980s at IRE NASU for 2-mm wavelength battlefield radars using a single antenna and circular polarization of transmitted/reflected signal [130].

The follow-on investigations showed that the electromagnetic characteristics of HDB were very attractive for radar cross-section (RCS) modeling. An HDB-based microcompact RCS testing range was built by V. Kiseliov, who succeeded Kuleshov as the head of the QO department at IRE NASU, in the early 1990s. This range was aimed at the indoor investigation of the scattering characteristics of various objects in the short-mm and sub-mm wave ranges [166]. Here, an object or its scaled model is placed inside the HDB, and the RCS (and also the forward scattering cross-section) is determined from the parameters measured for the reflected and transmitted HE_{11} mode signal. The conditions have been found (diameter not less than 5λ) under which HDB provides the amplitude and phase distributions of the incident field necessary to simulate free space scattering [167-169].

15.7.2   Metal-dielectric Waveguides

In the late 1960s, Y. Kazantsev and other researchers at IRE RAS in Moscow, who had at that time no idea of Kuleshov's HDB at that time, proposed another OSW technology with a metal-dielectric inner coating [170]. A detailed theoretical investigation of the basic idea can be found in [171]. A metal-dielectric waveguide (MDW) is commonly a large diameter (~10λ) metal tube of rectangular or circular cross-section. The inner surface of the tube is covered with a low-loss dielectric layer, thus having the features of an impedance boundary. Unlike HDB, the thickness of the lining layer here is small, and can cover only a part of the inner metallic surface. The principle of operation is based on the fact that if the partial plane waves encounter the waveguide wall at grazing incidence (this occurs for the lower-order modes), the ohmic losses in the metal are significantly smaller than for all-metal OSW. The higher-order mode losses in MDW are greater than in OSW – this effect is referred to as self-filtering of the modes.

In circular MDW, the working mode is the lowest hybrid mode HE_{11}, the same as in HDB. In 1970, Kazantsev demonstrated that a remarkably low attenuation could be achieved [171]; this effect was analyzed in detail in [172]. In particular, the losses in MDW having a diameter of 40 mm were estimated to be $3.7 \cdot 10^{-3}$ dB/m at $\lambda = 2$ mm ($S/\lambda^2 = 314.2$) with a dielectric coating thickness of 0.25 mm and complex dielectric permittivity characterized by $\varepsilon = 2.3$ and tgδ = $2 \cdot 10^{-4}$. In addition to the loss, Kazantsev considered MDW excitation problems, putting the design of MDW-based components on more solid ground than the theory available for HDB. MDW was quickly gaining a reputation of convenient and attractive low-loss sub-mm wave transmission line. Therefore, as an alternative to their own HDB-based designs, the same IRE NASU team of Kuleshov developed, in the early 1980s, a kit of QO components and instruments based on circular MDW having a diameter of 20 mm, whose inner surface was coated with a 0.2-mm thick Teflon layer. This waveguide enabled the designers to use a maximum of the already existing HDB technology without costly modifications. In particular, in [173] a QO circuit for a superheterodyne receiver was considered that used an intermediate frequency of 0.3 GHz, and operated in the wavelength range of 1.3 to 2.2 mm. The circuit included halfwave and quarterwave polarization interferometers, a wavemeter, a polarization attenuator, and was packaged in a very compact configuration. Similar integrated units were designed to serve as antenna feeds and for the mixers in the high-temperature plasma diagnostics facilities.

Rectangular cross-section MDW was also developed. It was designed having two options: one with the dielectric coating on two opposite walls, and the other with the coating on all four walls. The second case had the advantage of providing low-loss transmission of waves of two orthogonal polarizations, and was used in circuits implementing polarization conversion. In such a waveguide, the working mode is the longitudinal-magnetic mode LM_{11}. It is characterized by a practically flat phase front, linear polarization, and amplitude pattern having a maximum at the waveguide axis, smoothly decreasing towards its walls, and being symmetrical in the two orthogonal planes [174,175]. Experimental examination of the losses in MDW of rectangular cross-section was carried out at wavelengths between 8 mm and 2 mm. In particular, in the 2-mm range, the waveguide tested had a cross-section of dimensions 10 mm by 23 mm (S/λ^2 = 57.5), with 0.55 mm thick polyethylene layers ($\varepsilon = 2.3$, tg$\delta = 5 \cdot 10^{-4}$) placed on the narrower walls. The measured loss was $3.0 \cdot 10^{-2}$ dB/m. Based on such a waveguide, various components and devices were designed in the 1970s at IRE RAS, Moscow by Kazantsev and his team members O. Kharlashkin and M. Aivazyan. These included exciters, waveguide bends, higher-mode filters, rotary junctions, and ferrite devices [175,176]. They were used in the development of antenna feed circuits in the 2-mm wavelength range. In the late 1980s, development of systems based on rectangular MDW was continued at the Institute of Radio Physics and Electronics of the Armenian AS in Yerevan (now

IRE AAS) [177], by R. Avakyan, K. Agababyan, and Aivazyan, who worked there by that time. They developed detector sections, mixers, and an integrated receiver for a radiometer [178].

At the same time at IRE NASU, Kuleshov developed a complete set of QO components for building measurement circuits in the near-mm and sub-mm wavelength ranges, based on square section MDW. Here, the square waveguides of two wall dimensions were used: 14 mm in the range $\lambda = 1.15$ to 3 mm, and 10 mm for $\lambda = 0.7$ to 1.7 mm [179]. The set consisted of waveguide sections, exciters, angle-bend transitions, a rotating junction, nonreflecting loads, a linear polarizer, a polarization transformer, a polarization attenuator, a phase shifter, a resonant frequency meter, dielectric and polarization power dividers, a movable reflector, a two-sided angled reflector, an amplitude modulator, and a ferrite isolator. In 1994, a square MDW of 14 mm by 14 mm cross-section was used when developing a single-antenna radar in the 2-mm range employing a circularly polarized signal (Figure 15.40) [180]. This work was accomplished by the joint efforts of the IRE team and a team headed by Churilov, of the Institute of Radio Astronomy (IRA NASU), which spun off from IRE in 1985.

15.7.3 High Temperature Plasma Diagnostics in the Moscow Tokamaks

The high mechanical stability of HDB-based circuits together with their excellent electromagnetic characteristics enabled the IRE NASU team to develop beam waveguide multichannel interferometers for the measurement of the electron density in the high temperature *Tokamak* controlled plasma fusion machines. In particular, in Figure 15.41, one can see a parallel array of HDB-based arms of the 9-channel interferometer ($\lambda = 0.9$ mm) mounted on top of the *Tokamak* T-10 of the IAE in Moscow [181].

Figure 15.40. Single-antenna receiving-transmitting system employing a circularly polarized signal in the 2-mm wavelength range based on square MDW having cross-section of 14 by 14 mm developed at IRE NASU and IRA NASU (photo provided by V. Kiseliov, 1975).

Figure 15.41 The first QO nine-channel HDB-based interferometer (λ=0.9 mm) with a BWT source for hot plasma diagnostics of the Tokamak T-10 at IAE (photo provided by V. Kiseliov, 1975).

In fact, the diagnostics of the high temperature plasma in *Tokamaks* was the best funded, although almost "invisible", area of application of QO sub-mm wave circuitry and measurement techniques in the USSR between 1973 and 1993. It is necessary to keep in mind that the Tokamak principle of creating very hot and dense plasma in toroidal camera was proposed in the USSR in the early 1950s. All related work was concentrated at the IAE, which was a huge laboratory of the State Committee for Atomic Energy (SCAE) in Moscow, and heavily classified. Today it is unbelievable that the location of the IAE was selected to lie within the boundaries of a large city, only 20 km from the Kremlin. This was the personal choice of Beriya, Stalin's powerful security minister and skillful manager, who supervised the nuclear industry. It was clear from the beginning that realistic power-generating machines of this type had to have cameras of large diameter, around 1 m, because a dense plasma can "live" long enough only if it is concentrated far from the enclosure walls. When the first small-size machines were built and thoroughly studied, it was revealed that one needed continuous control of the plasma density and internal structure. The required monitoring could be realized with electromagnetic waves but the denser the plasma the shorter the wavelength that is required to penetrate the plasma, determine the phase shift and thus the electron column density in the beam. In the first *Tokamaks*, sensing with radiation of 4 mm wavelength and later 2 mm wavelength was adequate. To build the required measurement circuits, IAE scientists led by N. Yavlinsky and then Y. Gorbunov used standard-waveguide components and devices developed in the IRE NASU in Kharkov and worked together with Kuleshov and his team.

In 1968, the design of the larger machine T-10 was conceived, and in 1970 the government approved it. The plasma density in it was expected to reach 10^{14} electron/cm^3, hence it was estimated that the sensing should be carried out at a wavelength $\lambda = 1.5$ mm or shorter. IAE scientists remembered that Kuleshov at IRE NASU already had components based on the standard waveguide 1.1 mm by 0.55 mm in cross-section available. When asked about the new project, Kuleshov gladly agreed to assist, but said that his approach would be based on a different (not waveguide) technology. In fact, as described in section 6.1, he had been already working on hollow dielectric beam waveguide (HDB). Thus, in 1973, a project was initiated by the IAE on studying the feasibility of a nine channel interferometer in the wavelength range 0.5 mm to 1.7 mm [182].

Scientists working with the Tokamak highly appreciated the new instrument, which was not sensitive to the intense vibrations characteristic of their pulsed fusion machine, and thus enabled them to avoid errors in the phase measurements. This was a result of the fixed length of the beam waveguides and the possibility of mounting them directly on the plasma camera supports. T-10 was operated for the first time in June 1975, and in December of the same year the nine channel QO interferometer was commissioned and made its first measurements at $\lambda = 0.9$ mm, fed from a BWO (backward wave oscillator) source (Figure 15.41). Associated HDB-based circuitry reached dozens of meters in length. The follow-up stages of plasma diagnostics research were directed toward further shortening the working wavelength, an effort which required using laser sources instead of vacuum tubes. HCN lasers for $\lambda = 0.337$ mm were developed by S. Dyubko at the Kharkov State University and also by Kuleshov's team at IRE NASU.

In 1980, the latter source was tested in a prototype single-channel homodyne interferometer at T-10. In addition to the laser, new components included cryogenic n-InSb detectors and a remote control unit [183]. In the 1980s, an even more ambitious and extremely expensive fusion machine was put into operation at IAE, the T-15 (Figure 15.42). Its revolutionary new feature was not the plasma camera but the superconducting magnetic system. Plasma diagnostics here were configured in a nine channel vertical sounding tomographic configuration (Figure 15.43) [184] based on the same QO components as the single-channel prototype tested previously. This interferometer was built in 1985 by combined team efforts of IAE and IRE NASU. Its adjustment was performed with special laser-beam tuning devices integrated with the angled bends and beam splitters in HDB. As the beam waveguide could operate at even shorter wavelengths, in the late 1980s another interferometer was built and operated at T-15. It was based on the $\lambda = 0.119$ mm CO_2 laser with a microwave pump, and enabled one to measure electron densities up to $2.10^{14}/cm^3$ [185]. In the course of this work with the new instrument, it was

Figure 15.42. General view of the nuclear fusion machine Tokamak T-15 at IAE with a circle marking the placement of the QO interferometer (photo provided by V. Kiseliov, 1985).

Figure 15.43. The second QO nine-channel HDB-based interferometer (λ=0.337 mm) with an HCN-laser source for hot plasma diagnostics of Tokamak T-15 at IAE (photo provided by V. Kiseliov, 1985).

shown that when a sub-mm wave beam passed through the peripheral part of the plasma column it experienced a Faraday rotation of the plane of polarization. To quantify this effect, which could yield valuable information on the plasma magnetic field, the IRE NASU team developed a prototype single-channel interferometer-polarimeter. This was the last project performed for the IAE. It was released in 1992, the last year of the USSR [186]. That December, after the independence referendum in Ukraine, the Russian government cut off financing for R&D projects carried out there. In the following year, all experiments with fusion machines at the IAE were terminated because of power shortage, and today the chance of reviving T-15 is essentially zero. Instead, a small ITER-compatible machine is being built in Moscow, assembled from the parts remaining after dismantling the T-10.

Since the late 1970s, one more mm-wave transmission line closely related to HDB and MDW has been developed. This is a circumferentially corrugated circular OSW [187, 188] with a corrugation depth equal to a quarter-wavelength. Here, the effect of the periodic rectangular grooves on the inner wall is essentially the same as of the dielectric lining in HDB and MDW. The principal mode is again the hybrid HE_{11} mode, whose ohmic losses are low as a result of the small values of the field at the walls; the higher-order modes are filtered out thanks to the high losses. The operating wavelength, however, has to be far from the Bragg reflection regime. These beam waveguides are made from stainless steel and are attractive in the high-power application as they are able to guide up to 1 MW, if evacuated [75]. In the USSR, M. Petelin and G. Denisov and their colleagues from IAP RAS in Nizhny Novgorod have been active in this field, developing various mm-wave circuit components [189-191].

15.8   BRIEF SURVEY OF MODELING METHODS AND TOOLS USED IN QUASIOPTICS

In terms of theory, sometimes it is said that QO is based on physical optics (PO) for analysis and on GO for synthesis. In fact, several major ideas, concepts, and methods have played important roles in the establishment of QO theory. Some of them relate to the description of wave beams, while the others handle diffraction of short wavelength radiation by objects of large size. There are two points to emphasize about today's tools for modeling and simulation of various QO waveguides and devices. First, most of the theories are scalar ones and study the solutions to the Helmholtz equation (or its asymptotic form in the paraxial domain - the parabolic wave equation) separately for each Cartesian component of electromagnetic vector field. Second, because of the rather large electrical dimensions of QO components (size measured in terms of the wavelength of the radiation), approximate analytical techniques still have the upper hand over rapidly expanding numerical methods. The area of electromagnetic simulation where $D > 20\lambda$ is largely inaccessible even for the Golem of today's electromagnetics: FDTD. Other specialized analytical-numerical methods, such as the method of moments, have greater impact on the low-frequency part of QO.

GO. The oldest simulation tool, GO, is the instrument of classical optical technology and considers the limiting case of $\lambda \to 0$ [192]. Here, the electromagnetic vector field is assumed to locally behave as a homogeneous transverse plane wave. This leads to the eikonal equation for the field amplitude. With GO, a beam wave is modeled as a bunch of rays, and the ray-tracing algorithms propagate these rays through the optical system and determine the field distribution on the output surface. A very important modification of GO was developed, as is well known, by J. Keller [193], who introduced so-called diffracted rays appearing due to edges, tips, corners and other surface discontinuities of the scattering body. Note however, that the range of validity of GO is determined by the approximate character of the eikonal equation. Well-known examples of its failure are focal-domain analysis and long-distance beam propagation where the problem sometimes reveals itself at moderate distances when applied in clearly QO circuits (see [70]). Besides, as do all other scalar theories, GO neglects the coupling of the vector components of the electric and magnetic fields through the boundary conditions.

In the USSR, Keller's theory was not widely known. Nevertheless GO antenna simulation tools were well developed as a result of the fundamental works of B. Kinber and his colleagues [194]. In subsequent work, GO was widely applied in the analysis and synthesis of large QO antennas including polyfocal [195] and anisotropic ones [196]. The books [66,197] covered GO-based characterization of basic microwave antennas of the lens type.

Direct PO and angular-spectrum method. At the core of direct PO lies the Huygens–Kirchhoff method based on the Helmholtz-Kirchhoff integral theorem derived from Green's theorem with the choice of Sommerfeld's uniform scalar spherical wave as the Green's function. It divides these into three domains according to the features of the diffraction field: the near zone, the intermediate (Fresnel) diffraction zone, and the far zone. In the far zone, i.e. at distances $R \gg 2D^2/\lambda$, the electromagnetic field behaves as a spherical wave. In the near and intermediate zones it propagates as a beam. In the latter two zones, with a quadratic correction in the antenna aperture-field phase, one can achieve proper focusing and perform directive power transmission to the receiving antenna. The efficiency of beam forming in such a system is determined by the electrical size of the antenna's aperture and its focal distance. Besides the direct PO method, the plane-wave spectrum approach was also proposed [155] and developed [198] in the 1950s. The concept of the angular spectrum was introduced as a complex-valued function depending on the direction of the wave propagation that is the Fourier-image of the field in the aperture plane. Here, one can usually neglect the contribution of so-called "evanescent" plane waves, which are characterized by complex-valued wave numbers. To determine the power of a non-planar wave incident on a receiving antenna, simple and convenient engineering estimations were found almost simultaneously in the USSR [199] and in the USA [95] as early as in the late 1950s. This approach turned out to be useful in the calculation of the scattering losses not only in free-space Gaussian beam propagation [68] but also in QO beam waveguides, where discontinuities distort the amplitude and phase of the wave beam [141].

In the early 1960s, approximate solutions of the effects of discontinuities in OSW based on PO and GO were obtained by several scientists working in Moscow [194,200-202]. Subsequently, some of these QO circuit elements were computed at IRE NASU, Kharkov by using accurate numerical methods based on semi-inversion [203]. The comparison showed the applicability of the approximate methods, within an accuracy of 10%, if the waveguide size satisfied the condition $D > 5\lambda$. Using PO, diffraction of the HE_{11} mode from local scatterers in HDB was considered in [167,169].

Parabolic-equation technique. The "paraxial approximation" implies that any field component can be written as the product of a plane wave propagating along a definite direction and a slowly varying "wave amplitude". This amplitude has to satisfy the parabolic wave equation as a paraxial form of the Helmholtz equation. In the late 1960s and 1970s, many remarkable studies were published based on this asymptotic technique, most notable being the papers on the modes of open resonators. In fact, a satisfactory explanation of resonators made of two flat or curved mirrors had been the main driving force behind the early development of Gaussian beam theory. In the USSR, the book [204] authored by L. Vainshtein covered various infinite two-mirror resonators in the scalar paraxial approximation, and marginally treated "open waveguides" of reflection and lens type. The publications of Katsenelenbaum [24,25,205] and Talanov [206,207] had much greater impact on QO practice, as they directly developed the Gaussian beam approach.

PTD. In the mid-1960s P. Ufimtsev published the method of edge waves, also known as the physical theory of diffraction (PTD) [208]. He introduced into PO the concept of secondary waves originating from currents at the edges, bends and other discontinuities on the scatterer's surface. This empirical approach later was partially justified. It enabled researchers to obtain concise representations for the electromagnetic fields scattered by objects of large size and complex shape. In the 1990s, PTD became generally recognized as an efficient and economic approach in the design of stealth aircraft and ships having reduced RCS signatures.

Gaussian Beams and the Paraxial Approximation to the Angular Spectrum Method. As was discussed in [17], the theory of Gaussian beams may be obtained from various starting points. For example, a fundamental solution of the parabolic equation in the free space is the Gaussian beam. In this manner, in [139], Goubau studied the angular-spectrum PO approach in the paraxial approximation, enabling him to obtain the field as a sum of Gauss-Hermite functions (in a Cartesian coordinate system) or of Gauss-Laguerre functions (in a cylindrical coordinate system). The characteristic parameters - beamwidth (or the field spot size) and phase front radius of curvature have convenient and clear physical meaning. Even more important is that the field expansion as a Fourier integral enables one to consider the beam transformation in various QO systems by using a sort of the scattering matrix technique [209,210]. This approach to wave beam propagation, together with the GO ideas of phase correction, takes into quite accurate account the diffraction effects associated with large-sized lenses and reflectors. Later J. Arnaud [211] and A. Greynolds [212] greatly enhanced this approach by introducing the decomposition of a field distribution on an aperture into a set of spatially confined fundamental Gaussian beams of at least 100λ width. This eliminated the need for intermediate GO propagation, and enabled one to compute the wave propagation between two surfaces very efficiently.

Complex-Source-Point Concept. This is another, and very elegant, way to introduce the beam-like field solutions of the full-wave electromagnetic equations. Although many authors touched it in the late 1960s, G. Dechamps is frequently credited for suggesting that one could add an imaginary component to the radius vector of the source point and obtain waves that behave as directive beams [213]. Interestingly, a similar paper published in the USSR a year earlier [214] went unnoticed. In the paraxial domain of the near zone, the complex-source-point (CSP) beam behaves as a Gaussian beam, while in the far zone it is an outgoing spherical or cylindrical wave. The CSP field can be taken as an incident field in the analysis of any scattering problem. Then, given a solution to the similar problem for "real-source-point" excitation, one should modify it by inserting the corresponding complex-valued parameters. In the 1970s, Felsen and his colleagues published many analytical studies of this sort, treating beam scattering by flat material interfaces and slabs [215]. This idea was later implemented in numerical analyses of reflectors with the aid of asymptotic GO [216] and accurate analytical regularization [217] approaches. Despite mathematical complications (the appearance of finite-size real-space branch cut in the CSP field function), and the inability to simulate radiation blockage by the feed, the CSP beam is clearly a step ahead of the simplified Gaussian beams, which satisfy only the parabolic equation.

Vector Theories. It is quite clear that in order to obtain the most accurate results one must use a vector form of diffraction theory based on Maxwell equations, exact boundary and edge conditions, and exact distances instead of approximations. The first satisfactory vector approaches to the electromagnetic scattering problem were formulated by F. Kottler [218], and then by J. Stratton and L. Chu [219]. The Gaussian beam decomposition formulation was extended to the vector case in [220]. Vector forms of CSP beams are straightforward and equally easy to handle – they imply adding an imaginary term to the real space location of an elementary dipole or a Huygens source. The corresponding field functions still are exact solutions to Maxwell's equations in the whole space considered. This approach has been successfully applied to the analysis of open resonators [221] and spherical-reflector antennas [222].

Modal Expansions and Mode Matching. Expansion of the electromagnetic field in a closed metallic waveguide in terms of a modal series has been known since the pioneering work of Lord Rayleigh around 1900 [223]. The analysis of wave propagation leads here to the eigenvalue problems in which one has to find a discrete set of real or complex valued propagation constants for the natural modes of a regular waveguide. The guided mode theory of closed metallic waveguides was developed in the 1940s and 1950s. Stratton was probably the first to consider the natural propagation modes of a circular cylindrical channel in an infinite dielectric [155]. He used separation of variables and derived generic dispersion equations. In 1964, E. Marcatili and R. Schmeltzer studied the same structure [154] in more detail assuming that the radius of the channel was large. They determined the natural modes together with their propagation and attenuation constants. The self filtering effect was noticed and explained by the greater absorption of higher order natural modes. The lowest hybrid natural mode HE_{11} having the minimum attenuation was even recommended for possible application in longdistance optical communication. Although this idea was later discarded in view of obvious success of optical fibers, propagation of waves in hollow channels was studied for the other applications. Interesting numerical data on the modes in a lossy tunnel were published in [224].

The mode-matching technique (MMT) is a widely known numerical-analytical method of solving the scattering problems associated with closed resonators and waveguides. Since the 1970s, it has been applied to study the effects of irregularities such as bends and steps in OSW. Here, economic algorithms having guaranteed convergence were developed in the 1970s and 1980s in IRE NASU, Ukraine [203, 225] and in the Rostov State University, Russia [226]. They were based on the concept of semi-inversion or analytical regularization and enabled a rigorous check of the validity of previously known asymptotic GO and PO solutions for angled bends, U-turns, steps and OSW cavities.

Grating Theories. Theoretical analysis of electromagnetic wave scattering by periodic scatterers such as wire grids and diffraction gratings cut in a metal plane was carried out by many researchers and has a long history since the pioneering papers of H. Lamb [227] and Lord Rayleigh [228]. In QO components, as is clear from the review presented above, two types of periodic structures are commonly used: free standing wire grids and flat metal-strip gratings deposited on dielectric slabs. This is because the most valuable property of gratings for QO instruments is their polarization selectivity. This occurs provided that the grating elements are metallic and that their period is less than one fifth of the wavelength, i.e. they can produce only zero-order scattering. Higher order scattering does not see much practical use; hence the application of gratings in QO is significantly different from that in traditional optics. In consequence, quasistatic theories of single order scattering of plane waves from gratings are of primary importance. In the USSR, Vainshtein of the IPP RAS was a pioneer in this area when he published, in 1955, a paper on the Wiener-Hopf solution to the PEC flat strip grating [229] not knowing that the same solution was published just one year before in [230]. His later paper [231] and papers of V. Yampolsky [232] can be credited with presenting the theoretical estimates of the reflectance and transmittance of PEC wire grids having period p in the range $2b < p \ll \lambda$, where $2b$ is the wire diameter. These works duplicated in part similar publications in the West [233-235]. Later, similar and more accurate expressions (in the USSR, they were usually called Lamb-type formulas) were obtained by several researchers in Kharkov for a wider class of diffraction gratings including imperfect wire grids and PEC flat strip gratings [225,236]. Here, the technique was a semi-inversion method with analytical inversion of the static part of the scattering problem that led to explicit solutions in the small-period case. These analyses revealed that realistic wire grids made of copper or tungsten could be still simulated as PEC ones at sub-mm wavelengths provided that $p < 0.2\lambda$, as the absorption did not exceed 5% even with the E-field vector parallel to the wires. These conclusions were supported by experimental measurements [237]. Later the semi-inversion method was used in [238] to obtain Lamb-type formulas for the free-standing gratings made of resistive and dielectric flat strips. The theories that had been developed remained valid for higher-order scattering as well, where a numerical study was imperative. This led to several interesting proposals about using the gratings of the groove and echelette type in the blazing mode, e.g., to provide additional resolution in open resonator spectra. A later book [239] included several examples of Gaussian beam scattering by infinite perfectly conducting strip grids.

Numerical Methods. There exist many numerical methods that can potentially be used in the analysis and design of QO components and devices. We will not discuss them in detail. However, we would like to emphasize that almost all of them either loose accuracy in the range of the characteristic QO relationship between the wavelength and the size of the scatterer, or lead to high numerical complexity of the algorithms and hence prohibitively large computation times. Therefore, one will look in vain for the frequency dependence of the gain of a reflector or lens antenna of realistic size computed with the method of moments or FDTD. There is, however, at least one full-wave analytical-numerical approach that is able to reach the true QO range of the size-to-wavelength ratio while maintaining high accuracy and reasonable computation time. It is referred to as the semi-inversion or analytical regularization method (MAR), developed primarily in the USSR by the groups of scientists in Kharkov, Rostov, Lvov, and Moscow since the early 1960s. Several researchers in Greece, Italy, Japan, and the USA also made significant contributions - see [240]. At the core of this approach is the idea of analytical inversion of the "worst" part of a singular integral equation equivalent to the Maxwell problem. This invertible most singular part is associated either with the static limit or with a canonical regular shape (circular cylinder or sphere) or with the high-frequency limit (actually, again a canonic scatterer, however a specific one – the halfplane). The resulting matrix equations that are generated by such an approach are always very small – just slightly larger than electrical size of the scatterer, and have favorable convergence features. At first, this approach was extensively used in the analysis of gratings and discontinuities in OSW [203,225], and then it was applied to other problems [239]. MAR based on the static part or canonic shape inversion encounters no computational problems when solving 10-100 λ reflectors and lenses in the QO domain [222]. However, it is more economical if based on the inversion of the high frequency scattering operator parts relating to the semi-infinite fragments of scatterers. This variant of MAR is much less developed than the others, and the known solutions are restricted to the flat strip and flat disk scatterers [241-243]. If extended to curved and combined scatterers, it can become an extremely efficient and accurate tool for computational electromagnetics in QO applications.

15.9   NEW FRONTIERS OF THE XXI CENTURY: OPTICS GOES QUASIOPTICAL

QO has been one of the important areas in millimeter wave and sub-mm wave science and engineering from the time of Hertz's first experiments to the present. It includes the methods of transmission of electromagnetic power, the principles underlying functional elements, the experimental methods of physical research, and analytical and numerical simulation tools. The scientists of the former USSR made significant contributions to fundamental and applied research related to mm and sub-mm waves. From the information that has been presented here, one can see that the main QO laboratories were established by the USSR Academy of Sciences and concentrated in Moscow, Kharkov and Nizhny Novgorod. The *Microwave Pioneer Award* of the IEEE Microwave Theory and Techniques Society given in 2000 to the Ukrainian scientist Yevgeny Kuleshov caps the recognition by the world scientific community of this contribution. His citation reads, *"For the development of a hollow dielectric beamguide technology and measuring technique of the short-wave part of millimeter and sub-millimeter wave ranges"* [244].

In the XXI century, QO has new and exciting frontiers mainly in the shorter than millimeter wavelength portion of the electromagnetic spectrum. These frontiers are set, on the one hand, by the amazing opportunities given by the available computer hardware and software, and, on the other hand, by the rapid development of terahertz technologies and optoelectronics. It can be expected that more and more sophisticated QO systems will appear, with development aided by computer design tools based on more accurate vector solutions. For example, today terahertz-range antennas show an example of new synergy between QO and quasistatic principles (the first was the application of small-period grid polarizers in QO beam waveguides). Here, microsize printed elements and coupling lines are integrated with lenses to boost the efficiency and sensitivity, and to provide beam collimating and focusing [245,246].

Remarkably, even the areas traditionally governed by classical optics have come to the point where comprehensive account of diffraction phenomena is possible and desired in order to improve system performance. This is the case, e.g., for optical and near-infrared astronomical observations, with large integration times, of faint objects. In 1997, the work stared on the Very Large Telescope Interferometer (VLTI) international project, with the ultimate goal of computer-controlled coherent combination of signals from the four "unit telescopes" (diameter 8.2 m) and several movable "auxiliary telescopes" (diameter 1.8 m). In 2001, the first stellar observations were undertaken with this interferometer. At a wavelength of $2\mu m$, effective propagation distances reach 200 m resulting in significant diffractive spreading of the starlight beam several cm in diameter; hence beam clipping and general deviations from GO behavior are inevitable. The way to achieve the ultimate performance is seen in high-precision steering and tuning the variable-curvature (pressurized) mirrors included into a control loop and aided by computer analysis of beam propagation provided by the *BeamWarrior* software package [247]. This combines several different methods in the common framework of scalar diffraction theory: direct

PO field computation and angular-spectrum methods (both with and without the paraxial approximation), and a Gaussian beam decomposition technique able to compute "end-to-end" wave optical propagation. In the latter case, the typical waist size of the elementary beams considered is 1300λ. This software package is the most powerful tool available today for computing quasioptical fields, including polarization effects, if both the apertures and the observation distance from the apertures are large compared to λ. Thus the right-hand side boundary of "QO strip" in Figure 15.1 is shifting to at least $D \approx 10^4 \lambda$.

At the opposite end of the chain are light-emitting components, devices and instruments, having typical dimensions not much larger than a wavelength. Hence, accounting for the QO effects in the light propagation, confinement, and scattering is mandatory for improvement of existing technologies. For example, comprehensive modeling of vertical-cavity surface-emitting lasers (VCSELs) to enhance the performance of these complex light sources is an active R&D topic. Recently it was found that beam confinement and threshold reduction achieved in VCSELs with oxidized windows can be correctly explained by assuming that the mode field has the shape of a Laguerre-Gauss beam [248]. Therefore it is no surprise that one of the most efficient ways of polarization selection of the VCSEL modes is to use a metal grid placed on top of the output aperture [249]. There is little doubt that this approach will be further extended, and that finer diffraction effects are accurately taken into detailed account in the design of mm-wave, terahertz, infrared, and optoelectronic devices and systems satisfying QO criteria.

Exciting opportunities for the design of revolutionary new QO components and instruments are offered by emergent technological innovations, such as electromagnetic bandgap materials and metamaterials (also known as twice-negative and left-handed materials). For example, a QO prism made of bandgap material may display frequency and angular dependence of the incident beam deflection one hundred times stronger than that of a similar homogeneous prism [250]. Exotic designs of new QO lens antennas and beam waveguides made of metamaterials can be based on interesting effects including negative refraction [251]. Therefore the future of quasioptics is perfectly secure as long as electromagnetic waves are still used by the information society.

15.10 ACKNOWLEDGMENTS

The authors would like to express their sincere thanks to G. I. Khlopov, V. K. Kiseliov, Y. M. Kuleshov, V. M. Yakovenko, M. S. Yanovsky (Kharkov, Ukraine), L. V. Kasatkin (Kiev, Ukraine), G. P. Gorbunov, V. V. Meriakri, M. N. Shepelev, V. V. Shevchenko, A. V. Shishlov (Moscow, Russia), L. I. Fedoseev, L. V. Lubyako (Nizhny Novgorod, Russia), M. T. Aivazian (Yerevan, Armenia), B. Z. Katsenelenbaum (Nahariya, Israel) for their interest, criticism, useful recommendations and materials kindly provided. The support for this work by the IEEE Life Members Committee is acknowledged with gratitude.

REFERENCES

[1]   O. J. Lodge, J. L. Howard, "On the concentration of electric radiation by lenses", *Nature*, vol. 40, p. 94, 1889.

[2]   E. Karplus, "Communication with quasioptical waves", *Proc. IRE*, vol. 19, pp. 1715-1730, Oct., 1931.

[3]   F. Sobel, F. L. Wentworth, and J. C. Wiltse, "Quasi-optical surface waveguide and other components for the 100 to 300 Gc region", *IEEE Trans. Microwave Theory Tech.*, vol. MTT-9, no. 6, pp. 512-518, 1961.

[4]   E. Wolf, "Microwave optics", *Nature*, vol. 172, pp. 615-616, 1953.

[5]   N. Carrara, L. Ronchi, M. Scheffner, and G. T. di Francia, "Recent research on microwave optical systems", *Alta Frequenza*, vol. 26, pp. 116-158, 1957.

[6]   G. F. Hull, "Microwave experiments and their optical analogues", in J. Strong, Ed., *Concepts of Classical Optics*, San Francisco: Freeman, 1958.

[7]   J. F. Ramsey, "Microwave antenna and waveguide techniques before 1900", *Proc. IRE*, vol. 46, no. 2, pp. 405-415, 1958.

[8]   A. F. Harvey, "Optical techniques at microwave frequencies", *Proc. IEE* (London), vol. 106, part B, no. 26, pp. 141-157, 1959.

[9]   C. L. Andrews, *Optics of the Electromagnetic Spectrum*, New York: Prentice-Hall, 1960.

[10]   A. F. Harvey, *Microwave Engineering,* London: Academic Press, 1963.

[11]   Proceedings of the Symposium on Quasi-Optics, New York: Brooklyn Polytechnic Press, 1964.

[12]   R. Tremblay and A. Boivin, "Concepts and techniques of microwave optics", *Appl. Opt.*, vol. 5, no. 2, pp. 249-278, 1966.

[13]   G. Goubau, "Optical or quasi-optical transmission schemes", in *Millimetre and Submillimetre Waves*, F. A. Benson, Ed., London: Iliffe, pp. 337-367, 1969.

[14]   R. H. Garnham, "Quasi-optical components", in *Millimetre and Submillimetre Waves*, F. A. Benson, Ed., London: Iliffe, pp. 403-450, 1969.

[15]   P. F. Goldsmith, "A quasioptical feed system for radioastronomical observations at mm wavelength", *Bell Syst. Tech. J.*, vol. 56, no. 10, pp. 1483-1501, 1977.

[16]   P. F. Goldsmith, "Quasioptical techniques at mm and sub-mm wavelengths", in *Infrared and Millimeter Waves*, vol. 6, K. Button, Ed., New York: Academic, 1982.

[17]   P. F. Goldsmith, "Quasioptical techniques", *Proc. IEEE*, vol. 80, no. 11, pp. 1729-1747, 1992.

[18]   J. C. G. Lesurf, *Millimetre-Wave Optics, Devices, and Systems*, Bristol: Adam Hilger, 1990.

[19]   P. F. Goldsmith, *Quasi-optical Systems: Gaussian Beam, Quasioptical Propagation and Applications*, New York: IEEE Press, 1998.

[20]   L. I. Mandelstam, Ed., *From the Prehistory of Radio*, Moscow: AN SSSR Press, 1948 (in Russian).

[21]   A. E. Salomonovich, "Optics of mm waves and radio astronomy", *Uspekhi Fizicheskikh Nauk*, vol. 77, no. 4, pp. 589-596, 1962 (in Russian. Translated in *Soviet Physics-Uspekhi*).

[22]   V. P. Borisov and V. N. Sretinski, "Experiments, theories and discoveries before the era of radio", *Radiotekhnika*, no. 4-5, pp. 10-15, 1995 (in Russian).

[23]   I. V. Lebedev, "Microwaves – past, present and future", *Radiotekhnika*, no. 4-5, pp. 74-78, 1995 (in Russian).

[24]   B. Z. Katsenelenbaum, "Quasioptical methods of shaping and transmission of mm waves", *Uspekhi Fizicheskikh Nauk*, vol. 83, no. 1, pp. 81-105, 1964 (in Russian. Translated in *Soviet Physics-Uspekhi*).

[25]   R. B. Vaganov and B. Z. Katsenelenbaum, "Quasioptical transmission lines for mm waves", *Antenny*, no. 1, pp. 22-33, 1966 (in Russian).

[26]   R. A. Valitov, Ed., *Sub-Millimeter Wave Engineering*, Moscow: Sovetskoe Radio Publ., 1969 (in Russian).

[27]   R. A. Valitov and B. I. Makarenko, Eds., *Measurements on Millimeter and Sub-Millimeter Waves*, Moscow: Radio i Svyaz Publ., 1984 (in Russian).

[28]   Y. M. Kuleshov "Measurements in sub-mm wavelength band", Chapter 8, in A. Y. Usikov, Ed., *Electronics and Radio Physics of Millimeter and Sub-millimeter Waves*, Kiev: Naukova Dumka Publ., 1986, pp. 140-157 (in Russian).

[29]   H. Hertz, "Ueber strahlen elektrischer kraft", *Ann. Phys. Chem.* (Leipzig), Bd. 36, s. 769-783, 1889.

[30]   A. Righi, "Sulle oscillazioni elettriche a piccola lunghezza d'onda e sul loro impiego nella produzione di fenomeni anloghi ai principali fenomeni dell'ottica", *Mem. Acad.* (Bologna), vol. 4, pp. 487-591, 1894.

[31]   P. N. Lebedev, "Double refraction of electric waves", *Ann. Phys.* (Leipzig), Bd. 56, s. 1-17, 1895.

[32]   P. N. Lebedev, "The scale of electromagnetic waves in ether", *Physical Review* (Warsaw), vol. 2, no. 2, pp. 49-60, 217-230, 1901 (in Russian).

[33]   V. A. Fabrikant, Ed., *Scientific Correspondence of P.N. Lebedev*, Moscow: Nauka Publ., 1990 (in Russian).

[34]   J. C. Bose, *Collected Physical Papers*, New York: Longmans, Green and Co, 1927.

[35]   D. T. Emerson, "The work of Jagadis Chandra Bose: 100 years of mm-wave research", *IEEE Trans. Microwave Theory Tech.*, vol. MTT-45, no.12, pp. 2267-2273, 1997.

[36]   O. J. Lodge, "Signalling through space without wires", in: *The Electrician*, London: Printing and Publishing Co., Ltd., 1898.

[37]   J. A. Fleming, *The Principles of Electric Wave Telegraphy and Telephony*, New York: Longmans, 1906.

[38]   A. Righi, *The Optics of Electric Oscillations*, Bologna: N. Zanichelli, 1897.

[39]   J. C. Bose, "On a complete apparatus for the study of the properties of electric waves", *Electrician*, vol. 37, p. 788-791, Oct., 1896.

[40] J. C. Bose, "Detector for electrical disturbances", *U.S. Patent* 755840, March 29, 1904.

[41] D. L. Sengupta, T. K. Sarkar, and D. Sen, "Centennial of the semiconductor diode detector", *Proc. IEEE*, vol. 86, no. 1, pp. 235-243, 1998.

[42] G. F. Hull, "On the use of interferometers in the study of electric waves", *Phys. Rev.*, vol. 5, pp. 231-246, Oct., 1897.

[43] J. C. Bose, "On a new electro-polariscope", *Electrician*, vol. 36, pp. 291-292, Dec., 1895.

[44] J. C. Bose, "On rotation of plane of polarization of electric waves by twisted structures", *Proc. R. Soc.* (London), vol. 63, pp. 146-152, Dec., 1898.

[45] A. Garbasso and E. Aschkinass, "Brechnung und dispersion electricher strahlen", *Ann. Phys.* (Leipzig), Bd. 53, s. 534-541, 1894.

[46] K. Birkeland, "On magnetization produced by Hertzian currents: a magnetic dielectric", *Comp. Rend. Acad. Sci.* (Paris), vol. 118, pp. 1320-1323, June, 1894.

[47] A. D. Cole, "Measurement of short electrical waves and their transmission through water cells", *Phys. Rev.*, vol. 7, pp. 225-230, Nov., 1898.

[48] E. Branly, "The transmission of Hertz waves through liquids", *Compt. Rend. Acad. Sci.* (Paris), pp. 146-152, Oct., 1899.

[49] G. Marconi, "Improvements in transmitting electrical impulses and signals, and in apparatus therefore", *British Patent Specification*, no. 12.039, June 2, 1896.

[50] C. Hulsmeyer, "Device for detecting distant metal objects with electric waves" ("Verfahren, um entfernte metallische Gegenstände mittels elektrischer Wellen einem Beobachter zu melden"), *German Patent*, no. 165546, April 30, 1904.

[51] E. F. Nichols and J. D. Tear, "Short electric waves", *Phys. Rew.*, vol. 21, pp. 587-610, June, 1923.

[52] A. Glagolewa-Arkadiewa, "Short electromagnetic waves of wavelength up to 82 mm", *Nature* (London), vol.113, p. 640, May 3, 1924.

[53] R. C. Hansen, Ed., *Microwave Scanning Antennas*: New York, Academic Press, 1964.

[54] G. E. Southworth, "New experimental methods applicable to ultra short waves", *J. Appl. Phys.*, vol. 8, pp. 660-665, 1937.

[55] A. A. Kostenko, A. I. Nosich, and I. A. Tishchenko, "Development of the first Soviet 3-coordinate *L*-band pulsed radar in Kharkov before WWII", *IEEE Antennas and Propagation Magazine*, vol. 44, no. 3, pp. 28-49, 2001.

[56] A. F. Bogomolov and S. M. Verevkin B. A., Poperechenko, and I. F. Sokolov, "Antenna system of the ground station "Orbita", *Antenny*, 1969, no. 5, pp. 3-5 (in Russian).

[57] A. E. Salomonovich, Ed., *Survey of the History of Radio Astronomy in the USSR*, Kiev: Naukova Dumka Publ., 1985 (in Russian).

[58] A. E. Salomonovich, "Astronomy with sub-mm waves from Earth atmosphere", *Uspekhi Fizicheskikh Nauk*, vol. 99, no. 3, pp. 417-438, 1969 (in Russian. Translated in *Soviet Physics-Uspekhi*).

[59] A. G. Kislyakov, "The radioastronomical investigations in mm and sub-mm wave range", *Uspekhi Fizicheskikh Nauk*, vol. 101, no. 4, pp. 607-654, 1970 (in Russian. Translated in *Soviet Physics-Uspekhi*).

[60] A. A. Konovalenko, L. N. Litvinenko, and C. G. M. van't Klooster, "Radio telescope RT-70 in Evpatoria and space investigations", in *Proc. Int. Conf. on Antenna Theory and Techniques*, Sevastopol, 2003, vol. 1, pp. 57-62.

[61] A. F. Bogomolov, B. A. Poperechenko, and A. G. Sokolov, "Tracking radio telescope TNA-1500 with 64-m parabolic reflector", *Antenny*, 1982, no. 30, pp. 3-13 (in Russian).

[62] P. Herouni, "The first radio-optical telescope", in *Proc. Int. Conf. Antennas Propagation (ICAP-89)*, 1989, vol. 1, Part.1, pp. 540-546.

[63] A. A. Tolkachev, B. A. Levitan, G. K. Solovjev, V. V. Veytsel, and V. E. Farber, "A megawatt power mm-wave phased array radar", *IEEE Aerospace Electronic Systems Mag.*, no. 3, pp. 25-31, 2000.

[64] S. P. Morgan, "General solution of the Luneburg lens problem", *J. Applied Physics*, vol. 29, no. 9, pp. 1358-1368, 1958.

[65] R. G. Rudduck and C. H. Walker, "A general analysis of geodesic Luneburg lens", *IEEE Trans. Atennas Propagat.*, vol. AP-10, no. 4, pp. 444-450, 1962.

[66] E. G. Zelkin and R. A. Petrova, *Lens Antennas*, Moscow: Sovetskoe Radio Publ., 1974 (in Russian).

[67] http://www.com2com.ru/konkur

[68] A. M. Pokras, "Some problems of using of wireless transmission lines in the radio relay systems", *Voprosy Radoelektroniki*, Ser. 10, no. 8, pp. 49-55 (in Russian), 1960.

[69] A. M. Pokras, *Wireless Ttransmission Lines*, Moscow: Svyaz Publ., 1967 (in Russian).

[70] W. A. Imbriale, M. S. Esquivel, and F. Manshadi, "Novel solutions to low-frequency problems with geometrically designed beam-waveguide systems", *IEEE Trans. Antennas Propagat.*, vol. AP-46, no. 12, pp. 1790-1796, 1998.

[71] A. I. Gaponov, M. I. Petelin, and V. K. Yulpatov, "The induced radiation of excited classical oscillators and its in high frequency electronics", *Radiophysics and Quantum Electronics*, vol. 10, no. 9-10, pp. 794-814, 1967.

[72] V. S. Averbakh, S. N. Vlasov, E. M. Popova, and. N. M. Sheronova, "Experimental study of reflector beam waveguide", *Radiotekhnika i Elektronika*, vol. 11, no. 4, pp. 750-752, 1966 (in Russian. Translated in *Radio Engineering and Electronic Physics*).

[73] S. N. Vlasov, L. I. Zagryadskaya, and M. I. Petelin, "Transformation of a whispering-gallery mode of the circular waveguide into a wave beam",

Radiotekhika i Elektronika, vol. 20, no. 10, pp. 2026-2030, 1975 (in Russian. Translated in *Radio Engineering and Electronic Physics*).

[74]   A. Fernandez, K. M. Likin, P. Turullols, J. Teniente, R. Gonzalo, C. del Rio, J. Marti-Canales, M. Sorolla, and R. Martin, "Quasioptical transmission lines for ECRH at TJ-II stellarator", *Int. J. Infrared and Millimeter Waves*, vol. 21, no. 12, pp. 1945-1957, 2000.

[75]   M. Thumm, "Passive high-power microwave components", *IEEE Trans. Plasma Science*, vol. 30, no. 3, pp. 755-786, 2002.

[76]   R. G. Fellers, "Mm wave transmission by non-waveguide means", *Microwave J.*, vol. 5, no. 5, pp. 80-86, 1962.

[77]   N. A. Irisova, "Sub-mm wave metrics ", *Vestnik Akademii Nauk SSSR*, no. 10, pp. 63-71, 1968 (in Russian).

[78]   A. A. Volkov, B. P. Gorshunov, A. A. Irisov, and G. V. Kozlov, "Electromagnetic properties of plane wire grids", *Int. J. Infrared and Millimeter Waves*, vol. 3, no. 1, pp. 19-43, 1982.

[79]   W. Culshaw, "The Michelson interferometer at mm wavelengths", *Proc. Phys. Soc.* (London), Sect. B, vol. 63, pp. 939-954, 1950.

[80]   W. Culshaw, "A spectrometer for mm wavelengths", *Proc. IEE* (London), vol. 100, pt. II, pp. 5-14, 1953.

[81]   W. Culshaw, "The Fabry-Perot interferometer at mm wavelengths", *Proc. Phys. Soc.* (London), Sect. B, vol. 66, pp. 597-608, 1953.

[82]   E. G. Goodall and J. A. C. Jackson, "Focused spectrometer for microwave measurement", *Marconi Rev.*, vol. 20, no. 125, pp. 51-59, 1957.

[83]   B. A. Lengyel, "A Michelson-type interferometer for microwave measurements", *Proc. IRE.*, vol. 37, no. 11, pp. 1242-1244, 1949.

[84]   E. A. Vinogradov, N. A. Irisova, T. S. Mandelstam, and T. A. Shmaonov, "Wideband sub-mm radiospectrometer for investigation of solid state absorption at liquid helium temperatures", *Pribory i Tekhnika Eksperimenta*, no. 5, pp. 192-195, 1967 (in Russian).

[85]   M. B. Golant, R. L. Vilenskaya, E. A. Zyulina, Z. F. Kaplun, A. A. Negirev, V. A. Parilov, T. B. Rebrova, and V. S. Saveliev, "Series of wideband oscillators of low power for the mm waveband", *Pribory i Tekhnika Eksperimenta*, no. 4, pp. 136-139, 1965 (in Russian).

[86]   A. A. Volkov, Y. G. Goncharov, G. V. Kozlov, S. P. Lebedev, and A. M. Prokhorov, "Dielectric measurements in the sub-mm wavelength range", *Infrared Phys.*, vol. 25, no. 1-2, pp. 369-373, 1985.

[87]   A. A. Volkov, Y. G. Goncharov, G. V. Kozlov, and S. P. Lebedev, "Dielectric measurements and properties of solid states at the frequencies of 10^{11} to 10^{12} Hz", in *Trudy IOFAN*, vol. 25, Sub-mm spectroscopy of solid state, Moscow: Nauka Publ., pp. 3-51, 1990 (in Russian).

[88]   R. Edlington, and R. Wylde, "A multichannel interferometer for electron density measurements in COMPASS", *Rev. Sci. Instrum.*, no. 63(10), pp. 4968-4970, 1992.

[89]   R. Prentice, T. Edlington, R. T. C. Smith, D. L. Trotman, R. J. Wylde, and P. Zimmermann, "A two color mm-wave interferometer for the JET divertor", *Rev. Sci. Instrum.*, no. 66(2), pp. 1154-1158, 1995.

[90]   D. Veron, J. Certain, and J. P. Crenn, "Multichannel HCN Interferometers for Electron Density Profile Measurements of Tokamak Plasmas", in *Preprint EUR-CEA-FC-799*, France, Fontenay-au-Roses, 1975.

[91]   N. Tesla, *Experiments with Alternate Current of High Potential and High Frequency*, New York: McGraw Publ. Co., 1904.

[92]   S. I. Tetelbaum, "On long-distance wireless transmission of electric power with radio waves", *Elektrichestvo*, no. 5, pp. 43-46, 1945 (in Russian).

[93]   S. I. Tetelbaum, "On the problem of wireless power transmission with high efficiency", *Doklady Akadademii Nauk SSSR*, vol. 52, no 3, pp. 223-226, 1946 (in Russian).

[94]   S. I. Tetelbaum, "Wave channel with radiators of improved shape", *Zhurnal Tekhnicheskoi Fiziki*, vol. 18, no. 10, pp. 1181-1186, 1947 (in Russian).

[95]   A. F. Kay, "Near-field gain of aperture antenna", *IEEE Trans. Antennas Propagat.*, vol. AP-8, no. 11, pp. 586-593, 1960.

[96]   R. W. Bickmore, "Power transmission via radio waves", *Proc. IRE*, vol. 48, no. 3, pp. 366-367, 1960.

[97]   M. I. Willinski, "Beamed electromagnetic power as a propulsion energy source", *American Rocket Society J.*, vol. 29, no. 8, pp. 601-603, 1959.

[98]   P. F. Glazer, "Power from the Sun: its future", *Science*, vol. 162, pp. 857-866, 1968.

[99]   V. A. Vanke, L. V. Leskov, and A. V. Lukianov, *Space Power Systems*, Moscow: Mashinostroenie Publ., 1990 (in Russian).

[100]   K. E. Tsiolkovsky, "Spaceship", in *Collected Works*, vol. 2, Moscow: AN SSSR Press, 1954, pp. 158-159 (in Russian. First published in 1924).

[101]   M. I. Willinski, "Microwave powered ferry vehicles", *Spaceflight*, vol. 8, no. 6, pp. 217-225, 1966.

[102]   W. S. Brown, "History of microwave power transmission and the microwave-powered helicopter", *Wire and Radio Commun.*, vol. 82, no. 11, pp. 76-79, 88-89, 1964.

[103]   H. Matsumoto, "Microwave power transmission from space and related nonlinear plasma effects", *The Radio Science Bulletin*, no. 273, pp. 11-35, June, 1995.

[104]   http://www.zerkalo-nedeli.com/nn/show/445/38642/

[105]   L. B. Felsen, "Quasi-optic diffraction", in *Proc. Symp. Quasi-Optics*, New York: Polytechnic Inst. of Brooklyn, 1964, pp. 1-40.

[106]   S. E. Miller, "Low-loss waveguide transmission", *Proc. IRE*, vol. 41, no. 3, pp. 348-358, 1953.

[107]   V. B. Shteinshleiger, Ed., *Low-Loss Transmission Lines*, Moscow: IIL Publ., 1960 (in Russian).

[108]   Y. I. Kaznacheyev, *Wideband Long-Distance Communication with Waveguides*, Moscow: AN SSSR Press, 1959 (in Russian).

[109]   R. B. Vaganov, R. F. Matveev, and V. V. Meriakri, *Multimode Waveguides with Random Inhomogeneities*, Moscow: Sov. Radio Publ., 1972 (in Russian).

[110] R. H. Garnham, *Optical and quasi-optical techniques and components for mm wavelengths*, Royal Establishment, Malvern, England, R. E. E. Rpt. no. 3020, 1958.

[111] D. H. Martin, Ed., *Spectroscopic Techniques for Far Infrared, Sub-milliumetre and Millimetre Waves*, Amsterdam: North-Holland Publ., 1967.

[112] L. N. Vershinina and V. V. Meriakri, "Sub-mm waveguide channel", *Radiotekhika i Elektronika*, vol. 12, no. 10, pp. 1815-1817, 1967 (in Russian. Translated in *Radio Engineering and Electronic Physics*).

[113] J. S. Butterworth, A. L. Cullen, and P. N. Robson, "Over-moded rectangular waveguide for high-power transmission", *Proc. IEE* (London), vol. 110, pp. 848-858, April-June, 1963.

[114] J. J. Taub, H. J. Hindin, O. F. Hinckelmann, and M. L. Wright, "Sub-mm components using owersize quasi-optical waveguide", *IEEE Trans. Microwave Theory Tech.*, vol. MTT-11, no. 5, pp. 338-345, 1963.

[115] J. J. Taub, "The status of quasi-optical waveguide components for mm and sub-mm wavelengths", *Microwave J.*, vol. 13, no. 11, pp. 57, 60-62, 1970.

[116] 116. J. Bled, A. Bresson, R. Papoular, and J. G. Wegrowe, "Nouvelles techniques d'utilisation des ondes millimetriques et submillimetriques", *L'Onde Electrique*, vol. 44, no. 442, pp. 26-35, 1964.

[117] L. V. Lubyako, "Interferometer based on oversize waveguides", *Pribory i Tekhnika Eksperimenta*, no. 5, pp. 130-132, 1968 (in Russian).

[118] L. V. Lubyako, "Investigation of electric and magnetic properties of materials in mm and sub-mm wave ranges", *Izv. Vyssh. Uchebn. Zaved. Radiofizika*, vol. 14, no. 1, pp. 133-137, 1971 (in Russian. Translated in *Radiophysics and Quantum Electronics*).

[119] L. I. Fedoseev and Y. Y. Kulikov, "Superheterodyne radiometers of mm and sub-mm wavebands", *Radiotekhika i Elektronika*, vol. 16, no. 4, pp. 554-560, 1971 (in Russian. Translated in *Radio Engineering and Electronic Physics*).

[120] Y. A. Dryagin, L.M. Kukin, and L. V. Lubyako, "On heterodyne noise suppression in superheterodyne receivers with intermediate frequency", *Radiotekhika i Elektronika*, vol. 19, no. 8, pp. 1779-1780, 1974 (in Russian. Translated in *Radio Engineering and Electronic Physics*).

[121] L. I. Fedoseev and Y. Y. Kulikov, "The measurement of the Lunar radiowave emission at 1.42 mm", *Astronomicheskii Zhurnal*, vol. 45, no. 4, pp. 914-916, 1970 (in Russian. Translated in *Soviet Astronomy*).

[122] I. V. Kuznetsov, L. I. Fedoseev, and A. A. Shvetsov, "Observations of radio emission from Venus, Jupiter and some galactic sources at wavelength 0.87 mm", *Izv. Vyssh. Uchebn. Zaved. Radiofizika*, vol. 25, no. 3, pp. 247-256, 1982 (in Russian. Translated in *Radiophysics and Quantum Electronics*).

[123] V. N. Voronov, V. M. Dyomkin, Y. Y. Kulikov V. G., Ryskin, and V. M. Yurkov, "Spectrum analyzer and the results of investigation of ozone in the upper atmosphere", *Izv. Vyssh. Uchebn. Zaved. Radiofizika*, vol. 29, no. 2,

pp. 1403-1413, 1986 (in Russian. Translated in *Radiophysics and Quantum Electronics*).

[124] A. A. Metrikin, *Antennas and Waveguides of Radio Relay Links*, Moscow: Svyaz Publ., 1977 (in Russian).

[125] H. E. King and J. L. Wong, "Characteristics of oversize circular waveguides and transitions at 3-mm wavelengths", *IEEE Trans. Microwave Theory Tech.*, vol. MTT-19, no.1, pp. 116-119, 1971.

[126] G. I. Khlopov, "Radar based on diffraction radiation generator", Chapter 13, in V. P. Shestopalov, Ed., *Diffraction Radiation Generators*, Kiev: Naukova Dumka Publ., 1991 (in Russian).

[127] A. A. Kostenko, S. P. Martinyuk, and G. I. Khlopov, "Quasi-optical systems of the mm wavelength band", in *Proc. Int. Kharkov Symp. on Physics and Engineering of Millimeter and Sub-millimeter Waves (MSMW-94)*, Kharkov, 1994, vol. 5, pp. 472-475.

[128] G. I. Khlopov and V. P. Churilov, "Double-reflector axially symmetric antenna for mm and sub-mm wave ranges", *USSR Patent*, no. 315453, priority date 16.02.1970 (in Russian).

[129] G. I. Khlopov, V. P. Churilov, and A. I. Goroshko, "Radiation from open end of dielectric beamguide", *Radiotekhnika*, Kharkov, Kharkov State Univ. Press, no. 18, pp. 3-9, 1971 (in Russian).

[130] V. I. Bezborodov, A. A. Kostenko, G. I. Khlopov, and M. S. Yanovsky, "Quasi-optical antenna duplexers", *Int. J. Infrared and Millimeter Waves*, vol. 18, no. 7, pp. 1411-1422, 1997.

[131] A. A. Kostenko, "Matching of ferrite elements of non-reciprocal quasioptical devices", *Radiotekhika i Elektronika*, vol. 26, no. 10, pp. 2044-2052, 1981 (in Russian. Translated in *Radio Engineering and Electronic Physics*).

[132] A. A. Kostenko and G. I. Khlopov, "Quasioptical ferrite devices for mm and sub-mm wave bands", *Int. J. Infrared and Millimeter Waves*, vol. 17, no. 10, pp. 1593-1605, 1996.

[133] R. H. Dicke, "Molecular amplification and generation systems and methods", *US Patent* 2 851 652, September 9, 1958.

[134] A. M. Prokhorov, "Molecular amplifier and oscillator of sub-mm wavelength", *Zhurnal Eksperimentalnoi i Teoreticheskoi Fiziki*, vol. 34, no. 6, pp. 1658-1659, 1958 (in Russian. Translated in *Soviet Physics-JETP*).

[135] A. L. Shawlow and C. H. Townes, "Infrared and optical masers", *Phys. Rev.*, vol. 29, pp. 1940-1949, Dec., 1958.

[136] A. G. Fox and T. Li, "Resonant modes in optical masers", *Proc. IRE*, vol. 48, no. 11, pp. 1904-1905, 1960.

[137] G. D. Boyd and J. P. Gordon, "Confocal multimode resonator for mm through optical wavelength masers", *Bell Syst. Tech. J.*, vol. 40, no. 2, pp. 489-508, 1961.

[138] G. D. Boyd and H. Kogelnik, "Generalized confocal resonator theory", *Bell Syst. Tech. J.*, vol. 41, no. 7, pp. 1347-1369, 1962.

[139] G. Goubau and F. Schvering, "On the guided propagation of electromagnetic wave beams", *IRE Trans. Antennas Propagat.*, vol. AP-9, no. 3, pp. 248-256, 1961.

[140] I. P. Christian and G. Goubau, "Experimental studies on a beam waveguide for mm waves", *IRE Trans. Antennas Propagat.*, vol. AP-9, no. 3, pp. 256-263, 1961.

[141] A. N. Akhiyezer, "Devices with quasioptical beams", in *Trudy Metrologicheskikh Institutov SSSR*, Moscow: USSR Standard Committee Publ., no. 99(159), 1969 (in Russian).

[142] V. V. Shevchenko, "Optical transmission line for mm and sub-mm waves", *USSR Patent* no. 171453, priority date 2.09.1963 (in Russian).

[143] V. V. Shevchenko, "Nonreflecting lenses for quasioptical lines", *Radiotekhika i Elektronika*, vol. 14, no. 10, pp. 1764-1767, 1969 (in Russian. Translated in *Radio Engineering and Electronic Physics*).

[144] L. N. Vershinina, A. A. Lagunov, and V. V. Shevchenko, "The investigation of quasioptical lines in sub-mm band", *Radiotekhika i Elektronika*, vol. 13, no. 2, pp. 346-348, 1968 (in Russian. Translated in *Radio Engineering and Electronic Physics*).

[145] V. V. Meriakri, V. N. Apletalin, A. N. Kopnin, G. A. Kraftmakher, M. G. Semenov, E. F. Ushatkin, and E. E. Chigryai, "Sub-mm beam waveguide spectroscopy and its applications", in V. A. Kotelnikov, Ed., *Problems of Current Radio Engineering and Electronics*, Moscow: Nauka Publ., 1980, pp. 164-180 (in Russian).

[146] I. P. Christian and G. Goubau, "Some measurements of an iris beam waveguide", *Proc. IRE*, vol. 49, no. 11, p. 1941, 1961.

[147] B. Z. Katsenelenbaum, "Millimeter wave transmission using reflections from the chain of focusing reflectors", *Radiotekhika i Elektronika*, vol. 8, no. 9, pp. 1516-1522, 1963 (in Russian. Translated in *Radio Engineering and Electronic Physics*).

[148] J. E. Degenford, M. D. Sirkis, and W. H. Steier, "The reflectimg beam waveguide", *IEEE Trans. Microwave Theory Tech.*, vol. MTT-12, no. 4, pp. 445-453, 1964.

[149] R. B. Vaganov, A. B. Dogadkin, and B. Z. Katsenelenbaum, "Periscopic reflector transmission line", *Radiotekhika i Elektronika*, vol. 10, no 9, pp. 1672-1675, 1965 (in Russian. Translated in *Radio Engineering and Electronic Physics*).

[150] A. A. Dyachenko and O. E. Shushpanov, "Investigation of reflector periscopic transmission lines", *Izvestiya Vyssh. Uchebn. Zaved. Radiofizika*, vol. 11, no. 5, pp. 707-713, 1968 (in Russian. Translated in *Radiophysics and Quantum Electronics*).

[151] B. A. Claydon, "Beam waveguide feed for a satellite earth station antenna", *Marconi Rev.*, vol. 39, no. 201, pp. 81-116, 1976.

[152] A. Y. Miroshnichenko, "Beam waveguides for double-reflector antennas", *Zarubezhnaya Radioelektronika*, no. 7, pp. 28-62, 1981 (in Russian).

[153] A. N. Akhiyezer, A. I. Goroshko, B. N. Knyazkov, Y. M. Kuleshov, D. D. Litvinov, N. I. Tolmachev, V. A. Scherbov, and M. S. Yanovsky,

"Dielectric beamguide of the sub-mm wave range", *USSR Patent*, no. 302054, priority date 28.11.1970 (in Russian).

[154] E. A. Marcatili and R. A. Schmeltzer, "Hollow metallic and dielectric waveguides for long distance optical transmission and lasers", *Bell Syst. Techn. J.*, vol. 43, no. 4, pt. 2, pp. 1783-1809, 1964.

[155] J. A. Stratton, *Electromagnetic Theory*. New York: McGraw-Hill, 1941.

[156] Y. M. Kuleshov, Ed., "Exploration of opportunities of development of radiomeasuring equipment and devices in the sub-mm band based on quasi-optical principles", *Techn. Report "Ozero"*, Kharkov: IRE NASU Press, 1964 (intermediate report, in Russian) and 1966 (final report; in Russian).

[157] A. I. Goroshko and Y. M. Kuleshov, "Investigation of a hollow dielectric beamguide of the mm and sub-mm wavelength ranges", *Radiotekhnika*, Kharkov: Kharkov State Univ. Press, no. 21, pp. 215-219, 1972 (in Russian).

[158] Y. M. Kuleshov, Ed., "Exploration of opportunities of development of a kit of radiomeasuring equipment in the sub-mm band", *Techn. Report "Oliva"*, Kharkov: IRE NASU Press, 1971 (in Russian).

[159] Y. M. Kuleshov, M. S. Yanovsky, D. D. Litvinov, V. A. Scherbov, B. N. Knyazkov, A. I. Goroshko, N. I. Tolmachev, V. V. Stenko, and V. L. Shumeiko, "Quasioptical measuring devices for mm and sub-mm waves", in *Proc. USSR Symp. on Propagation of Millimeter and Sub-millimeter Waves in Atmosphere*, Moscow-Gorky, 1974, pp. 124-127 (in Russian).

[160] Y. M. Kuleshov, V. K. Kononenko, and V. N. Polupanov, "Ferrites and semiconductors in non-reciprocal quasioptical devices", *Elektronnaya Tekhnika*, Ser. 1, Elektronika SVCh, no. 7, pp. 90-99, 1976 (in Russian).

[161] Y. M. Gershenzon, G. N. Goltsman, and N. G. Ptitsyna, "Sub-mm spectroscopy of semiconductors", *Zhurnal Tekhnicheskoi Fiziki*, vol. 64, no. 2, pp. 587-598, 1973 (in Russian. Translated in *Soviet Physics-Technical Physics*).

[162] Y. M. Gershenzon, "Spectral and radio-spectroscopic investigations of semiconductors at sub-mm waves", *Uspekhi Fizicheskikh Nauk*, vol. 122, no. 1(500), pp. 164-174, 1977 (in Russian. Translated in *Soviet Physics-Uspekhi*).

[163] B. N. Knyazkov, Y. M. Kuleshov, D. D. Litvinov, V. P. Churilov, and M. S. Yanovsky, "Mm-wave duplex device", *USSR Patent*, no. 401278, priority date 07.06.1971 (in Russian).

[164] J. F. Ramsey and W. F. Gunn, "A polarized-mirror duplexer for use with a circularly-polarized lens aerial", *Marconi Rev.*, vol. 18, no. 1, pp. 29-33, 1955.

[165] H. Buizert, "Circular polarization at mm waves by total internal reflection", *IEEE Trans. Microwave Theory Tech.*, vol. MTT-12, no. 4, pp. 477-453, 1964.

[166] V. K. Kiseliov and T. M. Kushta, "Method for radar cross-section measurements in mm and sub-mm wave regions", *Int. J. Infrared and Millimeter Waves*, vol. 16, no. 6, pp. 1159-1165, 1995.

[167] V. K. Kiseliov and T. M. Kushta, "A spherical scatterer inside a circular hollow dielectric waveguide", *Int. J. Infrared and Millimeter Waves*, vol. 18, no. 1, pp. 151-163, 1997.

[168] V. K. Kiseliov and T. M. Kushta, "Ray representation of the electromagnetic field in a circular hollow dielectric waveguide", *IEEE Trans. Antennas Propagat.*, vol. AP-46, no. 7, pp. 1116-1117, 1998.

[169] V. K. Kiseliov, T. M. Kushta, and P. K. Nesterov, "Quasi-optical waveguide modeling method and micro-compact range for the mm and sub-mm wave bands", *IEEE Trans. Antennas Propagat.*, vol. AP-49, no. 5, pp. 784-792, 2001.

[170] L. N. Vershinina, Y. N. Kazantzev, V. V. Meriakri, and V. V. Shevchenko, "Sub-mm quasi-optical lines", in *European Microwave Conference Handbook*, Kent, 1969, pp. 174.

[171] Y. N. Kazantsev, "Eigenwave attenuation in a broad waveguide of thick dielectric coating", *Radiotekhika i Elektronika*, vol. 15, no. 1, pp. 207-209, 1970 (in Russian. Translated in *Radio Engineering and Electronic Physics*).

[172] Y. N. Kazantsev and O. A. Kharlashkin, "Circular waveguides of the "hollow dielectric channel" type", *Radiotekhika i Elektronika*, vol. 29, no. 8, pp. 1441-1450, 1984 (in Russian. Translated in *Radio Engineering and Electronic Physics*).

[173] V. I. Bezborodov, B. N. Knyazkov, Y. M. Kuleshov, and M. S. Yanovsky, "Quasioptical channel of the superheterodyne receiver with the separate reception of bands", *Izv. Vyssh. Uchebn. Zaved. Radioelektronika*, vol. 32, no 3, pp. 29-33, 1989 (in Russian. Translated in *Radiophysics and Quantum Electronics*).

[174] Y. N. Kazantsev and O. A. Kharlashkin, "Waveguides of rectangular cross section with small losses", *Radiotekhika i Elektronika*, vol. 16, no. 6, pp. 1063-1065, 1971 (in Russian. Translated in *Radio Engineering and Electronic Physics*).

[175] Y. N. Kazantsev and O. A. Kharlashkin, "Rectangular waveguides of the "hollow dielectric channel class", *Radiotekhnika i Elektronika*, vol. 23, no. 10, pp. 2060-2068, 1978 (in Russian. Translated in *Radio Engineering and Electronic Physics*).

[176] M. T. Aivazian, Y. N. Kazantsev, and O. A. Kharlashkin, "Set of the waveguide components based on the rectangular metal-dielectric waveguide for short-mm range", in *Proc. USSR Symp. on Millimeter and Sub-Millimeter Waves*, Gorky, 1980, pp. 106-107 (in Russian).

[177] R. S. Avakian, K. R. Agababian, M. T. Aivazian, Y. N. Kazantsev, and R. M. Martirosian, "Complete set of waveguides for the 120-180 GHz band", in *Proc. Int. Conf. Infrared and Millimeter Waves*, Lausanne, 1991, pp. 642-643.

[178] R. S. Avakian, K. R. Agababian, and G. G. Gabrielian, "Mm-wave radiometer complex", in *Proc. Int. Conf. Infrared and Millimeter Waves*, Lausanne, 1991, pp. 626-627.

[179] V. I. Bezborodov, V. K. Kiseliov, B. N. Knyazkov, E. M. Kuleshov, V. N. Polupanov, and M. S. Yanovsky, "Quasioptical radiometric devices of the mm and sub-mm ranges based on the rectangular metal-dielectric waveguides", in *Proc. Int. Conf. Microwave Engineering and Telecommunication Technology*, Sevastopol, 1997, vol. 2, pp. 630-631 (in Russian).

[180] V. I. Bezborodov, V. K. Kiseliov, B. N. Knyazkov, E. M. Kuleshov, V. P. Churilov, and M. S. Yanovsky, "Quasi-optical arrangements of transceiving circuits of mm and sub-mm waves", in *Proc. Int. Kharkov Symp. on Physics and Engineering of Millimeter and Sub-Millimeter Waves (MSMW-94)*, Kharkov, 1994, vol. 5, pp. 476-478.

[181] A. A. Bagdasarov, V. V. Buzankin, N. L. Vasin, E. P. Gorbunov, V. F. Denisov, Y. M. Kuleshov, V. N. Savchenko, V. V. Khilil, and V. A. Scherbov, "Nine-channel interferometer of the sub-mm range for measurements of electron concentration in Tokamak-10", in Pergament M.I., Ed., *Plasma Diagnostics*, no. 4(1), Moscow: Energoizdat Publ., 1981 (in Russian).

[182] Y. M. Kuleshov, Ed., "Exploration of opportunities of development and design of microwave circuit of multi-channel interferometer of the sub-mm range", *Techn. Report "Fakel"*, Kharkov: IRE NASU Press, 1973 (in Russian).

[183] Y. M. Kuleshov, Ed., "Development and manufacturing of a laser interferometer with wavelength of 0.337 mm for measuring the plasma density", *Techn. Report "Fider"*, Kharkov: IRE NASU Press, 1980 (in Russian).

[184] Y. M. Kuleshov, Ed., "Development of multi-channel laser interferometer for measuring the plasma density pattern in the vertical sounding in Tokamak-15", *Techn. Report "Farada"*, Kharkov: IRE NASU Press, 1985 (in Russian).

[185] Y. M. Kuleshov, Ed., "Development and manufacturing of a laser super-heterodyne interferometer with optical pump for the wavelength 0.119 mm", *Techn. Report "Fregat"*, Kharkov: IRE NASU Press, 1987 (in Russian).

[186] Y. M. Kuleshov, Ed., "Development of a single-channel interferometer-polarimeter for measuring the magnetic field of the plasma current in Tokamak-15", *Techn. Report "Filigran"*, Kharkov: IRE NASU Press, 1992 (in Russian).

[187] P. J. B. Clarricoats, A. D. Olver, and S. L. Chong, "Attenuation in corrugated circular waveguides", *Proc. IEE* (London), vol. 122, no. 11, pp. 1173-1186, 1975.

[188] J. L. Doane, "Low-loss propagation in rectangular waveguide at 1-mm wavelength", *Int. J. Infrared and Millimeter Waves*, vol. 8, pp. 13-27, 1987.

[189] N. I. Kovalev, I. M. Orlova, and M. I. Petelin, "Wave transformation in multimode waveguide with corrugated walls", *Izv. Vyssh. Uchebn. Zaved.*

Radiofizika, vol. 11, no. 5, pp. 783-786, 1968 (in Russian. Translated in *Radiophysics and Quantum Electronics*).

[190] V. L. Bratman, G. G. Denisov, N. S. Ginzburg, and M. I. Petelin, "FEL's with Bragg reflection resonators: Cyclotron autoresonance maser versus ubitrons", *IEEE J. Quantum Electronics*, vol. QE-19, no. 3, pp. 282-296, 1983.

[191] D. A. Lukovnikov, A. A. Bogdashov, and G. G. Denisov, "Properties of oversized corrugated waveguides at moderate diameter-wavelength ratio", in *Proc. Int. Kharkov Symp. on Physics and Engineering of Millimeter and Sub-Millimeter Waves (MSMW-98)*, Kharkov, 1998, vol. 2, pp. 595-597.

[192] M. Born and E. Wolf, *Principles of Optics*, Oxford, London, Edinburgh, New York, Paris, Frankfurt: Pergamon Press, 1966.

[193] J. B. Keller, "Geometrical theory of diffraction", *J. Optical Society America*, vol. 52, no. 2, pp. 116-130, 1962.

[194] V. A. Borovikov and B. Y. Kinber, *Geometrical Theory of Diffraction*, Moscow: Svyaz Publ., 1976 (in Russian).

[195] V. I. Klassen, B. Y. Kinber, A. V. Shishlov, and A. K. Tobolev, "Hybryd and polyfocal antennas", *Antenny*, 1987, no. 34, pp. 3-24 (in Russian).

[196] B. Y. Kinber, E. E. Gasanov, and M. M. Vainbrand, "Reflector antennas with anisotropic surface", *Radiotekhika i Elektronika*, 1988, vol. 33, no. 8, pp. 1590-1599 (in Russian. Translated in *Radio Engineering and Electronic Physics*).

[197] L. D. Bakhrakh and G. K. Galimov, *Scanning Reflector Antennas*, Moscow: Nauka Publ., 1981 (in Russian).

[198] H. G. Booker and P.C. Clemmow, "The concept of an angular spectrum of plane waves, and its relation to that of polar diagram and aperture distribution", *Proc. IEE* (London), vol. 97, pt. III, pp. 11-17, 1950.

[199] E. L. Burstein, "Power of non-plane wave received by an antenna", *Radiotekhika i Elektronika*, vol. 3, no. 2, pp. 186-189, 1958 (in Russian. Translated in *Radio Engineering and Electronic Physics*).

[200] V. A. Borovikov and A. G. Eidus, "Diffraction by the rectangular bend of a waveguide with a reflector", *Radiotekhika i Elektronika*, vol. 21, no. 1, pp. 47-56, 1976. (in Russian. Translated in *Radio Engineering and Electronic Physics*).

[201] A. V. Foigel, "Bend of a broad waveguide", *Radiotekhika i Elektronika*, vol. 18, no 11, pp. 2276-2283, 1973 (in Russian. Translated in *Radio Engineering and Electronic Physics*).

[202] B. Z. Katsenelenbaum, "Diffraction on a plane mirror in a wide waveguide bend", *Radiotekhika i Elektronika*, vol. 8, no. 7, pp. 1111-1119, 1963 (in Russian. Translated *Radio Engineering and Electronic Physics*).

[203] V. P. Shestopalov, A. A. Kirilenko, and L. A. Rud, *Resonant Scattering of Waves*. vol. 2, *Waveguide Inhomogeneities*, Kiev: Naukova Dumka Publ., 1986 (in Russian).

[204] L. A. Vainshtein, *Open Resonators and Open Waveguides*, Moscow: Sovetskoye Radio Publ., 1966 (in Russian).

[205] B. Z. Katsenelenbaum, *High-Frequency Electromagnetics: Foundations of Mathematical Methods*, Moscow: Nauka Publ., 1966 (in Russian).

[206] N. G. Bondarenko and V. I. Talanov, "Some aspects of the theory of quasioptical systems", *Izv. Vyssh. Uchebn. Zaved. Radiofizika*, vol. 7, no. 2, pp. 313-327, 1964 (in Russian. Translated in *Radiophysics and Quantum Electronics*).

[207] V. I. Talanov, "Operator method of the wave beam theory in complex quasioptical systems", *Izv. Vyssh. Uchebn. Zaved. Radiofizika*, vol. 8, no. 2, pp. 260-271, 1965 (in Russian. Translated in *Radiophysics and Quantum Electronics*).

[208] P. Y. Ufimtsev, *Method of Edge Waves in the Physical Theory of Diffraction*, Moscow: Sovetskoye Radio Publ., 1962 (in Russian. Translated by U.S. Air Force, Foreign Technol. Div., Wright-Patterson AFB, OH, 1971; Techn. Rep. AD no. 733203 DTIC, Cameron St., Alexandria, VA 22304-6145).

[209] G. A. Deschamps and P. E. Mast, "Beam tracing and application", in *Proc. Symp. Quasi-Optics*, New York: Polytechnic Inst. of Brooklyn, 1964, pp. 379-395.

[210] H. Kogelnik and T. Li., "Laser beams and resonators", *Appl. Opt.*, vol. 5, no. 10, pp. 1550-1567, 1966.

[211] J. A. Arnaud, "Non-orthogonal optical waveguides and resonators", *Bell Sysyem Techn. J.*, 1970, vol.49, no. 9, pp. 2311-2348.

[212] A. W. Greynolds, "Propagation of general astigmatic Gaussian beams along skew ray paths", in *Diffractive Phenomena in Optical Engineering Applications*, Proc. SPIE, 1985, vol. 560, pp. 33-50.

[213] G. A. Deschamps, "Gaussian beams as a bandle of complex rays", *Electron. Lett.*, vol. 7, pp. 684-685, 1971.

[214] A. A. Izmestiev, "One-parametric wave beams in free space", *Izv. Vyssh. Uchebn. Zaved. Radiofizika*, vol., no 9, pp. 1380-1388, 1970 (in Russian. Translated in *Radiophysics and Quantum Electronics*).

[215] L. B. Felsen, "Complex source point solutions of the field equations and their relation to the propagation and scattering of Gaussian beams", in *Symposia Mathematica*, 1975, vol. 18, pp. 39-56.

[216] G. A. Suedan and E. Jull, "Beam diffraction by planar and parabolic reflectors", *IEEE Trans. Antennas Propagat.*, vol. AP-39, no. 4, pp. 521-527, 1991.

[217] T. Oguzer, A. Altintas, and A. I. Nosich, "Accurate simulation of reflector antennas by the complex source – dual series approach", *IEEE Trans. Antennas Propagat.*, vol. AP-43, no. 8, pp. 793-801, 1995.

[218] F. Kottler, "Theory of diffraction by black screens", *Ann. Phys.* (Leipzig), 1923, vol. 70, pp. 405-456; vol. 71, pp. 457-508; vol. 72, p. 320.

[219] J. A. Stratton and L. J. Chu, "Diffraction theory of electromagnetic waves", *Phys. Review*, 1939, vol. 56, pp. 99-107.

[220] A. W. Greynolds, "Vector formulation of ray-equivalent method for general Gaussian-beam propagation", in *Current Developments in Optical Engineeringand Diffractive Phenomena*, Proc. SPIE, 1986, vol. 679, pp. 129-133.

[221] A. L. Cullen and P. K. Yu, "Complex source-point theory of the electromagnetic open resonator", *Proc. R. Soc.* (London), 1979, vol. A-366, pp. 155-171.

[222] S. S. Vinogradov, P. D. Smith, E. D. Vinogradova, and A. I. Nosich, "Accurate simulation of a spherical reflector front-fed by a Huygens CSP beam: a dual series approach", in *Proc. Int. Conf. on Antennas (JINA-98)*, Nice, 1998, pp. 550-553.

[223] Lord Rayleigh, "On the passage of electric waves through tubes, or the vibrations of dielectric cylinders", *Phil. Mag.*, vol. 43, pp. 125-132, Feb., 1897.

[224] C. L. Holloway, D. A. Hill, R. A. Dalke, and G. A. Hufford, "Radio wave propagation characteristics in lossy circular waveguides such as tunnels, mine shafts, and boreholes", *IEEE Trans. Antennas Propagat.*, vol. AP-48, no. 9, pp. 1354-1366, 2000.

[225] V. P. Shestopalov, L. N. Litvinenko, S. A. Masalov, and V. G. Sologub, *Diffraction of Waves by Gratings,* Kharkov: Vyscha Shkola Publ., 1973 (in Russian).

[226] G. F. Zargano, A. M. Lerer, V. P. Lyapin, and G. P. Sinyavsky, *Waveguides of Complicated Cross-Sections*, Rostov: Rostov State University Press, 1983 (in Russian).

[227] H. Lamb, "On the diffraction and transmission of electric waves by a metallic grating", *Proc. Math. Soc.* (London), 1898, vol. 29, pp. 523-544.

[228] Lord Rayleigh, "On the dynamical theory of grating", *Proc. R. Soc.* (London), Ser. A, vol. 79, pp. 399-416, 1907.

[229] L. A Vainshtein, "Diffraction of electromagnetic waves by the gratings of parallel conducting strips", *Zhurnal Tekhnicheskoi Fiziki*, vol. 25, no. 5, pp 847-852, 1955 (in Russian. Translated in *Soviet Physics-Technical Physics*).

[230] G. L. Baldwin and A. E. Heins, "On the diffraction of a plane wave by an infinite plane grating", *Math. Scand.* vol. 2, pp. 103-118, 1954.

[231] L. A. Vainshtein, "Electrodynamic theory of gratings", in *High-Power Electronics*, Moscow: AN SSSR Press, no. 2, pp. 26-74, 1963 (in Russian).

[232] V.G. Yampolsky, "The diffraction of plane waves by a wire grid situated inside a dielectric slab", *Radiotekhika i Elektronika*, vol. 3, no. 12, pp. 1516-1518, 1958 (in Russian. Translated in *Radio Engineering and Electronic Physics*).

[233] E. A. Lewis and J. Casey, "Electromagnetic reflection and transmission by gratings of resistive wires", *J. Applied Physics*, vol. 23, no. 6, pp. 605-608, 1952.

[234] J. R. Wait, "Reflection at arbitrary incidence from a parallel wire grid", *Appl. Sci. Research*, vol. B-4, pp. 393-400, 1955.

[235] T. Larsen, "A survey of the theory of wire grids", *IRE Trans. Microwave Theory Tech.*, vol. MTT-10, no. 3, pp. 191-201, 1962.

[236] Z. S. Agranovich, V. A. Marchenko, and V. P. Shestopalov, "Diffraction of a plane electromagnetic wave from plane metallic lattices", *Zhurnal Tekhnicheskoi Fiziki*, vol. 32, no. 4, pp. 381-394, 1962 (in Russian. Translated in *Soviet Physics-Technical Physics*).

[237] A. A. Kostenko, "Engineering characteristics of one-dimensional small-period diffraction gratings in mm and sub-mm wave range", *Microwave Opt. Technol. Lett.*, vol. 19, no. 6, pp. 438-444, 1998.

[238] T. L. Zinenko, A. I. Nosich, and Y. Okuno, "Plane wave scattering and absorption by resistive-strip and dielectric-strip periodic gratings", *IEEE Trans. Antennas Propagat.*, vol. AP-46, no. 10, pp. 1498-1505, 1998.

[239] L. N. Litvinenko and S. L. Prosvirnin, *Spectral Scattering Operators in the Problems of Wave Diffraction by Flat Screens*, Kiev: Naukova Dumka Publ., 1987 (in Russian).

[240] A. I. Nosich, Method of analytical regularization in the wave-scattering and eigenvalue problems, *IEEE Antennas Propagat. Magazine*, vol. 41, no. 3, pp. 34-49, 1999.

[241] K. Kobayashi, "Wiener-Hopf and modified residue calculus techniques", in E. Yamashita, Ed., *Analysis Methods for Electromagnetic Wave Problems*, Norwood: Artech House, pp. 245-301, 1990.

[242] V. G. Sologub, "Solution of an integral equation of the convolution type with finite limits of integration", *Soviet J. Comput. Maths. Mathem. Physics*, vol. 11, no. 4, pp. 33-52, 1971.

[243] V. G. Sologub, "Short-wave asymptotic behavior of the solution of the problem of diffraction by a circular disk", *Soviet J. Comput. Maths. Mathem. Physics*, vol. 12, no. 2, pp. 135-164, 1972.

[244] Microwave Pioneer Awards, *IEEE Microwave Magazine*, vol. 1, no. 1, p. 75, March, 2000.

[245] D. F. Filipovic, G. P. Gauthier, S. Raman, and G. M. Rebeiz, "Off-axis properties of silicon and quartz dielectric lens antennas", *IEEE Trans. Antennas Propagat.*, vol. 45, no. 5, pp. 760–766, 1997.

[246] J. Rudd and D. Mittelman, "Influence of substrate-lens design in terahertz time-domain spectroscopy", *J. Opt. Soc. Am. B*, vol. 19, no. 2, pp. 319-329, 2002.

[247] R. Wilhelm, "Comparing geometrical and wave optical algorithms of a novel propagation code applied to the VLTI", *Proc. SPIE*, vol. 4436, no. 11, pp. 89-100, 2000.

[248] S. Ryapoulos, D. Dialetis, J. Inman, and A. Phillips, "Active-cavity VCSEL eigenmodes with simple analytic representation", *J. Opt. Soc. Am. B*, vol. 18, no. 9, pp. 1268-1284, 2001.

[249] J. G. Ju, J. H. Ser, and Y. H. Lee, "Analysis of metal-interlaced-grating VCSELs using the modal method", *IEEE J. Quantum Electronics,* vol. 33, pp. 589-595, 1997.

[250] H. Kosaka, T. Kawashima, A. Tomita, M. Notami, T. Tomamura, T. Sato, and S. Kawakami, "Superprism phenomena in photonic crystals: toward

microscale lightwave circuits", *J. Lightwave Technol.*, vol. 17, no. 11, pp. 2032-2038, 1999.

[251] J. B. Pendry, "Negative refraction index makes a perfect lens", *Phys. Rev. Lett.*, vol. 85, pp. 3966-3969, 2000.

16

THE EVOLUTION OF ELECTROMAGNETIC WAVEGUIDES: FROM HOLLOW METALLIC GUIDES TO MICROWAVE INTEGRATED CIRCUITS

ARTHUR A. OLINER, *Polytechnic University (Emeritus), 11 Dawes Road, Lexington, MA 02421*

16.1   HOLLOW METALLIC WAVEGUIDES

16.1.1   Early Investigations on Guided Waves

The theory presented by James Clerk Maxwell, in his original memoirs and in his *Treatise on Electricity and Magnetism*, published about 140 years ago, showed that electromagnetic waves were possible. The theoretically predicted velocity of those waves was about the same as the measured velocity of light, so that his theory excited much interest but also much skepticism. The finite velocity of those electromagnetic waves also contradicted the generally accepted theory of action at a distance, which presumed an infinite velocity of propagation. The Prussian Academy of Sciences (Berlin) established a prize, known as the *Berlin Prize,* challenging anyone to validate or invalidate Maxwell's theory experimentally [1] within a period of approximately three years, ending in March 1882. However, there were no takers since there were no sources or detectors available at that time, and no evident way to proceed.

The Berlin Prize was established under the urging of Hermann von Helmholtz, during the time that *Heinrich Rudolf Hertz* was a student of his at the University of Berlin. Helmholtz urged Hertz to think about the problem, and Hertz did so but concluded that the available measurement equipment was inadequate. Nevertheless, Hertz continued to explore how to proceed, and, in particular, critically examined both Maxwell's theory and competing theories. By November 1886, a little more than 20 years after Maxwell published his well-known paper on his theory [2], Hertz began his series of measurements. In all that time no one else had tried to perform such experiments.

In 1988, the IEEE and its Microwave Theory and Techniques (MTT) Society celebrated the Centennial of the year in which Heinrich Hertz made his most important contributions, proving that Maxwell's theory is indeed correct, and that the velocity of electromagnetic waves is finite, and, by extension, that light is an electromagnetic wave. The MTT Society collected some of Hertz's original equipment and displayed it during its International Microwave Symposium in New York City, May 25-27, 1998. In honor of this Centennial, the IEEE also published an excellent monograph [3], "Heinrich Hertz, the Beginning of Microwaves," by John H. Bryant. This publication presents all of Hertz's experiments in chronological order and in substantial detail, from his key initial invention of his novel method for generating and detecting electromagnetic waves in the VHF and UHF ranges, to many sophisticated experiments.

Many of these experiments involved guided waves, which included using standing waves in space to measure guide wavelengths and to examine polarization effects. His experiments included creating a transmission line consisting of a wire over a ground plane, constructing a coaxial line with an outer conductor composed of parallel wires so that a probe could be moved along the wires, forming the first slotted line, and, while exploring the skin effect, making measurements using thin metallic sheets of various thicknesses and observing their shielding effects. Finally, it is worth adding that he created a dipole antenna and a parabolic mirror and with them made focused-beam experiments, and he drew impressively accurate sketches of how the fields radiate outward from a dipole antenna. Hertz first gained public recognition when he published a paper referring to "waves in air". It was published in English in 1893 [4].

Although he is best known for his remarkable experiments, Hertz made important theoretical contributions as well. Perhaps the most important of these is Hertz's reformulation of Maxwell's equations "in the compact form that rapidly achieved widespread influence and has been standard ever since [5]". It is known that Albert Einstein referred to Maxwell's equations as the Hertz-Maxwell equations. In 1894, soon after Hertz completed his series of experiments in electromagnetics, but as he was entering a new group of challenges in physics, he became ill and died when he was only 36. During the late 1960s, various professional societies discussed the awkwardness of the term "cycles per second," which was then employed as the unit of frequency. It was agreed that the new unit of frequency should become the Hertz (Hz). In view of the nature and importance of Heinrich Hertz's contributions, the term is eminently suitable and the honor totally deserving.

After it was recognized from Hertz's experiments that electromagnetic waves were a reality, research activity in those waves exploded, and over the next decade or so many significant contributions were made, at frequencies even reaching into the microwave range. Some chapters in the present book treat some of the activities conducted during that period.

Guglielmo Marconi was one of the important contributors, but he was concerned primarily with long-distance wireless communication. He conducted experiments at various frequencies, but he concentrated on waves at lower frequencies when it was recognized that such waves could propagate over much longer distances and that the equipment required was less costly and more efficient. Because of Marconi's success, and in view of the need for wireless communication from ship to shore, for example, the amount of research activity at the higher frequencies decreased substantially. Some important basic studies, mostly theoretical, continued to be conducted, but the motivations for practical reasons dwindled essentially to zero.

Another major contributor to waveguide theory was John William Strutt, *Lord Rayleigh*, who succeeded Maxwell as Cavendish Professor at Cambridge. The prolific Lord Rayleigh, who seems to be first with nearly everything, made basic contributions to all sorts of topics in classical physics, including the resolving power of gratings, an explanation of why the sky is blue, a host of new results on the theory of sound, and the discovery of argon, for which he received the Nobel Prize. In waveguide theory, he is the first to discuss in detail (in 1897) the electromagnetic modes that can propagate through metallic tubes [6], and the scattering of electromagnetic waves by circular apertures and by ellipsoidal obstacles [7]. The latter work provided the foundation for the highly useful "small aperture" and "small obstacle" methods which were revived and developed further during World War II and later. The former work actually contains the fundamental ideas of mode propagation and cutoff in waveguides. However, this information became lost because it came out just as the research interest in electromagnetic waves was shifting from the higher frequencies to lower frequencies. In fact, when interest in hollow metallic waveguides was renewed in the 1930s, the contributors at that time were not at first aware of Rayleigh's work, and these concepts needed to be rediscovered.

16.1.2 The 1930s Period: The Real Beginnings of Waveguides

The first *systematic* development of waveguide theory occurred during the 1930s, in connection with hollow metallic waveguides. The possible use of hollow waveguides for guiding electromagnetic waves was investigated independently during the early 1930s by two groups, one at the Bell Laboratories under *George C. Southworth*, and the other at the Massachusetts Institute of Technology under *Wilmer A. Barrow*.

Most of the information presented in this section is taken from the article [8] that I wrote for the Special Centennial Issue of the *IEEE Trans. MTT*, devoted to "Historical Perspectives of Microwave Technology". For those who are interested in further details relating to the 1930s period and the contributions of Southworth and Barrow to waveguide theory, I recommend reading the article [9] by Karle S. Packard in that same Special Issue. That article served as the basis for Packard's M.S. in the History of Science from the Polytechnic Institute of New York (now Polytechnic University), and it contains much unpublished source material.

We are very fortunate that Dr. Southworth has written a book [10] detailing his personalized history of that period (and also his earlier work); it is very revealing not only to technical details but also the attitudes of the time. He points out that since Marconi found that longer wavelengths were more effective for long-distance transmission, shorter waves were neglected. However, by the end of the 1920s, the ship-to-shore and transoceanic telephone projects became a practical reality, and the techniques below 25 MHz became commonplace, so "there was an urge everywhere to explore the frequencies beyond". Accordingly, in 1931, Southworth began a few "homespun experiments".

Southworth first built two oscillators which gave wavelengths tunable from 123-200 cm, but these wavelengths were too long for experiments on air-filled waveguides. He therefore filled his waveguides with water ($\varepsilon' \approx 80$) so that the guide diameters could be reduced by a factor of about nine. The first measurements were made on water-filled copper pipes of circular cross section, and also bakelite pipes of the same cross-sectional size. Soon afterwards, he ordered some triodes from France which provided Barkhausen oscillations at wavelengths as short as 15 cm, thereby permitting measurements on air-filled pipes only five or six inches in diameter.

In 1920, Otto Schriever [11] had performed experiments on guided modes on dielectric rods, verifying the earlier theoretical paper [12] by Hondros and Debye. Southworth had to rely at first on the paper by Schriever as a guide to his work, even though he concentrated on hollow metal guides, because he was unaware of Lord Rayleigh's analysis [6].

Southworth's experiments came to the attention of the mathematicians, particularly John R. Carson and Sallie P. Mead, who were intrigued and began their own investigations. About that time, Sergei A. Schelkunoff also joined these studies. However, they were also unaware of Rayleigh's work, so they (especially Mead) studied waves on dielectric wires to begin with, and later analyzed dielectric cylinders with a metallic sheath. They then realized that the dielectric was unnecessary, and they moved directly to hollow metallic guides. After that, the mathematicians rederived Rayleigh's original results and also located his original paper. They also extended Rayleigh's results to include metal attenuation losses, and had discovered that for one of these modes, later designated the TE_{01} mode, the attenuation *decreased* as the frequency is increased. This discovery led to the low-loss, circular-electric mode, oversized-waveguide project undertaken after World War II by the Bell Laboratories and others around the world.

Somewhat earlier, the Bell Laboratories were skeptical of not only the value but even the validity of his experiments. He reports that "one of the leading mathematicians of the company … had doubted its feasibility." As a result, he was ordered to "be assigned to more constructive work". Since they were slow in carrying out these orders, Southworth continued his measurements, and in the meantime the mathematician found an error in his own earlier results and sent a correcting memorandum. Because of the change in the mathematician's opinion, and the success shown by Southworth in his experiments, he was eventually transferred to the Research Department.

After further successful work, Southworth wanted to publish his results but his superiors were reluctant because of a fear that the work was fallacious and that "the Company might be made to appear ridiculous". It was only after it was learned that Barrow of MIT was doing similar work and that he was about to publish it that Southworth was given permission.

Although both Southworth and Barrow independently indicated that their investigations overlapped and that the other person was also about to publish, the Bell Laboratories material appeared in print several months earlier than Barrow's paper. It appeared as a pair of companion papers in the *Bell System Technical Journal* in April 1936, the first by Southworth [13] on general considerations and experiments, and the second by Carson, Mead, and Schelkunoff [14] on the theory. Both papers referred to Rayleigh's much earlier paper [6] as the first to present the idea of "critical frequencies," above which propagation occurs and below which there is no transmission. They also referenced the work on dielectric rods by Hondros and Debye [12].

The stress in both papers was on modes in circular hollow metal guides, for the fields, the cutoff frequencies, the guide wavelengths (expressed as relative velocity), and the attenuation constants. More briefly, they also considered cylindrical dielectric guides, for the fields and the variation of guide wavelength with frequency. Southworth's paper [13] presented the field distributions, and curves comparing theory and measurement; the theory paper [14] presented the equations and their derivations. Southworth also described some of his measuring equipment: oscillator, tunable resonator, crystal detector, wavemeter, etc.

W. L. Barrow's paper [15] was published in the *Proceedings of the IRE* in October 1936. Although Barrow treated only circular hollow metal pipes, the overlap with the Bell Labs papers [13, 14] was surprisingly large. Barrow's paper [15] contained both theory and measurement, with the information usually presented in a more useful engineering format. Some material was not included in the Bell Labs papers; for example , Barrow also discussed radiation from the open end of the guide, and excitation from a coaxial line feed.

Following the publication of these papers, and corresponding presentations at a joint meeting of the Institute of Radio Engineers and the American Physical Society, Southworth and his colleagues delivered several semi-popular talks on waveguides. The first of these was given on February 2, 1938 before the Institute of Radio Engineers in New York, and it stressed different modes of transmission and their respective cutoff frequencies. A photograph taken on that occasion appears in Figure 16.1.

K. S. Packard, in his article [9] mentioned above, points out that "If Rayleigh's work had not been forgotten, the course of events leading to waveguide technology would have been quite different, and could well have followed a linear sequential process...". In the case of Southworth and the Bell Laboratories' theorists, they were misled for about a year by the belief that their work could be described in terms of waves on dielectric cylinders. In Barrow's case, "He did not immediately recognize the low-frequency cutoff property of electromagnetic waves in hollow tubes..." As a result, "His first experiment in May 1935 failed to produce transmission through the tube, and this result would have been expected given the source wavelength and tube diameter used, had the theory been available to him".

Figure 16.1. Photograph of George C. Southworth (foreground) demonstrating different waveguide modes of transmission and their cutoff frequencies on February 2, 1938. This was the first demonstration of waveguides before the Institute of Radio Engineers in New York (from [10], courtesy Gordon and Breach).

It is most interesting that all of the experiments and also all of the theoretical analyses performed on hollow metallic waveguides during the first half of the 1930s period involved circular cross sections. On the other hand, rectangular hollow waveguides turned out later to be much more practical and also seem so much simpler to us, because the field distributions are simpler, and the analyses involve trigonometric functions rather than Bessel functions. When was attention first paid to *rectangular* hollow metallic guides?

It turns out that Lord Rayleigh was again the first [6]. The second seems to be L. Brillouin, who independently published an analysis [16] in 1936 that considered lossless waveguides only, as did Rayleigh [6]. The first consideration of wall attenuation in rectangular waveguides is due to Schelkunoff in 1937, as part of a larger paper [17]. Finally, a comprehensive paper [18] by L. J. Chu and W. L. Barrow appeared in 1938, again performed independently and in fact part of Chu's doctoral thesis. Their paper contained detailed theoretical results for the attenuation properties, as well as for the fields in lossless guides, and also presented field patterns, structures for exciting various modes, and results of experiments.

During the second half of the 1930s period, a variety of additional papers on waveguides were published. They include several more experimental papers by Southworth, an analysis of hollow waveguides of elliptical cross section by Chu, and several noteworthy papers by Schelkunoff, who was a very active contributor during this period.

16.1.3   The World War II Period

The development of the magnetron in Great Britain furnished a reliable source of centimeter waves and made radar feasible. It was the tremendous push to improve radar during World War II that led to striking advances in such a short time for the microwave field.

Many laboratories in the United States, such as those at Harvard, Stanford, Columbia, and Brooklyn Polytechnic, contributed to some phase of the overall radar program, but the center of the activity and the most famous of these laboratories was the Radiation Laboratory at the Massachusetts Institute of Technology. Also active in radar development was the Telecommunications Research Establishment in England and McGill University in Canada. The grouping together of prominent and highly capable individuals under these special war-time circumstances produced an unusually stimulating and productive working environment.

Waveguide theory at that time recognized that microwave circuits consisted of lumped control elements connected together by lengths of waveguide. The representation of the guiding regions by transmission lines was already quite well understood by 1941 or so. The representation of waveguide discontinuities by lumped elements, however, was only in its infancy.

It was understood *qualitatively* that waveguide discontinuities could be viewed in terms of lumped elements since higher-order modes were excited at the geometrical discontinuities which could not propagate away because they were below cutoff. It was even recognized that some lumped elements were capacitive and some inductive, depending on the discontinuity geometry and the incident mode. But it was not known how to evaluate the lumped elements *quantitatively.* Some small progress in the direction of quantitative calculations was begun around 1940 or so, but the real advances were made during World War II, partly in England and Canada, but primarily in the United States at the MIT Radiation Laboratory.

At MIT's Radiation Laboratory, one of the prime activities was to obtain *precise measurements on waveguide discontinuities.* At the Polytechnic Institute of Brooklyn, Ernst Weber, then a Professor there (later the President of the University, and the first President of the IEEE, when the IRE and the AIEE merged in 1963), had just formed the Microwave Research Institute at the Polytechnic, and was attempting to establish a collaborative effort with the MIT Radiation Laboratory. In this connection, Weber negotiated in January 1942 an arrangement with F. W. Loomis and I. I. Rabi, Associate Directors of the Radiation Laboratory, to trade *Nathan Marcuvitz*, his prize graduate student, for a contract with MIT. At MIT, Marcuvitz was made responsible for the measurements on waveguide discontinuities.

Marcuvitz was in the Fundamental Development Group, headed by E. M. Purcell; others in the group included C. G. Montgomery and R. H. Dicke. Their function was to supply information to the various applications groups, who then used the network parameter results in their designs. They would usually make rough calculations based on these results and then adjust or optimize by cutting and trying. It was felt, however, that the major contribution made by these network parameter results was in creating a way of thinking, so that waveguide plumbing could be designed in network terms.

Marcuvitz' responsibility involved two requirements: first, the development of an accurate measurement setup, and second, the evolution of a measurement procedure that would permit the network parameters of geometrical discontinuities to be determined with great precision.

Much creative attention had to be paid to the precision measurement procedure, but the method finally adopted turned out to be taken from a paper published in 1942 in Germany by A. Weissfloch [19]. That method has been called the "*D* versus *S* procedure," where *S* is the distance from the discontinuity output reference plane to a short circuit in the output, and *D* is the distance from the input standing wave minimum to the discontinuity input reference plane. It has also been referred to as the "tangent method," since the *D* and *S* values are related by tangent functions. A description of the method appears in a publication [20] and in the *Waveguide Handbook* [21, sec.3.4.]. The method was systematically used, and it furnished very precise results for many discontinuity structures.

I had wondered, since this method was clearly valuable, why it was published during the war and whether or not the Germans actually used it. I learned that after the war Weissfloch had moved to France, and that he was working near Paris in a laboratory of a French affiliate of ITT. In connection with a trip to Europe in 1956, I wrote to Weissfloch and arranged to visit him. He informed me that during the war his colleagues in Germany felt his method to be uninteresting and did not care whether or not he published it; he was in fact not permitted to do anything further with it on the job. He also commented that not until much later did the Germans understand the value of radar, and that they had discouraged war-time research in microwaves.

Verification of this surprising situation is given in Southworth's book [10], pp. 174-175], where several paragraphs are quoted from a 1959 letter to Southworth from Dr. H. Mayer, then Vice-President of Siemens and Halske. Mayer's laboratory in Germany had received, in 1943, some equipment that came from "an English plane shot down near Amsterdam, Holland". The letter continues:

> *For a considerable time this piece of equipment was quite a riddle to us, especially the strange components such as waveguides, magnetrons and the like, indicating that microwaves were used. But for which purpose? At that time, microwave techniques were badly neglected in Germany. It was generally believed that it was of no use for electronic warfare, and those who wanted to do research work in this field were not allowed to do so.*

In contrast to the German view, Winston Churchill was widely quoted as saying appreciatively that "Radar won the Battle of Britain," referring to the defense of Britain against the Nazi bombings which were so devastating until radar was sufficiently well developed that the British radars could locate the incoming Nazi planes in the dark and fog and destroy them before they came close enough to do much damage. In my courses on microwaves, I used to quote

Churchill's remark during my first class to motivate the students' interest. It appeared to work well until a class in the late 1970s, at which I overheard one student lean over to his neighbor and ask "Who's Churchill?" I then realized that I was dealing with the next generation, and I never told that story again.

Marcuvitz is best known as an outstanding theorist, rather than an experimentalist. This transition from experimentalist to theorist was made easier because of his close association with *Julian Schwinger*. They both rented rooms in the same house near Harvard Square, and they became friends.

Schwinger worked during the night and slept all day. Marcuvitz would wake him up at 7:30 PM, and they would go to dinner. After that they would often discuss their research problems until midnight, after which Marcuvitz would go home to bed and Schwinger would begin his work.

Schwinger was in the Theory group, which was headed by G. Uhlenbeck. Despite the brilliance of many in that group, Schwinger's contributions stood out above the rest. The principal challenge faced by Schwinger and the others was how to quantitatively characterize waveguide discontinuities in terms of lumped elements. As indicated earlier, people already understood the lumped-element concepts in a qualitative sense, but methods had to be developed which yield numerical values as a function of the geometrical parameters. It was necessary to solve the "diffraction" problem posed when the wave incident on the discontinuity excited the higher-order modes in unknown proportions. Schwinger's contribution was that he established an *integral equation formulation* of the field problem, and then developed *various methods for its solution*. It was a giant step forward.

Various *approximate* methods were developed for solving the integral equations for different geometrical discontinuities. One of these methods involves techniques for manipulating the static kernel rather than the dynamic one, based on the recognition that for almost all the higher-order modes the cutoff frequencies are much larger than the operating frequency, so that for those higher-order modes, the operating frequency can be set to zero. A second method, termed the equivalent static method, employs the same static kernel but then view the problem as an electrostatic one that can be solved by a conformal mapping to a simpler geometry. Another procedure recasts the integral equation into a variational form, from which the normalized susceptance is obtained via a judicious field assumption and appropriate integrations. In addition, more accurate results were obtained in some cases by finding the field first by means of one of the first two methods and inserting that field into a variational expression.

Some time later, Schwinger realized that certain types of discontinuities could be solved using the *Wiener-Hopf technique*, which provided exact solutions for those geometries. The previous solutions, while accurate, were only approximate. Here was a breakthrough of another type. Weiner-Hopf solutions for different structures were first obtained by Schwinger and later by Carlson, and then by Heins.

In order to interact with others in the group, Schwinger reluctantly got up a little earlier and presented seminar lectures during the afternoon. Figure 16.2

shows a photograph of Schwinger at the blackboard. Notes on these lectures were taken by D. S. Saxon [22]. These notes became famous, and were reprinted or copied privately by various groups.

The theoretical results derived by Schwinger and others in that group, together with experimental data taken by Marcuvitz' group on structures for which no theory was available, were systematically arranged and edited by Marcuvitz and published as the *Waveguide Handbook* [21], volume 10 of the MIT Radiation Laboratory Series. This book was very widely used in the design of microwave components.

It was characteristic of this period that everything was done so rapidly that no one paid attention to who did what first; they were not concerned with publication but with getting the task done. As a result, many people have never been properly credited for their contributions.

Another very useful approximate technique for certain classes of waveguide discontinuities is *small aperture theory*, and its dual, small obstacle theory. Although the initial ideas were outlined by Lord Rayleigh [7] in 1897, H. A. Bethe [23] revived and generalized them, in effect introducing another method. The approach is based on the recognition that a wave incident on a small hole in a conducting wall produces a field in the hole equivalent to the sum of an electric and a magnetic dipole, the polarizabilities of which are given to a good approximation by electrostatic expressions if the hole diameter is small relative to wavelength and the hole is not near the waveguide wall. The great virtue of this approach is that very simple closed-form expressions can be obtained quickly. Later in the war period, Bethe left to head up the theoretical group at Los Alamos working on the atomic bomb.

After the war, Marcuvitz returned to the Polytechnic Institute of

Figure 16.2. Photograph of J. S. Schwinger at the blackboard during one of his lectures at the MIT Radiation Laboratory during World War II (from the official MIT Radiation Laboratory Book, *Five Years*).

Brooklyn to complete his doctorate, and his thesis dealt with a reformulation of small aperture theory [24]. He showed that it was derivable from originally rigorous expressions, and he rephrased the results in practical network terms, thereby eliminating the need for the integrations appearing in Bethe's formulation if normalized mode functions are employed. Before issuing the *Waveguide Handbook* [21], Marcuvitz derived analytical expressions for a large number of waveguide discontinuities that were not considered at the Radiation Laboratory, in order to produce a more useful final volume. All of these additional derivations were based on small aperture or small obstacle theory; in fact, almost one-third of the solutions appearing in the *Waveguide Handbook* were obtained using this theory.

Others also contributed to the utility of this method after the war. For example, Seymour B. Cohn performed careful electrolytic tank **measurements** [25] to obtain the electrostatic polarizabilities for aperture shapes for which no theory was available. I extended Marcuvitz' results to apply to *longitudinal* small obstacles [26]; previously, only transverse obstacles had been treated. L. B. Felsen and W. K. Kahn [27] generalized the procedure further to cover *multimode* situations. Recently, there has been a revival of this approach. Several people, principally R. F. Harrington [28], have extended it to apply to thick walls, to conductance expressions, and to coupling to cavities.

During the war period, contributions to the quantitative description of waveguide discontinuities were also made by others not affiliated with the Radiation Laboratory. In the United States, for example, analyses for discontinuities in parallel plate guide [29] and in coaxial line [30] were performed by J. R. Whinnery and his colleagues. Whinnery coauthored with S. Ramo an excellent and widely used book [31] in this field. In England, at the Telecommunications Research Establishment, G. G. MacFarlane developed a quasi-stationary field approach for a variety of discontinuities, such as capacitive and inductive irises and strips, and periodic strip gratings. The work was conducted in 1942, but published [32] only in 1946. As a third example, a group at McGill University in Canada under W. H. Watson [33] also worked on such problems, but concentrated on resonant slots. Had it not been for Schwinger's more extensive, more systematic, and more accurate contributions; these other results, which were certainly significant, would have been better recognized.

Note should also be made of a book [34] by L. Lewin published in 1951. Lewin was aware of work done by others, including Schwinger [22] and MacFarlane [32], but, working alone at the Standard Telecommunications Laboratories, Ltd., in England, he rederived results for many of these discontinuities in somewhat different ways, employing different modal separations, and he obtained slightly different final solutions. He also presented solutions for new structures, such as the tuned post and tuned window.

Lastly, some comments should be made on similar investigations in the Soviet Union. I do not know whether the work was performed during World War II or immediately afterwards, but papers were published shortly after the war. The most noteworthy contributions on waveguide discontinuities were made by L. A. Weinstein (or Wainstein, or Vainshteyn – different spellings have appeared, but the last one shown indicates how to pronounce his name), who applied the Wiener-Hopf technique to obtain rigorous solutions. He went beyond the Schwinger group in the sense of considering different incident modes and also multimode situations. His original papers in 1948, for example [35,36], were in Russian, but a translation of his book into English was available later [37].

16.1.4   The Microwave Research Institute (MRI)

To those who worked in microwaves in the two or three decades after the end of World War II, the name "Microwave Research Institute" commanded great respect, but, because of several factors, including the disuse of the name during the past two decades, the name is hardly known today, and certainly not to the current generation. Because of its earlier world-wide fame, and the important role it played then, it is an important part of microwave history, and it deserves to be discussed, at least briefly.

Many of the contributors to microwave theory at the MIT Radiation Laboratory were physicists before the war, and most of them went back to physics research after it. Notable among those who returned to Electrical Engineering departments at universities was N. Marcuvitz, who became an Assistant Professor at the Polytechnic Institute of Brooklyn. Although Dr. E. Weber had earlier lost Marcuvitz to the Radiation Laboratory, the contract he received from it furnished the basis for the successful establishment in 1942 of the later world-famous Microwave Research Institute (MRI) at Brooklyn Polytechnic (now Polytechnic University). Dr. Weber was its Director for more than a dozen years, being succeeded in that position by Dr. Marcuvitz and later by me. During that period, Weber and Marcuvitz provided the leadership in the areas of microwave components and microwave field theory, respectively, raising MRI to a position of world-wide prominence in both areas.

MRI had established the reputation of being perhaps the most prominent university activity in microwave theory in the world. For many years, it attracted post-doctoral researchers from around the world to spend a year or more, coming from such countries as Japan, France, U.S.S.R., Israel, Italy, England, Denmark, Sweden, Hungary, Poland, and Finland. Many of those researchers have since become famous in their own right. MRI was also well known for its series of annual symposia on topics in the forefront of the electronics field, and for the symposium proceedings volumes, 24 in all, that accompanied them.

Not only did MRI produce much important research in microwave theory and on basic microwave components, but it also trained a whole generation of microwave engineers. The journal, *MicroWaves* (now *Microwaves and RF*) in an interview with many microwave engineers in 1968, asked them various questions, including from what school they received their microwave education. One of the article's conclusions was that more microwave engineers graduated from Brooklyn Polytechnic than from any other school, and that the second was MIT, with only half as many microwave graduates.

The most important contribution made by MRI to microwave field theory, in my opinion, is the rigorous, systematic *reformulation of field theory in microwave network terms,* leading to a solid body of microwave network methods. These microwave network approaches greatly simplified both the setting up and the solving of many microwave field problems, and they helped to make the problems more transparent, thus also providing enhanced physical insight. The major credit for the systematic development of this reformulation into network terms goes to Marcuvitz, who pursued that goal with enormous zeal and effectiveness. Of course, much had already been accomplished along these lines by others, at the Radiation Laboratory and elsewhere.

For about a decade, we held a weekly seminar at which every applicable field problem was examined in these network terms. The organizer and prime mover was Marcuvitz, who presented most of the seminars himself. I joined MRI in 1946, just when all this began, so that I was fortunate enough to have experienced it all.

Unfortunately, very little of all this was published. One basic paper [38], which Marcuvitz coauthored with Schwinger, appeared in 1951; it incorporated many of the simplifications in the phrasing employed, but it was compactly presented. Marcuvitz and Schwinger had agreed to coauthor a book with the title, *Theory of Guided Waves,* but it was never written. Marcuvitz and I had agreed to write a book called Microwave Network Theory; several chapters were written and some parts of them were issued as reports, but unfortunately the rest was never completed. Another book, planned as a sequel to the others, did actually reach a publisher, but many years later (1973) and in a rather different form from that originally intended. That book [39], coauthored by L. B. Felsen and N. Marcuvitz, covers an enormous number of topics in a highly compressed form. On the other hand, these methods were included in the many graduate courses we presented in electromagnetics generally and microwaves more specifically. Many students took these courses over many years, so that the methodology was widely disseminated. As a result, these network methods have subsequently appeared in many research papers, by us and by many others.

Marcuvitz formed a group, and obtained funding for an extensive joint research project concerned with ***equivalent circuits for slots*** located in various positions in rectangular waveguide. The investigations were both experimental and theoretical, and served to teach us many techniques as well as to obtain many research results. We derived simple closed-form theoretical expressions and performed precise measurements on rectangular slots, resonant and nonresonant, located at the end of rectangular guide, transverse inside the guide, coupling E-plane tees, coupling H-plane tees, and radiating from the top or side of the guide. The study was comprehensive and was summarized in two large (no longer available) reports [40,41].

The project extended from 1946 into the early 1950s, with Marcuvitz as the principal investigator for the first three or four years, until he began to spend much of his time with the mathematicians at the NYU Courant Institute. I then took over that function. The others in that group were H. M. Altschuler, J. Blass, L. B. Felsen, H. Kurss, and A. Laemmel. Again, unfortunately, very little of this work ever got published. It seemed there were too many of us, and by the time we agreed some had left.

By the early 1950s the group broke up, but we were joined by some brilliant new colleagues and students, who helped us make many significant contributions on other microwave topics. Among these topics were surface waves, leaky waves, periodic structures, stripline discontinuities, high-frequency scattering, etc. These new colleagues included A. Hessel, T. Tamir, I. Palocz, S. T. Peng, and many others as time progressed, including H. J. Carlin and D. C. Youla, who made outstanding contributions to network theory.

In 1955, Brooklyn Polytechnic, through MRI, became the fifth university to join the elite Joint Services Electronics Program (JSEP). The existing member universities at that time were MIT, Stanford, Columbia and Harvard, and the next to join after Brooklyn Polytechnic was the University of Illinois, in 1959. Afterward, the program expanded to comprise nine universities, and ultimately twelve. With MRI included in JSEP, the scope of MRI expanded to include programs in optics and physics.

MRI was originally part of the Electrophysics Department, separate from the Electrical Engineering Department. Then, in 1971, the university administration merged these two departments, and MRI had to include other research areas within its scope, such as communication theory, systems theory, and so on. These additions were accepted by the JSEP with respect to the funding provided to MRI, but the new faculty members who joined MRI considered that the name MRI was too narrow as a description of the expanded scope of activities. After a few years of controversy, the name was changed to Weber Research Institute (WRI), since Weber was the founder of MRI. This new name did not have wide appeal, and the net effect was essentially to cause the research institute to lose its cohesiveness and its identity.

16.2   THE TRANSFORMATION TO MICROWAVE INTEGRATED CIRCUITS

16.2.1   The Competition Between Stripline and Microstrip Line

In the years immediately after World War II, rectangular waveguide became the dominant waveguide structure largely because good components could be designed using it. By 1950, however, people sought components that could provide greater bandwidth, and they therefore examined other waveguides. Ridge waveguide offered a step in that direction, and a neat theoretical paper by S. B. Cohn appeared on it in 1947 [42], but it was not the answer. Coaxial line would have been very suitable, since it possessed a dominant mode with zero cutoff frequency, thereby yielding two important virtues: a very wide bandwidth, and the capability of miniaturization. The lack of a longitudinal component of field, however, made it more difficult to create components using it, although various novel suggestions were put forth. In addition, those components would be expensive to fabricate.

In an attempt to overcome these fabrication difficulties, the center conductor of coaxial line was flattened into a strip and the outer conductor was altered into a rectangular box. Components with such interiors where then fitted with connectors for use with regular coaxial line. After a few years, some excellent components became commercially available employing this approach.

At about the same time, others took a much bolder step; they removed the side walls altogether, and extended the top and bottom walls sideways. The result was called strip transmission line, or ***stripline***. Different methods were used by different companies to support the center strip, but in all cases the region between the two outer plates was filled (or effectively filled) with only a single medium, either dielectric material or air. A modification that emerged at roughly the same time involved removing the top plate also, leaving only the strip and the bottom plate, with a dielectric layer between them to support the strip. That structure was termed ***microstrip***. The two different basic structures are illustrated in Figure 16.3.

Sketches of the structures of the three most important commercial microwave printed circuits are shown in Figure 16.4. Structures (b) and (c), registered as "Tri-Plate" and "Stripline", respectively, are both of the basic type seen in Fig. 16.3(a), but differ in that the structure in (b) is clearly dielectric-filled, whereas that in (c) is essentially air-filled because very little field is present in the thin center sheet that supports the center strip.

The inventor of the stripline concept was *Robert M. Barrett* of the Air Force Cambridge Research Center. [This organization has gone through many name changes over the years. It is now called the Air Force Research Laboratory (AFRL) at Hanscom Field.] He was also the prime mover in its development, not simply furnishing contract money but also encouraging the various researchers in different laboratories to exchange information. Among the organizations he supported were the Polytechnic Institute of Brooklyn, Tufts College, and the Airborne Instruments Laboratory Inc. (now AIL). I interacted very closely with both of these organizations, particularly with W. E. Fromm and E. G. Fubini of AIL. Barrett also wrote a popular article [43] early on, in 1952, to encourage

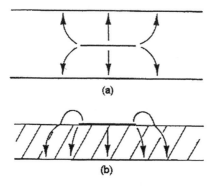

(a)

(b)

Figure 16.3. Cross-sections of (a) stripline and (b) microstrip line, showing the basic electric field lines (from [8], p. 1033, Figure 5).

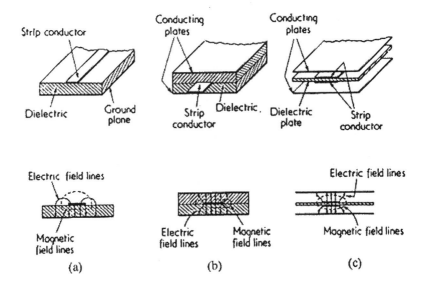

Figure 16.4. Typical commercial microwave printed systems; (**a**) Microstrip, Federal Telecommunication Laboratories, ITT. (**b**) Tri-Plate, Sanders Associates. (**c**) Stripline, AIL. (From [45], p. 985, Fig. 4.)

interest in this new type of waveguide. He stressed the simplicity of the structure, its printed circuit nature, and its many other virtues including the fact that circuits based on stripline could be trimmed by applying a razor blade to the center strip. He also wrote a short historical survey of work on these lines as of 1954 [44].

In recognition of Barrett's vital contributions to the early stages of microwave printed circuits, he received in 1992 the Microwave Pioneer Award of the IEEE Microwave Theory and Techniques Society.

Shortly after the appearance of Barrett's article [43], a group of engineers from the Federal Telecommunications Laboratories of ITT presented a series of three papers [46-48] on microstrip, the competing printed circuit line. They presented the concept, an approximate theory, and various components. The basic point of view and the intended virtues were similar to those propounded by Barrett; they differed in their choice of waveguiding structure.

Progress on stripline and microstrip proceeded so rapidly that a full-scale symposium on Microwave Strip Circuits was held in October, 1954, at Tufts College under the sponsorship of AFCRC. At this symposium, the Proceedings of which were published as a special issue of the *IRE Transactions on Microwave Theory and Techniques* in March 1955 [49], the extensive developments on microstrip were summarized in a paper by M. Arditi [50], but it became clear that only ITT was working on microstrip and that everyone else was using stripline.

There were good technical reasons for the preference for stripline. Because the region between the two outer plates of stripline contains only a single medium, the phase velocity and the characteristic impedance of the dominant (TEM) mode do not vary with frequency; because of the symmetry of the structure, all discontinuity elements in the plane of the center strip are purely reactive. In contrast, the two-media nature of microstrip causes its dominant mode to be hybrid, not TEM, with the result that the phase velocity, characteristic impedance, and field variation in the guide cross section all become mildly frequency dependent. Because of the symmetry unbalance in microstrip, all discontinuity elements possess some resistive content and therefore radiate to some extent.

AIL stressed this last-mentioned point by calling their stripline "High-Q Stripline". In fact, AIL's measurements showed Q values of over 4000. The term "High-Q Stripline" was used both in their advertising and as part of technical presentations made by Fubini. ITT became very sensitized on this issue, particularly after Dr. Fubini made some sharp comments during the presentation of a paper. After that talk, Dr. A. Clavier, a Vice-President of ITT (who was a well-known and highly respected contributor to wireless telephony in Europe during the 1930s), confronted Fubini (who became an Assistant Secretary of Defense a decade later) on this matter. Fubini then backed off and offered to modify AIL's terminology.

A very interesting additional remark was made by Fubini in his attempts to undermine the value of microstrip line as the basis for microwave components. He drew a sketch of a half-wave microstrip dipole, and, as an intended joke, he said that the dipole would probably serve as a pretty good antenna. That "joke" particularly irritated Clavier. No one at the time realized that Fubini's remark should have been treated seriously, least of all Fubini himself. It wasn't until 1970 or so, about 15 years later, that the microstrip patch antenna was proposed, which was based on precisely the same concept.

By the mid-1950s, however, the symmetrical form of stripline was the clear winner, and microstrip stayed in the background for another decade, except for development efforts.

The only fairly systematic compilation of available information on stripline was published in 1956 by Sanders Associates, and titled "A Handbook of Tri-plate Microwave Components". A much more complete and useful book [51], by the name "Stripline Circuit Design," by Harlan Howe, became available in 1974.

16.2.2 Theoretical Research on Stripline

During the first half of the 1950s there was also significant *theoretical* research on stripline.

The first theoretical challenge faced by researchers during the early 1950s in connection with stripline involved the *characteristic impedance* of the dominant mode. Many investigators worked on this problem. Early on, it was recognized that conformal mapping could yield a rigorous answer because the mode was TEM. The solution involved the ratio of elliptic functions, however, and in those days computers were not yet available. It was desirable, therefore, to have at one's disposal a simple formula (or formulas) that would permit an easy calculation of the characteristic impedance over the complete range of dimensional parameters.

Such a result was published by S. B. Cohn [52]. In that solution, a wide strip was approximated by accounting for the fringing fields at the sides, and a narrow strip by choosing an equivalent circular rod. The two solutions overlapped to better than 1 percent at some intermediate geometry.

Subsequent to that, Cohn proceeded to calculate the attenuation constant for stripline, but I concentrated on deriving equivalent circuits for *discontinuities*, stimulated by measurements being taken at AIL, and encouraged by R. M. Barrett of AFCRC.

Although it was widely understood that reactive effects were associated with junctions and other discontinuities in rectangular waveguide or coaxial line most designs in stripline at that time completely ignored such reactive effects. Some people did not realize that these effects could be important, and others simply did not know how to characterize them. With the aim of characterizing some common stripline discontinuities and assessing their significance, the group at AIL took careful measurements on two discontinuity elements in the center strip: a gap and a round hole. The group at Tufts College took measurements on other discontinuities, such as bends.

Realizing that a Green's function analysis of these discontinuity elements would present formidable difficulties, I concentrated instead on possible approximate approaches. I began with stored power considerations combined with an approximate model of the strip line, in which the fringing fields at the sides were compensated by extending the strip width and placing magnetic walls at the sides. That approach yielded good results for several discontinuities, but an even more successful approximate method involved a novel application of Babinet equivalences. By taking the Babinet dual of the approximate model mentioned above, the resulting structure became a parallel plate line of finite width; it was then possible to obtain the equivalent circuits for several additional discontinuities by taking the appropriate duals of existing solutions for corresponding discontinuities in parallel plate guide. By using these two methods in appropriate contexts, I derived simple expressions for the equivalent network parameters of a large variety of discontinuities in the center conductor of stripline, including gaps, holes, bends, changes in width, and tee junctions. Very good agreement was found with the measured results obtained by AIL and Tufts College.

These approximate approaches and network expressions, together with some comparisons with measurements, were presented at the Tufts College symposium mentioned above. A paper was then published in the IEEE Trans. Microwave Theory Tech. [53], in 1955. I was awarded the Microwave Prize of the IEEE MTT Society for that paper. It is also my understanding that the equivalent circuit results presented there was widely applied in stripline circuit designs, and that the results were quoted in various books and reports. Since no one else derived comparable theoretical expressions, the only alternative would have been to compensate empirically for the reactive effects of the discontinuities.

16.2.3   Microwave Integrated Circuits

Once the concepts of stripline and microstrip line were announced to the microwave community, they were recognized as "milestones," and practical components based on them were systematically developed. Those milestones can be identified with the year 1952. The circuits that were designed with these transmission lines were built as individual separate components that would then be connected together as desired, similar to the way circuits were formed with rectangular waveguide. Roughly fifteen years later the circuits were designed primarily in a very different form, that of *microwave integrated circuits*. The transformation between these two circuit forms occurred in various steps, and contributions were made by various groups, so that the steps were many but it was difficult to identify when they occurred. There were few identifiable milestones. A very apt description of this situation in connection with one phase of this transformation was presented in a paper [54] by a group from Texas Instruments. They remarked, "The professional historian is well aware of the blurring of events in tracing a particular series of developments. We all tend to measure a history by milestones and to forget or overlook the in-between transitions which led to the milestone events. These transitory events often are more important in acquiring a real perspective and true knowledge than the milestone events". This description is not applicable to all developments, but it certainly is the case in the transformation to integrated circuits.

Although we may not know precisely when certain key modifications occurred, or who is responsible for them, we do understand what the changes were and what their consequences are. One major change, which seems obvious today, was the recognition that it is not necessary to create a circuit by connecting individual separate components, with each component terminated at each end with a connector, usually in coaxial form. Once the *connectors were removed*, and the components were connected together directly, it was obvious that advantage could be taken of the *miniaturization* property of both stripline and microstrip. The components themselves could then be (and were) reduced in size, so that the subsystems became smaller in size and lighter in weight (and eventually less expensive).

The reduction in the size of the cross section also had very important *electrical* consequences. The reactance (or susceptance) of *discontinuities* on these transmission lines is, to first order, proportional to (dimension/wavelength), or to its reciprocal, depending on the type of discontinuity, so that a reduction in the relevant dimension would cause a corresponding reduction in the importance of the discontinuity on the component design. A really large reduction in the cross-sectional size of the line would usually allow one to neglect the effect of the discontinuity altogether.

The reactance aspect of discontinuities holds for both stripline and microstrip, but *microstrip* discontinuities have a resistive aspect in addition because microstrip discontinuities also *radiate*. This radiative aspect was a major concern for microstrip before the lines were miniaturized. After the miniaturization, the radiative contribution was also greatly reduced, thereby removing one of the prime objections to the original microstrip circuits. As a result, microstrip circuits became very popular, because for many situations microstrip then behaved essentially like a pure TEM line. An additional virtue was that for microstrip one need not remove a top cover in order to see the circuit.

Today, there are certain applications for which stripline *must* be used, and microstrip would not be suitable. A prime example is furnished by components designed to operate over very wide frequency ranges. Stripline supports a true TEM mode, whereas the properties of the microstrip dominant mode vary to some extent as the frequency increases. For example, the propagation constant increases with frequency, and the field distribution in the cross section changes somewhat with frequency. Until about 1980 or so, microstrip circuits were operated at the lower microwave frequencies, where the variations with frequency were hardly noticed, but as the operating frequencies rose various corrections had to be taken into account.

In the transformation to microwave integrated circuits, we have so far taken into account the following changes, with the associated consequences:

(a) Originally, circuits were produced by connecting together individual components, each of which had its own connectors at each end. After the connectors were removed, the components were connected together directly.

(b) Once the components were connected together directly, it became clear that the transmission-line cross sections could be greatly reduced in size because the dominant modes were TEM (in stripline) or quasi-TEM (in microstrip). As a result, the circuits became greatly reduced in size and weight.

(c) The reduction in cross-section had electrical consequences as well. The influence of discontinuity elements on circuit design became greatly reduced, thereby allowing circuit performance to more directly conform to the initial designs, even when discontinuity effects were neglected altogether.

(d) The reduction in the importance of discontinuity elements also applied to microstrip radiation effects, thereby reducing the differences in the electrical behaviors of microstrip and stripline, and as a result increasing the popularity of microstrip. Eventually, microstrip exceeded stripline in popularity, but stripline is still essential for applications requiring operation over very wide frequency ranges.

The above considerations apply directly to passive circuits, but also are key aspects of the transformation to active circuits. The first additional step for active circuits was the selection of a *semiconductor* as the substrate for the circuits. The semiconductors employed consisted of silicon most of the time, but later moved to gallium arsenide.

A further key step was the development of the *monolithic microwave integrated circuit (MMIC)*. To quote from an excellent and detailed paper on such circuits [54], "Monolithic literally means 'one rock' and, in electronics, has come to mean the processing of active and passive components *in situ* on a semiconductor slab and providing interconnections to the components to form an integrated circuit (IC). This method is substantially the concept Jack Kilby of Texas Instruments originated in July of 1958 and had reduced to practice by September 12, 1958 [55]". Major advances were made within the MERA program (Molecular Electronics for Radar Applications), which was encouraged and supported primarily by the Air Force at Wright Patterson AFB. The historical sequence of the various developments that emerged from that program at Texas Instruments is presented in detail in Reference [54] cited above.

REFERENCES

[1] Monthly Report of the Prussian Academy of Sciences (Berlin), pp. 519, 528 and 529, July 1879 (in German).

[2] James Clerk Maxwell, "A dynamical theory of the electromagnetic field," *Phil. Trans. Royal Soc. (London),* vol. 155, pp. 459-512, 1865.

[3] John H. Bryant, *Heinrich Hertz, the Beginning of Microwaves*, The Institute of Electrical and Electronics Engineers, Inc., New York, NY, 1988.

[4] Heinrich Hertz, "On electromagnetic waves in air and their reflection," *Electric Waves,* Chap. 8, D. E. Jones translation, London, Macmillan and Co., 1893, and New York, Dover, 1962.

[5] Reference [3], p. 48.

[6] Lord Rayleigh, "On the passage of electric waves through tubes, or the vibration of dielectric cylinders," *Phil. Mag.,* vol. 43, pp. 125-132, Feb. 1897.

[7] Lord Rayleigh, "On the incidence of aerial and electric waves on small obstacles in the form of ellipsoids or elliptic cylinders; on the passage of electric waves through a circular aperture in a conducting screen," *Phil. Mag.,* vol. 44, p. 28, 1897.

[8] Arthur A. Oliner, "Historical Perspectives on Microwave Field Theory," *IEEE Trans. on Microwave Theory Tech.,* vol. MTT-32, No. 9, pp. 1022-1045, September 1984.

[9] Karle S. Packard, "The Origin of Waveguides: A Case of Multiple Rediscovery," *IEEE Trans. on Microwave Theory Tech.,* vol. MTT-32, No. 9, pp. 961-969, September 1984.

[10] G. C. Southworth, *Forty Years of Radio Research*, New York: Gordon and Breach, 1962.

[11] Otto Schriever, "Elekromagnetische Wellen an dielektrischen Drähten," *Ann. d. Phys.,* vol. 63, ser.4, p. 645, 1920.

[12] D. Hondros and P. Debye, "Elektromagnetische Wellen an dielektrischen Drähten," *Ann. Der Physik*, vol. 32, pp. 465-476, June 1910,

[13] G. C. Southworth, "Hyper-frequency wave guides – General

consideration and experimental results," *Bell Syst. Tech. J.*, vol. 15, pp. 284-309, April 1936.

[14] J. R. Carson, S. P. Mead, and S. A. Schelkunoff, "Hyper-frequency wave guides – Mathematical theory," *Bell Syst. Tech. J.*, vol. 15, pp. 310-333, April 1936.

[15] W. L. Barrow, "Transmission of electromagnetic waves in hollow tubes of metal," *Proc. IRE*, vol. 24, pp. 1298-1398, October 1936.

[16] L. Brillouin, "Propagation d'ondes électromagnetiques dans un tuyan," *Rev. Gen. De l'Elec.*, vol. 40, pp. 227-239, August 1936.

[17] S. A. Schelkunoff, "Transmission theory of plane electromagnetic waves," *Proc. IRE*, vol. 25, pp. 1457-1492, November 1937.

[18] L. J. Chu and W. L. Barrow, "Electromagnetic waves in hollow metal tubes of rectangular cross section," *Proc. IRE*, vol. 26, pp. 1520-1555. December 1938.

[19] A. Weissfloch, *Hochfrequenz u. Elektro.*, vol. 60, p.67, 1942.

[20] N. Marcuvitz, "On the representation and measurement of waveguide discontinuities," *Proc. IRE*, vol. 36, pp. 728-735, June 1948.

[21] N. Marcuvitz, *Waveguide Handbook* (MIT Radiation Laboratory Series vol. 10). New York: McGraw-Hill, 1951.

[22] D. S. Saxon, "Notes on lectures by Julian Schwinger: Discontinuities in waveguides," February 1945.

[23] H. A. Bethe, "Theory of diffraction by small holes," *Phys. Rev.*, vol. 66, no. 7/8, pp. 163-182, October 1944.

[24] N. Marcuvitz, "Waveguide circuit theory: Coupling of waveguides by small apertures," Ph.D. dissertation, Polytechnic Institute of Brooklyn, New York, June 1947.

[25] S. B. Cohn, "Determination of aperture parameters by electrolytic-tank measurements," *Proc. IRE*, vol. 39, pp. 1416-1421, November 1951.

[26] A. A. Oliner, "Equivalent circuits for small symmetrical longitudinal apertures and obstacles," *Trans. IRE Microwave Theory Tech.*, vol. MTT-8, pp. 72-80, January 1960.

[27] L. B. Felsen and W. K. Kahn, "Network properties of discontinuities in multi-mode circular waveguide, Part 1," *Proc. Inst. Elec. Eng.*, part C, monograph no. 503E, pp. 1-13, February 1962.

[28] R. F. Harrington, "Resonant behavior of a small aperture backed by a conducting body," *IEEE Trans. Antennas Propagat.*, vol. AP-30, no. 2, pp. 205-212, March 1982.

[29] J. R. Whinnery and H. W. Jamieson, "Equivalent circuits for discontinuities in transmission lines," *Proc. IRE*, vol. 32, pp. 98-114, February 1944.

[30] J. R. Whinnery, H. W. Jamieson, and T. E. Robbins, "Coaxial line discontinuities," *Proc. IRE*, vol. 32, pp. 695-709, November 1944.

[31] S. Ramo and J. R. Whinnery, *Fields and Waves in Modern Radio*. New York: Wiley, 1944.

[32] G. G. MacFarlane, "Quasi-stationary field theory and its application to diaphragms and junctions in transmission lines and wave guides," *J. Inst.*

Elec. Eng., pt. IIIA, vol. 93, no. 4, pp. 703-719, 1946.

[33] W. H. Watson, *The Physical Principles of Wave Guide Transmission and Antenna Systems*. Oxford at the Clarendon Press, 1947.

[34] L. Lewin, *Advanced Theory of Waveguides*, published for the Wireless Engineer. London:Iliffe and Sons, Ltd., 1951.

[35] L. A. Weinstein, "A rigorous solution of the problem of the plane waveguide with an open end, *Bull. Acad. Sci. USSR, series Phys.* vol. 12, no. 2, pp. 144-165, March/April 1948 (in Russian).

[36] L. A. Weinstein, "Theory of symmetrical waves in a circular waveguide with an open end," *Zh. Tekh. Fiz.*, vol. 18, pp. 1543-1564, December 1948 (in Russian).

[37] L. A. Weinstein, *The Theory of Diffraction and the Factorization Method*. Boulder, CO: Golem Press, 1969. Translated from the Russian.

[38] N. Marcuvitz and J. Schwinger, "On the representation of electric and magnetic fields produced by currents and discontinuities in waveguides," *J. Appl. Phys.*, vol. 22, pp. 806-819, June 1951.

[39] L. B. Felsen and N. Marcuvitz, *Radiation and Scattering of Waves*. Englewood Cliffs, NJ: Prentice Hall, 1973.

[40] N. Marcuvitz, "The representation, measurement, and calculation of equivalent circuits for waveguide discontinuities with application to rectangular slots," *MRI, PIB Report*, under Contract No. WLENG W28-099-ac-146 with the Air Force, 1949. The report was a group effort.

[41] A. A. Oliner, "Equivalent circuits for slots in rectangular waveguide," *MRI, PIB Report*, under Contract No. AF19(122)-3 with the Air Force Cambridge Research Center, August 1951. The report was a group effort.

[42] S. B. Cohn, "Properties of ridge wave guide," *Proc. IRE*, vol. 35, pp. 783-788, August 1947.

[43] R. M. Barrett, "Etched sheets serve as microwave components," *Electronics*, vol. 25, pp. 114-118, June 1952.

[44] R. M. Barrett, "Microwave printed circuits – A historical survey," *IRE Trans. Microwave Theory Tech.*, vol. MTT-3, pp. 1-9, March 1955. In Special Issue: Symposium on Microwave Strip Circuits.

[45] R. M. Barrett, "Microwave printed circuits – The early years," *IEEE Trans. Microwave Theory Tech.*, vol. MTT-32, no. 9, pp. 983-990, September 1984.

[46] D. D. Grieg and H. F. Engelmann, "Microstrip – A new transmission technique for the kilomegacycle range," *Proc. IRE*, vol. 40, pp. 1644-1650, December 1952.

[47] F. Assadourian and E. Rimai, "Simplified theory of microstrip transmission systems," *Proc. IRE*, vol. 40, pp. 1651-1657, December 1952.

[48] J. A. Kostriza, "Microstrip components," *Proc. IRE*, vol. 40, pp. 1658-1663, December 1952.

[49] *Proceedings of Symposium on Microwave Strip Circuits*, Tufts College, October 11-12, 1954, Special Issue of *IRE Trans. Microwave Theory Tech.*, vol. MTT-3, no. 2, March 1955.

[50] M. Arditi, "Characteristics and applications of microstrip for microwave wiring," *IRE Trans. Microwave Theory Tech.*, vol. MTT-3, pp. 31-56, March 1955. In Special Issue: Symposium on Microwave Strip Circuits.

[51] H. Howe, *Stripline Circuit Design*. Dedham, MA: Artech House, 1974.

[52] S. B. Cohn, "Characteristic impedance of the shielded-strip transmission line," *IRE Trans. Microwave Theory Tech.*, vol. MTT-2, pp. 52-57, July 1954.

[53] A. A. Oliner, "Equivalent circuits for discontinuities in balanced strip transmission line," *IRE Trans. Microwave Theory Tech.*, vol. MTT-3, pp. 134-143, March 1955. In Special Issue: Symposium on Microwave Strip Circuits.

[54] D. N. McQuiddy, Jr., J. W. Wassel, J. B. LaGrange, and W. R. Wisseman, "Monolithic microwave integrated circuits: An historical perspective," *IEEE Trans. Microwave Theory Tech.*, vol. MTT-32, no. 9, pp. 997-1008, September 1984.

[55] J. S. Kilby, "Invention of the integrated circuit," *IEEE Trans. Electron Devices*, vol. ED-23, pp. 648-654, July 1976.

17

A HISTORY OF PHASED ARRAY ANTENNAS

ROBERT J. MAILLOUX, AFRL, *Hanscom AFB, MA*
(Currently with the University of Massachusetts, Amherst, MA)

17.1   INTRODUCTION

Electronically scanned arrays were a natural outgrowth of the fixed beam array technology that was well understood in the late 1920s. Among other examples are the well-known Yagi array and the vertical array of half wave dipoles used in the RAF Chain Home (CH) air defense radar system. Early scanning systems used mechanical devices to produce the required phase changes and beam motion, but by the 1950s there was a technology that is the direct ancestor of our present day phased array systems.

A number of important developments have been necessary to bring phased arrays to the present state of refinement. These include the development of good synthesis tools, the understanding of mutual coupling phenomena and array blindness, the development of powerful modeling techniques to accurately describe arrays with elements on and within dielectric substrates, the development of adaptive optimization techniques, and the computer aided design that has made precision fabrication possible. This chapter will review some of the developments that have brought us this useful electromagnetic technology.

The benefits of enhanced resolution led researchers in the 1920s and 1930s to investigate ways of combining the output signals from simple elements in order to produce narrower, more directive antenna patterns. One of the earliest of these was the end-fire array called the Yagi or the Yagi-Uda array, [1] which used several "director" elements in addition to a feed dipole and a reflector element to form a directional beam that radiated in the end-fire direction. Other fixed array combinations formed narrowed broadside beams and were used for both radar and communication during the 1930s, when the first mechanically scanned array was reported.

Although the basic principles of array scanning were known in the 1920s, and electromechanical scanning was shown to be practical for some applications in the 1930s, it was not until the 1950s that the first fully electronic scanning was realized. In the interim there were numerous mechanical "scanners" invented and used, but device technology didn't support electronic scan until the first ferrite phase shifters in 1954 – 1955.

Since the 1950s, and powered by substantial military funding in the United States and Europe, there have been significant advances in scanning array technology. These have included large ground and ship-based radar, and more recently airborne radar arrays. Communication systems, which provided the first impetus to array developments at HF frequencies, later stimulated research into adaptive techniques and to new frequency bands up through the EHF range. Many of the various components and techniques that were developed to further array development were chronicled in the classic 3 volume text, edited by Hansen and entitled Microwave Scanning Antennas [2]

The cost of arrays has been a major factor in slowing their use in more system applications. In the process of performing the system calculations that start with required gain and noise temperature, and working back through the comparatively higher array loss, increased control power and added complexity, the array usually is second choice to a mechanically positioned aperture unless there is some over-riding capability that can only be provided by the array. Recent history is changing this, and the change is built upon solid state array developments that began in the 1960s with the advent of printed circuit technology and later the first T/R module developments. Another major step in this evolution was the invention of the microstrip patch antenna, which, in its numerous variations, has come to be the element of choice for many low-cost array developments.

The electromagnetic foundations of array theory have developed along with the needs of the technology. Accurate calculations of dipole and dipole array currents and fields were first obtained from Hallén's equation and later from other forms of the integro-differential equations for wires and (later) slots. More general formulations were based upon the free space and half-space Green's functions, or equivently the magnetic and electric potentials, and used for wire arrays and waveguide slot arrays, all solved in the frequency domain. Printed circuit microstrip antennas and arrays presented some initial difficulty, and necessitated the use of more complex Sommerfeld-type Green's Functions, followed by the use of numerical techniques for solving the resulting electromagnetic boundary value problems. Presently there is a mix of frequency domain Method of Moment and Finite Element approaches, along with a variety of time domain methods being applied to element and array studies as the geometries continue to become more complex.

Cost is still the most important factor inhibiting the growth of this technology, but the constantly expanding list of functions now expected of modern radar and communication systems is making the phased array an invaluable tool.

17.2   THE EARLY HISTORY

Although radar was later to become the more significant impetus to array technology development, it was short wave radio that first utilized the flexibility of scanned arrays. As early as 1925, Friis [3] studied the reception of waves with directional antennas made up of several loop antennas whose signals were combined electronically, but not scanned. At wavelengths of 492 meters and 600 meters, loops of 1.76 meter diameter, placed 34.4 meters apart, were used to demonstrate the enhanced directivity of the antenna pair, and these were rigidly joined and rotated together to change the direction of the combined beam. Friis described a similar experiment at long wave (λ = 5000 – 6000 meters) with square loops eight feet on a side, and located 400 meters apart. This larger structure was not rotated. In a third experiment, he used a single loop and a co-located "condenser" antenna, which consisted of two parallel flat plates excited by a tuned circuit. This combination resulted in a cardoid pattern with a null in the "back" direction, but it was not as directive as the two loop system.

Friis' array was not the first experiment with directive antennas, for since the time of Marconi's paper entitled "Directive Antenna" [4] researchers had been investigating means for improving long range communication with enhanced directivity. A measure of the sophistication of these studies is the publication by Beverage et al. [5] describing the "wave antenna" now commonly called the "Beverage" antenna, which is a long horizontal wire of about one wavelength mounted over earth, and loaded with appropriate resistances. The Beverage studies for RCA (Radio Corporation of America) were also conducted to provide communication from the United States to Europe (England, France and Norway). Among other issues, Beverage noted that the received wavefronts were not perpendicular to the ground, but tilted, and it is this tilt that excites the longitudinal current in the horizontal wire. The studies revealed choices of length, height and terminating resistance to optimize the received signal.

The need for a scanning array became apparent as a result of further research on propagation effects. Although other investigations continued exploring enhanced directivity to optimize intercontinental communications [6,7] using various tilted long wire antennas, it became apparent [8,9] that increased directivity alone did not lead to improved quality of reception over these distances for several reasons. First the received signal did not arrive from one elevation angle, but from waves which arrive at different vertical angles, and which have traveled from the transmitter over paths of different lengths. Secondly, these angles change (although slowly) with time. Increased vertical angle directivity alone can reduce channel fading by favoring waves arriving from a certain path to the exclusion of others, but eventually further increases in vertical directivity result in a pattern so narrow that the signal arrives outside of the angular range of the antenna much of the time.

At the conclusion of their paper on the direction of arrival of short radio waves, Friis et al. [8] proposed constructing a more elaborate antenna with a single lobe directional pattern and steerable in elevation. In 1937 Friis and Feldman [9] reported the results of a multiple unit steerable antenna (MUSA) for short wave reception. This antenna system was to be a test-bed for a more elaborate system built later, and so it supported a very sophisticated test of angle and time diversity combining. The test array consisted of 6 rhombic antennas arrayed along a three-quarter mile path in the direction of propagation. To accommodate the angle diversity the received signals were amplified and down-converted to an intermediate frequency where they were split to form three independent beams, each controlled in elevation by three sets of mechanically rotated phase shifters geared together to form the desired progressive phase Δ, 2Δ, 3Δ, etc.. Of the three beams thus formed, one was used to search for the principal returns, and the other two were set at the angles with strongest signals. These outputs were then time delayed, passed through a network to equalize loss over the frequency band, and added to complete the combination of angle and time diversity. The system achieved nearly 8dB increase in average signal to noise power, relative to that of a single rhombic and proved the value of being able to select the optimum group of angles at any time then combine multiple time delayed output signals. A commercial installation of this type, consisting of 16 rhombic antennas and forming four beams, was built to provide single-sideband voice telephone communication between Manahawkin, N.J. and London, England. This system was approximately 1.5 miles long and achieved an average of 12–13 dB signal to noise improvement [10] relative to a single rhombic. The three beam phasing and combining was done automatically using the fourth beam as monitor.

This early demonstration of the value of electronic (or more accurately electromechanical) scanning certainly verified the electromagnetic foundations and at least one very practical application of the technology, but did not lead to fully electronic scan without many more years of technology development.

Early radar systems took advantage of increased directivity of arrayed elements. The Chain Home radar of Great Britain was the first radar to be organized into a complete defense system [11]. The radar at Dalby, on the Isle of Man, is shown in Figure 17.1. Built between 1935 and 1942, the Chain Home system consisted of a number of sites, and several different radar designs, but the one described here is the "East Coast" version, such as the one built on the Isle of Wight. Operating between 20 and 30 MHz, the transmitter array consisted of a stack of eight half-wave dipoles, spaced a half wave apart and excited in-phase. Each transmitting dipole was slung between a set of 360 foot towers, and had a

Figure 17. 1. Chain Home Radar of Great Britain.
http://www.castletown.org.im/heritage/chl_nl5_p3.html

tuned reflector spaced 0.18 wavelength behind it. The mean height of the array was 215 feet, and the combined forward and ground-reflected radiation had its peak at 2.6° and first null at 5.2°. The azimuth beamwidth was about 100°.

To fill the gap at 5.2°, a second array was slung between several of the towers. This "Gapfiller" array had a mean height of 95 ft and so a peak at 6°. The signal from this auxiliary array was selected by a remote operator in order to provide an elevation pattern without gap. The system achieved azimuth direction finding by phase comparison using a nulling system, and height finding by comparing the signals received by the basic array and the gap-filler array.

The receiver antennas were mounted in shorter (215 ft.) wooden towers, arranged in pairs and switched to control the azimuth receiving pattern of each tower.

In April 1947, Friis and Lewis [12] published a detailed account of radar antenna technology, including the basic electromagnetics of antenna radiation and the effects of amplitude taper and linear, square law and cubic aperture phase variation. They included a short catalog of military land based and airborne radar antennas, most of which were basic apertures, like parabolas or metal plates lenses but including some details of an electromechanically controlled phased array, the Polyrod Fire Control antenna developed by Bell Laboratories and based on the MUSA experience [12,13,14]. This array, shown in Figure 17.2a, consisted of fourteen S-band elements, with each element an in-phase vertical array of three polyrods. The inter-element spacing was about two wavelengths

because the array was only required to scan ± 15° in azimuth. The high gain and narrowed pattern of the polyrod enabled the two wavelength spacing and reduced element count while suppressing some of the grating lobes. Rotary phase shifters invented by Fox [13] provided a reliable mechanical phase shift, but not truly electronic scan. The phasing was accomplished by an ingenious arrangement of cascaded rotary phase shifters as shown in Figure 17.2b. The array was center-fed and used extra fixed line lengths to equalize phase across the array with all phase shifters set to zero. The phase shifters were ganged together, all rotating at

Figure 17.2a. Polyrod Array.

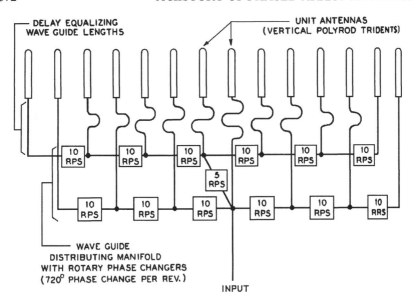

Figure 17.2b. Phase Control for a Polyrod Array.

10 revolutions per second, with 720° phase change per revolution. The series arrangement of phase shifters shown in the figure made the outer elements phase vary as 35 revolutions per second, or 25,200 degrees per second and so scanned the beam continuously with inter-element progression of 3600 degrees per second. This system was successfully used in Word War II as the Mark 7 and Mark 8 fire control radars. Figure 17.3 shows the Mark 8 radar (along with two others) that was put into production in 1942 by Western Electric Company (a Bell Laboratory subsidiary). Additional details about this array are given by Kummer in the reference [14] and by Fowler [15].

One of the earliest scanning arrays that saw service in the Second World War was the German Wullenweber array. The basic Wullenweber, shown in Figure 17.4 is a set of vertical dipoles arranged in a circle with receiving ports connected to a sector of the array, and that sector commutated around the circle by means of a capacitive coupling. The received signals were fed to a radiogoniometer [16] that combined the signals and produced a response proportional to angular deflection. The device was used for azimuth direction finding, and was said to be more accurate in bearings than American systems. After reverse-engineering and studies at the University of Wisconsin, the United States Navy built a Wullenweber near Corea, Maine. Using sets of concentric receiving rings over 200 feet high and diameter 1300 feet. It operated from very low VLF frequencies up to VHF bands. Another was built for the Navy by the University of Illinois in the 1950s and 1960s [17]. This system had 120 vertical monopole antennas and operated between 2 and 28 MHz, with one third of the array connected to the receiver. Later, the Navy AN/FRD-10 became the most

Figure 17.3. Mark 8 Shipboard Fire Control Radar.
http://www.vectorsite.net/ttwiz4.html

Figure 17. 4. Wullenweber Array.
http://www.mindspring.com/~cummings7/wullen.html

common developed system, and consisted of two concentric rings of 120 elements in an 850-foot diameter circle covering the higher frequency range, and 40 elements on a 750-foot diameter covering the lower frequency range (2-8 MHz). [18].

17.3   ELECTROMECHANICAL AND FREQUENCY SCANNING

The success of the polyrod and Wullenweber antennas still did not lead to any obvious transition to the technology of today. The lack of an electronic phase shifter was to limit scanning systems to electromechanical scan or frequency scan until the later invention of the microwave ferrite phase shifter. The impossibility of cascading large numbers of Fox rotary phase shifters led to ingenious electromechanically scanned mechanisms, and some of these are documented in the chapter by Kummer [14] and in the article by Fowler [15].

The series arrangement of phase shifters, as in the polyrod antenna, enabled the use of mechanical phase shift devices that provided only about a wavelength of phase variation. These devices included coaxial "trombone" phase shifters that consisted of telescoping coaxial transmission lines that were fitted with sliding outer and inner conductors that effectively lengthened or shortened the line without changing the outer network connections. These mechanical phase shifters originally used sliding mechanical contacts, but later versions used noncontacting joints with microwave choke circuits or electromagnetically coupled concentric helix lines [19]. Four port hybrid power dividers, terminated in sliding waveguide or coaxial short circuits and introduced in the late 1940s made excellent mechanical phase shifters. This basic concept is used today in diode and MMIC phase shifter circuits.

In addition to mechanical devices for changing the physical length of the transmission lines, there were a number of mechanical devices developed to change the electrical length of the lines without changing the physical length, by mechanically changing the propagation constant. Kummer [14] describes several of these devices including a corrugated coaxial line due to Cady with propagation constant that varies as the center conductor is rotated [19]. The 1947 Friis and Lewis article also described the AN/APQ-7 Radar Bombsight Antenna, later called the Eagle Scanner, operating at X-band, which was a line source array of 250 dipoles excited at half-wave spacing by a variable width waveguide feed line. The change in waveguide propagation velocity and resulting progressive phase change produces scan over a ± 30° horizontal scan sector, while the vertical pattern has a fixed cosecant like pattern. This scanning system, sometimes called Alvarez scanner, is also described in an MIT Radiation Laboratory report in 1946 [20], and in the reference by Kummer.

The several schemes mentioned above were mechanical means for replacing the phase shifters of an array, but these assumed a discretized, more or less conventional, phased array. In addition to these, there have been a number of electromechanical techniques for scanning lenses and reflectors as documented by Kelleher [21], Fowler [15], Shnitkin [22], and Johnson [23]. Too numerous to mention, these techniques date back to the work of Luneberg [24] in 1946 and to many other inventors referenced in the above [21-23] surveys.

17.4   THE TECHNOLOGY OF ARRAY CONTROL

17.4.1   Phase Shift and Time Delay

The 1937 paper by Friis and Feldman [9] left no doubt about the tremendous potential benefits of scanned arrays. In many ways, the paper exposed much of what was to be the subject of development throughout the rest of the century. The array formed independently scanned beams, controlled by sets of phase shifters at an IF frequency and re-combined using appropriate time delays. The development thus anticipated developments in multiple beam arrays, IF (and digital) beamforming, and even space-time-adaptive processing (STAP). Unfortunately there was no technology to explore these opportunities, and progress in electronic scan was to wait until the 1950s when the ferrite phase shifter was introduced. A detailed history of early microwave ferrite devices is given by Button [25]. Relevant studies leading directly to phase shifter devices began with measurements of transversally magnetized ferrite slabs in rectangular waveguide [26], which produce non-reciprocal phase shift. Single ferrite slabs located at the plane of the waveguide circular magnetic polarization were found to produce significant non-reciprocal phase shift, as did dual slab configurations (Figure 17.5a) with antisymmetric applied H-fields. The toroid phase shifter of Treuhaft and Silber [27] provided a very practical method of applying the antisymmetric H-field using direct current pulses through separate drive wires. The toroidal phase shifter of Figure 17.5b, after many practical upgrades in design and material, remains one of the most important components for array control.

Reciprocal Phase shift can be produced by using a longitudinally biased ferrite rod in the waveguide. Bush [28] first reported a practical phase shifter of this type, and later Reggia and Spencer [29] presented experimental results that showed over 200° of phase shift per inch of ferrite at X-band with 30 Oersteds change in longitudinal magnetic field.

The first phased array developed in Russia was a four element dielectric rod array scanned by reciprocal phase shifters [30] and built in 1955. This experimental array was part of an intensive program under Professor Yuri Yurov, of Leningrad Electrical Engineering Institute. The work culminated in 1960 with construction of a 61 element lens filled with ferrite phase shifters.

A latching reciprocal "Dual-Mode Ferrite Phase Shifter" sketched in Figure 17.5c was developed [31] using ferrite quarter-wave plates with fixed bias that launch a circularly polarized wave into a small, square or circular waveguide plated on the ferrite. The magnetic circuit that provides a longitudinal magnetic field is completed external to the waveguide. The device can be switched in 50 - 100 microseconds with insertion loss between 0.5 – 0.7dB. Another development is the rotary field phase shifter [32], which is the electrical analog of the mechanical Fox phase shifter that was used in the early polyrod array. This device uses a transverse magnetic field and results in extremely precise

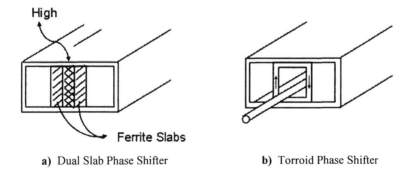

a) Dual Slab Phase Shifter   **b)** Torroid Phase Shifter

Figure 17.5 Ferrite Phase Shifter Configurations.

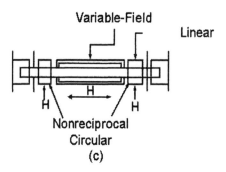

Figure 17.5c. Ferrite Phase Shifter Configuration: Dual Mode Reciprocal Phase Shifter.

reciprocal phase control (on the order of one degree) but slower switching time. Typical switching times can be one hundred to several hundred microseconds, depending on the control circuitry. Average power varies from 600 watts at S-band to 40 watts at Ku-band. Insertion loss is 0.6 – 0.7 dB.

Non-reciprocal coaxial line phase shifters were introduced by Sucher and Carlin [33] and later demonstrated experimentally by Rowley [34]. Other TEM phase shifters have been described for planar circuits. Simon et al. describe a stripline ferrite phase shifter [35].

The ferrite phase shifter became the enabling technology for phased arrays in the late 1950s and 1960s, and is the preferred technology to date for high power arrays at frequencies above 2 GHz. A recent survey is given by Adam et al. [36].

The next step in array control followed from the insight and pioneering work of Robert Barrett, who introduced the concept of the microwave printed circuit [37] in 1951. Prior to Barrett's work there had been transmission lines consisting of flat-strip coaxial power dividers and rods of various cross sections as center conductors between flat plates. Barrett's insight was to realize that "not only could flat-strip coaxial lines in a flat plate form be employed to carry energy from point to point, but they could also be used to make all types of microwave components, such as filters, directional couplers, hybrids etc." He also realized that they could be fabricated by the same methods as low frequency printed circuits and so could include distributed elements to replace the lumped circuit elements used at lower frequencies. These new microwave printed circuits became the essential ingredients of diode phase shifter technology and the foundation for later developments in monolithic microwave integrated circuits.

During the 1960s, the technology for diode phase shifters brought a second option for array control. Summaries of the relevant developments are given in references [38] and [39]. P-I-N [40,41] and varactor [42] diodes have both been used as microwave phase shifters, although P-I-N diodes are preferable for high power applications. Throughout the years P-I-N diode phase shifter power handling capacity has grown, so that peak power of kilowatts and average power in the hundreds of watts [39] can be handled with only two diodes per phase bit. Figure 17.6 shows the three basic types of microwave phase shifters, with Figure 17.6a illustrating a switched line phase shifter that adds short sections of transmission lines to simulate the various phase bits. Figure 17.6b shows a reflection phase shifter and Figure 17.6c shows a loaded line phase shifter. The appropriateness of each of these types depends upon the available space for the circuit as well as loss and matching constraints. In general, diode phase shifters offer several advantages when compared with ferrite devices. They are far lighter weight, and switch in nanoseconds as compared to microseconds for the fastest ferrite devices, but they tend to have higher loss, especially at higher frequencies. In general, they are lower power devices than ferrites, but at lower microwave frequencies, where there is adequate interelement space, they are capable of very high power operation [39].

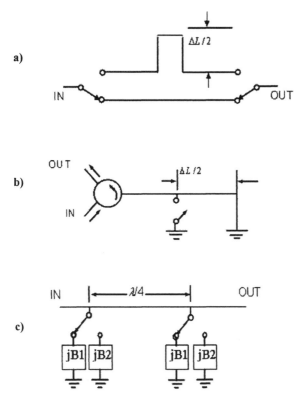

Figure 17.6. Electronic Phase Shifter Circuits **a)** switched line phase shifter **b)** reflection phase shifter **c)** loaded line phase shifter.

Time delay control has been provided primarily by switching lengths of transmission line (primarily coaxial line) into the microwave circuits. Early discussions of this technology are found in [43, 44].

17.4.2   Digital and Optical Control of Arrays

The process of digitally sampling the output signals from a number of antenna elements or subarrays, and combining these samples appropriately to an angular filter response that is a virtual antenna pattern with arbitrarily small beamwidth, is termed digital beamforming. Figure 17.7 is a sketch that illustrates this process for an array in the receive mode. The transmit configuration is similar, but using power amplification at each element and a digital synthesizer as the D/A device. The figure also illustrates the multiple beam feature available by using different sets of weights W^P for each beam. This capability is one of the major advantages of the technology.

In the late 1950s, the use of underwater acoustic arrays for underwater detection systems began to show the advantages of digital processing [45,46]. This technology was ideally suited for the low kilohertz frequencies and narrow bandwidth of acoustic sensors, and it was to be many years before it transitioned to radio frequency electromagnetic systems. Acoustic technology developments in signal processing and the obvious flexibility inherent in digital beamforming made it inevitable that digital control would eventually play a significant role for arrays at RF and microwave frequencies. That role is still developing, [47,48] and largely stimulated by advances in digital signal processing that have brought the initial analog sidelobe canceller concepts of Howells [49,50] and others [51,52] to ever larger problems of multiple interference cancellation. In addition to interference cancellation, digital beamforming provides the flexibility to record multiple beams, modify radiation patterns at will, and provide the ultimate in precise sidelobe control.

Figure 17.8 shows the most basic of optical control circuit for array steering. On transmit the RF signal (including whatever modulation is to be transmitted) is used to modulate the optical source (or sources). This optical signal power is then divided to address each antenna port, and the various lines

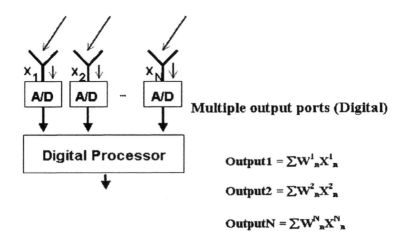

Figure 17.7.   Digital Beamformer In Receive Mode.

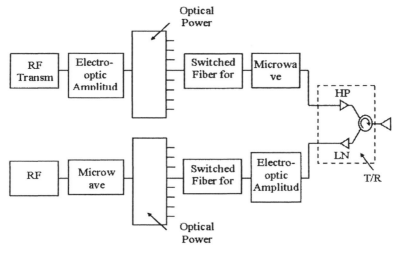

Figure 17.8. Optical Control Network.

are then time-delayed by switching appropriate lengths of fiber transmission line into the circuit. Each optical signal is then passed through a detector circuit to produce the time delayed RF signals, which are then amplified and radiated. The received signal is handled in a similar manner. This RF/optical path is inefficient and requires amplification elsewhere in the network, but the technology can provide accurate time delay with little dispersion, as required for large arrays with wide bandwidth.

Actual networks that are configured for photonic array control are often far more complex than the simple one shown in the figure, and may use independent optical sources for each control port. The first demonstration of the use of optical delay in an array was by Ng et al. [53], who built an array using individual optical sources for each element. A recent survey by Seeds [54] describes this effect in addition to several recent contributions to this technology [55,56].

Implementation of optical time delay steering in arrays has been slowed because of the inefficiencies in modulation, detection and fiber switching. Recent advances in multiple interconnect networks for forming multiple independent beams with MEMS mirror switches [57] may provide a capability that cannot be equaled by microwave analog multiple beam networks.

17.5   PHASED ARRAY ANALYSIS AND SYNTHESIS

17.5.1   Mathematical Developments

Phased array technology was preceded by many years of fundamental antenna technology, and so the powerful analytical techniques previously developed were readily applied to array problems.

The integral equation formulations of Pocklington [58] and Hallén [59] for wire elements led to practical solutions that included the mutual coupling between elements in arrays. King and others [60] presented iterative solutions of these equations as well as solutions made up of a finite number entire domain basis functions. Storer, in 1950 [61] introduced a variational method of solution for dipole impedance. Generalized Green's Function formulations were introduced by Levine and Schwinger [62] and applied to numerous antenna and array studies. Waveguide elements were analyzed using similar integral equation techniques by Lewin [63] and later Stevenson [64].

Thus established, integral equation formulations for element and array problems became the norm for research studies throughout the 1960s culminating in the methodical procedure that was to be called the Method of Moments, and presented by Harrington [65,66]. In the intervening years this procedure has been extended and developed and now is the source matter for a number of large, efficient, generalized software codes for antenna and array analysis.

Surface integral equation methods based on the free-space Green's Functions (or the equivalent potential functions) cannot be applied to finite antennas or arrays of elements printed on dielectric substrates, and for these important problems it is necessary to formulate the boundary value problems in the transform domain, or equivalently to use a Sommerfeld [67] type kernel. More recently, antenna and array problems have also been studied using finite difference time domain [68] or finite element [69] formulations.

17.5.2 Antenna Pattern Synthesis

The relationship between antenna sidelobes and array aperture distribution was understood during the 1930s and 1940s, and discussed in the 1949 text by Silver [70]. Early synthesis methods relied on the use of well known functions for the aperture distributions, like the $(\cos)^n$ and cosine on pedestal distributions. In 1943 Schelkunoff [71] published a synthesis theory based on the linear array polynomial that led to numerous practical applications and inspired additional researchers throughout the rest of the century. Fourier series techniques were also first used in the 1940s. The polynomial nature of the linear array pattern factor was again used in the 1940s by Dolph [72] who adapted a Chebyshev polynomial representation of the array factor and thus obtained a set of equal sidelobe distributions that allowed one to explicitly set the desired sidelobe level for an array and obtain the narrowest main beam achievable. Taylor [73] improved upon the Chebyshev distributions by recognizing their gain limiting characteristics for large arrays, and proposing a modification that maintains a small set of sidelobes at roughly the Chebyshev design level, while allowing the wider angle sidelobes beyond the nth pattern zero to decay as the inverse of the sine of the angle from the peak. This technique became the most important procedure for practical array synthesis. Figure 17.9a shows a Taylor pattern for a 50 element array with 40 dB sidelobes and $n=8$. A similar approach was later used by Bayliss [74] (Figure 17.9b) for the antisymmetric excitation required of monopulse radar. The pattern synthesis technique of Woodward [75,76], based on synthesizing with orthogonal sinc beams, has proven extremely useful for pattern shaping, as well as for direct application with multiple beam systems. Recently there have been numerous examples of the use of numeric techniques including gradient methods [77], simulated annealing [78], genetic algorithms [79] and an extremely important application of iterative projection methods [80,81].

17.5.3   Array Mutual Coupling and Blindness

Although the phenomenon of antenna mutual coupling was understood earlier, and relatively simple mathematical models for dipoles existed as early as 1934 [60], it was not until the 1960s that coupling effects were included regularly in

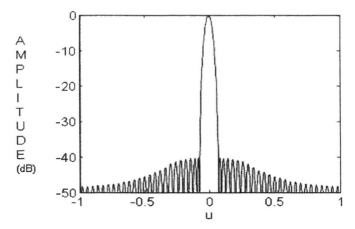

Figure 17. 9a. Low Sidelobe Taylor [73] Radiation Pattern $(n = 8)$.

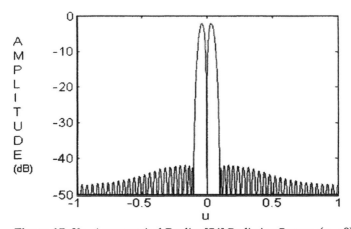

Figure 17. 9b. Asymmetrical Bayliss [74] Radiation Pattern $(n = 8)$.

antenna and array calculations. The inclusion of coupling made it possible to explain the "surface wave-like" behavior of Yagi arrays [82] and the impedance of element pattern behavior of multiple antenna phased arrays. Throughout this period, and to the present, there have been two traditional methods for including coupling. One is the explicit calculation of element to element interaction using Green's Function or potential methods which has been commonly used for relatively small arrays, while the infinite array approach is often used to characterize elements and behavior near the center of large arrays. The availability of advanced acceleration techniques and vast computer speed and storage is making it possible to use the element to element modeling methods for arrays of many thousands of elements.

The phenomenon of array blindness is a condition that results from array mutual coupling and can bring about essentially complete cancellation of the antenna-radiated beam at certain scan angles. This result is accompanied by near-utility reflection coefficient at most of the central elements of the array. From the element pattern point of view, it is seen as a zero in the array element pattern. Figure 17.10 and shows element pattern data due to Farrell and Kuhn [83,84] for an array of waveguides on a triangular grid. The figure shows experimental H-plane scan data (solid line) for a finite array of 95 waveguide elements compared with the infinite array modal expansion (dashed line) and single-mode (dotted line) theories. The experiment and infinite-array model theory show good correlation, while the single-mode theory (called grating lobe series in the figure) does not exhibit the blindness.

The logic that explains this phenomenon is now apparent, but required substantial experimental and computational proof during the 1960s and 1970s. From a mathematical point of view, the array formulation is in terms of a set of inhomogeneous coupled Fredholm equations of the first kind, and the blindness condition arises at frequencies and scan angles when the associated homogeneous set has a solution.

Array blindness results when the array geometry with short-circuited input ports would support a normal mode (lossless nonradiating propagation) along the structure at some given scan angle. At the angle of array blindness with the array excited at all input ports, the input impedance at all ports is identically zero, with the structure supporting a nonradiating lossless mode. Mathematically, this is analogous to a resonant L-C circuit. In the L-C circuit case, at resonance, the input current is unbounded ($I = V/Z_{in}$) because of the zero in the impedance ($Z_{in} = 0$). The resonance can be defined as the condition at which a nonzero current is supported with no input signal. The resonant frequency is the solution of the eigenvalue problem, and is that frequency of undamped oscillation of the circuit with input terminals shorted (by the zero resistance path of the ideal voltage source).

The array, however, is more complex than the L-C circuit because it has a distributed set of input ports, with signals applied to each port. However, in a manner entirely analogous to the resonant circuit, if there is a propagating nonradiating solution that would satisfy the boundary conditions of the shorted array structure, then applying a set of signals with that phase progression would

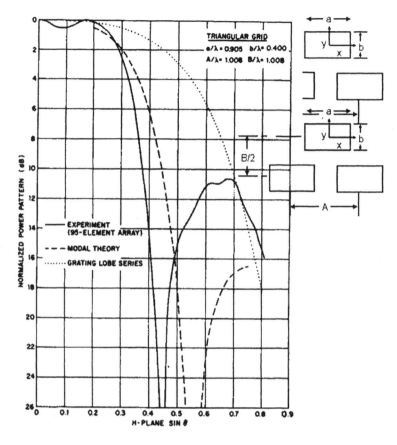

Figure 17. 10. Array Element Pattern Illustrating "Blindness," after Farrell and Kuhn [83,84].

result in zero input impedance at all input ports. If the input impedance is zero, one can place short circuits at the terminals without changing the solution.

Extension of the above logic to the array case for scan angles in real space and with spacing greater than one-half wavelength is not obvious. For such spacings, the inter-grating lobe separation is less than 2 in $\sin\theta$-space, and there is always at least one beam in real space $|\sin\theta| < 1$. Thus, at least one array beam should radiate, and the combined network should have loss. If one were to look at the equivalent shorted array, one would argue that it cannot support a normal mode solution because the radiation would preclude a lossless solution. Yet the blindness phenomenon is caused by the existence of a normal mode solution that exists precisely because it allows no radiation.

Part of the answer to this intuitive dilemma came from the study of Farrell and Kuhn [83,84], who provided an essential key to understanding blindness and performed rigorous analysis of a waveguide array with a blind spot. They were the first to observe that waveguide higher order modes play a dominant role in achieving the cancellation necessary for a null. The null occurs when radiation contributions from the lowest order symmetric and antisymmetric modes cancel to produce a zero in the element pattern. In addition, Wu and Galindo [85], demonstrated that the only radiating (fast) wave of the periodic structure spatial harmonic spectrum is identically zero at a null (because of cancellation by the odd mode), and that for this reason a normal mode can exist even for a structure with a period greater than one-half wavelength.

Knittel et al. [86] reinforced this point of view, describing the blind spot as associated with the normal mode solution of an equivalent, reactively loaded passive array, and the condition of a complete null on the real array occurs when the elements are phased to satisfy the boundary conditions for the equivalent passive array.

Although dielectric layers have been used for many years to improve scan match, they too can be the source of array blindness. This phenomenon is not new and was observed in the early work of Wu and Galindo [85]. More recently, blindness has been observed to occur in microstrip patch arrays or microstrip dipoles when the combination of dielectric constant and substrate thickness is such as to support a tightly bound surface wave, one with a phase velocity that is sufficiently slow so that it couples to an array grating lobe. A particular case is illustrated by arrays of microstrip patches etched on dielectric substrates. Pozar and Schaubert [87] have correlated the TM surface wave propagation constant with the observed blindness angle. The dielectric layer itself supports a surface wave, and although the boundary conditions are perturbed by the array patch or dipole structure, the location of the blindness is often predicted very accurately by the surface wave propagation constant. The mechanism for coupling into the surface wave is that the periodic perturbations of the surface wave traveling (slowly) in, for example, the negative "x" direction, form a grating lobe that would radiate into the positive θ angle, but that it is exactly canceled by the array beam, also at the $+\theta$ direction.

Array blindness, though now understood, remains an issue that must be accounted for in any array development. It can usually be avoided by using smaller element spacings and very careful use of dielectric substrates or superstrates, or any coupling structures like transmission lines and baluns at the array face.

17.5.4   Some Major Historical Array Developments Since 1950

Aside from the brilliant work of a small number of researchers, most of the important developments in array technology have occurred since 1950. Some of the major developments related to array control, like phase shifters and optical and digital array control, have been discussed earlier and are not mentioned below. One way to describe "major historical array developments" would be to list the significant "system" accomplishments, and looking back from 2003, one could produce a huge list of important military and commercial systems that have incorporated electronic scanning technology in one or both scan planes. With enough research it would become clear that most major countries have made contributions to this technology or have ongoing developments of some significance. Instead, however, the following group of developments lists very few of these systems and concepts, and those because they are examples of the major historical developments in the underlying technology, the breakthrough technologies themselves.

17.5.5   Frequency Scanning

The high cost of phase control led to the use of frequency scanning for arrays that scanned in elevation and were rotated to provide azimuthal steering. [88] One of the earliest of these systems to see application to a major system was the AN/SPS-39, a United States Navy shipboard radar that used a parabolic cylinder feed by forty elements connected to a serpentine delay line. Developed by Hughes Aircraft Company with it first prototype delivered in 1957, it was used to stabilize the elevation pointing of the shipboard radar while the ship rolled and pitched in rough seas. This very successful application led to numerous other frequency scanned arrays for Naval systems, and to 3-D radars using a combination of azimuthal phase scanning and frequency scanning in elevation. Of these, the earliest was the AN/SPS-32. [89]

17.5.6   Retrodirective Array

Figure 17.11 shows the concept of the retrodirective array as attributed to Van Atta [90] who invented the array configuration that bears his name. The Van Atta array collects signals from an incident wavefront and re-radiates each received signal from an element that is at the mirror-image location on the other side of the array, as shown in the figure. Because of the array symmetry, the net result is that each element re-transmits a signal with the complex conjugate of the phase it

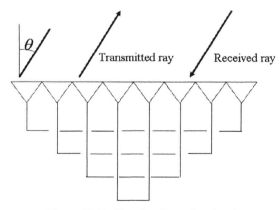

Figure 17.11. Van Atta Retrodirective Array.

received. The circuit is the array analog of the well-known corner reflector. Not only does this array exhibit a large scattering cross section, but it has major application as a beacon or transponder wherein the received signals are amplified and perhaps modulated before re-transmission back to the source of the received signal. Retrodirective arrays have also been implemented by electronic phase conjugation using up-conversion and down-conversion mixer circuits operating on a received pilot signal.

The concept has been used for communication with satellites as well as ground communication, but its import extends beyond this, since it was also recognized as the first venture into signal processing and adaptive arrays.

17.5.7   Adaptive Arrays

The primary reason for using electronically scanned arrays is to move a directional beam throughout some chosen scan sector without physically moving the antenna. A secondary reason has always been to minimize interference from unwanted sources, and this was done using a degree of pattern shaping, but in the 1960s it became apparent that using sophisticated signal processing techniques the additional degrees of freedom provided by arrays could also provide a new capability for pattern control.

Although the retrodirective array was meant to control the radiation of a transmit array, most of the adaptive arrays have been built to control the receive pattern. In the 1960s P.W. Howells [49] invented a sidelobe canceller analog circuit that used correlation loops to adaptively null interference coming into a receiver through the antenna sidelobes. This work stimulated a flurry of activity at General Electric and Syracuse Research Corporation [50] and led to mathematical development by Applebaum [51] for optimizing the signal to noise in quite general arrays. Widrow et al. [52] looked at self-optimizing algorithms based on a gradient search method that optimized array outputs within least mean square (LMS) accuracy. Early adaptive arrays used analog processing circuits and feedback loops because the bandwidth and speed requirements of digital processing were prohibitive (unlike those for acoustic systems). More recently it has become common practice to use digital beamforming technology for adaptive processing. In STAP (Space-Time Adaptive Processing) the intention is to cancel clutter and external interference using all both angular and doppler degrees of freedom. STAP involves huge amounts of digital processing, and implies digital beamforming at the aperture or subarray level [91].

Adaptive processing has become extremely important to wireless system design, with several important references addressing that current topic [92,93].

17.5.8   Multiple Beam Lenses and Networks

An important class of arrays, the multiple beam array, is shown schematically in Figure 17.12 where each input port excites an independent beam in space. Multiple beams can be produced with a digital beamformer as in Figure 17.7, but in addition there are a variety of antenna hardware concepts that produce

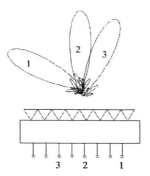

Figure 17.12. Multiple Beam Array.

multiple beams, including Butler matrices [94,95] and other networks [96]. Butler matrices (Figure 17.13) are a circuit implementation of the Fast Fourier Transform technique and radiate orthogonal sets of beams with uniform aperture illumination. Lens and reflector systems can produce wide-band multiple beams whose beam locations do not vary with frequency. A particularly convenient implementation is the Rotman lens [97] of Figure 17.14, a variant of the earlier Gent bootlace lens [98] that has the special feature of forming three points of perfect focus for one plane of scan. The Rotman lens can provide good wide-

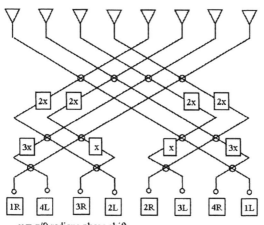

x = π/8 radians phase shift

⊠ Hybrid coupler convention: Straight through arms have no phase shift
while coupled arms have 90° phase shift

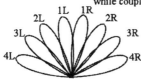

Figure 17.13. 8 × 8 Butler Matrix.

angle scanning out to angles exceeding 45°. Multiple beam lenses and reflectors have been chosen for satellite communication systems, and in that application serve to produce either switched individual beams or use clusters of beams to cover particular areas on the earth. Figure 17.14 shows a sketch of a Rotman lens, illustrating the several ray paths through the lens, and the associated radiating wavefront. In 1975 Archer proposed a microstrip implementation of Rotman Lens [99]. This compact, inexpensive printed circuit antenna has found a number of important system applications.

An early multiple beam array, described by Schrank [100] was the Navy TYPHON system radar array which consisted of a transmit spherical array and three Luneburg lens receiving antennas. The transmit array consisted of 3240 elements with 900 elements activated to form each beam in space. This system appears to be the first switched beam spherical array, and was developed in 1960 as part of the SPG-59 TYPHON experimental radar.

A major technological contribution reported by Blass in 1960 [101] is depicted in Figure 17.15. This network used combinations of directional couplers to combine the signals of multiple constrained (waveguide) feeds, each exciting a single radiating beam with a unique scan angle. This network was implemented in the FAA experimental AHSR-I three dimensional radar that operated at S-band with elevation beamwidths between 0.2° with 1.2° [15,22]. The array was 150 feet high and formed 111 beams. Following the Blass work, Lopez [102] presented a monopulse network using directional couplers and demonstrated the formation of independent lossless, orthogonal sum and difference beams.

17.5.9   Subarray Systems for Wideband Scanning

During the 1960s there was much concern about the high cost of large arrays that required time delay scanning. Scanning with variable time delays is required to eliminate array beam peak motion of a phase scanned array as a function of frequency, which can result in the array beam moving off the target as the frequency is moved away from the chosen center frequency. This bandwidth

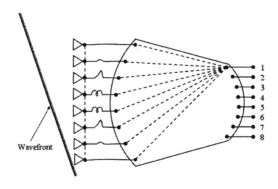

Figure 17.14. Rotman Lens, ray traces, and radiating wavefront.

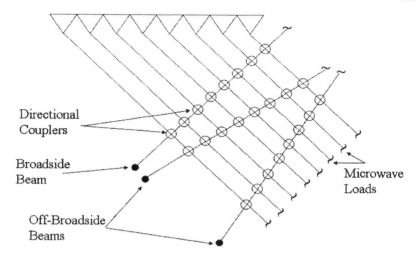

Figure 17.15. Time Delayed Version of Blass Matrix.

limitation is significant for large arrays, when the beamwidth is so narrow that a relatively small beam pointing error can result in a major loss in target response.

The obvious solution to this "beam squint" problem is to use time delay devices of some sort at each of the elements. Conceptionally the simplest approach is just to switch coiled lengths of line in series with each element to provide time delay. Unfortunately this approach is costly, lossy and often requires more space than available, although it has been implemented at lower frequencies. Some alternative approaches are given in the writings of Tang [103] and Butler [95], and consist of using the time delayed version of the Blass Matrix, providing each element with time delayed pulses, and providing time delay at an intermediate frequency. Since these approaches are all very costly if implemented at the element level, researchers have investigated a number of ingenious subarray schemes so that phase shift could be provided at the element level with time delay at many fewer subarrays where the element signals are grouped.

The most obvious way to break the array into subarrays is using a set of corporate feeds, one for each subarray, and then to connect contiguous subarrays to feed the whole array. With phase shift at each element and time delay at each subarray, the bandwidth based upon "beam squint" loss is multiplied by the number of subarrays. Unfortunately this is accompanied by very high sidelobes, called "quantization lobes" that occur off center frequency when the array is scanned.

A number of subarraying schemes have been devised to avoid this problem, most of which are described by Tang [103], and shown in Figure 17.16. This figure suggests general classes of interlacing and overlapping the signals from groups of subarrays. Arguably the most original and promising of these for

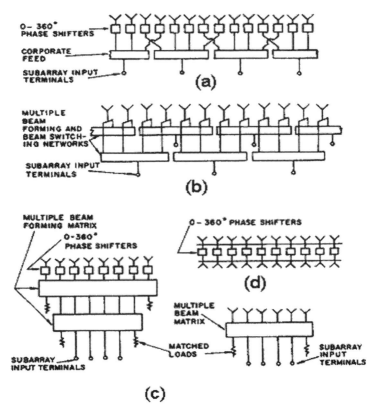

Figure 17.16. Subarray schemes for minimizing gain reduction and grating lobe level. (after Tang [104]).

    (a) Interlaced subarray scheme
    (b) Overlapped subarray scheme
    (c) Completely overlapped subarray scheme. (Constrained feed approach)
    (d) Completely overlapped subarray scheme (Space feed approach)

large arrays are the constrained feed and space feed versions of completely overlapped subarrays. A space fed antenna of this class is the HIPSAF antenna, shown schematically in Figure 17.17. This structure uses a multiple beam array as a space feed for an objective lens. Each input to the feed array radiates to produce a (sin x)/x type illumination that radiates to the objective lens back face. The secondary far field radiation is a pulse shaped subarray pattern of an angular width such that it suppresses the quantization lobes. The HIPSAF antenna itself was built and tested around 1964, and the concept of overlapped subarrays has been implemented in the Australian Jindalee Over-the-Horizon Radar and other systems. Southall and McGrath [104] give a comprehensive description of an experimental overlapped subarray development.

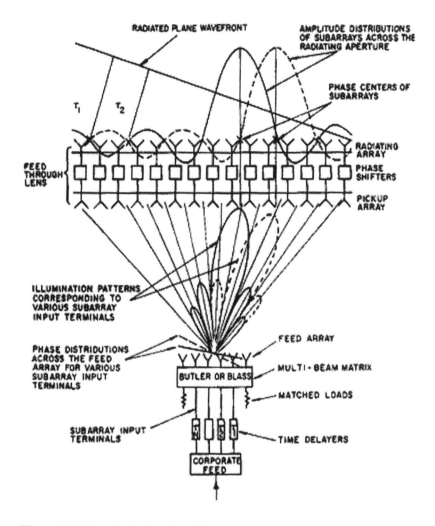

Figure 17.17. Completely Overlapped Subarray Antenna System (after Tang [104]).

17.5.10 Subarrays and Space Fed Arrays for Limited Field of View Systems

The subarraying concept that led to the wideband HIPSAF antenna was implemented with a radiating space feed, but there have been a number of constrained feed geometries that have produced overlapped subarrays with pulse shaped radiation patterns. Among these are the networks of Dufort [105] and Skobelev [106] that use overlapped power divider networks to achieve low sidelobe pulse shaped patterns. These constrained networks have much broader subarray patterns than the space-fed versions, and so are useful either for very wide band systems or to reduce the number of phase controls for arrays that scan over limited field of view (LFOV). Skobelev [107] describes a number of these systems and their historical relevance.

In addition to the subarraying feeds for LFOV, there have been other LFOV systems that used arrays as feeds for reflectors, but placed closer than a focal length to the objective aperture. The array in this case is adjusted to best match the converging rays (on receive). Early examples of this concept include papers by Winter [108] and Tang [109]. This technology was later to become the basis of the AN/TPS- 25, a precision approach radar antenna.

17.5.11 The Advent of Printed Circuit Antennas

The insights of Barrett and others who brought about the age of microwave printed circuits were destined to impact antenna technology during the latter half of the century. Among the innovators of this technology was Georges Deschamps who, in 1953, authored the paper with William Sichak entitled "Microstrip Microwave Antennas" [110]. The Deschamps/Sichak paper described a variety of microwave circuit structures, but the actual microstrip antennas they considered were an open horn structure and a lens fed by microstrip. In both cases they built small arrays and recorded pattern and impedance data. Microstrip patch antennas (and multi-feed strip antennas) were first described by Munson [111,112] and by Howell [113,114] who referenced Munson as the originator of the concept. The first microstrip array data was published by Munson in 1974. During that period there was no accurate theory to describe the radiation or impedance characteristics of any of the microstrip configurations, but in 1978 Y.T. Lo and colleagues presented a cavity based theory [115] which, though approximate, allowed engineering practice to go forward for a number of years. Alexopoulos and colleagues [116] and Pozar and Schaubert [117] developed full wave electromagnetic solutions appropriate for the numerical analysis of elements and arrays.

Figure 17.18a shows a proximity coupled patch designed to have wider bandwidth than an individual patch. At this time the microstrip patch and its many printed variants have been incorporated into low profile passive and scanning arrays of modest bandwidth. They are the radiators that have first reached Barrett's goal of fully printed arrays and circuits.

Among the "printed circuit" antennas, there is one class of elements that is not flush mounted like the patch, but is often made using printed circuit technology on substrates mounted perpendicular to the aperture. Called by various names and excited by a variety of transmission lines, the stripline fed flared notch antenna of Figure 17.18b is a wide band antenna element that was first developed as an individual element, and later saw application to arrays [118].

This element, along with the slotline fed "Vivaldi" element [119] have undergone significant developments in recent years to broaden their bandwidth in an array environment [120-123]. Results to date show up to 10:1 operational bandwidth is achievable at the expense of very tight element spacing (0.05λ) at the lowest frequency.

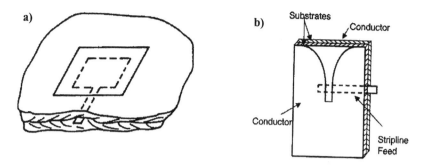

Figure 17.18. Microstrip Patch and Flared Notch Elements: a) proximity coupled microstrip patch (ground screen beneath lower substrate); b) Strip-line fed Flared Notch Element.

17.5.12 Solid State Modules

Throughout the 1960s, array antennas continued to be powered by microwave tubes, magnetrons, klystrons and traveling wave tubes, all providing relatively high power single source feeds. The feed signal was then distributed to transmission lines with phase or time delay control, and radiated through coaxial or waveguide elements. This process was relatively inefficient because of power divider, transmission line and phase shifter losses. In addition, the array reliability was limited by the single point failure rate of the transmitter. By the end of the 1960s advances in solid state amplifiers and switches made it possible to consider distributing relatively lower power solid state sources and controls throughout the array. It was expected, and has turned out to be true in many instances, that the solid state "module" array would have improved reliability, better efficiency, and ultimately lower cost.

The next and continuing development in the history of array control was the integration of solid state modules behind, and eventually in the array aperture. The early history of this technology is reviewed in references [124] and [125]. The concept of an integrated circuit on a block consisting of layers of insulating, conducting, rectifying and amplifying materials with all electrical functions connected by cutting out areas of the layers was first proposed by G. A. W. Dummer of the Royal Radar Establishment in 1952 in England [126]. The processing of active and passive components on the semiconductor slab is attributed to J. Kilby [124] of Texas Instruments in 1958. This technology saw its first application to arrays through the MERA (Molecular Electronics for Radar Applications) program [100,127,128] sponsored by the United States Air Force Avionics Laboratory at Wright-Patterson Air Force Base, and contracted in 1964. The MERA module included some monolithic integration on silicon, but all of the microwave components were fabricated using these chips on alumina microstrip substrate. The array had 604 elements, produced 32 dB gain on transmit and 30 on receive, transmitted 352 watts peak at 9 GHz, and had a system noise of 12.5 dB. The array had multi-mode radar modes for ground mapping, terrain following/avoidance, and air to ground ranging. Although the technology was successful, the module approach was far more expensive than traditional passive (tube and phase shifter) technology, so in another Avionics Laboratory effort, the Radome Antenna and RF Circuitry (RARF) program, sought simple, lightweight, economical ferrite phase shifter scanning arrays. An Electronically Agile Radar (EAR) followed. It employed mass-produced phase-shifter/radiator elements that cost approximately $300 each, considerably lower than technology developed under MERA. The EAR influenced design of the B-1B avionics.

Despite the achievements of these early "passive" arrays, they suffered reliability problems because of their continued reliance on tubes and high power voltages. More solid-state technology became the remedy. The reason was obvious: solid-state array modules not only required less power, they delivered higher mean time between failures (MTBF). Solid-state antenna arrays tolerated hundreds of module failures before needing repair, a fact that gave them a crucial edge over their tube-based cousins.

The last of the airborne solid state demonstrator radar programs, the Solid State Phased Array (SSPA) program ended in 1988 using traditional microwave integrated circuit technology, which was again to prove the capability and reliability of modern solid state arrays, but not its affordability. The breakthrough in affordability for airborne arrays came with the availability of Microwave Monolithic Integrated Circuit (MMIC) technology, sponsored and developed by DARPA to advance solid state technology.

In 1972 the first all solid-state radar (AN/TPS-59) was fielded for the United States Marines by the General Electric Company. This mobile 3-D radar system used an array of 54 row feeds, each with a T/R module to provide elevation scanning. The array is mounted on a rotating platform for azimuth scan.

The first phase solid state radar array was the Pave Paws UHF array. This system, shown in Figure 17.19 has two array faces, each consisting of 1972 bent dipoles with T/R modules and 885 dummy elements. The radar was installed in 1978.

These and other numerous advances in the solid state technology of T/R modules have made that technology the state of the art for modern radars. A recent article [129] by Kopp et al. discusses the present state of T/R modules including performance and cost aspects of the semiconductor, packaging and assembly technologies associated with T/R modules. Parker and Zimmerman [130] list a number of solid state active arrays as typical of present developments.

17.6   THE FUTURE

The history of phased arrays has depended on two technologies. Progress has alternately accelerated by and slowed by device technology and electromagnetic technology.

The lack of electronic phase control before the 1960s severely limited applications to more advanced radar and communication systems. However the

Figure 17.19. PAVE PAWS Array.
http://www.boeing.com/defense-space/space/gmd/gallery/rcgal/7ce8699.html

availability of ferrite and diode phase shifters since the 1960s, coupled with major recent developments in solid state module technology has brought us to the age of multi-function radar and communication systems.

Further advances in these technologies are close at hand but tied to the price of solid state monolithic integrated circuit modules and subarrays. For some applications passive arrays based on RF MEMS and ferroelectric phase shifters may become the enabling technologies.

The electromagnetics that has slowed array developments when poorly understood and accelerated them when accommodated will probably lead the way in a manner never seen in the past. The electromagnetic phenomenon of array blindness resulting from the mutual interaction of elements once led to a temporary halt in development programs, but better electromagnetic tools made it possible to refine designs that were free of blindness. Similarly, the difficulty of analyzing wide band elements has slowed progress in broadband arrays. The encouraging aspect of this issue is that very powerful numerical electromagnetic solvers have recently been developed, and yet more advanced packages are being produced. These general solvers will allow engineers to produce very complex electromagnetic surfaces and to analyze geometries that were beyond analysis just a few years ago.

With all of these developments in technology and theory, we can look forward to arrays with multiple independent wideband scanning beams, high quality pattern performance, flexibility, low cost and compact light weight configurations. Most of these attributes were anticipated by Friis and Feldman [9] in 1937, but not supported by the technology. They aren't fully supported by today's technology either, so there remains much device and array architectural work to make this goal a reality.

17.7   AUTHOR'S COMMENTS

It was not my intention to spend so much of this history describing American technology. I would rather have given more equal treatment to development from around the world. However, the first scanning arrays were developed in the United States, and much of the rest of the early literature is difficult to access because of national and industrial security classifications. I encourage others who may know additional details to publish them and to tell me of them. Uncovering the history of this stimulating area is a worth-while goal, and I offer this contribution as a first iteration.

17.8   ACKNOWLEDGMENTS

My sincere thanks are given to Mrs. Marjorie Hobbs for typing this manuscript and for the diligent web-searches that have been indispensable to this work.

REFERENCES

[1]   H. Yagi and S. Uda, "Projector of the Sharpest Beam of Electric Waves," *Proc. Imp. Acad.* (Tokyo) Vol. 2, p. 49 (1926).

[2]   R. C. Hansen, *Microwave Scanning Antennas*, Academic Press, N.Y., 1964.

[3]   H. Friis "A New Directional Receiving System," *IRE Proc.*, Vol. 13, No. 6, Dec. 1925, pp. 685-708.

[4]   G. Marconi, "Directive Antenna," *Proc. Royal Soc.,* London, 77A, p. 413, 1906.

[5]   H. H. Beverage, C. W. Rice and E. W. Kellogg, "The Wave Antenna," *Journal of the American Institute of Electrical Engineers,* Vol. XLII, March 1923, pp. 258-269; April 1923, pp. 372-381;May 1923, pp. 510-519; June 1923, pp. 636-644; July 1923, pp. 728-738.

[6]   E. Bruce, "Developments in short-wave directive antennas," *Proc. I.R.E.*, Vol. 19, pp. 1406-1433, Aug 1931.

[7]   E. Bruce and A. C. Beck, "Experiments with directivity steering for fading reduction," *Proc. I.R.E.*, Vol. 23, pp. 357-371, April 1935.

[8]   H. T. Friis, C. B. Feldman and Sharpless, "The determination of the direction of arrival of short radio waves," *Proc. I.R.E.*, Vol. 22, pp. 45-78, Jan 1934.

[9]   H. T. Friis and C. B. Feldman, "A Multiple Unit Steerable Antenna for Short-Wave Reception," *Proc. I.R.E.*, Vol. 25, No. 7, July 1937, p. 841, 1937.

[10]   F. A. Polkinghorn "A Single-Sideband Musa Receiving System for Commercial Operation of Transatlantic Radiotelephone Circuits," *Proc.*

I.R.E., Vol. 28, April 1940, pp. 157-170; also *Bell System Technical Journal*, April 1940.

[11]   B. T. Neale, "CH – The first operational radar," http://www.radarpages.co.uk/mob/ch/chainhome.htm, Copyright by Dick Barrett.

[12]   H. T. Friis and W. D. Lewis, "Radar Antennas," *Bell System Technical Journal*, Vol. XXVI, April 1947, No. 2, pp. 219 -317.

[13]   A. G. Fox, "An adjustable wave-guide phase changer," *Proc. I.R.E.*, Vol. 35, pp. 1489-1498, Dec 1947.

[14]   W. H. Kummer, "Feeding and Phase Scanning" in *Microwave Scanning Antennas*, Vol. III, R.C. Hansen, Ed., Academic Press, NY, p. 82.

[15]   C. A. Fowler, "Old Radar Types Never Die; They Just Phased Array," *IEEE AES Systems Magazine*, Sept 1998, pp. 24A – 24L.

[16]   R. A. Watson Watt and J. F. Herd, "An Instantaneous Direct Reading Radiogoniometer," *J. IEE* (London) Vol. 64, 1926, pp. 611-622.

[17]   G. W. Swenson Jr, "Wullenweber Direction Finder," University of Illinois, 1 Sept 1994, http://www.ece.vivc.edu/pubs/spotlight/wullnart.htm

[18]   Wullenweber/CDDA Antenna Homepage, http://www.mindspring.com/ncummings7/wullen.html.

[19]   W. M. Cady, C.V. Robinson, F.B. Lincoln, and F.J. Mehringer, "Antennas, Scanners and Stabilization" in *Radar Systems Engineering*, MIT Radiation Lab. Series, Vol. 1, Ch. 9, pp. 303-304, McGraw-Hill Book Company, New York 1948.

[20]   R. M. Robertson, "Variable width waveguide scanner for Eagle (AN/APQ-7) and GCA (AN/MPN-1) Rept. No. 840, MIT Radiation Laboratory, Cambridge Massachusetts, 1946.

[21]   K. S. Kelleher, "Electromechanical Scanning Antennas," Chapter 18 in *Antenna Engineering Handbook*, Third Edition, R.C. Johnson, Editor, McGraw-Hill Inc., 1993.

[22]   H. Shnitkin, "Survey of Electronically Scanned Antennas," Parts I and II, *Microwave Journal*, Part I, Dec 1960, pp. 67-72. Part 2, Jan 1961, pp. 57-64.

[23]   R. C. Johnson, "Optical Scanners," Chapter 3 in *Microwave Scanning Antennas, Volume 1, Apertures*, R.C. Hansen, Ed., Academic Press, New York, 1964, pp. 213-261.

[24]   R. K. Luneburg, *The Mathematical Theory of Optics*, pp. 208-213, Brown University Press, Rhode Island, 1944.

[25]   K. J. Button, "Microwave Ferrite Devices: The First Ten Years," *IEEE Trans. on Microwave Theory and Techniques*, MTT-32, No. 9, Sept 1984, pp. 1088-1096.

[26]   B. Lax, K. J. Button, and L. M. Roth, "Ferrite phase shifters in rectangular waveguide," *J. Appl. Phys.*, Vol. 25, p. 1413, 1954.

[27]   M. A. Truehaft and L. M. Silber, "Use of microwave ferrite toroids to eliminate external magnets and reduce switching power," *Proc. I.R.E.*, Vol. 46, pp. 1538, Aug. 1958.

[28] D. Bush, "Discussion on microwave apparatus," *Proc. Inst. Elec. Eng.,* Vol. 104B, Suppl. 6, p. 368, 1956.

[29] R. Reggia and E.G. Spencer, "A new technique in ferrite phase shifting for beam scanning of microwave antennas", *Proc. I.R.E.,* Vol. 45, p. 1510, 1957.

[30] O. G. Vendik and Y. V. Yegorov, "The First Phased-Array Antennas in Russia: 1955-1960, *IEEE Antennas and Propagation Magazine,* Vol. 42, No. 4, Aug 2000, pp. 46-52.

[31] C. R. Boyd, Jr., "A dual-mode latching, reciprocal ferrite phase shifter," in 1970 *IEEE G-MTT Int. Microwave Symp. Dig,* pp. 337-340.

[32] C. R. Boyd and G. Klein, "A precision analog duplexing phase shifter," in *IEEE MTT-S Int. Microwave Symp. Dig.,* 1972, pp. 248-250.

[33] M. Sucher and H. J. Carlin, "Coaxial line nonreciprocal phase shifter," *J. Appl. Phys.* Vol. 28, p. 921, 1957.

[34] J. J. Rowley, "Phase shift studies in ferrite-dielectric loaded coaxial lines at 2200 Mc," *J. Appl. Phys.,* Vol. 32, p. 321S, 1961.

[35] J. W. Simon, W. K. Alverson, and J. E. Pippin, "A reciprocal TEM latching ferrite phase shifter," in *IEEE MTT-S Int., Microwave Symp. Dig.,* 1966, pp. 241-247.

[36] J. D. Adam, L. E. Davis, G. F. Dionne, E. F. Schloemann, and S. N. Stitzer, "Ferrite Devices and Materials," *IEEE Trans. on Microwave Theory and Techniques,* Vol. 50, No. 3, Mar 2002, pp. 721-737.

[37] R. M. Barrett, "Microwave Printed Circuits – the Early Years," *IEEE Trans. MTT-32,* No. 9, Sept 1984, pp. 983-990.

[38] J. F. White, "Review of Semiconductor Microwave Phase Shifters," *Proceedings of the IEEE,* Vol. 56, No. 11, Nov. 1968, pp. 1924-1931.

[39] J. F. White, "Origins of High-Power Diode Switching," *IEEE Trans. MTT-32,* No. 9, Sept 1984, pp. 1105-1117.

[40] K. E. Mortenson et al., "High-Power semiconductor phase shifting devices," Quarterly Progress Reports, Navy Contract NObsr-81470, 1960-1962, also "Microwave semiconductor control devices," *Microwave Journal,* Vol 7, pp. 49-57, May 1964.

[41] W. J. Ince and D. H. Temme, "Phasers and time delay elements," MIT Lincoln Laboratory, Lexington MA, Proj. Rept RDT-14, July 11, 1967.

[42] R. H. Hardin, E. J. Downey, and J. Munoshian, "Electronically variable phase shifter utilizing variable capacitance diodes," *Proc. IRE* (letters) Vol. 48, pp. 944-945, May 1960.

[43] L. Stark, "Microwave Theory of phased arrays – A Review," *IEEE Proc.,* Vol. 62, 1974, pp. 1661-1701.

[44] W. T. Patton, "Array Feeds" in *Practical Phased Array Antenna Systems,* Eli Brookner Editor, Artech House, 1991, pp. 6-18.

[45] V. C. Anderson, "Digital Array Phasing, " *Journal of Acoustical Society of America,* Vol. 33, No. 7, July 1960, pp. 867-870.

[46] T. E. Curtis and R. J. Ward, "Digital beam forming for sonar systems", *Proc. I.R.E.,* Vol. 127, pt. F, No. 4, August 1980, pp. 257-265.

[47] J. Litva, T.K-Y. Lo, *Digital Beamforming in Wireless Communications*, Artech House Inc., 1996.

[48] P. Barton, "Digital beam forming for radar", *Proc. I.R.E.*, Vol. 127, pt F, No. 4, Aug 1980, pp. 266-277.

[49] P. W. Howells, "Intermediate Frequency Sidelobe Canceller," U.S. Patent 3202990, August 24, 1965.

[50] P. W. Howells, "Explorations in Fixed and Adaptive Resolution at GE and SURC," *IEEE Trans. Antennas Propag.*, Special Issue on Adaptive Antennas, Vol. AP-24, No. 5, pp. 575-584, Sept 1976.

[51] S. P. Applebaum, "Adaptive Arrays," Syracuse University Research Corporation, Rep. SPL TR66-1, Aug 1966.

[52] B. Widrow, "Adaptive Filters I: Fundamentals," Stanford University Electronics Laboratories, System Theory Laboratory, Center for Systems Research, Rep. SU-SEL-66-12, Tech. Rep. 6764-6, Dec 1966.

[53] W. Ng, A. A. Walston, G. L. Tangonan, J. J. Lee, I. L. Newberg, and N. Bernstein, "The first demonstration of an optically steered microwave phased array using true time delay," *J. Lightwave Technol.*, Vol. 9, pp. 1124-1132, 1991.

[54] A. J. Seeds, "Microwave Photonics", *IEEE Trans. Microwave Theory Tech.*, MTT-50, No. 3, Mar 2002, pp. 877-887.

[55] J. J. Lee, R. Y. Loo, S. Livingston, V. I Jones, J. B. Lewis, H-W. Yen, G. L. Tagonau and M. Wechsberg, "Photonic Wideband Array Antennas", *IEEE Trans. Antennas and Propagation*, AP-43, No. 9, pp 966-982, 1995.

[56] M.Y. Frankel., P. J. Matthews and R. D. Esman, "Fiber optic true time steering of an ultra wide band receive array", *IEEE Trans. Microwave Theory Tech.*, Vol. 45, pp. 1522-1526, 1997.

[57] A. Morris III, "In search of transparent networks," *IEEE Spectrum*, Vol 38, No. 10, Oct 2001, pp. 47-51.

[58] H. C. Pocklington, "Electrical Oscillations in Wire," *Cambridge Phil. Soc. Proc.*, Vol. 9, 1987, pp. 324-332.

[59] E. Hallén "Theoretical Investigations into Transmitting and Receiving Qualities of Antennas," *Nova Acta Regiae Soc. Sci.*, Upsaliensis, Jan 1938, pp. 1-44.

[60] R. W. P. King, *The Theory of Linear Antennas*, Harvard University Press, Cambridge Massachusetts, 1956.

[61] J. E. Storer, "Variation Solution to the Problem of the Symmetrical Cylindrical Antenna," Cruft Laboratory Report No. 101, Harvard University, 1950.

[62] H. Levine and J. Schwinger, "On the Theory of Electromagnetic Wave Diffraction by an Aperture in an Infinite Conducting Scan," *Comm. Pure and Applied Math*, Vol. 44, 1950-51, pp. 355-391.

[63] L. Lewin, *Advanced Theory of Waveguides*, London, Iliffe and Sons, 1951.

[64] A. F. Stevenson, "Theory of Slots in Rectangular Waveguides," *J. Appl. Phys.*, Vol. 19, 1948, pp. 24-38.

[65] R. F. Harrington, "Matrix Methods for Field Problems," *IEEE Proceedings*, Vol. 55, No. 2, pp. 136-149, Feb 1967.

[66] R. F. Harrington, *Field Computation by Moment Methods*, New York: Macmillan Co., 1968.

[67] A. Sommerfeld, "Uber die Ausbreitung der Wellen in der drahtlosen telegraphie," *Ann. Physik*, Vol. 28, pp. 665-737, 1909.

[68] A. F. Peterson, S. L. Ray and R. Mittra, *Computational Method for Electromagnetics*, Ch 12, IEEE Press, New York, Oxford University Press, Melbourne, 1998.

[69] J. L. Volakis, A. Chatterjee and L. C. Kempel, *Finite Element Method for Electromagnetics*, IEEE Press, Oxford University Press, 1998.

[70] S. Silver, *Microwave Antenna Theory and Design*, MIT, Rad. Lab., Series, Vol. 12, McGraw-Hill, New York, 1979.

[71] S. A. Schelkunov, "A Mathematical Theory of Linear Arrays," *Bell System Tech. J.*, 1943, pp. 80-107.

[72] C. L. Dolph, "A Current Distribution for Broadside Arrays Which Optimizes the Relationship Between Beamwidth and Sidelobe Level," *Proc. IRE*, Vol. 35, June 1946, pp. 335-345.

[73] T. T. Taylor, "Design of Line Source Antennas for Narrow Beamwidth and Low Sidelobes," *IEEE Transactions on Antennas Propagation*, AP-3, Jan. 1955, pp. 16-28

[74] E. T. Bayliss, "Design of Monopulse Antenna Difference Patterns with Low Sidelobes," *Bell System Tech. J.*, Vol. 47, 1968, pp. 623-640.

[75] P. M. Woodward, "A Method of Calculating the Field Over a Plane Aperture Required to Produce a Given Polar Diagram," *Proc. IEE*, Part IIIA, Vol. 93, 1947, pp. 1554-1555.

[76] P. M. Woodward and J. P. Lawson, "The Theoretical Precision With Which an Arbitrary Radiation Pattern May be Obtained From a Source of Finite Size," *Proc. IEEE*, Vol. 95, P1, Sept. 1948, pp. 362-370.

[77] T. J. Peters, "A Conjugate Gradient-Based Algorithm to Minimize the Sidelobe Level of Planar Arrays with Element Failures," *IEEE Transactions on Antennas Propagation*, AP-39, No. 10, Oct. 1991, pp. 1497-1504.

[78] V. Murino, A. Trucco and C. S. Ragazzoni, "Synthesis of Unequally Spaced Arrays by Simulated Annealing," *IEEE Transactions on Signal Processing*, Vol 44, No. 1, Jan 1996, pp. 119-123.

[79] R. L. Haupt, "Genetic Algorithm design of antenna arrays," *Proceedings of Aerospace Applications Conference*, Vol. 1, pp 3-10, Feb 1996.

[80] O. M. Bucci, G. Franceschetti, G. Mazzarella and G. Panierello, "The Intersection approach to array synthesis," *Proc. Inst. Elec. Eng.*, pt H, Vol. 137, pp. 349-357, 1990.

[81] O. M. Bucci, G. D'elia and G. Romito, "A Generalized Projection Technique for the Synthesis of Conformal Arrays," *IEEE-AP-S International Symp.*, 1995, pp. 1986-1989.

[82] L. Brillouin, *Wave Propagation in Periodic Structures*, Dover Publications, Inc., New York, 1953.

[83] G. F. Farrell, Jr., and D. H. Kuhn, "Mutual Coupling Effects of Triangular Grid Arrays by Modal Analysis," *IEEE Transactions on Antennas Propagation*, AP-14, 1966, pp. 652-654.

[84] G. F. Farrell, Jr., and D. H. Kuhn, "Mutual Coupling Effects in Infinite Planar Arrays of Rectangular Waveguide Horns", *IEEE Transactions on Antennas Propagation*, AP-16, 1968, pp. 405-414.

[85] C. P. Wu and V. Galindo, "Surface Wave Effects on Dielectric Sheathed Phased Arrays of Rectangular Waveguide," *Bell Syst. Tech. J.*, Vol. 47, 1968, pp. 117-142.

[86] G. H. Knittel, A. Hessel and A. A. Oliner, "Element Pattern Nulls in Phased Arrays and Their Relation to Guided Waves," *Proceedings of the IEEE*, Vol 56, 1968, pp. 1822-1836.

[87] D. M. Pozar and D. H. Schaubert," Scan Blindness in Infinite Arrays of Printed Dipoles," *IEEE Transactions on Antennas Propagation*, AP-32, No. 6, June 1984, pp. 602-610.

[88] N. A. Begovich, "Frequency Scanning," Chapter 2 in *Microwave Scanning Antennas*, R.C. Hansen, Editor, Academic Press, New York, 1966

[89] M. I. Skolnik, "Survey of Phased Array Accomplishments and Requirements for Navy Ships," in *Phased Array Antennas*: Proceedings of the 1970 Phased Array Antenna Symposium, Artech House, 1972, A. A. Oliner and G.H. Knittel, Editors .

[90] L. C. Van Atta, "Electromagnetic Reflection," U.S. Patent 2908002, October 6, 1959.

[91] L. C. Godora, "Application of Antenna Arrays to Mobile Communications. Part II: Beamforming and Direction-of-Arrival Considerations," *IEEE Proc.*, Vol. 83, No. 8 (August), pp. 1195-1245, 1997.

[92] G. V. Tsoulos (Editor), *Adaptive Antennas for Wireless Communication*, IEEE Press, 2001.

[93] R. Klemm, *Space-Time Adaptive Processing*, The Institute of Electrical Engineers, London, 1998.

[94] J. Butler and R. Lowe, "Beam Forming Matrix Simplifies Design of Electronically Scanned Antennas," *Elect. Design*, Vol. 9, April 1961, pp. 170-173.

[95] J. L. Butler, "Digital Matrix and Intermediate-Frequency Scanning," Chapter 3 in *Microwave Scanning Antennas*, R.C. Hansen, Editor, Academic Press, New York, 1966.

[96] J. P. Shelton and K. S. Kelleher, "Multiple Beams from Linear Arrays," *IRE Trans Antennas Propagat.*, AP-9, 1961, pp. 154-161.

[97] W. Rotman and R. F. Turner, "Wide Angle Microwave Lens for Line Source Applications," *IEEE Transactions on Antennas Propagation*, AP-11, 1963, pp. 623-632.

[98] H. Gent, "The Bootlace Aerial," *Royal Radar Establishment Journal*, pp 47-57, October 1957.

[99]   D. Archer, "Lens Fed Multiple Beam Arrays," *Microwave Journal*, Oct 1975, pp. 37-42.

[100] H. E. Schrank, "Some Notable First in Array Antenna History," *Digest, 2001 Antenna Applications Symposium*, Sept 19-21, 2001, pp. 231-249.

[101] J. Blass, "Multidirectional Antenna – A New Approach to Stacked Beams," *IRE Conv. Record* 1960, pt 1, pp. 48-50.

[102] A. R. Lopez, "Monopulse Networks for Series Feeding an Array Antenna, *IEEE Antennas Propagat. Int. Symp. Dig.*, 1967.

[103] R. Tang, "Survey of Time-Delay Beam Steering Techniques," *Proc. 1970 Phased Array Antenna Symp.*, Artech House, Inc., Dedham, Mass., 1972, pp. 254-260.

[104] H. L. Southall and D. T. McGrath, "An Experimental Completely Overlapped Subarray Antenna," *IEEE Transactions on Antennas Propagation*, AP-34, No. 4, April 1986, pp. 465-474.

[105] E. C. DuFort, "Constrained Feeds for Limited Scan Arrays," *IEEE Transactions on Antennas Propagation*, AP-26, May 1978, pp. 407-413.

[106] S. P. Skobelev, "Analysis and Synthesis of an Antenna Array with Sectoral Partial Radiation Patterns," *Telecommunications and Radio Engineering*, 45, Nov 1990, pp. 116-119.

[107] S. P. Skobelev, "Methods of Constructing Optimum Phased Array Antennas for Limited Field of View," *IEEE Antennas and Propagation Magazine*, Vol. 40, No. 2, April 1998, pp. 39-49.

[108] C. E. Winter, "Phase Scanning Experiments With Two Reflector Systems," *Proc. IEEE*, Vol. 56, 1968, pp. 1984-1999.

[109] C. H. Tang, "Application of Limited Scan Design for the AGILTRAC-16 Antenna," *20th Annual USAF Antenna Research and Development Symp.*, Univ. of Illinois, 1970.

[110] G. Deschamps and W. Sichak, "Microstrip Microwave Antennas," *Proceedings of the Third Symposium on the USAF Antenna Research and Development Program*, October 18-22, 1953.

[111] R. E. Munson, "Microstrip Phased Array Antennas," *Proc. of the Twenty-Second Symposium on the USAF Antenna Research and Development Program*, October 1972.

[112] R. E. Munson, "Conformal Microstrip Antennas and Microstrip Phased Arrays," *IEEE Transactions on Antennas Propagation*, Vol. 22, pp. 74-78, 1974.

[113] J. Q. Howell, "Microstrip Antennas," in *IEEE AP-S Int. Symp. Digest*, 1972, pp. 177-180.

[114] J. Q. Howell, "Microstrip Antennas," *IEEE Transactions on Antennas Propagation*, Vol. 23, pp. 90-93, Jan1975.

[115] Y. T. Lo, D. Solomon and W. F. Richards, "Theory and Experiment on Microstrip Antennas," *Proc. of the 1978 Antenna Applications Symposium*, September 20-22, 1978.

[116] N. G. Alexopoulos and D. R. Jackson, "Fundamental superstrate (cover) effects on printed circuit antennas," *IEEE Transactions on Antennas Propagation*, AP-32, pp. 807-816, Aug. 1984.

[117] D. Pozar and D. Schaubert, "Analysis of an Infinite Array of Rectangular Microstrip Patches with Idealized Probe Feeds," *IEEE Transactions on Antennas Propagation*, Vol. AP-32, pp. 1101-1107, Oct. 1984.

[118] L. R. Lewis, M. Fassett, and J. Hunt, "A Broadband Stripline Array Element," *IEEE AP-S Symp. Dig.*, Atlanta, GA, June 1974, pp. 335-337.

[119] P. J. Gibson, "The Vivaldi Aerial," *9th European Microwave Conf.*, Brighton, U.K., 1979, pp. 101-105.

[120] T. H. Chio and D. H. Schaubert, "Parameter study and design of wide-band widescan dual-polarized tapered slot antenna arrays," *IEEE Transactions on Antennas Propagation*, Vol. 48, pp. 879-886, June 2000.

[121] H. Holter and H. Steyskal, "Some experiences from FDTD analysis of infinite and finite multi-octave phased arrays," *IEEE Transactions on Antennas Propagation,* AP-50, No. 12, Dec 2002, pp 1725-1731.

[122] N. Schuneman, J. Irion, and R. Hodges, "Decade bandwidth tapered notch antenna array element," *2001 Antenna Applications Symposium*, Sept 18-20, 2001, pp 280-294

[123] K. Trott, B. Cummings, R. Cavener, M. Deluca, J. Biondi and T. Sikina, "Wideband Phased Array Radiator," *IEEE Int. Symposium on Phased Array Systems and Technology 2003, Symposium Digest*, p. 383.

[124] D. N. McQuiddy, J. W. Wassel, J. B. Lagrange, and W. R. Wisseman, "Monolithic Microwave Integrated Circuits: An Historical Perspective," *IEEE Trans. of Microwave Theory and Techniques,* MTT-32, No. 9, Sept 1984, pp 997-1008.

[125] D. N. McQuiddy, R. L. Glassner, P. Hull, J. S. Mason and J. M. Bedinger, "Transmit/Receive module Technology for X-Band Active Array Radar," *Proc. IEEE*, Vol. 79, No. 3, March 1991, pp. 308-341.

[126] J. S. Kilby, "Invention of the integrated circuit," *IEEE Trans. Electron Devices*, Vol. ED-23, pp. 648-654, July 1976.

[127] J. F. Rippin, Jr., "Survey of Airborne Phased Array Antennas," in Phased Array Antennas, *Proceedings of the 1970 Phased Array Antenna Symposium*, June 2-5, 1970, A. Oliner and G. H. Knittel, Eds., Artech House Inc. Dedham, MA.

[128] T. E. Harwell, "Airborne Solid State Radar Technology," Ch. 20 in *Radar Technology*, E. Brookner, Artech House Inc., Dedham MA, 1977, pp. 275-287.

[129] B. A. Kopp, M. Borkowski and G. Gerinic, "Transmit/Receive Modules," *IEEE Trans. of Microwave Theory and Techniques,* MTT-50, No. 3, Mar 2002, pp. 827-834.

[130] D. Parker and D. C. Zimmermann, "Phased Arrays-Part II: Implementations, Applications and Future Trends," *IEEE Trans. of Microwave Theory and Techniques*, MTT-50, No. 3, March 2002, pp 688-698.

INDEX

A

Abraham, Max, 90
Abraham, Henry, 110, 112
absorption spectra, 493
AC, 75, 77, 80, 84, 86, 112, 113,
    116, 121, 131, 142, s*ee also*
    alternating current
acceptor, 146
accumulator, 16
Ackerman, S. L., 144
actinoscope, 105
action at a distance, 2, 9, 11, 13, 44,
    46, 543
active circuit, 562
Ada, 59
Adams, John Couch, 32
Adams, P. R., 142
adaptive array, 586
adaptive equalization, 152-3, 158
adaptive optimization, 567
Adcock, F., 109
adding machine, 79
Adler, Robert, 143
admittance, 138
Advanced Research Projects
    Agency Network, 154, 155
    see also ARPANET
advertising media, 115
AEG Company, 90, 125
aeolight lamp, 128
Aepinus, Franz Maria Ulrich
    Theodor Hoch, 4, 9, 11, 166
Aerial for Wireless Signaling, 385
aerial system, Clifden station, 398
aerial, 69, 79, 85, 87, 88, 92-4, 97,
    100, 102, 105, 109, 110-2, 115,
    118, 120, 122, 125, 128, 140-1,
    355-6, 368, 370, 377, 382, 384,
    388, 392, 403, 408;
    Brant Rock, 103, 400, 400f;
    Marconi's fan, 393-4
    Telegraph, 69, 70
aeroplane, 103, 113, 127

aether, 9, 16-9, 23-4, 27-9, 34, 38,
    41, 45, 48-51, 78
AFCRC, 560
Affel, H. A., 126, 139
AFRL, 157, 557, *see also* Air Force
    Research Laboratory
Agababyan, K., 514
Agustar, 154
AIEE, 78, 87, 100, 152, 549, *see
    also* American Institute of
    Electrical Engineers
Aiken, Howard, 135
Aiken, William Ross, 148
AIL, 557, 560, *see also* Airborne
    Instruments Laboratory Inc.
air defense, 567
Air Force Research Laboratory, 557
    see also AFRL
air navigation, 126
Airborne Instruments Laboratory
    Inc., 557, *see also* AIL
aircraft, 104, 109, 110-3, 127-8,
    132, 135-6, 139-43, 147, 151,
    157;
    anti-, 135, 141, 142, 520;
    blind landing, 128, 466;
    detection of, 127, 133;
    guidance, 104;
    locating, 132, 135;
    pilot-less, 111
airline pilot, 112
airplane, 110, 133, 143, 148, 495
airport, 133, 146
Airy, George Biddell, 28
Aisentein, S. M., 109
Aivazyan, M., 513-4
Akhiyezer, A., 504
alcohol, 483
Alexanderson, Ernst Frederik
    Werner, 100, 102, 106, 109,
    351, 377-8
Alexopoulos, N. G., 592
Alferov, Zhores Ivanovich, 344
Alford, Albert, 139
algebra, Boolean, 64, 138
alkali, 82, 138

605